ANNUAL REVIEW OF
EARTH AND
PLANETARY SCIENCES

EDITORIAL COMMITTEE (1986)

Responsible for the organization of Volume 14
(Editorial Committee, 1984)

Production Editor   KEITH DODSON
Subject Indexer   CHERI D. WALSH

ANNUAL REVIEW OF EARTH AND PLANETARY SCIENCES

VOLUME 14, 1986

GEORGE W. WETHERILL, *Editor*
Carnegie Institution of Washington

ARDEN L. ALBEE, *Associate Editor*
California Institute of Technology

FRANCIS G. STEHLI, *Associate Editor*
University of Oklahoma

ANNUAL REVIEWS INC. 4139 EL CAMINO WAY PALO ALTO, CALIFORNIA 94306 USA

ANNUAL REVIEWS INC.
Palo Alto, California, USA

International Standard Serial Number: 0084-6597
International Standard Book Number: 0-8243-2014-X
Library of Congress Catalog Card Number: 72-82137

Annual Review and publication titles are registered trademarks of Annual Reviews Inc.

Annual Reviews Inc. and the Editors of its publications assume no responsibility for the statements expressed by the contributors to this *Review*.

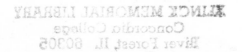
TYPESET BY AUP TYPESETTERS (GLASGOW) LTD., SCOTLAND
PRINTED AND BOUND IN THE UNITED STATES OF AMERICA

Annual Review of Earth and Planetary Sciences
Volume 14, 1986

CONTENTS

GEOPHYSICS ON THREE CONTINENTS, Anton L. Hales 1

TRIGGERED EARTHQUAKES, D. W. Simpson 21

EL NIÑO, Mark A. Cane 43

MOLECULAR PHYLOGENETICS, Jerold M. Lowenstein 71

CONODONTS AND BIOSTRATIGRAPHIC CORRELATION, Walter C. Sweet and Stig M. Bergström 85

OCCURRENCE AND FORMATION OF WATER-LAID PLACERS, Rudy Slingerland and Norman D. Smith 113

EARTHQUAKES AND ROCK DEFORMATION IN CRUSTAL FAULT ZONES, Richard H. Sibson 149

GENESIS OF MISSISSIPPI VALLEY–TYPE LEAD-ZINC DEPOSITS, Dimitri A. Sverjensky 177

CARBON DIOXIDE INCREASE IN THE ATMOSPHERE AND OCEANS AND POSSIBLE EFFECTS ON CLIMATE, Chen-Tung A. Chen and Ellen T. Drake 201

COASTAL PROCESSES AND THE DEVELOPMENT OF SHORELINE EROSION, Paul D. Komar and Robert A. Holman 237

FORECASTING VOLCANIC ERUPTIONS, Robert W. Decker 267

RUPTURE PROCESS OF SUDUCTION-ZONE EARTHQUAKES, Hiroo Kanamori 293

GEOCHEMISTRY OF TEKTITES AND IMPACT GLASSES, Christian Koeberl 323

CLIMATIC RHYTHMS RECORDED IN STRATA, Alfred G. Fischer 351

TEMPERATURE DISTRIBUTION IN THE CRUST AND MANTLE, Raymond Jeanloz and S. Morris 377

PETROGENESIS OF ANDESITES, Timothy L. Grove and Rosamond J. Kinzler 417

GEOLOGIC SIGNIFICANCE OF PALEOZOIC AND MESOZOIC RADIOLARIAN CHERT, David L. Jones and Benita Murchey 455

CHEMICAL GEODYNAMICS, Alan Zindler and Stan Hart 493

INDEXES

Subject Index 573

Cumulative Index of Contributing Authors, Volumes 1–14 584

Cumulative Index of Chapter Titles, Volumes 1–14 587

SOME RELATED ARTICLES IN OTHER *ANNUAL REVIEWS*

From the *Annual Review of Astronomy and Astrophysics*, Volume 23 (1985)

Sunspots, Ronald Moore and Douglas Rabin

Cold Outflows, Energetic Winds, and Enigmatic Jets Around Young Stellar Objects, Charles J. Lada

Big Bang Nucleosynthesis, Ann Merchant Boesgaard and Gary Steigman

From the *Annual Review of Biochemistry*, Volume 54 (1985)

Origin of Immune Diversity: Genetic Variation and Selection, Tasuku Honjo and Sonoko Habu

From the *Annual Review of Ecology and Systematics*, Volume 16 (1985)

Approaches in Evolutionary Morphology: A Search for Patterns, L. B. Radinsky

Longevity of Individual Flowers, Richard B. Primack

Computer-Aided Reconstruction of Late-Quaternary Landscape Dynamics, Allen M. Solomon and Thompson Webb, III

Hybrid Zones and Homogamy in Australian Frogs, Murray J. Littlejohn and Graeme F. Watson

Analysis of Hybrid Zones, N. H. Barton and G. M. Hewitt

Patterns of Species Diversity on Coral Reefs, M. A. Huston

Phenological Patterns of Terrestrial Plants, Beverly Rathcke and Elizabeth P. Lacey

Ecology of Kelp Communities, Paul K. Dayton

Ontogeny and Systematics, Arnold G. Kluge and Richard E. Strauss

Predation, Competition, and Prey Communities: A Review of Field Experiments, Andrew Sih, Phil Crowley, Mark McPeek, James Petranka, and Kevin Strohmeier

Speciation in Cave Faunas, Thomas C. Barr, Jr. and John R. Holsinger

Feeding and Nonfeeding Larval Development and Life-History Evolution in Marine Invertebrates, Richard R. Strathmann

Gene Flow in Natural Populations, Montgomery Slatkin

Biological Aspects of Endemism in Higher Plants, Arthur R. Kruckeberg and Deborah Rabinowitz

From the *Annual Review of Fluid Mechanics*, Volume 18 (1986)

Eddies, Waves, Circulation, and Mixing: Statistical Geofluid Mechanics, Greg Holloway

Wind-Wave Prediction, Rodney J. Sobey

The Continental-Shelf Bottom Boundary Layer, William D. Grant and Ole S. Madsen

Vorticity Dynamics of the Oceanic General Circulation, Peter B. Rhines

From the *Annual Review of Materials Science*, Volume 15 (1985)

    Secondary Ion Mass Spectrometry, Peter Williams

    Sol-Gel Processing of Silicates, L. C. Klein

    Defects and Order in Perovskite-Related Oxides, D. M. Smyth

From the *Annual Review of Microbiology*, Volume 39 (1985)

    Marine Microbiology Far From the Sea, Robert A. MacLeod

    Sulphate-Reducing Bacteria and Anaerobic Corrosion, W. A. Hamilton

    Microbial Desulfurization of Fossil Fuels, D. J. Monticello and W. R. Ginnerty

    The Distribution and Evolution of Microbial Life in the Late Proterozoic Era,
      Andrew H. Knoll

Ann. Rev. Earth Planet. Sci. 1986. 14 : 1–20

GEOPHYSICS ON THREE CONTINENTS

Anton L. Hales

Research School of Earth Sciences, Australian National University, Canberra, Australia

Let me say at the outset how much I have enjoyed the journey down memory lane occasioned by the invitation of the Editors of the *Annual Review of Earth and Planetary Sciences* to write about my experiences as a student, teacher, and investigator in the Earth sciences. As I have written this account, it has been driven home to me how fortunate I have been in the students and the colleagues with whom I have been associated over the past 50 years.

After matriculation from the Sea Point Boys High School, I enrolled for a BSc degree at the University of Cape Town, intending to become a science teacher. After completing an MSc in Applied Mathematics, I was appointed a Junior Lecturer at the University of the Witwatersrand but resigned after six months to go to St. John's College, Cambridge, where I completed the Mathematical Tripos in 1933.

At that time (1931) quantum physics was the glamor field, and at first I intended to choose quantum physics for Schedule B (now Part III) of the Mathematical Tripos. However, following a suggestion of Basil Schonland (Senior Lecturer in Physics at Cape Town), I finally decided to take a sequence of courses in geophysics, mostly given by Harold Jeffreys. These lectures (as might be expected) covered a very wide range of topics in geophysical hydrodynamics and elasticity. I found them most stimulating, and indeed most of my activities since then have their origin in those lectures, with the significant exception of paleomagnetism.

Before completing the Tripos, I was offered a Junior Lectureship in Pure Mathematics at the University of the Witwatersrand. This was 1933, a time when jobs were not very plentiful, so I chose to return to South Africa. I started some work on convection in the Earth and later in geysers. The

1

latter proved most interesting theoretically because I could not find a solution with flow up the center and down the outside (or, alternatively, down the center and up the outside). The only solution I could find that would grow was one in which the upgoing and downgoing currents were interleaved spirals. These papers constituted my PhD thesis at Cape Town.

In 1937 Bernard Price of the Victoria Falls Power Company provided funds for the establishment of the Bernard Price Institute of Geophysical Research (BPI) at the University of Witwatersrand. Schonland, who had met Bernard Price because of their shared interest in lightning research, was appointed Director. Philip Gane was appointed to do research in seismology. Schonland invited me to become an Honorary Research Associate of the Institute. The creation of the Bernard Price Institute and my association with it had a very important influence on my future career.

The first seismology project was the creation of a network for locating the foci of the rockbursts, or earth tremors, that occurred daily in Johannesburg. The network operated for about a year in 1938 and 1939. In September 1939 the BPI and its staff became the nucleus of a South African radar manufacturing and operations group. I was given the task of completing a report on the rockburst location project for the Chamber of Mines. A paper describing this work was published in 1946. Other projects discussed in 1939 were the cause of the reversed magnetization of the Pilanesberg dikes and the measurement of stress underground, but World War II intervened before these got past the discussion stage.

After completing the rockburst report, I joined the South African Engineer Corps and served in East Africa, Abyssinia, and Egypt in a water-finding unit. Later I served with the War Supplies organization in South Africa, returning to teach at the University of the Witwatersrand in 1945. In 1946 I transferred to the Bernard Price Institute as a Senior Research Officer.

The 1939 study had shown that the rockbursts frequently had local magnitudes of 3 to 4. Thus they had sufficient energy to be used for seismic refraction studies of crustal structure. The first requirement was for a network from which location and time of origin could be found. Gane and Hugh Logie developed a radio telemetry system that used frequency modulation of an audio tone for the seismic signal. They also built a memory system consisting of an endless wire loop that delayed the signals for 13 s; the signals were then recorded on photographic paper only when the recording system was triggered by a tremor. There remained the problem of high-speed recording in the field. The Gane solution was to transmit the output from a high-Q tuning fork by radio when a tremor occurred. The field receiver output was fed to a matched tuning fork filter, and the time marks were superimposed on the trigger signal. By 1948 the system was ready to go. P. Willmore came from Cambridge to participate, bringing with him three stations recording at lower speed. I ran an array with three sets of seismometers spaced at 0.25 km. The experiment was very successful. The P_n and S_n velocities were high, and so far as I was concerned, the existence of an "intermediate" layer in the crust was established. At this time also a gravity survey using the Cambridge pendulums was planned. Ian Gough was appointed to the Council for Scientific and Industrial Research to assist me with the measurements. After the first leg from Cape Town to Pretoria, Gough and his assistant carried on with observing while I read the records.

When I joined the BPI in 1946, I was interested in seeing whether one could detect reflection from the M discontinuity at near-vertical incidence using rockbursts as the source. I ran a small array in the University grounds for this purpose. There were indeed phases with the high apparent velocities that would correspond to reflections from the M discontinuity, but not just one at 11–13 s as I expected; instead, there were several between 8 and 13 s. I tried again at a site north of Johannesburg after I became Professor of Applied Mathematics at Cape Town in 1949, but with the same result.

At Cape Town, teaching and departmental administration left me little time for research other than writing up the gravity survey and my contribution to the 1948 seismic refraction program. In 1952, I spent six months on sabbatical leave in Cambridge, with most of my time there devoted to writing a paper on thermal contraction. However, I did spend some time with S. K. Runcorn's paleomagnetic students at Madingley Rise. On my return to Cape Town I continued to work on thermal contraction.

Much to my surprise, in 1954 Schonland accepted the Deputy Directorship of the Atomic Energy Research Establishment at Harwell; even more to my surprise, he suggested that I should be a candidate for the Directorship of the BPI. I was offered the appointment and went back to Johannesburg after attending the General Assembly of the International Union of Geodesy and Geophysics at Rome in 1954.

The seismology program at the BPI was well looked after by Gane, and so I looked for a new field. I decided that it should be paleomagnetism. On a suggestion of Eric Simpson, I offered a Research Assistantship to Kenneth Graham, who had just completed his BSc (Honors) in Geology. I told him that I thought that if enough measurements were made sufficiently carefully, one would find that the continents had not drifted. Graham accepted the offer but told me he was convinced that drift had occurred. Ken collected samples while I built an airspinner magnetometer of the pattern used by John Graham, although with a single-stage narrow-band-pass filter instead of the matched pair used by Graham. Later on, through the kindness of Merle Tuve of the Department of Terrestrial Magnetism,

Carnegie Institution of Washington, we were loaned two matched narrow-band-pass filters. We collected a suite of specimens of the Karroo sediments, but these showed scatter. We also collected surface samples of dolerite; scatter was apparent in these samples too, though much less than with the sediments. Underground samples from an East Rand mine shaft and a railway tunnel under construction in Natal showed consistent results in both the normal and reverse senses. Furthermore, baked sediments in general agreed with the dike baking them. To Kenneth Graham's great pleasure, the paper we presented at the symposium organized by P. M. Blackett in 1956 contained the statement that "the simplest way of reconciling the divergent pole positions [from Europe, North America, Australia, and South Africa] is to postulate relative displacement of the continents, i.e. Continental drift" (Graham & Hales 1957). From that time on I called myself a "reluctant drifter," reluctant for much the same reason, I suppose, as why H. H. Hess called his 1962 paper "An Essay in Geopoetry." In the 1950s it simply was not proper to acknowledge a belief in drift. The extent of the antidrift feelings can best be judged from the published reports of the 1928 symposium organized by the American Association of Petroleum Geologists or the 1948 meeting of the British Association for the Advancement of Science.

There were, of course, ardent supporters of drift (for example, Sam Carey from Tasmania and Lester King in South Africa). Carey organized a symposium on drift in Tasmania at about the same time as the London one organized by Blackett. Chester Longwell was there and wrote the Introduction to the proceedings of the symposium. Longwell wrote, "If the fit between South America and Africa is not genetic then surely it was devised by Satan for our confusion," but he was still unconvinced. It was several years before it was generally accepted that Satan was not responsible (Carey 1958).

I was also reasonably certain that there were real reversals of the Earth's field, for otherwise it seemed improbable that the baked sediments should be magnetized in the same sense as the dikes baking them.

This was not the only paleomagnetism work being done at the BPI. Gough had earlier built a magnetometer to measure samples from the Pilanesberg dikes, which were known to be reversely magnetized. The first results from surface outcrops scattered all over the stereoplot. The project was nearly abandoned, but Gough took the opportunity to take samples from a Pilanesberg dike in one of the deep Witwatersrand mines, and these gave a consistent direction, with the pole located in Ethiopia.

The scatter of the surface outcrop directions was puzzling. We thought that lightning was excluded, because magnetization due to lightning would be dominantly horizontal. Kenneth Graham and the Institute's instrument maker, Jock Keiller, designed and built a drill that could take oriented samples to a depth of 2 m. Graham showed that the magnetization in one outcrop was due to a network of currents a centimeter or two above the present outcrop surface, one of which was 50,000 A. Jan van Zijl, a graduate student working with Graham, sampled two sections of the Stormberg lavas very carefully. There was a reversal in the section. During the transition, the intensity of the field was reduced by a factor of four.

As is happened it was not long before I again became involved in seismology, for Gane moved to a post in industry in 1955. I inherited a graduate student, Selwyn Sacks, whose thesis topic was an attempt to use a signal with continuously varying frequency for seismic reflection. The source, devised by Sacks, was a water hammer, the frequency of which was varied by moving a male cone inside a female cone, thus varying the acoustic impedance. A small one worked well to a depth of about 10 m, and a larger one to about 100 m. We wanted to try to get reflections from a depth of 10 km and hopefully from the M discontinuity. The water hammer concept did not work because of corrosion fatigue in the spring, so a hydraulic piston device was built, which operated successfully.

Sacks and I decided to carry out a seismic refraction profile to the east of Johannesburg similar to the profile taken in 1948 to the west of Johannesburg. To get more data, we built satellite stations that were triggered by a VHF relay of the trigger signal from the Institute. We thus got three records for each tremor. The existence of the lower crust layer was confirmed. Working with Rodleigh Green, we then began to make observations at greater distances, to 900 km to the south and up to 1500 km to the north. The northern survey was less successful, because a station with a frequency close to ours had poor frequency control and thus heterodyned with our signal to produce false triggers that ran the cameras out of recording paper before they had caught a tremor. The arrivals beyond 400 km were 2 or 3 s earlier than the Jeffreys-Bullen travel times or the times extrapolated from our P_n and S_n travel times.

Before I became Director of the BPI, the Department of Terrestrial Magnetism (DTM) had given the Institute parts for a mass spectrometer. Deneys Schreiner, a chemist, made measurements on lepidolites, but he was concerned about possible diffusion effects when measurements were made on mineral separates from the granite. He thought that if it was possible to make measurements on the total rock, the diffusion difficulty could be avoided. Initially there were problems with the ion exchange column separation, but these were solved while working on the Bushveld granite. The age was found to be 1920 Myr (Schreiner 1958).

By reason of staff shortage I became more deeply involved in isotope geochemistry than I had expected. The Witwatersrand System is of considerable significance in South Africa. Some of the quartzites of the System contain as much as 1.37% potassium and very little calcium. A research assistant, Hazel Roberts, and I made age measurements on the sericite (thought to be authigenic) and found the age to be 2710 Myr. This is close to the present-day estimates, though in 1960 it was thought to be too high.

The more precise location of rockbursts was of interest to the Institute and also to the Chamber of Mines. Observations with high timing accuracy were needed on a scale of 2 or 3 km instead of the 20- or 30-km spatial coverage of the surface array. Preferably, the seismometers should be underground. After long discussion with Sacks it became clear that the best way to achieve this was by recording on a slow-running magnetic tape recorder, with the tape being changed once a day. Keiller built the recorder and playback systems to Sacks' design. Neville Cook joined the Institute to process the records and analyze them. He also started on some rock mechanics experiments. John Jaeger was still a regular visitor to the Institute, and he and Cook continued to collaborate after this project was completed.

I spent 1960 in the United States, staying several months at DTM in Washington, three months at the Seismological Laboratory at the California Institute of Technology, and a month at the University of Wisconsin, Madison. I wrote a short paper on the attenuation of short period S-waves and planned a small surface-wave array in the neighborhood of Johannesburg. Frank Press, then Director of the Seismological Laboratory, had a quite extraordinary group of students, who have had a considerable impact on seismology in the United States.

In early 1961 I returned to Johannesburg. We were fortunate to obtain the cooperation of the South African Navy for a program of seismic refraction and gravity measurements in April and June of 1962 and began to plan a two-ship operation. However, in January 1962 we were told that only one ship would be available. We hurriedly planned a one-ship operation using free-floating buoys, with commercial tape recorders triggered from the shooting ship using the same equipment used to trigger land stations. Matters became more complicated when in February I received a letter from Lloyd Berkner asking me whether I would be prepared to lead a geophysics program in the newly created Southwest Center for Advanced Studies. Quickly I arranged to spend a weekend in Dallas in March, after a crustal studies meeting in Paris and a meeting to establish the International Year of the Quiet Sun (also in Paris). Before I left Dallas to return to South Africa, I was committed to return to head the geophysics program at the Southwest Center.

It was an appropriate time to hand over the Bernard Price Institute to new leadership. The long-distance seismic refraction studies had shown that the travel times at distances between 500 and 1000 km were considerably shorter than either the Jeffreys-Bullen times or the extrapolated times of P_n from our own crustal studies to 400 km. The program for precisely locating rockbursts using the underground network had been very successful, with nearly all events being located within 50 m of a working face. Neville Cook was writing up his thesis on these results. The study of the paleomagnetism of the Stormberg lavas had shown a reversal during the sequence, and after considering a number of self-reversal hypotheses (van Zijl et al 1962), we concluded that there was a real reversal of the Earth's field in Jurassic times. I assumed that Sacks would continue the seismic program at the Institute, but he accepted a post at the Department of Terrestrial Magnetism not long after I moved to Dallas. Unfortunately, the vibrator work was not written up for publication, and the only published reference to it is in a paper of mine reviewing geophysical research at the BPI (Hales 1960).

In late 1962, I went to a Vela program meeting in Colorado, and at a cocktail party Charlie Bates remarked to me that there was a lot of good data from the Long Range Seismographic Monitoring (LRSM) program. He wondered whether anything useful could be done with it. I replied immediately that a study of travel times and their regional variation would be very worthwhile. Within a few weeks, I had written the proposal and it had been funded. John Cleary, who had just handed in his PhD thesis at the Australian National University, joined the Southwest Center in April 1963. The LRSM network was extremely well run and the time control excellent. By the end of the year we knew that travel times to stations in the western United States were 1 to 2 s later than those to stations in the central or eastern United States. In order to avoid regional bias, we used a series of events from the west, another from the east, and still another from the south and weighted the three profiles as equally as the events made possible. Here we ran into a difficulty. The travel times from all profiles were earlier than the Jeffreys-Bullen times, but the differences for the south and west profiles were tilted with respect to one another. We spent some time looking for more events from the south, but the effect persisted. Gene Herrin and James Taggart were also studying travel times, using much more data. Their travel time curve was tilted with respect to ours but was very similar to our western event profile. Herrin and I discussed this quite often, each looking for bias in the other's procedure. Eventually I took the Herrin and Taggart data and weighted it in a similar manner to the way in which Clearly and I had weighted ours. The result was a mean curve not dissimilar from that we had found earlier. One possible explanation was that although the major part of the regional variation in travel times clearly originated in the upper mantle, there was a significant contribution from the lower mantle.

Having shown that there were regional variations in P travel times, it was natural to investigate whether it was possible to measure the variations that obviously would occur in S travel times. The S times would necessarily be read from long-period records, and I was not as certain that it would be possible to determine the onsets with sufficient accuracy. Hugh Doyle, also from the Australian National University (ANU), undertook this task and was successful. The S station delays correlated well with the corresponding P delays. The regression coefficient was high (3.72 instead of the expected value of 1.75). It is probable that partial melting in the upper mantle is responsible. Later, Jeannie Roberts, my research assistant in the teleseismic studies for all the years I was in Dallas, and I extended the analysis to 100° and also S-diffracted around the core boundary. In addition, we determined a lower-mantle S velocity distribution. A study of ScS travel times followed. In this study, differential travel times of S and ScS were used in order to avoid upper-mantle variations. There was evidence of velocity variation in the lower mantle.

The first observed PKP phases bottom about 1600 km below the core-mantle boundary, so PKP travel times provide only limited constraints on the velocity distribution in the outer core. We therefore embarked on a study of SKS travel times and derived a velocity distribution for the outer core. This study should be repeated when sufficient broadband stations well distributed over the surface of the globe are available, for I think it will be found that there are SKS phases with periods of about 5 s. This will enable more accurate readings of onsets.

Once an S velocity distribution for the mantle and ScS times were available, it was possible to determine the radius of the core. Our estimates (Hales & Roberts 1970a) were 3486 and 3489 km. These were higher than the estimate made from PcP measurements by Taggart & Engdahl (1968) of 3477 km. At the time I thought that the PcP estimates would be more accurate. I think now that the greater values are more likely to be correct.

Seismic work at sea continued to interest me, and a number of free-floating buoys were built by Keiller. These recorded on magnetic tape running at 15–20 mil s^{-1} and ran for 15–20 hr. They were used at sea in the Caribbean, in the Gulf of Mexico, off the east coast of the United States, and off the south coast of Africa. They were also used on land, with seismometers replacing the hydrophone in South America and the United States (in particular, the Early Rise and Greeley observations). Initially only analog replays were used, but from the time of Early Rise in 1966 on, all analyzed records have been digitized. My research assistant for these studies was Joe Nation, who is still at the University of Texas at Dallas.

Seismology, isotope geochemistry, paleomagnetism, and experimental petrology were the initial programs of the geophysics group. John Graham and Charles Helsley were recruited early, so when the question of an appointment for Ian Gough arose, it was clear that he should do something other than paleomagnetism. He decided to come, and we explored possible programs.

I had been impressed by the work that J. Bartels and U. Schmucker had done using magnetic variometers, and after some discussion Gough began to design a variometer that could be buried in the ground as were the seismometers we used in our field programs. It seemed to us that one needed to have at least 15 or 20 instruments in an array to make the program efficient, and that was the target Gough set himself. It was achieved very quickly with a simple instrument of conventional design that has proved very effective in the field. Thus, deep magnetic sounding became the fifth program of the group. After Gough and J. S. Reitzel left, the deep magnetic sounding program was continued by Hartmut Porath and David Bennett. Later, micropaleontology and organic geochemistry were added.

There are few properties of the mantle that can be estimated from observations at the surface of the Earth. Electrical conductivity is one of these few, and I regard it as of prime importance that the electromagnetic methods should be fully exploited. There are some intriguing questions. It is known that electrical conductivity increases by several orders of magnitude between the base of the crust and a depth of 700 km. The question is, does the increase occur in steps or zones of rapid increase as some models (e.g. those of Woods 1979) suggest, and if so, at what depths do these steps occur? Even if the changes are relatively smooth, more precise information on the behavior of the conductivity would complement the seismological information significantly. More observations are needed, especially perhaps in programs of long-term magnetotelluric measurements on the continents of the kind that Jean Filloux has made so successfully in the oceans. One other requirement is for more and better laboratory measurements of the behavior of the conductivity of upper-mantle minerals at high pressure and temperature.

Mapping of the regional variations of conductivity would also be significant. It can be done. Schmucker, Porath, Reitzel, and Gough have shown this in the Rocky Mountain region, and others have done so elsewhere.

In 1963 and 1964 I took part in seismic refraction experiments organized by DTM and the University of Wisconsin. Later, in 1966, the United States

Geological Survey (USGS) organized the Early Rise experiment. Many organizations participated in the observing program, among them the Southwest Center for Advanced Studies (SCAS). Our profiles ran from the Texas-Mexico border to Lake Superior, and from Little Rock, Arkansas, to Lake Superior. In addition, Rodleigh Green, who was in Dallas to write up the South African seismic work at sea, observed along an arc from Colorado to the Smoky Mountains. Green and I worked on the analysis of the SCAS observations. As in South Africa, the arrivals beyond 500 or 600 km were earlier than the Jeffreys-Bullen times and earlier than would have been predicted from the P_n times observed to 300 or 400 km. We ascribed the early arrivals to an increase in velocity at a depth of 89 km. While the analysis of the Early Rise data was underway, we were told by the USGS that a nuclear explosion was to be fired in Nevada within a few days. This offered the opportunity of observing over an even greater distance range than Early Rise. We quickly planned to set up stations from Oklahoma to North Carolina. The observations of the Greeley explosion showed clearly that there was a discontinuity in the travel times at about 20° and another at a distance of 23° to 24° corresponding to discontinuities, or rapid increases of velocity, at depths of about 400 and 650 km. However, Greeley was fired at the Nevada Test Site, and Inge Lehmann had shown that there was a low-velocity zone in that area. The arrivals from Greeley from 1000 to 2000 km were later by up to 6 s than those of Early Rise, although the observations had been made at stations of the arc profile of Early Rise. We interpreted these arrivals as coming from below the low-velocity zone at a depth of 156 km. Although the delay is relatively well determined, the depth is not because of the difficulty in estimating the velocity in the low-velocity zone (Green & Hales 1968).

At about the time that the analyses of Early Rise and Greeley observations were complete, Helsley and I decided that a long profile in the Gulf of Mexico would complement the results on land. With the cooperation of the Institute of Geophysics at the University of Mexico, we obtained permission to set up stations in Mexico. Meanwhile, the University of Wisconsin was setting up stations in Florida. The shots were fired at sea from a US Coast Guard ship. This Gulf of Mexico profile also showed a significant increase of velocity in the upper mantle, in this case at a depth of 57 km. Toward the end of the profile, there were arrivals that we interpreted as coming from below the low-velocity layer at a depth of 170 to 200 km. We attempted to use amplitudes to discriminate between models of the velocity distribution in the low-velocity zone. The discriminating power was not sufficient to set very close limits on the depth to the base of the low-velocity zone. I was now convinced that this discontinuity at about 200 km was a global feature.

By the mid-1960s sufficient data had been collected by the small surface-wave array in the Witwatersrand region to justify analysis, and Selwyn Bloch came to Dallas for this purpose. At about the same time, Adam Dziewonski joined the Southwest Center as a Polish Academy scholar to work with Mark Landisman. Both Bloch and Dziewonski were working on surface-wave problems, the immediate one being the determination of phase and group velocities. As a result of discussions with Stefan Mueller, they developed a narrow-band filter technique and employed it successfully in a number of studies. Both Bloch and Dziewonski like to start work around noon and continue until the early hours of the morning, and most of our fruitful discussion of their progress took place very late at night. Unfortunately, more often than not I had meetings early the next morning, to which I arrived somewhat jaded.

Dziewonski soon graduated from surface-wave studies to free oscillation analysis and began his productive association with Freeman Gilbert. At this time all of the free oscillation data were from the Chile (1960) and Alaska (1964) earthquakes, and these only from a few stations. Dziewonski set about digitizing the World-Wide Standardized Seismograph Network records from the Alaska earthquake, an earthquake on the Peru-Bolivia border, and a Colombian earthquake. This was tedious work and had to be carried out with meticulous care. As much credit is due to his research assistants of that time as to Dziewonski himself. In fact, these three earthquakes and the Chile earthquake provided the data on which all free oscillation models were based until the advent of the digital seismic stations in the 1970s. In 1969 the Southwest Center for Advanced Studies became the University of Texas at Dallas (UTD), and a UTD Earth model UTD 124A was developed by Dziewonski. It was later superseded by models 1066A and 1066B. All of this data analysis required large amounts of computer time and could not have been achieved but for the generosity of Sun Oil in allowing Dziewonski use of spare night-shift time on the CDC6400 at their Richardson laboratories.

Other seismic activities of the 1960s were concerned with the continental margin: the East Coast Onshore Offshore Experiment, and a similar experiment, the Gulf Coast Onshore Offshore experiment, carried out in cooperation with the US Coast and Geodetic Survey.

In 1970 I became Acting Vice President for Academic Affairs and for a year was involved with the development of the graduate programs at UTD. In late 1971, soon after my return to the Geoscience Program, Bob Meyer called me to tell me that he and his graduate student Joseph Gettrust had been discussing new programs in seismology and had come to the conclusion that the time was ripe for a serious effort to apply seismic reflection methods to the study of the deep crust and the M discontinuity.

He suggested that this should be a joint venture. We discussed this idea further at a Seismological Society of America meeting, came to the conclusion that it would only be viable as a national program, and decided to invite Jack Oliver and Bob Phinney to join us in promoting it. The four of us met and decided on the name COCORP, the Consortium for Continental Reflection Profiling. It was agreed that a proposal should be written by Oliver. The National Science Foundation funded the initial program, and its subsequent success is well known.

By the mid-1960s, continental drift and seafloor spreading had evolved into plate tectonics and had become respectable. A spate of modeling of the convection processes ensued. It seemed to me that the convection modelers took insufficient account of the evidence from seismology and from studies of the thermal regime of the oceanic plates that the bases of the oceanic plates were not level or equipotential surfaces. I wrote a short paper (Hales 1969) suggesting that gravity sliding was at least a plausible alternative to the convection models then in vogue. Since the bases of the oceanic plates slope toward the trenches, if one resolves the forces acting on the plates parallel to the base of the plate, there is a component of gravity acting toward the trench. This is offset to some degree by viscous drag, but numerical calculations showed that the forces were adequate to move the plates. Another way of looking at the forces is to resolve them radially. There is then a component of the pressure on the base of the plate acting toward the trench.

More recent work (for example, the deep magnetic soundings of Schmucker, Porath, Reitzel, and Gough) in the western United States and the interpretations of these soundings by Porath and Gough have shown that on the continents, the base of the plates or lithosphere is a much more irregular surface than in the oceans. The anelasticity, or effective viscosity, is very temperature dependent, as is the electrical conductivity. Thus one can expect tension over regions where the isothermals approach closer to the surface of the Earth. I am reasonably sure that rifting is initiated by a thermal pulse. The pressure on the base of the plates will have components acting outward from the hot region, thus producing tension. Of course, the base of the plates is not a sharply defined surface, but nevertheless there must be tension over a hot plume.

I was impressed by John Verhoogen's 1960 paper suggesting that the heat that drove convection in the core was provided by the latent heat of solidification of the inner core. It seemed to me then that the problem of convection in the core needed to produce the magnetic field was solved. Later, S. Braginski and D. Gubbins showed that gravitational differentiation was a more significant source of energy than the latent heat of solidification.

In 1977, G. Jones, a student of Verhoogen, wrote a paper pointing out that convection in the core would result in the creation of a thermal boundary layer at the base of the mantle, and that episodic breakdown of this boundary layer was responsible for plate tectonics. I regard this suggestion and its later development by H. Schloessin and Jack Jacobs as attractive. In fact, I think that plate tectonics does begin at the boundary of the inner core.

Other activities that began in the Dallas years were the development of a cooperative PhD program between Southern Methodist University (SMU) and the Southwest Center and my involvement in the affairs of the International Seismological Centre and the Geodynamics Program.

I had always thought of the training of graduate students and research as going hand in hand. Although SCAS had a charter permitting it to offer degrees, Lloyd Berkner did not plan to do so. In the course of a discussion with Gene Herrin, it became clear that he wanted to see a PhD program at SMU, and that together SMU and SCAS had the resources to establish it. Berkner and James Brooks, then Chairman of the Department of Geological Sciences at SMU, were enthusiastic about the idea, and the program was started in 1963. It turned out some good students in those early years and was the foundation on which the present PhD programs at SMU and UTD developed.

Before World War II, global seismological data collection was provided by the International Seismological Summary in Britain and the Bureau Central International Seismologique in France. As a result of World War II, a large backlog in analysis had developed, and the question was raised of the need for a new organization to collect and analyze data. The International Seismology Centre (ISC) was set up by the International Union of Geodesy and Geophysics in 1963. I became a member of the ISC committee of the International Association of Seismology and Physics of the Earth's Interior (IASPEI) in 1967 and served on the Governing Council of the ISC successively as IASPEI, Australian, and United States representative from then until 1983 (with a short break in 1978), the last five years as Chairman of the Executive Committee.

I became a member of the Bureau of the Interunion Commission of Geodynamics in 1973 and President in 1975. The Earth sciences are essentially global in character, and thus there is a need for programs that cross political boundaries or for complementary programs on either side of the political boundaries. For many countries, the establishment of their own programs and cooperation with neighboring countries is greatly facilitated if carried out under some kind of international umbrella. Furthermore, it is clear that cooperation between geologists, geophysicists, and geochemists is necessary if we are to understand the Earth and the processes that have shaped it.

I regard the time I have spent on international programs as time well spent. In the course of the Geodynamics Program, the really hard work was done by the Secretary-General Don Russell. Frances Delaney, former Secretary-General, was most helpful, especially in liaison with the Asian countries, and Chuck Drake gave more of his time to Geodynamics than would ordinarily be expected of a past President. I had a fairly easy time in consequence.

I had expected that I would serve out my time to retirement in Dallas. However, unexpectedly in October 1972, John Jaeger wrote to me asking whether I would consider being a candidate for the foundation Directorship of the about-to-be-created Research School of Earth Sciences at the Australian National University. Whether the challenge of creating the new Research School or the great opportunities for seismological research in Australia (which had been revealed to me in a discussion with Ken Muirhead on a January 1973 visit to Australia) were the more attractive, I do not know, but I agreed to become a candidate, was offered and accepted the position, and took up my new duties in June 1973.

The committee that recommended the creation of the Research School of Earth Sciences at ANU suggested four new programs: Economic Geology, Environmental Geochemistry, Geophysical Fluid Dynamics, and Global Geodynamics. It was decided that we should try to get three of these started during my term as Director. The first step was to recruit senior staff to act as group leaders for the new programs. We were successful in persuading Stewart Turner, an Australian working in the Department of Applied Mathematics and Theoretical Physics at Cambridge, to return to Australia. We also recruited Lewis Gustafson for Economic Geology, and later Kurt Lambeck for the Global Geodynamics Group. The development of these new programs placed very considerable demands on the resources of the new School, and consequently expansion of existing programs was only possible to a limited degree.

One major development of an existing program was a start on construction of an ion probe. During my January 1973 visit to Australia, I talked with Bill Compston and Glen Riley about the ion probe. They convinced me that high resolution and high sensitivity required a larger secondary mass analyzer than was provided by any of the machines then available. After discussion with Compston, it was decided to appoint Steve Clement, a former student of his, as a Research Fellow to undertake design of the system under Compston's guidance. After some setbacks the probe was completed, and significant results have since been achieved with it. A notable result is the identification of zircons from Western Australia with an age of more than 4000 Myr.

Before I left Dallas, Kenneth Muirhead and I began to plan a seismic program. One major problem was that the seismic array at Warramunga was in disarray, with less than half the stations being operative. Clearly a major refurbishment of the array was required if it was to be of any use as a research tool. We decided to replace the cables from the seismometers by radio telemetry, as had already been done in Canada, and later to power the stations by solar panels. We also decided to build new electronics of Muirhead design for the 14-track magnetic tape recorders built at UTD. These systems used analog recording on magnetic tape running at 4 mil s^{-1} and could run unattended for 2 months on a set of dry batteries. Muirhead started planning a digitizing system and began development of the software for it and for the processing of the records. As a complement to the fifteen 14-track recorders, ten new 6-track recorders of Muirhead type were to be built.

All of this took time, and so while the equipment was being built, I looked at the problem of the averaging of oceanic and continental crust and upper mantle in free oscillation inversion. I also pursued the notion that the smoothing essential in any such inversion was most efficiently achieved by using parameterized models, such as what Cleary and I had done for the lower-mantle velocity. The use of parameterized models was approved at the Lima meeting of the Standard Earth Model Committee in 1973 (Hales et al 1974). Adam Dziewonski spent three months in Canberra in 1974, and during this time the parametric Earth model (PEM) was developed and the PEM paper written. As usual with Dziewonski, most of the work was done late at night or in the early hours of the morning.

In all of the free oscillation inversion models of the late 1960s and early 1970s, the period of the radial mode $_0S_0$ showed a residual several times larger than the estimated error in the observed period, whereas the residuals for the fundamental spheroidal mode period had the same sign for far more consecutive order numbers than was probable if the errors were random. All the models also required baseline corrections of 2 to 3 s to the P body wave travel times of Cleary & Hales and the Herrin 1968 Standard Travel Time Tables, and of 4–6 s to the S travel times of Hales & Roberts (1970b). These corrections were larger than I thought probable.

During the development of PEM, Dziewonski found that the residual for $_0S_0$ and the baseline correction to the body wave travel times were correlated, and that both could be reduced to zero at the expense of a small increase in the residuals of the fundamental modes. It was known at the time that the Q of the radial mode $_0S_0$ was very high, and I think now that we should have realized that there was a Q-effect on the periods of the fundamental modes. As it was, it was not until the paper of Randall (1976) that it was accepted that Q affected the periods of the fundamental modes, and that if this effect was allowed for, no baseline correction was needed for the body wave travel times.

Dziewonski returned to Canberra for a short visit in mid-1974. On the long flight to Australia, he had mulled over the possibility of trying to determine lateral variation of the velocities in the lower mantle using body wave travel times. Clearly, if this were to be possible, a very large body of travel time data was necessary. Even so it was possible that lower-mantle velocity variations would be masked by the lateral variations of the travel times through the upper mantle. However, Dziewonski was enthusiastic and was convinced that he could handle the complex numerical analysis required, so we called the Director of the International Seismological Centre and ordered the ISC tapes from 1964 onward. Dziewonski carried out the analysis at Harvard and (as reported by Dziewonski, B. H. Hager & R. J. O'Connell) found that there were long-wave anomalies in the lower mantle that correlated negatively with the long-wave anomalies in the gravity field found from satellite studies. Hager later gave the explanation for the negative correlation.

In order to extract these long-wave anomalies, it was, of course, necessary to establish global travel times. Dziewonski pursued the refinement of the global body wave travel times over the next several years, experimenting with the effect of relocating the earthquakes and with various ways of weighting the earthquakes used so as to minimize the effects of the nonuniform distribution of earthquakes and stations over the surface of the globe. I was an interested onlooker in these experiments and was particularly happy when Dziewonski and D. L. Anderson published the preliminary reference Earth model (PREM) paper, which found that the baseline tilt was not very different from the Cleary & Hales version of 1966.

The PREM model has a transversely anisotropic upper mantle. There is, of course, evidence of anisotropy in the upper mantle. However, I am not convinced that in the PREM model a pseudo-anisotropy of the kind described by K. Aki is not being lumped in with whatever real anisotropy exists. The point is of considerable importance, for by introducing transverse anisotropy Dziewonski and Anderson have removed the low-velocity zone from the upper mantle. The transversely isotropic model predicts differences of several seconds between SV and SH times at arc distances of 15–19°. This is a distance range where short-period S data are very poor and broadband seismograms covering the 4–10 s period range are needed. However there are few such data. Muirhead has looked at intermediate-period records from Canberra, but not enough suitable events were recorded while the intermediate-period instruments were operative for a definitive answer.

Two of my Dallas students, J. M. Mills and J. R. Clements, moved to Canberra to complete their PhD studies. Mills looked at group velocities and Q in the period range 80–300 s, which covers the minimum in Rayleigh-wave group velocities.

One of the things that impressed me about the Warramunga records was the large number of ScP phases routinely recorded with periods of 2–3 s. Simple calculations showed that if the free oscillation Q structure were used, ScP should have an amplitude smaller than that of P by more than a factor of 10. Such was not the case. Clearly, Q was frequency dependent. Clements' PhD thesis was concerned with efforts to determine Q for P, S, PcP, and ScP from the Warramunga records. The difficulty in determining transient body-wave Q's is to insure that other effects do not introduce systematic errors. I think Clements avoided most of these, and his conclusions that

$$Q_p(\omega) = {_pQ_0}[1 + \tau_p\omega],$$

$$Q_s(\omega) = {_sQ_0}[1 + \tau_s\omega],$$

(where the subscript 0 indicates the value of Q at the free oscillation periods, and τ_p, τ_s are constants) are valid. These equations are consistent with the equation proposed by Azimi et al (1968) to preserve causality [see Aki & Richards (1980), Equation 5.75]. The Q for S becomes equal to that for P at about 2 Hz, but the attentuation for S only approaches that for P asymptotically. The high-frequency P and S phases reported by Daniel Walker from the Wake Island hydrophone array can be accounted for by a combination of frequency-dependent Q of the type proposed by Clements and a whispering gallery effect.

Most of my own research time at Canberra was spent analyzing the records of the portable array system. The first subarray of this system was installed in December 1974, and the results from it showed that the apparent velocity for P waves in the distance range 800–1800 km was greater than 8.7 km s^{-1} and for S of the order of 4.75 km s^{-1}. A larger-scale array was installed in 1975, and analysis of the records from this array led to the construction of a model CAP8 considerably more complex than most upper-mantle models.

Shortly after I became Director of the Research School of Earth Sciences, Bill Best suggested that ANU should take over an infrasonic array operated by the U.S. Air Force in Australia. Before accepting this suggestion, I had to find out what kind of research was possible with the infrasonic array. Herrin and John Macdonald operated a similar array near Dallas, and I talked with them about it. One of Herrin's students, Gene Smart, had done his MSc research in Colorado on the propagation of atmospheric waves associated with thunderstorms and gave me a copy of his thesis. I was very interested in this work, and thus the infrasonic array was moved to

Warramunga and configured to determine the propagation velocity of the kind of wave studied by Smart. We hoped also to investigate the relation between long-period microseisms and atmospheric pressure variation, but the long-period seismometers at Warramunga never functioned satisfactorily.

Almost as soon as the array became operational, long-period solitary waves were observed, and the study of these has become the focus of the research carried out with the infrasonic array by Douglas Christie. He has been able to show, using portable infrasonic units, that these solitary waves propagate coherently over distances of several hundred kilometers.

My term as Director of the Research School of Earth Sciences ended in June 1978. I spent the next three months as a Fairchild Scholar at Caltech. As in 1960, there was a bright and enthusiastic group of graduate students at the Seismological Laboratory. It was a most enjoyable and relaxing change from administration. I returned to UTD at the beginning of the academic year and there continued to work on travel times.

I retired from UTD in January 1982; shortly before my leaving, my colleagues Richard Mitterer and Ronald Ward organized a symposium in my honor. One of the papers of that symposium (which was not published) was by Robert Meyer, who made a very strong case for many-element portable arrays.

During the next two or three years I visited Southern Methodist University twice. On one of these visits I spent a day at Caltech on the way to Dallas. It was one of the most exciting days of my life. Dziewonski, then a Fairchild Scholar, showed me color pictures of the upper mantle based on Ichito Nakanishi and Anderson's surface-wave analyses and J. H. Woodhouse and Dziewonski's free oscillation analyses. The sets of events were not the same and the methods of analysis different, but the mappings were very similar, given that the resolution was only of the order of 1000 km. Clearly, it was now possible to develop global maps of the upper mantle. Dziewonski also showed me mappings of the lower-mantle P velocity based on a new analysis of travel time anomalies. Together we spent the afternoon making sections through the Earth in regions that interested me. It was a fascinating afternoon. I later saw more of these pictures at a meeting in San Diego to create an organization for a global digital seismic network, which eventually crystallized as IRIS.

I believe that in order to achieve the full potential of the mapping now possible, it is necessary that there should be available portable arrays of broadband instruments with at least 50 elements. In fact, returning now to upper-mantle structure (which has been one of my primary interests of the past 20 years), I feel that the details of the S velocity distribution and the Q

variation in the upper mantle will only be resolved by the study of records in the period range 2–10 s.

One of the noteworthy features of the Central Australia study was that the refractions from below the 200-km discontinuity started with large amplitudes and petered out toward the end of the refraction branch. This behavior is most easily explained by a gradient below the discontinuity to give the large amplitudes at the start of the branch, followed by a region of decreasing velocity or a low-Q zone. Since both velocity and Q are temperature dependent, it is likely that both decreasing velocity and low-Q contribute to the decreasing amplitudes toward the end of the refracted branch.

Similar effects occur below the other upper-mantle discontinuities, which suggests that the temperature gradients between 200 and 700 km are greater than in most of the current temperature models. Thus, the effect of increasing temperature on the velocities overrides, or at any rate greatly reduces, the increase in velocity due to increasing pressure.

There are still some features of the lower-mantle velocity distribution that require further study. In the Jeffreys velocity distribution (Bullen 1963), the values of dV_p/dp (where V_p is the P velocity and p is the pressure) were high both in the transition zone and below it. The introduction of the velocity discontinuities at depths of 400 and 600–650 km resulted in much lower values of dV_p/dp in models such as PREM. Nevertheless, in PREM dV_p/dp decreases from 0.007 km s^{-1} kbar^{-1} at a depth of 746 km to 0.0022 km s^{-1} kbar^{-1} at a depth of 2121 km, thereafter increasing slightly to the onset of the thermal boundary layer. It is possible that the introduction of minor discontinuities in the lower mantle, such as were suggested as a result of the $dt/d\Delta$ studies (e.g. Johnson 1969) would lead to still lower values of dV_p/dp between depths of 700 and 2000 km. There is also some support for a velocity discontinuity about 300–400 km above the core-mantle boundary.

In summary, one can expect that the new global digital seismic network will, if properly configured, contribute significantly to the refinement of the present rather crude (although very exciting) models of lateral variations in the upper mantle. As I have remarked, the fact that broadband information will be available will be a significant addition to the seismological data bank. Given the Global Digital Seismic Network and the many-element portable arrays, I am confident that seismologists will in the next decade map the deep Earth in sufficient detail to be able to contribute significantly to a fuller understanding of the processes that have shaped the Earth. An exciting prospect!

Literature Cited

Aki, K., Richards, P. G. 1980. *Quantitative Seismology.* San Francisco: Freeman. 932 pp.

Azimi, Sh. A., Kalinin, A. V., Kalinin, V. V., Pivovarov, B. L. 1968. Impulse and transient characteristics of media with linear and quadratic absorption laws. *Izv. Acad. Sci. USSR Phys. Solid Earth* 42:88–93

Bullen, K. E. 1963. *An Introduction to the Theory of Seismology.* Cambridge: Cambridge Univ. Press. 381 pp.

Carey, S. W., ed. 1958. *Continental Drift: A Symposium.* Hobart, Aust: Univ. Tasmania. 374 pp.

Graham, K. W. T., Hales, A. L. 1957. Paleomagnetic measurements on Karroo dolerites. *Adv. Phys.* 6:149–61

Green, R. W. E., Hales, A. L. 1968. The travel times of P waves to 30° in the central United States and upper mantle structure. *Bull. Seismol. Soc. Am.* 58:267–89

Hales, A. L. 1960. Research at the Bernard Price Institute of Geophysical Research. *Proc. R. Soc. London Ser. A* 258:1–26

Hales, A. L. 1969. Gravitational sliding and continental drift. *Earth Planet. Sci. Lett.* 6:31–34

Hales, A. L., Roberts, J. L. 1970a. Shear velocities in the lower mantle and the radius of the core. *Bull. Seismol. Soc. Am.* 60:1427–36

Hales, A. L., Roberts, J. L. 1970b. The travel times of S and SKS. *Bull. Seismol. Soc. Am.* 60:461–89

Hales, A. L., Lapwood, E. R., Dziewonski, A. M. 1974. Parameterization of the spherically symmetrical earth model with special reference to the upper mantle. *Phys. Earth Planet. Inter.* 9:9–12

Johnson, L. R. 1969. Array measurements of P velocities in the lower mantle. *Bull. Seismol. Soc. Am.* 59:973–1008

Randall, M. J. 1976. Attenuative dispersion and frequency shifts of the Earth's free oscillations. *Phys. Earth Planet. Inter.* 12:1–4

Schreiner, G. D. 1958. Comparison of the ^{87}Rb → ^{87}Sr ages of the red granite of the Bushveld complex from measurements on the total rock and separated mineral fractions. *Proc. R. Soc. London Ser. A* 245:112–17

Taggart, J. M., Engdahl, E. R. 1968. Estimation of PcP travel times and the depth to the core. *Bull. Seismol. Soc. Am.* 58:1293–1303

van Zijl, J. S. V., Graham, K. W. T., Hales, A. L. 1962. The paleomagnetism of the Stormberg lavas of South Africa. I. Evidence for a genuine reversal of the Earth's field in Triassic-Jurassic times. *Geophys. J.* 7:23–39

Woods, D. V. 1979. *Geomagnetic depth-sounding studies in Central Australia.* PhD thesis. Aust. Natl. Univ., Canberra

Ann. Rev. Earth Planet. Sci. 1986. 14 : 21–42

TRIGGERED EARTHQUAKES

D. W. Simpson

Lamont-Doherty Geological Observatory of Columbia University, Palisades, New York 10964

INTRODUCTION

Over the past 30 years, it has become increasingly apparent that some of man's engineering activities can have a measurable influence on the way in which crustal stresses are released in earthquakes. Increases in seismic activity have resulted from the impounding of reservoirs behind high dams; large-scale surface quarrying; deep underground mining; the injection of fluids under high pressure into the ground in solution mining, waste disposal, geothermal power generation, and secondary oil recovery; the removal of fluids in petroleum operations; and the aftereffects of large underground explosions.

A number of adjectives have been used to describe this type of seismicity—man-made, induced, artificial, triggered—some of which give the erroneous impression that the human activities are the primary cause of the earthquakes, rather than just the trigger that acts to release preexisting stress of tectonic origin. In this paper, "triggered" and "induced" are used as the terms that best convey the triggering nature of the human influence.

Because of their association with major engineering projects, triggered earthquakes have significant social and economical implications. Since the very nature of induced earthquakes implies that they will occur near the engineering activity responsible for triggering them, even earthquakes as small as magnitude 5 are cause for concern. The largest generally accepted example of a triggered earthquake was the magnitude 6.5 earthquake at Koyna Reservoir in India in 1967, which caused 200 deaths, 1500 injuries, and major damage to the nearby town (Gupta & Rastogi 1976). The earthquakes at Koyna and Hsinfengkiang (China) dams caused significant damage to the dams themselves, and other triggered earthquakes (e.g. Kremasta, Greece; Oroville, California) have caused damage in surrounding populated areas. The occurrence of triggered earthquakes has resulted

21

0084–6597/86/0515–0021$02.00

in major modification to the engineering activities at some sites (e.g. the halting of fluid injection into a well at the Rocky Mountain Arsenal), and the potential for induced seismicity was an important factor in stopping the construction of the proposed Auburn Dam in California (Allen 1978).

Induced earthquakes were first associated with the initial filling of Lake Mead in the late 1930s (Carder 1945), but it was not until the late 1960s, after earthquakes larger than magnitude 5.5 had occurred at four large reservoirs in China, Rhodesia, Greece, and India, that major interest developed in the field of reservoir-induced seismicity. Early work by Rothé (1968, 1970), Archer & Allen (1969), Gupta et al (1969), and Gough & Gough (1970a,b) drew attention to the correlation of increased seismicity with loading at these and other sites and led to early attempts to determine the mechanism responsible for reservoir triggering (Gough & Gough 1970b, Snow 1972). A meeting sponsored by UNESCO on Seismic Phenomena Associated with Large Reservoirs in Paris in 1970, a publication of the US National Academy of Sciences (1972), and two symposia on induced seismicity organized by UNESCO in London, England, in 1973 (Judd 1974) and Banff, Canada, in 1975 (Milne 1976) were important in focusing the attention of various branches of the Earth sciences (seismology, geology, rock mechanics, hydrology, etc) on this problem.

In recent years, as more reservoir sites have been instrumented with seismic networks, numerous other examples of induced seismicity have been identified. There are now at least 50 cases where changes in earthquake activity have been detected following the filling of reservoirs.

While an interest in the effects of reservoir impoundment on seismicity was growing in the mid-1960s, an important example of a different type of induced seismicity had occurred at Denver, Colorado (Evans 1966, Healy et al 1968). High-pressure injection of fluid wastes into a 3.8-km-deep well at the Rocky Mountain Arsenal began in 1962 and was immediately followed by an anomalous increase in seismicity. Even though pumping was halted because of this seismicity in 1966, earthquakes continued to occur for several years afterwards, with the largest earthquake (of magnitude 5.5) in 1967.

Generated by interest in these early cases of reservoir- and injection-induced seismicity, a number of studies soon began on the mechanism responsible for triggering these earthquakes. At first, two independent effects were studied: Gough & Gough (1970a,b), as part of their work on the Lake Kariba (Africa) earthquakes, emphasized the influence of the elastic load; while Healy et al (1968), in studies related to fluid injection seismicity, stressed the influence of increased pore pressure. Later studies of reservoir-induced seismicity, including those of Snow (1972), Withers & Nyland (1976), and Bell & Nur (1978), considered the full consolidation problem and incorporated the coupled interaction between elastic and fluid stress. McGarr et al (1975), Cook (1976), and others have modeled the elastic stress changes near deep underground excavations and their relationship to triggered earthquakes.

Numerous reviews have summarized the case histories at specific sites of triggered earthquake activity, and the details are not presented here. Gupta & Rastogi (1976), Gough (1978), and the two symposia volumes edited by Judd (1974) and Milne (1976) provide summaries of all types of triggered seismicity. Mining-induced seismicity is described by Cook (1976). Early cases of reservoir-induced activity are reviewed by the National Academy of Sciences (1972), Rothé (1970), Simpson (1976), Packer et al (1977), Stuart-Alexander & Mark (1976), and Snow (1982). Important recent examples of reservoir-induced seismicity that are not covered in these earlier reviews include those at Manic-3, Quebec (Leblanc & Anglin 1978); Oroville, California (Bufe et al 1976, Toppozada & Morrison 1982); Nurek, USSR (Simpson & Negmatullaev 1981, Keith et al 1982); Jocassee and Monticello, South Carolina (Talwani 1979); Toktogul, USSR (Simpson et al 1981); and Aswan, Egypt (Kebeasy et al 1981, Simpson et al 1982).

ROCK FAILURE AND INDUCED STRESS CHANGES

Engineering works can influence the stress regime through increases in solid (elastic) stress or fluid (pore) pressure. The Mohr-Coulomb failure criterion (Jaeger & Cook 1971) provides a simple representation of the failure process in rocks and the influence of induced stresses. The criterion for shear failure (Figure 1a) in dry rock is $\tau = \tau_0 + \mu\sigma_n$, which relates the shear stress necessary for failure on a plane (τ) to the inherent shear strength of the material (τ_0), the normal stress across the plane (σ_n), and the coefficient of friction of the surface (μ). The stresses on such a system are usually specified with respect to an orthogonal set of axes, referred to as the principal stress axes, along which the stresses are purely compressive, there being no shear stress. The stress components relative to these axes are called the principal stresses and are designated σ_1 (maximum), σ_2 (intermediate), and σ_3 (minimum). It can be shown that failure will occur along a plane containing the intermediate stress and is independent of the value of the intermediate stress. Thus, for the evaluation of failure criteria, it is necessary to consider only the values of the maximum and minimum principal stresses. In order to study the failure of rock under a given set of principal stresses, it is necessary to resolve these stresses into shear and normal components along planes other than the principal stress directions. This can be done using the Mohr circle representation (Figure 1b). A circle is drawn containing points σ_1 and σ_3 on the compressive stress axis. This

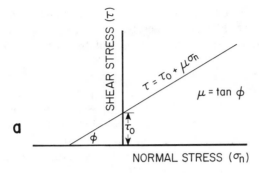

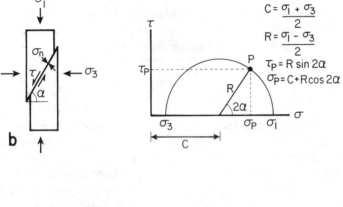

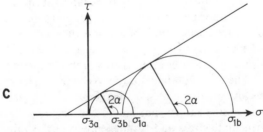

Figure 1 (a) Coulomb's law for failure in dry rock, showing the relationship between the shear stress (τ) required for failure and the normal stress σ_n across the plane. Here τ_0 is the cohesion and μ is the coefficient of friction. (b) The Mohr circle diagram, which provides a graphical method by which the principal (compressive) stresses (σ_i) can be resolved into shear (τ) and normal (σ_n) components on a plane at angle α to the σ_3 direction. (c) The Mohr-Coulomb failure criterion. Given maximum (σ_1) and minimum (σ_3) principal stresses, failure will occur on a plane containing the intermediate stress (σ_2) and at an angle α to σ_3 if the circle containing points σ_1 and σ_3 intersects the failure curve defined in (a).

circle is then the locus of normal (σ_n) and shear (τ) stress components on a suite of planes at an angle α to the minimum stress direction (σ_3). When the Mohr circle diagram is combined with the Coulomb-Navier failure criterion (Figure 1c), the relative values of σ_1 and σ_3 necessary for failure (i.e. for a circle tangent to the failure envelope) and the orientation α of the plane along which failure will occur can be determined.

Since the inherent shear strength of intact rock is considerably larger than that for preexisting failure surfaces, failure in previously fractured rock may occur first on preexisting planes other than at the angle α (Raleigh et al 1972). If the rock is subjected to internal fluid pressure P, the compressive stresses will be opposed by the hydrostatic fluid pressure, and the effective principal stress components will be $\sigma_i - P$. In terms of the Mohr circle representation, this is equivalent to a translation of the circle to the left by an amount equal to the magnitude of the fluid pressure. Thus, in the presence of fluid pressure, the effective normal stress is the elastic stress minus the pore pressure, and the failure criterion becomes $\tau = \tau_0 + \mu(\sigma_n - P)$.

The orientation of the principal stresses in nature varies with the tectonic environment, with the minimum compressive stress vertical in regions of thrust faulting, the maximum compressive stress vertical in regions of normal faulting, and the intermediate stress vertical in regions of strike-slip faulting (Anderson 1951). Snow (1972) has shown how the influence of a reservoir load varies with tectonic environment. The triggering effect of the induced elastic stress will be greatest in regions where the incremental induced stress adds to the tectonic shear stress by increasing the radius of the Mohr circle, either by increasing the maximum principal stress or by decreasing the minimum (Figure 2). Thus, an increase in vertical stress from the elastic load of a reservoir will have the greatest impact in regions of normal faulting, where the vertical stress is the maximum (Snow 1972, Simpson 1976); whereas a decrease in vertical stress (e.g. from the removal of surface load in quarrying operations) will have its greatest effect in regions of thrust faulting, where the vertical stress is the minimum (Pomeroy et al 1976). Increasing pore pressure (e.g. from fluid injection), while not changing the radius of the Mohr circle, translates the circle to the left (Figure 2) and thus always moves the circle toward the failure envelope (Hubbert & Rubey 1959), regardless of the tectonic stress environment.

In terms of this simple representation of failure, earthquakes are triggered when one of the following situations occurs:

1. The preexisting stress conditions on a fault are such that the magnitude of the induced stresses are sufficient to produce failure.
2. The triggering source acts to drive the natural stress condition closer to

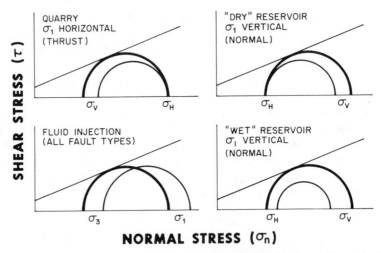

Figure 2   Examples of induced changes in stress that can trigger failure in different faulting environments. The thin circle is the original stress state and the heavy circle the state after the stress change. For simplicity, the small change in σ_H due to a lateral expansion caused by an increase in σ_V is ignored.

the failure envelope, either (*a*) by a purely elastic effect, increasing the deviatoric stress (increasing the radius of the Mohr circle); or (*b*) through increased pore pressure, decreasing the effective stress (translating the Mohr circle toward the failure envelope); or (*c*) a combined elastic and pore pressure effect.

TYPES OF TRIGGERED EARTHQUAKES

The three main types of triggered seismicity—injection, mining, and reservoir loading—can be distinguished in terms of the way in which they influence the stress field. Mining and quarrying operations have their major impact on the local elastic stress. Fluid injection affects primarily the fluid pressure distribution in the crust. The impounding of large reservoirs can affect both the elastic and the fluid stresses.

Figure 2 compares the influence of isolated changes in pore pressure (fluid injection), elastic stress (mining), and the combined elastic and pore pressure effects (reservoir loading).

Fluid Injection

The injection of fluids under high pressure into the crust has provided some of the best-documented and best-understood cases of triggered seismicity.

Since the injected fluid represents a relatively small mass and cannot substantially alter the elastic stresses, the major influence on the stress field is by increasing the pore pressure within the formation into which the fluids are injected. The resulting changes in strength can be expressed in terms of a single parameter—the excess of the injection pressure over the preexisting formation pressure. In the case of the Denver earthquakes (Healy et al 1968), seismicity started when the downhole pressure of 389 bars exceeded the original reservoir pressure (estimated at 269 bars) by 120 bars. When injection was stopped and the reservoir pressure at the well dropped to 311 bars, seismic activity near the well ceased, but earthquakes continued to occur up to 6 km from the well for at least two years as the anomalous pressure front continued to migrate outward from the injection point.

Following the Denver earthquakes, a controlled experiment was carried out in an oil field in Rangely, Colorado (Raleigh et al 1972, 1976), where fluid injection during secondary oil recovery had triggered minor seismicity. In this experiment, it was possible to measure all of the parameters necessary to establish the connection between the injection and the induced seismicity. A dense array of seismometers was used to accurately determine the locations and focal parameters of the induced earthquakes. Hydrofracturing was used to measure the state of stress in the rock at the depth of injection. Injection pressures were both measured and controlled. Laboratory tests provided estimates of the strength of the rock into which the injection took place. In their earlier paper, Raleigh et al (1972) combined the stress determinations from hydrofracturing, the orientations of fault planes, and the frictional properties of Weber sandstone to show that a pore pressure of approximately 260 bars (an increase of 90 bars from the original formation pressure of 170 bars) would be sufficient to induce slip. This value was in agreement with the pressure of 275 bars that existed in the well field at the time of the induced seismicity. In the later paper (Raleigh et al 1976), they were able to show that seismicity could be controlled by raising and lowering the pressure about the predicted value of 260 bars.

Mines and Quarries

At the other end of the elastic stress–pore pressure spectrum are those cases where seismicity is triggered primarily by changes in elastic stress caused by the removal of large masses of rock in mining and quarrying operations (Cook 1976). Two types of triggered earthquakes have resulted from the excavation of rock from the crust: those in the immediate working area of deep underground mines, and those triggered at depth beneath shallow mines and large surface quarries.

Various types of rock failure are an inherent problem in underground mines. Some of these failures, including "outbursts," "rock-falls," and gas-related "bumps" in coal mines, result from processes fundamentally different from those producing earthquakes (Osterwald 1970). True "rockbursts," however, do result from shear failure and, especially in very deep mines, have many characteristics in common with natural earthquakes (Cook 1976, Spottiswoode & McGarr 1975). Although the association of rockbursts with mining activity has been apparent for centuries, Gane et al (1946) were the first to begin seismological studies to establish the direct relationship of seismicity to deep gold mining in the Witwatersrand area, South Africa. More recent studies (Cook 1976, McGarr et al 1975, Spottiswoode & McGarr 1975) have shown that these events, of up to magnitude 5, occur within a few hundred meters of the active mine face, at depths of nearly 3 km from the surface. Because of the great depths of these mines and the high lithostatic stresses, the removal of mass in the mining operations can lead to induced stress changes of a kilobar or more. Since these events apparently result from the initiation of new shear failures in intact rock (Spottiswoode & McGarr 1975), this type of seismicity may be the closest to true "man-made" earthquakes.

In addition to seismic events generated in the immediate area of mine workings in deep mines, earthquakes have also occurred beneath shallow mines (Smith et al 1974) and large surface quarries (Pomeroy et al 1976, Gibowitz 1982). In these cases, the removal of mass from near the surface causes stress changes sufficient to trigger the release of preexisting stress at depth. In some mining operations, pumping of water from the mine workings may have a significant effect on the regional groundwater system (Gibowitz et al 1981), but the dominant effect in most mining operations will be the influence on the elastic stress field.

Reservoir Impounding

The most common type of triggered earthquake, and as yet the most poorly understood, is related to the impounding of large reservoirs. In these cases, the large mass of the reservoir represents an imposed load that causes a significant increase in elastic stress (Gough 1969), while increases in pore pressure can be generated both directly (by the infiltration of reservoir water) and indirectly (through the closure of water-saturated pores and fractures in the rock beneath the reservoir load). The surface load produced by a reservoir is 0.1 bar for each meter of water depth, or a maximum of 20 bars for all but the deepest of the world's large reservoirs. However, while the magnitude of the stress changes produced by reservoirs is smaller than for most mining and fluid injection sites, the larger physical dimensions of reservoirs allow their influence to extend over much broader areas. It is the coupling between the elastic and fluid effects and the major heterogeneities in rock properties influenced by the large physical dimensions of man-made lakes that make modeling the impact of reservoirs much more difficult than for cases of fluid injection or mining.

The most complete models of the effects of reservoir impounding have been presented by Withers & Nyland (1976) and Bell & Nur (1978). Both are based on Biot's (1941) consolidation theory (Rice & Cleary 1976). As pointed out by Bell & Nur (1978), the theory accounts for the three main effects of reservoir loading:

1. The elastic effect—the rapid increase of elastic stress that follows the loading of the reservoir;
2. Compaction—the increase in pore pressure, in saturated rock, caused by the decrease in pore volume from compaction due to the increased elastic stress;
3. Diffusion—the diffusion of pore pressure, related to fluid migration, from both (a) the reservoir itself and (b) the redistribution of pore fluids in response to the anomalous pore pressure changes caused by compaction.

In regions where the local groundwater table is deep prior to the filling of the reservoir, a fourth effect must be added (Snow 1972, Bell & Nur 1978):

4. Flow—the flow of reservoir water into previously unsaturated layers, raising the groundwater table.

The redistribution of pore pressure from compaction and diffusion in saturated rock involves relatively little fluid flow. Significant amounts of water, however, can be involved in flow into unsaturated rock, which substantially increases both the effective head and the effective volume of the reservoir. This effect is most pronounced in flat, arid regions where the final reservoir extends well beyond the original confines of the river channel.

Bell & Nur (1978) consider two types of reservoirs: one in which the water of the reservoir is isolated from the rock underneath by an impermeable bottom, and one in which the bottom is permeable and thus permits the flow of water from the reservoir into the underlying rock. The initial loading of a reservoir always leads to an instantaneous increase in pore pressure due to the elastic compaction of saturated pore space. For the case of an impermeable reservoir bottom, this initial increase in pore pressure disappears with time as the fluids migrate outward in adjustment to the new elastic stress field. For the permeable reservoir bottom, steady-state flow from the reservoir can develop, maintaining or exceeding the initial pore pressure increase. Thus, the impermeable bottom can lead to a transient weakening due to increased pore pressure on first filling of a reservoir, whereas the permeable case produces an effect in which the weakening increases with time. Bell & Nur (1978) also point out (cf Snow 1972, Simpson 1976) that rapid unloading of a reservoir can always lead to relative weakening, since the load stresses are removed more rapidly than the pore pressures can diffuse away.

Bell & Nur (1978) considered the case of spatially homogeneous hydraulic and elastic parameters. If the rock beneath the reservoir is inhomogeneous or fractured, zones of more compressible rock may give rise to localized pockets of higher pore pressure as a result of compaction following initial loading. In these zones, the pore pressure may decay away with a short time constant related to the spatial dimensions of the inhomogeneities, in constrast to the longer time constant required for the effect of the reservoir at the surface to reach to the depths of the inhomogeneities (Leith et al 1984). In this case, compaction may lead to weakening that is dependent on the rate of loading: The pore pressure will rise if the time for loading is short compared with the time constant for diffusion out of the localized zones of high compressibility; if the loading rate is slow, the pore pressure within the localized zones may be able to dissipate before critical values are reached. This may provide an explanation for the apparent correlation between induced seismicity and loading rate at some reservoir sites (Simpson & Negmatullaev 1981, Toppozada & Morrison 1982, Gupta 1983).

The difference between the time scales for compaction- and diffusion-controlled changes in pore pressure may be responsible for two distinct types of reservoir-induced seismicity (Leith et al 1984). Compaction results from increased elastic stress and thus acts with little time delay after the surface load is imposed. In contrast, increased pore pressure transmitted by diffusion from the surface may require months to years to reach hypocentral depths (Howells 1974, Talwani 1981). Figure 3 shows the water levels in some of the well-studied reservoirs that have shown induced seismicity. The times of significant earthquakes or bursts of seismicity are indicated. At Nurek, Manic-3, and Monticello reservoirs, the seismicity initiated immediately after the first rapid increase in water level, and at Nurek further rapid changes in water level were also immediately followed by increases in seismicity. In contrast, the large earthquakes at Oroville, Kariba, Aswan, and Koyna occurred many years after the reservoirs began to fill. At Oroville and Koyna, the water levels had been close to the maximum for 4 to 6 yr prior to the time of the large earthquakes. Leith et al (1984) suggest that the "rapid response" in seismicity in cases like Nurek, Manic-3, and Monticello is related to elastically induced changes in pore pressure due to compaction, whereas the "delayed response" at Oroville, Kariba, Aswan, and Koyna is more closely related to the diffusion of pore pressure to depth along fault zones. There are also differences in the

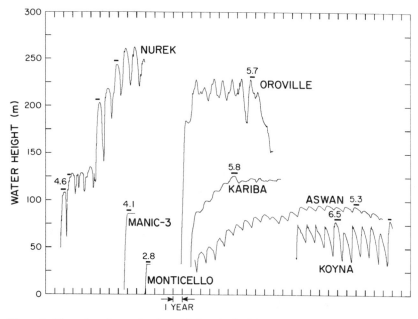

Figure 3   Examples of water levels and induced seismicity. Vertical scale is absolute water depth at the dam. The horizontal scale is the same for all curves, but the absolute positions have been shifted for clarity. Numbers above the water level curves are magnitudes of the largest earthquakes. Bars indicate times of prominent bursts of seismicity. For details of seismicity at each site, see references in text.

characteristics of the seismicity within these two types of responses. The rapid response is generally related to shallow, low-magnitude seismicity, directly beneath or adjacent to the reservoir and hence within the main influence of the elastic stress increase. Cases of delayed response are often deeper, of larger magnitude, and up to 20 km from the reservoir, but have occurred on fault zones that intersect the reservoir.

STRESS CYCLES ON FAULTS AND TRIGGERED EARTHQUAKES

Through various tectonic processes, the state of stress in the Earth's crust changes as a function of time. The most direct manifestation of these changes is the abrupt release of stress during earthquakes. While the entire crust is in a stressed state, the release of stress is spatially concentrated along zones of lowest strength, or faults. The temporal variations of stress changes along faults is cyclic, with a gradual build-up of stress until a failure level is reached, producing a drop in stress during an earthquake. This is followed by a recovery of the stress drop until the next earthquake occurs and the cycle repeats. This regularity in the earthquake cycle has formed the basis for recent models of earthquake prediction (Shimazaki & Nakata 1980, Sykes & Quittmeyer 1981). In regions of high strain accumulation, the repeat time between major earthquakes along the same fault segment is measured in tens to hundreds of years. In more stable regions, the repeat times for large earthquakes may be thousands to hundreds of thousands of years.

It is through the modification of the cycle of stress release that human activities have their major impact in triggered earthquakes. Figure 4 shows a simplified schematic of the temporal variations in stress along a fault zone. The natural cycle of earthquakes is represented by an idealized constant repeat time and constant stress drop [see Shimazaki & Nakata (1980) for a discussion of the influence of variations in stress drop on the time to the subsequent earthquake]. An increase in the shear stress (e.g. from the

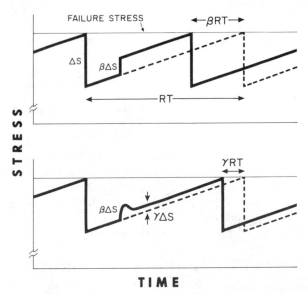

Figure 4 Schematic representation of stress on a fault zone. The normal earthquake cycle is shown as the dashed line. The stress drop (ΔS) in an earthquake is recovered during an interseismic period with repeat time RT. When the failure level is reached, rupture occurs and the cycle repeats. The failure stress may be on the order of kilobars, while the stress drop may be tens of bars or less. If an induced stress change of $\beta\Delta S$ occurs, the repeat time is shortened by βRT. In the lower diagram, the induced stress change includes a transient component. Note that in both cases the induced stress change will trigger rupture if it occurs during the last βRT yr of the cycle.

addition of a reservoir load or removal of overburden in mining) and/or a decrease in the effective normal stress (e.g. from increased pore pressure) modifies the natural process by rapidly increasing the stress and advancing the time at which failure occurs. How much the anomalous stress increase will influence the seismicity depends on the magnitude of the induced stress changes compared with the total amount of stress remaining to be recovered between earthquakes.

While the ambient stresses at hypocentral depths are on the order of kilobars, crustal earthquakes relieve only part of this stress, and recent determinations (Archuleta et al 1982, Sykes & Quittmeyer 1981) indicate stress drops of tens of bars or less. Thus, for faults exhibiting repeated earthquakes, the state of stress may never be more than the equivalent of one stress drop (tens of bars) away from failure, and for a substantial part of the interseismic period even smaller changes in stress (a few bars) may be sufficient to trigger failure. Thatcher (1982) has suggested that episodes of accelerated stress increase may be responsible for triggering natural seismicity and notes the similarity with the types of triggered seismicity being considered here. He has also suggested (Thatcher 1983) that in some regions the rate of stress accumulation may be more rapid immediately following earthquakes, with a significant part of the stress being recovered in a short time, which leads to even longer fractions of the interseismic period in which the stress may be just below failure. The triggering of earthquake swarms by water level changes of less than one meter (equivalent to less than 0.1 bar) at some sites of reservoir-induced seismicity (e.g. Simpson & Negmatullaev 1981) in itself supports the idea that seismicity can be triggered on faults that otherwise remain stable even at stress levels extremely close to failure (Leith & Simpson 1985).

A preexisting state of stress that is both high and close to failure has been given as a prerequisite for the occurrence of triggered seismicity (e.g. Gough & Gough 1970b, Simpson 1976); the assumption is that induced stress changes will be small in comparison to both the level of ambient stress at hypocentral depths and the failure strength of rock. As Figure 4 shows, however, it is not the absolute levels of stress that are important, but only the differences between the magnitude of the induced stresses and the stress recovery required to initiate failure. Since the ambient tectonic stress field can maintain a level that is high compared with the temporal modulation of these stresses that occurs during earthquakes, it is not unreasonable that induced stress changes on the order of a few bars or less can trigger seismicity on preexisting faults. The failure of intact rock requires a much higher level of shear stress (to overcome the cohesive strength) than is needed to overcome frictional forces on preexisting fault surfaces. Thus, except for seismicity in deep mines, where induced stress changes on the order of kilobars may occur, it is unlikely that significant triggered earthquakes will occur in the absence of preexisting faults.

It has been suggested that one method of assessing the potential for induced seismicity at a particular reservoir site would be to measure the in situ state of stress and determine how close it is to failure. However, the uncertainties involved in both the stress measurements and the failure criteria may be large compared with the stress change required to initiate failure. There may also be substantial differences between the magnitudes of the stresses and failure conditions extrapolated from near-surface measurements in accessible boreholes and those that actually exist on faults at hypocentral depths. A much more important application of in situ stress measurements is in determining the orientation of the current stress field. Especially in regions of low seismicity, the current stress regime may be different than that which existed when major faults in the area were produced. The orientation of the principal stresses determined from in situ measurements can aid in identifying those faults that have orientations conducive to failure under the current tectonic stress field.

Arguments have also been made that the triggering of earthquakes (by advancing the time at which they would occur naturally) may alleviate the seismic hazard by causing smaller earthquakes than would eventually occur. As suggested by Figure 4, however, the time at which the earthquake is triggered may have little effect on the magnitude of the stress drop. What will be more important in determining the size of a triggered earthquake is the length of rupture over which failure occurs. This will be controlled primarily by the spatial extent along the fault over which the induced stresses act; this, in turn, will be strongly influenced by the geometry of the fault system with respect to the source of induced stress changes. Faults that parallel and intersect a large reservoir may be much more prone to failure than those that lie outside the direct influence of the reservoir. Fluid injection is most likely to trigger earthquakes when the injection is directly into a fault zone.

While a prestressed state is still a prerequisite for the occurrence of triggered seismicity, it may be that fault zones in many parts of the Earth's surface lie in a state of stress that is susceptible to triggering by stress changes of relatively small magnitude such as are produced by the engineering works of man. The more important diagnostics as to whether significant triggered earthquakes will occur at a particular site may be (a) whether or not potential fault surfaces exist, (b) whether these faults are within the influence of the inducing agent, (c) whether the induced stress changes can be focused on these faults, and (d) whether the fault geometry is conducive to failure within the current tectonic stress regime.

TRIGGERED EARTHQUAKES AND INCREASED SEISMIC RISK

The development of any major engineering activity, especially in areas of known seismicity, requires that the seismic hazards be assessed and that proper seismic criteria be incorporated into the design of structures.

There has been a general reluctance in parts of the engineering community to accept the significance or even existence of induced seismicity. In a recent publication of the US National Academy of Sciences concerning safety of dams (Committee on the Safety of Existing Dams 1983, p. 274), a cursory section on induced seismicity states:

> There is a question of whether a reservoir may induce earthquakes. To date there is no universally accepted proof that this can occur, but it is a possibility that should be given consideration.... A recent study (Mead 1982) indicated that the possibility of reservoir-induced earthquakes is very limited and probably should not be considered except for extremely large and deep reservoirs.

This statement is made in spite of the fact that the largest induced earthquake occurred at Koyna Reservoir, which, with a dam height of 85 m above the riverbed and a reservoir volume of 2.8 km^3, is not large in comparison with many of the world's dams. A number of well-documented cases of induced seismicity (albeit with considerably smaller magnitudes) have occurred at even smaller reservoirs (e.g. Manic-3, Monticello). It is true that many reservoirs, mines, quarries, and injection operations have not triggered any detectable increases in seismicity and that most of the reported cases of induced seismicity involve small earthquakes that in themselves are not of engineering significance. However, it is clear that at Koyna, Kariba, Hsinfengkiang, and Oroville the resulting induced seismicity far exceeded that which might have been anticipated on the basis of standard risk assessment techniques that do not incorporate the possibility of induced seismicity.

Allen (1982) summarizes and refutes many of the arguments presented as to why reservoir-induced seismicity is not of significant importance in engineering design. One of the common arguments is that a relatively small percentage of large reservoirs have caused significant induced seismicity. Depending on the definition of "large reservoir" and "significant seismicity," percentages of from 2% (Allen 1982) to 15% (Gough 1978) can be obtained. If the sample were large enough and both the process of triggering and the geological conditions were the same at all sites, a purely statistical approach might provide a reasonable method for assessing the potential for induced seismicity. As discussed above, however, the specific geological structures in the reservoir area, their spatial relationship to the reservoir, and the prior seismicity of the region are far more important in controlling the occurrence of induced seismicity than the physical size of the reservoir itself (Castle et al 1980).

Most of the standard methodologies used for the determination of seismic hazard and the associated risk rely on the basic uniformitarian principle of geology—that "the past is the key to the present." The seismicity of the region is evaluated on the basis of available geological, seismological, and historical data to determine the expected maximum size of an earthquake, the rate of occurrence of these earthquakes, and the spatial distribution of seismicity with respect to the planned structure. Using these parameters, along with the magnitude distribution of seismicity, it is possible to determine the expected annual occurrence of earthquakes for significant faults near the structure. Given the source characteristics of the earthquakes and the rate of attenuation of seismic waves, the annual probability for various levels of shaking at the site can be obtained. While these techniques are appropriate for most passive structures such as large buildings or nuclear power plants, they require modification for the types of engineering activities described here because the structures themselves are capable of significantly *changing* the seismic regime and thus violating the basic principle on which the methods are based.

Engineering structures have finite lifetimes, and most risk assessments are based on expected lifetimes measured in tens to hundreds of years. The determination of design criteria involves an assessment of the acceptable level of risk. In an area of relatively low seismicity and long repeat times between large earthquakes, a very low annual probability of strong ground shaking may be an acceptable risk in terms of the expected lifetime of the structure under consideration. Thus, it is not only the maximum size of the earthquakes that is important in risk assessment but also their distribution in time. As noted above, it is by changing the temporal distribution of seismicity that triggered earthquakes have their greatest impact.

Two inherent characteristics of triggered earthquakes must be considered during the assessment of seismic risk:

1. Triggered earthquakes will occur during the lifetime of the activity triggering them.
2. Triggered earthquakes will occur in close proximity to the activity triggering them.

A methodology for the proper assessment of seismic risk at potential sites of triggered seismicity will require the incorporation of the probability of occurrence of triggered earthquakes expressed not as an annual probability based on the previous natural rate of seismicity, but as the probability that triggered earthquakes will occur at that site during the lifetime of the structure. Especially for the case of seismicity related to reservoirs, we do not yet have sufficient understanding of the mechanism by which seismicity is induced at any particular site to provide the specific criteria necessary to quantitatively assess the potential for triggering.

It is in areas of relatively low natural seismicity that triggered earthquakes can have their greatest impact on engineering design, and it is in these areas that the largest induced earthquakes have occurred. In active tectonic regions, a reservoir may have less impact in changing the seismic regime (Simpson 1976, Lomnitz 1974), and in any case the occurrence of natural earthquakes would normally play a major role in the engineering design criteria because of the relatively short repeat times between earthquakes. In relatively aseismic regions, where the repeat times of natural earthquakes may be thousands of years or longer, an increased probability of triggering the maximum expected earthquake during the lifetime of the structure can significantly alter the risk estimate. To illustrate this difference, consider the implications for a probabilistic risk estimate at a reservoir site for a simple model suggested by Figure 4. For simplicity, we assume that the stress drops, and hence the repeat times, for earthquakes are constant. Variations from this assumption add statistical uncertainty to the probability of earthquake occurrence but do not alter the general conclusions of the following example. We also assume various ratios β of the magnitude of the induced stress to the stress drop in earthquakes. While this parameter cannot be assessed with any great certainty given our current knowledge of the mechanism of induced seismicity, it is useful for illustration purposes and suggests a possibly productive avenue for future study.

In the absence of induced seismicity, the annual probability of the occurrence of an earthquake in this model (assuming the temporal position within the stress recovery curve is unknown) is $P_a = 1/RT$, where RT is the repeat time between earthquakes (i.e. there is a 1 in RT chance that the given year is the last point in the stress recovery curve).

If an induced stress increase raises the stress level by a value of $\beta\Delta S$, there is now a β chance that this will lead to failure (i.e. the induced stress change at any time during the final βRT yr of the stress recovery curve will lead to failure).

Note that in estimating the risk over the lifetime of a structure, we are calculating the total probability that rupture will occur and not an annual probability. Figure 4 also shows how the short-term risk is greatest during the early stages of reservoir impoundment. Depending on the rate of water level rise and the local geological conditions, the initial stress increase may take months to years. After this initial increase has passed, however, the *annual* probability is the same as in the case with no induced stress, if we assume that no further changes in stress are induced. If there are annual or other types of rapid changes in water level, these will be associated with additional periods of short-term increased risk.

If the lifetime of the structure under consideration is LT yr, then the total probability of occurrence during the structure's lifetime (assuming that the induced stress change is permanent and the same natural rate of stress increase continues after the induced stress change) is $P_t = (\beta RT + LT)/RT = \beta + LT/RT$, where the case of $\beta = 0$ (no induced stress increase) corresponds to the normal case of $P_t = LT/RT$. In other words, the induced stress has produced the equivalent of βRT yr of natural stress recovery and has increased the "effective lifetime" to $LT + \beta RT$ yr.

Figure 5*a* shows the total probability as a function of the ratio of the lifetime to repeat time (LT/RT) for various values of β. Figure 5*b* shows the increase in the probability over that without the assumption that induced changes in stress occur.

If the induced stress changes include a transient component at the initiation of filling of the reservoir (Figure 4), the natural and induced components in the probability are no longer additive and the total probability will be $P_t = LT/RT$ (for $LT/RT > \beta$) and $P_t = \beta$ (for $LT/RT < \beta$). As can be seen from Figure 5, this produces only a minor deviation from the curves shown in the vicinity of $LT/RT = \beta$, so that to a first approximation the influence of any transient component in the induced stress change can be added to the effect of the permanent component.

This simple example clearly shows that the impact of a reservoir on the increased risk is relatively small in areas where the repeat time is short compared with the lifetime of the structure, but that it has a major impact, increasing the probability by up to a factor of 1000 or more, in areas of much longer repeat times. Even in areas of long repeat times, the total probability approaches a minimum value equal to the ratio β of the induced stress to stress drop. There are a number of factors that would make these estimates of increased risk a minimum. If there is an initial period of rapid stress recovery following an earthquake (Thatcher 1983, Castle et al 1980), then the probabilities for later times will be considerably higher than for the linear recovery of stress indicated in Figure 4. In addition, this model assumes that the position on the stress recovery curve is unknown; however, it is likely that the position is not located during the early part of the curve, since an earthquake in the recent past would have been apparent and have been given special consideration in the risk analysis.

Mitigation of Risk From Triggered Earthquakes

There are indications that the risk posed by triggered earthquakes can be mitigated by careful control of the activity responsible for the triggering. In

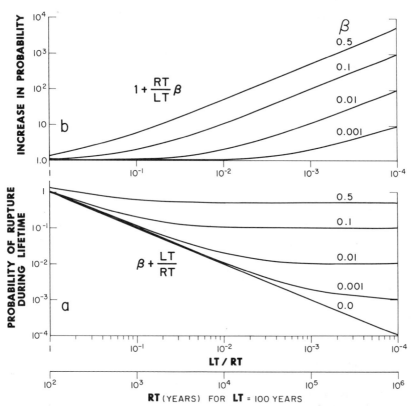

Figure 5   (*a*) Probability that the induced stress change of $\beta \Delta S$ in Figure 4 will trigger rupture during the lifetime LT of a structure, as a function of the ratio LT/RT of the lifetime to the repeat time for various values of β. For simplicity, the model in Figure 4 assumes a constant repeat time and hence is deterministic, so that as LT/RT approaches unity the "probability" (better considered as the expected number of events) can exceed 1. (*b*) The increase in probability for the case shown in (*a*) over that when there is no induced stress change ($\beta = 0$).

the cases of seismicity triggered by fluid injection at Denver and Rangely, it was clearly shown that seismicity eventually can be stopped by ceasing the injection or using lower pumping pressures. The occurrence of the largest earthquakes at Denver a year after pumping had stopped, however, indicates that the process, once started, may not be completely controlled. In deep mines, careful planning of the excavation process can minimize the extent of failure or at least control the timing of major rockbursts so that precautionary measures can be taken. Monitoring of microseismic activity in mines can be used to predict the occurrence of larger failures.

At some reservoir sites it appears that the triggering of earthquakes is controlled by the rate at which the water level is changed (Simpson & Negmatullaev 1981, Toppozada & Morrison 1982, Gupta 1983). In those environments where compaction-induced changes in pore pressure are important, gradual changes in the water level of the reservoir may allow excess pressures to dissipate before critical levels are reached and thus may decrease the transient component (Figure 4) in the induced stress change.

ACKNOWLEDGMENTS

I would like to thank Chris Scholz and Wayne Thatcher for stimulating my interest in the relationship between stress drop, induced stresses, and increased risk. Bill Leith and Dan Davis provided helpful reviews of the manuscript. The water level data in Figure 3 were provided by a number of agencies, including the Tadjik Institute of Seismoresistant Construction and Seismology (Nurek), the California Division of Mines and Geology (Oroville), the Central African Power Corporation (Kariba), the Aswan and High Dam Authority (Aswan), and the Central Water and Power Research Station (Koyna). This work was supported by the US Geological Survey under contract USGS 14-08-0001-22002 and is Lamont-Doherty Geological Observatory Contribution Number 3847.

Literature Cited

Allen, C. R. 1978. Evaluation of seismic hazard at the Auburn damsite, California. *U.S. Bur. Reclam. Rep.*, Denver, Colo. 10 pp.

Allen, C. R. 1982. Reservoir-induced earthquakes and engineering policy. *Calif. Geol.* 35:248–50

Anderson, E. M. 1951. *The Dynamics of Faulting.* Edinburgh: Oliver & Boyd. 206 pp. 2nd ed.

Archer, C. B., Allen, N. J. 1969. *A Catalogue of Earthquakes in the Lake Kariba Area, 1959–1968.* Salisbury: Meteorol. Serv. 35 pp.

Archuleta, R. J., Cranswick, E., Mueller, C., Spudich, P. 1982. Source parameters of the 1980 Mammoth Lakes, California, earthquake sequence. *J. Geophys. Res.* 87:4595–4607

Bell, M. L., Nur, A. 1978. Strength changes due to reservoir-induced pore pressure and stresses and application to Lake Oroville. *J. Geophys. Res.* 83:4469–83

Biot, M. A. 1941. General theory of three-dimensional consolidation. *J. Appl. Phys.* 12:155–64

Bufe, C. G., Lester, F. W., Lahr, K. M., Lahr, J. C., Seekins, L. C., Hanks, T. C. 1976. Oroville earthquakes: normal faulting in the Sierra Nevada foothills. *Science* 192:72–74

Carder, D. S. 1945. Seismic investigations in the Boulder Dam area, 1940–1944, and the influence of reservoir loading on earthquake activity. *Bull. Seismol. Soc. Am.* 35:175–92

Castle, R. O., Clark, M. M., Grant, Z. A., Savage, J. C. 1980. Tectonic state: its significance and characterization in the assessment of seismic effects associated with reservoir impounding. *Eng. Geol.* 15:53–99

Committee on the Safety of Existing Dams. 1983. *Safety of Existing Dams: Evaluation and Improvement.* Washington, DC: Natl. Acad. Press. 345 pp.

Cook, N. G. W. 1976. Seismicity associated with mining. *Eng. Geol.* 10:99–122

Evans, M. D. 1966. Man-made earthquakes in Denver. *Geotimes* 10:11–17

Gane, P. G., Hales, A. L., Oliver, H. O. 1946. A seismic investigation of Witwatersrand earth tremors. *Bull. Seismol. Soc. Am.* 36:49–80

Gibowitz, S. J. 1982. The mechanism of large mining tremors in Poland. *Proc. Int. Congr. Rockbursts and Seism. in Mines, 1st,* ed. N. C. Gay, E. H. Wainwright, pp. 17–28. Johannesburg: S. Afr. Inst. Min. Metall.

Gibowitz, S. J., Droste, Z., Guterch, B., Hordejuk, J. 1981. The Belchatow, Poland, earthquakes of 1979 and 1980 induced by surface mining. *Eng. Geol.* 17:257–71

Gough, D. I. 1969. Incremental stress under a two-dimensional artificial lake. *Can. J. Earth Sci.* 6:1067–1151

Gough, D. I. 1978. Induced seismicity. In *The Assessment and Mitigation of Earthquake Risk*, pp. 91–117. Paris: UNESCO

Gough, D. I., Gough, W. I. 1970a. Stress and deflection in the lithosphere near Lake Kariba, 1. *Geophys. J. R. Astron. Soc.* 21:65–78

Gough, D. I., Gough, W. I. 1970b. Load-induced earthquakes at Lake Kariba, 2. *Geophys. J. R. Astron. Soc.* 21:79–101

Gupta, H. K. 1983. Induced seismicity hazard mitigation through water level manipulation at Koyna, India: a suggestion. *Bull. Seismol. Soc. Am.* 73:679–82

Gupta, H. K., Narrain, H., Rastogi, B. K., Mohan, I. 1969. A study of the Koyna earthquake of December 10, 1967. *Bull. Seismol. Soc. Am.* 59:1149–62

Gupta, H. K., Rastogi, B. K. 1976. *Dams and Earthquakes*. Amsterdam: Elsevier. 299 pp.

Healy, J. H., Rubey, W. W., Griggs, D. T., Raleigh, C. B. 1968. The Denver earthquakes. *Science* 161:1301–10

Howells, D. A. 1974. The time for a significant change in pore pressure. *Eng. Geol.* 8:135–38

Hubbert, M. K., Rubey, W. W. 1959. Role of fluid pressure in mechanics of overthrust faulting, 1. *Geol. Soc. Am. Bull.* 70:115–66

Jaeger, J. C., Cook, N. G. W. 1971. *Fundamentals of Rock Mechanics*. London: Chapman & Hall. 515 pp.

Judd, W. R., ed. 1974. *Eng. Geol.* Vol. 8. 212 pp.

Kebeasy, R. M., Maamoun, M., Ibrahim, E. 1981. Aswan lake induced earthquake. *Bull. IISEE* 19:15–160

Keith, C., Simpson, D. W., Soboleva, O. V. 1982. Induced seismicity and style of deformation at Nurek reservoir, Tadjik SSR. *J. Geophys. Res.* 86:4609–24

Leblanc, G., Anglin, F. 1978. Induced seismicity at the Manic-3 reservoir, Quebec. *Bull. Seismol. Soc. Am.* 68:1469–85

Leith, W., Simpson, D. W. 1985. Seismic domains within the Gissar-Kokshal seismic zone, Soviet Central Asia. *J. Geophys. Res.* In press.

Leith, W., Simpson, D. W., Scholz, C. H. 1984. *Two types of reservoir-induced seismicity.* Presented at IASPEI Reg. Assem., Hyderabad, India

Lomnitz, C. 1974. Earthquakes and reservoir impounding: state of the art. *Eng. Geol.* 8:191–98

McGarr, A., Spottiswoode, S. M., Gay, N. C. 1975. Relationship of mine tremors to induced stresses and to rock properties in

the focal region. *Bull. Seismol. Soc. Am.* 65:981–93

Mead, R. B. 1982. The evidence for reservoir-induced macroearthquakes. *Misc. Pap. S-73-1*, US Army Corps Eng., Waterw. Exp. Stn., Vicksburg, Miss. 194 pp.

Milne, W. G., ed. 1976. *Eng. Geol.* Vol. 10, pp. 83–388

National Academy of Sciences/National Academy of Engineering, USA. 1972. *Report: Earthquakes Related to Reservoir Filling.* Washington, DC: Div. Earth Sci., Natl. Res. Counc. 24 pp.

Osterwald, F. W. 1970. Comments on rockbursts, outbursts and earthquake prediction. *Bull. Seismol. Soc. Am.* 60:2083–88

Packer, D. R., Lovegreen, J. R., Born, J. L. 1977. Reservoir induced seismicity. In *Woodward Clyde Consultants, Earthquake Evaluation Studies in the Auburn Dam Area*, Vol. 6. *Natl. Tech. Inf. Ser. Rep. PB-283 531/GA, PC A15/A01.* 327 pp.

Pomeroy, P. W., Simpson, D. W., Sbar, M. L. 1976. Earthquakes triggered by surface quarrying—Wappingers Falls, New York sequence of June, 1974. *Bull. Seismol. Soc. Am.* 66:685–700

Raleigh, C. B., Healy, J. H., Bredehoeft, H. D. 1972. Faulting and crustal stress at Rangely, Colorado. *Geophys. Monog. Am. Geophys. Union* 16:275–84

Raleigh, C. B., Healy, J. H., Bredehoeft, H. D. 1976. An experiment in earthquake control at Rangely, Colorado. *Science* 191:1230–37

Rice, J. R., Cleary, M. P. 1976. Some basic stress-diffusion solutions for fluid-saturated elastic porous media with compressible constituents. *Rev. Geophys. Space Phys.* 14:227–41

Rothé, J. P. 1968. Fill a lake, start an earthquake. *New Sci.* 39:75–78

Rothé, J. P. 1970. Seismic artificiels (man-made earthquakes). *Tectonophysics* 9:215–38

Shimazaki, K., Nakata, T. 1980. Time-predictable recurrence model for large earthquakes. *Geophys. Res. Lett.* 7:279–82

Simpson, D. W. 1976. Seismicity changes associated with reservoir impounding. *Eng. Geol.* 10:371–85

Simpson, D. W., Negmatullaev, S. Kh. 1981. Induced seismicity at Nurek reservoir, Tadjikistan, USSR. *Bull. Seismol. Soc. Am.* 71:1561–86

Simpson, D. W., Hamburger, M. W., Pavlov, V. D., Nersesov, I. L. 1981. Tectonics and seismicity of the Toktogul Reservoir region, Kirgizia, USSR. *J. Geophys. Res.* 86:345–58

Simpson, D. W., Kebeasy, R. M., Maamoun, M., Albert, R., Boulous, F. K. 1982. Induced seismicity at Aswan Lake, Egypt.

Eos, Trans. Am. Geophys. Union 63:371 (Abstr.)

Smith, R. B., Winkler, P. L., Anderson, J. G., Scholz, C. H. 1974. Source mechanisms of microearthquakes associated with underground mines in eastern Utah. *Bull. Seismol. Soc. Am.* 64:1295–1317

Snow, D. T. 1972. Geodynamics of seismic reservoirs. *Proc. Symp. Percolation Through Fissured Rocks*, T2-J:1–19. Stuttgart: Ges. Erd- und Grundbau

Snow, D. T. 1982. Hydrogeology of induced seismicity and tectonism: case histories of Kariba and Koyna. *Geol. Soc. Am. Spec. Pap. No. 189*, pp. 317–60

Spottiswoode, S. M., McGarr, A. 1975. Source parameters of tremors in a deep-level gold mine. *Bull. Seismol. Soc. Am.* 65:93–112

Stuart-Alexander, D. E., Mark, R. K. 1976. Impounding-induced seismicity associated with large reservoirs. *US Geol. Surv. Open-File Rep. 76-770*

Sykes, L. R., Quittmeyer, R. C. 1981. Repeat times in great earthquakes along simple plate boundaries. In *Earthquake Predic-tion—An International Review, Maurice Ewing Ser.*, ed D. W. Simpson, P. G. Richards, 4:217–47. Washington, DC: Am. Geophys. Union. 680 pp.

Talwani, P. D. 1979. Induced seismicity and earthquake prediction studies in South Carolina. *8th Ann. Tech. Rep., Contract 14-08-0001-14553*, US Geol. Surv., Reston, Va.

Talwani, P. D. 1981. Hydraulic diffusivity and reservoir-induced seismicity. *Final Tech. Rep.*, US Geol. Surv., Reston, Va. 48 pp.

Thatcher, W. 1982. Seismic triggering and earthquake prediction. *Nature* 299:12–13

Thatcher, W. 1983. Nonlinear strain buildup and the earthquake cycle on the San Andreas fault. *J. Geophys. Res.* 88:5893–5902

Toppozada, T. R., Morrison, P. W. 1982. Earthquakes and lake levels at Oroville, California. *Calif. Geol.* 35:115–18

Withers, R. J., Nyland, E. 1976. Theory for the rapid solution of ground subsidence near reservoirs on layered and porous media. *Eng. Geol.* 10:169–85

Ann. Rev. Earth Planet. Sci. 1986. 14:43–70

EL NIÑO

Mark A. Cane

Lamont-Doherty Geological Observatory of Columbia University,
Palisades, New York 10964

INTRODUCTION

In the year 1891, Señor Dr. Luis Carranza, President of the Lima Geographical Society, contributed a small article to the Bulletin of that Society, calling attention to the fact that a counter-current flowing from north to south had been observed between the ports of Paita and Pacasmayo.

The Paita sailors, who frequently navigate along the coast in small craft, either to the north or to the south of that port, name this counter-current the current of "El Niño" (the Child Jesus), because it has been observed to appear immediately after Christmas.

As this counter-current has been noticed on different occasions, and its appearance along the Peruvian coast has been concurrent with heavy rain in latitudes where it seldom, if ever, rains to any great extent, I wish, on the present occasion, to call the attention of the distinguished geographers here assembled to this phenomenon, which exercises, undoubtedly, a very great influence over the climatic conditions of that part of the world.

These remarks are taken from an 1895 address to the Sixth International Geographical Congress by Señor Federico Alfonso Pezet of the Lima Geographical Society (Pezet 1896). The El Niño countercurrent was apparently first reported by the Frenchman Lartigue in 1822, but the Peruvian climate anomalies that are a part of the larger pattern we now refer to as El Niño are documented at least as far back as 1726 (Quinn et al 1978).

Carranza's article and Pezet's talk were prompted by the "tremendous rains" associated with the El Niño of 1891, probably the strongest in the past 100 years—until superseded by the El Niño of 1982/83, the El Niño preceding the present article. Concerning 1891, Pezet goes on to note that

it was then seen that, whereas nearly every summer here and there there is a trace of the current along the coast, in that year it was so visible, and its effects were so palpable by the fact that large dead alligators and trunks of trees were borne down to Pacasmayo from the north, and that the whole temperature of that portion of Peru suffered such a change owing to the hot current which bathed the coast.

43

0084–6597/86/0515–0043$02.00

The connection between El Niño and climate is not restricted merely to the region directly touched by the El Niño current; rather, it is a *global* pattern of anomalies referred to as the Southern Oscillation. Major El Niño–Southern Oscillation (ENSO) events, such as those of 1891 or 1982, have profound global ecological, social, and economic consequences (e.g. Barber & Chavez 1983, Canby 1984, Glantz 1984). ENSO is also of scientific interest as the best-defined, most prominent signal in year-to-year climate variability. As such, it is widely perceived as an entrée to a broader understanding of the atmosphere and ocean as a coupled climate system.

The seminal figure in delineating the Southern Oscillation was Sir Gilbert Walker, the Director-General of Observatories in India. Walker assumed his post in 1904, shortly after the famine resulting from the monsoon failure of 1899 (an El Niño year). He set out to predict the monsoon fluctuations, an activity begun by his predecessors after the disastrous monsoon of 1877 (also an El Niño year). Walker was aware of work indicating a swing of sea-level pressure from South America to the India-Australia region and back, with a period of several years (cf Rasmusson 1985).

Over the next 30 years Walker added correlates from all over the globe to this primary manifestation of the Southern Oscillation; among these were rainfall in the central equatorial Pacific and in India, and temperatures in southeastern Africa, southwestern Canada, and the southeastern United States (Walker 1924, Walker & Bliss 1932). No conceptual framework supported the patterns he found; Walker's methods were strictly empirical. Together with the short duration of the records then available, this made it easier for others to dismiss his findings as mere artifacts.

By the 1960s interest in the Southern Oscillation began to revive (cf Rasmusson 1985), and Walker's correlations were found to hold when reexamined with decades of new data (see especially Horel & Wallace 1981). Although both El Niño and the Southern Oscillation had been known at the turn of the century, it was only now that the connection between them was finally recognized, principally through the work of Bjerknes (1966, 1969, 1972). Figure 1 shows how closely these two are related.

Bjerknes did more than just point out the empirical relation between the two; he also proposed an explanation that depends on a two-way coupling between the atmosphere and ocean. His ideas were motivated by observations of large-scale anomalies in the atmosphere and tropical Pacific Ocean during 1957–58, the International Geophysical Year.

As it happened, a major El Niño occurred in those years. (Coincidentally, they were also the last two years of Sir Gilbert Walker's life.) It is implausible that a local coastal warming could cause global changes in the atmosphere, but the 1957 data showed that the rise in sea-surface

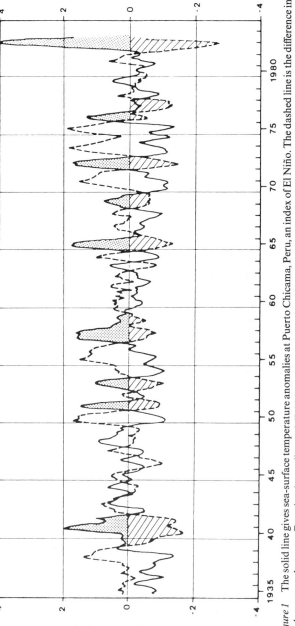

Figure 1 The solid line gives sea-surface temperature anomalies at Puerto Chicama, Peru, an index of El Niño. The dashed line is the difference in sea-level pressure between Darwin, Australia, and Tahiti, an index of the Southern Oscillation. Both are normalized by their long-term standard deviations. Major El Niño events are shaded (from Rasmusson 1985).

temperature (SST) extended along the equator from the South American coast to the date line (cf Figure 2). Bjerknes suggested that this feature was common to all El Niño events; he was correct, and the term "El Niño" is now often used to denote the basin-scale oceanic changes. In his account of the connection between the ocean and atmosphere, the coastal events constituting the narrow definition of El Niño are incidental to the important oceanic change: the warming of the tropical Pacific over a quarter of the circumference of the Earth.

Bjerknes suggested a tropical coupling between El Niño and the Southern Oscillation; he also hypothesized a link between tropical Pacific SST and midlatitude circulation anomalies. This teleconnection idea is consistent with the global nature of Walker's Southern Oscillation pattern. It is a subject of intense interest among theoreticians, observationalists, and

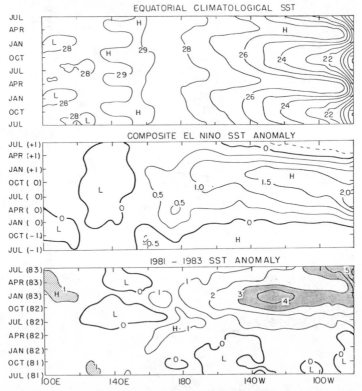

Figure 2   Time-longitude sections of sea-surface temperature (SST). The sections follow the equator to 95°W and then follow the climatological cold axis to its intersection with the South American coast at 85°S. (*top*) Mean climatology. (*middle*) Composite El Niño anomalies. The El Niño year is year 0. (*bottom*) Anomalies from 1981 to 1983. Note the larger contour interval.

long-range forecasters, but it is beyond the scope of this article (see Rasmusson & Wallace 1983, and references therein). The tropical coupling is central to our theme, for we regard El Niño and the Southern Oscillation as the oceanic and atmospheric components of a single phenomenon, referred to as ENSO.

The Bjerknes Hypothesis

Bjerknes (see especially the 1969 paper, from which we quote freely) begins his account by pointing out that the eastern equatorial Pacific is unusually cold among low-latitude oceans. He attributes this to equatorial upwelling and horizontal advection of cold waters driven by the easterly trade winds prevailing along the equator. Since the western Pacific is very warm, there is a large SST gradient along the equator in the Pacific [cf Figure 2 (top)]. As a result there is a direct thermal circulation in the atmosphere along the equator: The relatively cold, dry air above the cold waters of the eastern equatorial Pacific flows westward along the surface toward the warm western Pacific. "There, after having been heated and supplied with moisture from the warm waters, the equatorial air can take part in large-scale, moist-adiabatic ascent" (Bjerknes 1969). Some of the ascending air joins the poleward flow at upper levels associated with the Hadley circulation, and some returns to the east to sink over the eastern Pacific. There is a zonal surface pressure gradient associated with this equatorial circulation cell (high in the east and low in the west).

Bjerknes named this the "Walker Circulation" because he felt that fluctuations in this circulation initiated pulses in Walker's Southern Oscillation. It can have such global consequences because "it operates a large tapping of potential energy by combining the large-scale rise of moist air and descent of colder dry air" (Bjerknes 1969).

The Walker Circulation is the link between eastern Pacific SST anomalies and the Southern Oscillation. As Bjerknes (1969) states:

A change toward a steeper pressure slope at the base of the Walker Circulation is associated with an increase in the equatorial easterly winds and hence also with an increase in the upwelling and a sharpening of the contrast of surface temperature between the eastern and western equatorial Pacific. This chain reaction shows that an intensifying Walker Circulation also provides for an increase of the east-west temperature contrast that is the cause of the Walker Circulation in the first place. Trends of increase in the Walker Circulation and corresponding trends in the Southern Oscillation probably operate in that way. On the other hand, a case can also be made for a trend of decreasing speed of the Walker Circulation, as follows. A decrease of the equatorial easterlies weakens the equatorial upwelling, thereby the eastern equatorial Pacific becomes warmer and supplies heat also to the atmosphere above it. This lessens the east-west temperature contrast within the Walker Circulation and makes that circulation slow down.

There is thus ample reason for a never-ending succession of alternating trends by air-sea interaction in the equatorial belt, but just how the turnabout between trends takes place is not yet quite clear.

Bjerknes' elegant scenario, in which both the tropical ocean and atmosphere are active participants, has been the principal stimulus for subsequent research on ENSO. The framework he proposed still serves as the underpinning for the more complete structure that has been built since. An enhanced observational picture has been constructed, and Bjerknes' "admittedly somewhat tenuous reasoning" about causal connection has been buttressed by a more solid theoretical foundation.

The greatest advances since Bjerknes' day have been on the oceanographic side of the problem. Although the oceanographic component of his scenario is not very specific, he did recognize that the variations in SST during El Niño are related to ocean dynamics and not to changes in surface heat flux. The key to these dynamics was to switch attention from SST to sea-level variations. The work of Wyrtki (1975, 1979) in collecting and interpreting sea-level data from a network of tide gauges in the tropical Pacific is the basis of our present understanding of the oceanography of El Niño.

Our goal in the remainder of this article is to provide the background for a coupled model of ENSO with strong emphasis on the oceanography (i.e. on El Niño). We begin with a highly selective review of the observations of the normal annual cycle in the Pacific before moving on to observations of the evolution of a typical El Niño event. The theory for the oceanography of El Niño follows. After briefly considering the influence of SST anomalies on the tropical atmosphere, we present results from a numerical model for the coupled system able to generate El Niño events. A discussion of the implications for the real ENSO cycle concludes this review.

CLIMATOLOGY OF THE TROPICAL PACIFIC

Where the lower-level winds converge in the tropics, there is upward motion, condensation, and release of latent heat. Such regions provide much of the thermal driving for the atmospheric circulation, both tropical and extratropical. By far the most powerful of these is the Indonesian Low over the "maritime continent" between the Indian and Pacific Oceans; it is the western terminus of the easterly trade winds that constitute the lower branch of the Walker Circulation. Extending southeastward from this zone into the Southwest Pacific is the South Pacific Convergence Zone; a second band of convergence, the Intertropical Convergence Zone, is generally found north of the equator and extends eastward from the central Pacific to the American coast.

As noted above, equatorial Pacific SST is warm in the west and cold in the east [Figure 2 (top)]. This surface picture reflects the distribution of oceanic heat content. Almost everywhere in the ocean the surface waters are well mixed, primarily by wind stirring. Along the equator in the Pacific this surface mixed layer typically exceeds 100 m in depth, becoming shallower to the east until it almost disappears at the South American coast. The depth of the thermocline, the thin layer of high temperature gradient separating the warm waters of the upper ocean from the cold abyssal waters, has a similar zonal variation. Sea level is also higher in the west. The western tropical Pacific is the largest pool of very warm water in the world ocean. It is maintained by the oceanwide trade winds, which drive currents westward along the equator under the tropical sun.

Figure 2 shows that there is very little temperature variation—annual or interannual—west of the date line, where the warm pool lies. The annual variation is largely in the east, roughly in the region bounded to the east and west by the South American coast and longitude 140°W and to the north and south by latitudes 3°N and 15°S. The interannual variability associated with El Niño is also largest in this region (Figures 2, 3). As noted by Bjerknes (and others) this area is the coldest in the low-latitude oceans. Heat exchange with the atmosphere is not the cause: The surface heat flux in this

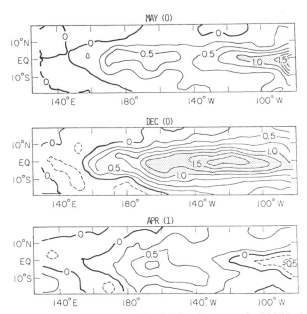

Figure 3 Sea-surface temperature anomalies (°C) for the composite El Niño for May and December of the El Niño year and April of the following year.

region is in excess of 50 W m^{-2} into the ocean, enough to increase the temperature of a 50-m-thick surface mixed layer by 1°C in less than a month. Most of the surface flow into this cold tongue is the relatively cold water brought in from the south in the Peru Current. There is a net inflow of colder water, since the outflow (westward in the South Equatorial Current and poleward to the north and south) is warmer, having been heated by the Sun.

Coastal and equatorial upwelling are other sources for the cold SST in this area. Winds along the South American coast are southerly, and the Coriolis effect turns the surface currents offshore. The waters leaving the coast are replaced by colder waters from below. Similarly, easterlies induce equatorial upwelling because the Coriolis effect turns the waters poleward in both hemispheres, making the surface flow divergent at the equator. Since the thermocline in the eastern Pacific is so close to the surface, the waters brought up to the surface mixed layer by both forms of upwelling are unusually cold. Two recent estimates (Wyrtki 1981, Bryden & Brady 1985) indicate that while the horizontal divergence of heat in the surface layer is not negligible, the flux of cold water from below associated with upwelling is the dominant process maintaining the cold tongue.

Beginning late in the boreal fall there is an annual warming of the usually cold tongue in the eastern equatorial Pacific [Figure 2 *(top)*]. SST variations are a consequence of basin-wide equatorial ocean dynamics, not of changes in surface heat flux. As the Sun retreats south, the Intertropical Convergence Zone migrates equatorward and the southeast trade winds relax, which reduces coastal and equatorial upwelling and slows the flow in the Peru and South Equatorial Currents. All these effects combine to reduce the flux of cold water into the surface layer, and since the surface heating rate is maintained, the surface waters warm.

These local factors are aided by remote influences. The warm waters to the west are advected eastward as a consequence both of the weakening of the easterlies in the western and central Pacific that occurs during the fall and winter and of the seasonal transition of the Asian monsoon (cf Cane 1983). A related oceanic response is a deepening of the thermocline in the east, so that water upwelled to the surface is warmer than before.

THE COMPOSITE EL NIÑO

As background for our discussion of the mechanisms governing the ENSO cycle, it is useful to describe the "composite" El Niño event. It is based on data from the 1950s to the 1970s from a number of sources [especially Rasmusson & Carpenter (1982) and Wyrtki (1975, 1979)]. It is possible to composite data from many past events because El Niño tends to be locked to the annual cycle [cf Figure 2 *(top, middle)*, Figure 4].

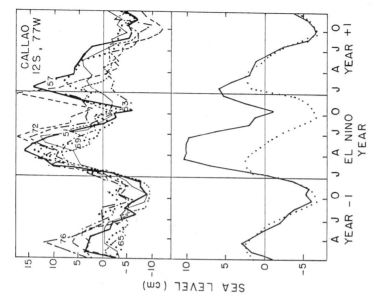

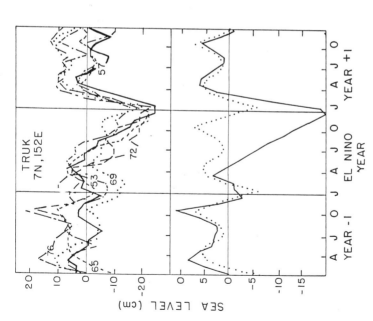

Figure 4 El Niño signatures: sea level at Truk and Callao for indicated El Niño events (top panels), for the composite El Niño (bottom panels, continuous line), and for the annual mean in non–El Niño years, 1953 to 1976 (bottom panels, dotted line). At Truk, note the similarity among El Niño events and their collective difference from the semiannual cycle of non–El Niño years. In the eastern Pacific (e.g. Callao), El Niño events typically appear as an enhancement of the annual cycle (after Meyers 1982).

Prelude

Typically, there are stronger than average easterlies in the western equatorial Pacific preceding an El Niño event, especially a strong event. These winds move water westward, and consequently sea level is higher than normal in the west and lower in the east. Equatorial SST is slightly warm in the west and somewhat cold east of 160°E.

Onset

In the fall preceding El Niño, the warm anomaly in the South Pacific develops a northward extension across the equator in the vicinity of the date line. This is associated with a northeast shift of the South Pacific Convergence Zone, which brings it closer to the equator than normal. The easterlies west of the date line have started to diminish, and the sea-level slope along the equator has begun to relax. There are positive precipitation anomalies west of the date line (e.g. Nauru and Ocean Islands), but no discernible pattern to the convergence anomalies.

Event

The anomalous warming off the coast of South America begins in January or February and increases until June (Figure 3). For the first several months it is difficult to distinguish it from the normal warming that occurs every winter [Figure 2 (top)]. At the same time, the sea level rises in a narrow region along the South American coast (see Callao in Figure 4) and the thermocline deepens. There is strong southward flow at the coast (the El Niño Current mentioned by Pezet). There is also evidence for a sea-level rise north of the equator, at least as far north as San Diego (Enfield & Allen 1980). The SST anomaly at the equator in the vicinity of the date line persists (and perhaps expands) throughout this period (Figure 2). At this time there are westerly wind anomalies along the equator from 110°W to 170°E, with the maximum near the date line. The Intertropical Convergence Zone has shifted equatorward in the east, so there is enhanced convergence and precipitation all along the equator from Peru to 175°E.

During the next half year, the warm anomaly spreads northwestward and then westward along the equator at a speed of about 1 m s^{-1}. By late fall the eastern anomaly has merged with the one in the central Pacific: Warm water now girdles a quarter of the Earth (Figures 2, 3). At this time, SST at the coast is only slightly anomalous, although the thermocline is still substantially deeper than normal there.

The large drop in sea level at Truk shown in Figure 4 is characteristic of the western Pacific. Thus there is a redistribution of mass from west to east during an event; in the 1976 El Niño this occurred at an average rate of 27×10^6 m^3 s^{-1}, about half the strength of the westward-flowing South Equatorial Current (Wyrtki 1979).

Mature Phase

Something more or less like the normal annual warming takes place at the coast beginning in December of the El Niño year. After reaching a peak, SST drops off very sharply, and by the end of the normal warm season in March-April it is slightly colder than normal (Figure 3). Sea level follows a similar pattern (Figure 4); note how quickly the sea level at Truk returns to normal. The positive SST anomaly farther to the west remains through the early part of the year, disappearing from east to west [Figure 2 (*middle*)]. During this time the winds relax to their normal pattern, and the westward sea-level slope is reestablished. During the second half of the year, the system overshoots its mean state: The trade winds are stronger than normal at the equator, as are the easterlies in the far western Pacific; sea-level slope is greater than normal; and eastern equatorial SST is below normal.

Discussion

This description of the composite event omits features judged to be extraneous to the question of how El Niño events are created and evolve. For example, events outside the tropical Pacific have been neglected. [Cane & Sarachik (1983) point out some possibly significant omissions.]

In Bjerknes' scheme, SST changes can induce wind changes, and conversely, wind changes can produce SST changes. The question has often been raised, which is the prime mover setting off the ENSO event—the atmosphere or the ocean? Our description is true to the data in its inability to sort out lag–lead relationships; to within the temporal resolution of the observations, many of the critical changes in the atmosphere and ocean are simultaneous.

Although SST anomalies, a surface expression of deeper changes, are conspicuous only in the east, El Niño is a basin-wide change in the equatorial Pacific that involves a substantial east-to-west redistribution of upper-ocean waters. If only the eastern Pacific Ocean were considered, it might seem proper to characterize El Niño as an enhancement of the usual annual cycle. On the other hand, El Niño in the western part of the Pacific Ocean is markedly different from the normal cycle (e.g. Figure 4). El Niño events are not just the tail of the distribution of annual events; the distribution is bimodal, with El Niño and non–El Niño years being distinctly different (Meyers 1982).

The atmospheric changes may be succinctly characterized by saying that the convergence zone normally centered over the maritime continent

observational evidence has accumulated to show that a signal much like the theoretically conceived Kelvin wave does indeed cross the breadth of the Pacific at speeds of almost 3 m s^{-1}.

In pre–El Niño conditions the prevailing easterlies, whether or not stronger than normal, tend to pile up warm water at the western side of the Pacific. A relaxation of the winds along the equator in the central or western Pacific excites packets of Kelvin waves, which cross the ocean to the South American coast. Their effect is to raise sea level in the east.

In principle, the coastal changes could depend on the local setup in response to longshore winds. In fact, as shown by Wyrtki (1975; see also Enfield 1981), there is almost no change in the coastal winds during El Niño. Hence, changes in currents and in all aspects of the thermal structure at the coast, including sea-level displacement and thermocline depth, depend solely on the amplitude of the incident Kelvin waves (Cane 1984).

This amplitude is determined by its initial value at the western end of the Pacific plus the amount added by wind forcing as it propagates along the equator. Model calculations show the latter to be the principal influence (BOB, Cane 1984). Furthermore, all that matters is the zonal wind stress within a Kelvin wave width of the equator—a few hundred kilometers. Figure 5 shows this forcing for the gravest baroclinic mode, based on the composite winds of Rasmusson & Carpenter (1982). The dashed lines show the path of a Kelvin wave through the longitude-time plane. For the composite El Niño, the primary cause of the rise in sea level at the beginning of the El Niño year is the change from easterly to westerly anomalies in the vicinity of the date line. For the four major events encompassed in the BOB calculations, the locale and nature of this initial wind signal varied: Sometimes it was east of the date line and sometimes west; sometimes it was more of a slackening of anomalously strong easterlies, and sometimes an actual westerly anomaly. The second peak late in the year is a response to the massive collapse of the trade winds that begins in the summer (Figure 5), a feature common to all events.

The most quantitative comparison of linear theory with El Niño sea-level changes to date is Busalacchi & Cane's (1985) study of the 1982/83 event. They found excellent agreement in the eastern Pacific (i.e. toward the end of the equatorial-coastal waveguide), with correlations between model and data above 0.9. A test of a different sort is provided by an ocean general circulation model simulation of the 1982/83 event (Philander & Seigel 1985). This model includes as complete an account of ocean physics as is presently possible, and the mechanisms described above are clearly operative.

Perhaps it is surprising that the linear dynamical theory works as well as it does. The dynamics governing currents along the equator (e.g. the

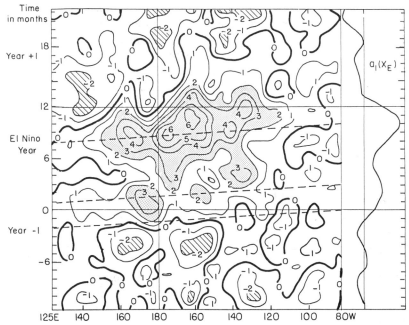

Figure 5 Forcing for the gravest baroclinic Kelvin wave, based on the composite El Niño wind anomaly field. Dashed lines indicate the path of a Kelvin wave. The curve on the right gives the Kelvin wave amplitude at the eastern boundary (from Cane 1984).

equatorial undercurrent) are highly nonlinear and viscous. However, numerical experiments with nonlinear models (Cane 1979, Philander 1979) indicate that variations in such essentially integral quantities as sea level and dynamic topography are well predicted by linear dynamics. [On the other hand, Cane (1984) suggests some important discrepancies in a linear calculation, and Philander & Seigel (1985) describe other mass redistribution mechanisms present in their general circulation model simulation.]

SEA-SURFACE TEMPERATURE

Despite their central role in ocean-atmosphere interaction, there have been few heat budget or modeling studies of El Niño SST anomalies (but see Barnett 1977). Although there may be local exceptions, the preponderant evidence is that surface heating does not contribute to the El Niño warming (Bjerknes 1969, 1972, Ramage & Hori 1981, Weare 1983). The data indicate an inverse relation between SST and heat flux into the ocean because of increased evaporation. This is the expected sense if SST anomalies are to lead to the heating that drives the atmospheric circulation during ENSO. If it is assumed that anomalous surface flux into the ocean depends only on local SST anomalies, then, as with sea level, the influence of the atmosphere reduces to the anomalous surface wind stress.

Based on this idea, Zebiak & Cane (1985) constructed a simple perturbation model for SST anomalies, with mean temperature structure and surface currents specified on the basis of climatological data. The dynamics of the model begin with the linear reduced-gravity model that is so successful in simulating thermocline depth anomalies and surface pressure changes during El Niño events. Such models produce only depth-averaged baroclinic currents, but the surface current is usually dominated by the frictional (Ekman) component. Therefore, a shallow frictional layer of constant depth is added to simulate the surface intensification of wind-driven currents in the real ocean. The dynamics of this layer are also kept linear, but only by using Rayleigh friction to stand in for nonlinear influences at the equator [see Zebiak & Cane (1985) for a discussion]. Upwelling velocity is computed as the divergence of the surface-layer transport. Inclusion of this surface layer allows a strong response to local winds; models that omit it understate upwelling effects as a consequence.

In contrast to the dynamics, the evolution equation for perturbation SST is complete and nonlinear, including three-dimensional temperature advection by both mean and anomalous currents. The temperature of upwelled water is parameterized as a function of the total thermocline depth, varying most rapidly when the thermocline is near the surface. Since the mean thermocline is shallower in the east, the model is more sensitive to depth changes there, in accord with observations. The only significant simplification in this equation is that the surface heat flux is taken to be linearly proportional to the SST anomaly. This Newtonian cooling formulation is the simplest embodiment of the inverse relation discussed above.

The Zebiak & Cane (1985) model was forced with the composite wind anomalies; results are shown in Figures 6 and 7. The first of these figures indicates the extent of the model's ability to reproduce the time evolution averaged over key regions. The second shows the anomaly patterns at key times and should be compared with Figure 3. Given the model's simplicity, the results are encouraging, especially in view of the large uncertainties in the forcing and verification data. Evaluation of the contribution of the various terms in the SST equation leads to the conclusion that all make a significant contribution at some time or place. Mean upwelling (i.e. warming of upwelled water due to a deepening of the thermocline) is the strongest influence at the South American coast (NINO1). In the central Pacific (NINO4), horizontal advection dominates because the reduction in the upwelling strength is offset by a rise in the thermocline. In between, in

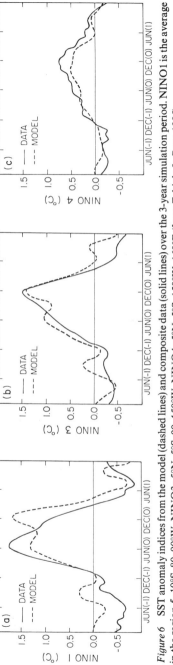

Figure 6 SST anomaly indices from the model (dashed lines) and composite data (solid lines) over the 3-year simulation period. NINO1 is the average in the region 5–10°S, 80–90°W. NINO3: 5°N–5°S, 90–150°W. NINO4: 5°N–5°S, 150°W–160°E (from Zebiak & Cane 1985).

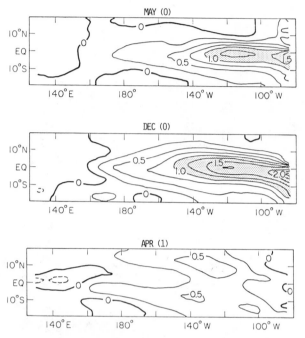

Figure 7 Model SST anomaly fields for composite wind forcing. Shown are May and December of the El Niño year, and April of the year after El Niño (from Zebiak & Cane 1985). Compare with Figure 3.

the eastern equatorial Pacific (NINO3), all the dynamical terms make a substantial contribution to the warming.

The discussion of SST has focused here on the equatorial waveguide. Meridional currents are important in spreading the anomalies off the equator and away from the coast (e.g. Leetmaa 1983, Zebiak & Cane 1985). The North Equatorial Countercurrent strengthens during El Niño, transporting more warm water eastward north of the equator (Wyrtki 1975). In the powerful 1982/83 event, some of this warm water may even have spilled across the equator (Philander & Seigel 1985).

TROPICAL ATMOSPHERIC RESPONSE DURING ENSO

We have seen that the ocean state can be accounted for if the wind stress is specified. Similarly, if SST anomalies characteristic of El Niño are given, then the principal changes in the tropical circulation may be calculated. This has been amply demonstrated by simulations with atmospheric general circulation models. [Shukla & Wallace (1983) contains a thoughtful review of general circulation model studies up to that time.] The most convincing is the set of 15-year integrations reported by Lau & Oort (1985) and Lau (1985). Runs made with observed tropical Pacific anomalies for the years 1961 to 1976 were contrasted with runs with climatological SSTs. The former simulated the ENSO signal, the latter did not.

The outstanding change observed during an ENSO event is the migration of the zone of convergence normally found in Australasia to the central Pacific. The general circulation models simulate this, and Bjerknes' ideas on the variations of the Walker Circulation explain the changes. Beyond that, observations show that the tropical anomalies have a simple vertical structure with a universal form, namely, a reversal of polarity between the lower and upper troposphere (e.g. regions of low-level convergence lie below regions of upper-level divergence). Linear dynamical models with a single degree of freedom in the vertical have proven remarkably adept at reproducing the horizontal structure of the atmosphere (Matsuno 1966, Gill 1980), but the physical interpretation of these models is uncertain [on this issue, see especially Geisler & Stevens (1982) and Zebiak (1982)].

In their purest form, these are models for the dynamical response to heating, not to SST variations (cf Shukla & Wallace 1983), and they give the most impressive results when the heating is specified (e.g. Gill & Rasmusson 1983). Anomalous heating need not be closely related to SST anomalies. First, the data show that the maximum convergence, and hence the maximum heating, tends to be near the SST maximum. Thus, the maximum heating *anomaly* need not be where the maximum SST *anomaly* is. Second, heating is determined less by local evaporation (a function of local SST) than by the convergence of moisture in the boundary layer. Thus the heating depends on the circulation and does so with a potentially positive feedback: Stronger moisture convergence gives more heating; more heating gives more convergence. General circulation models include parameterizations of all the physical processes involved in this complex chain, while the simple models usually do not. There are exceptions; for example, Webster (1981) and Zebiak (1985) include convergence feedbacks. Zebiak's is a model for the perturbations about the climatological state, which takes explicit account of the effect of mean convergence and total SST. The resulting model is capable of reproducing most of the principal features of the surface wind field over the tropical Pacific during an El Niño event.

COUPLED MODEL RESULTS

The atmospheric and oceanic responses during El Niño have been considered separately, but it has not yet been demonstrated that the mechanisms discussed above are enough to account for the evolution of the coupled system. Furthermore, investigation of the coupled system is essential to understanding the initiation and termination of events, the turnabout between El Niño and non–El Niño states that puzzled Bjerknes.

In the past decade there have been numerous restatements, modifications, elaborations, and enhancements of the Bjerknes hypothesis, all with the goal of providing a scenario for the complete ENSO cycle. The more interesting descriptive ones include Julian & Chervin (1978), Wyrtki (1981), and Philander (1983). Based on Bjerknes' ideas, McWilliams & Gent (1978) developed a nonlinear model with only five degrees of freedom. The model's most El Niño–like oscillations were rapidly damped. A number of simplified linear or nearly linear dynamical coupled models are reviewed in McCreary (1985). All exhibit long-period regular oscillations and/or growth of unstable modes. The characterization of the essential air-sea interaction as an instability (Philander 1983) taps a rich hydrodynamical literature; we return to this point below.

The results described here are more realistic than those obtained previously. They derive from the ocean-atmosphere interaction model constructed by coupling the tropical atmosphere model of Zebiak (1985) to the upper-ocean model of Zebiak & Cane (1985). The most significant differences from earlier coupled models are the attention to mean climatological fields and the explicit inclusion of an ocean mixed layer. The atmospheric heating is strongly influenced by the mean convergence pattern; SST can be changed by advection of mean temperature fields and by the advection of anomalies by mean currents and upwelling. Inclusion of surface-layer dynamics allows a mode of response that is local and rapid, granting upwelling its full measure of influence.

A numerical experiment with the coupled model was initiated with an imposed 2 m s^{-1} westerly wind anomaly of 4 months duration beginning in December of the year designated -1. There was no external forcing thereafter: Aside from the model physics, evolution of anomalies in SST, winds, etc, depends only on this initial condition and on the monthly mean climatological fields specified in the component ocean and atmosphere models. Furthermore, because of the damping in the model, the initial conditions are largely forgotten within a decade.

A 90-year time series of model SST anomalies averaged over the eastern equatorial Pacific is shown in Figure 8. There are peaks of varying amplitude occurring at irregular intervals but typically 3 to 4 years apart.

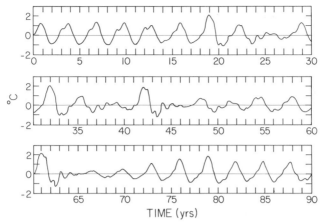

Figure 8   SST anomalies averaged over the eastern equatorial Pacific (the region NINO3—see Figure 6) for 90 years of coupled model integration.

They tend to be phase locked to the annual cycle, with major events reaching maximum amplitude at the end of the calendar year and decaying rapidly thereafter. All of these features are characteristic of observed El Niño events, as described earlier. The amplitudes of model events are similar to observed ones, although the model did not produce anything as extraordinary as the 1982/83 event. Note that the mean anomaly is close to zero (i.e. the model climatology is correct). Although it is a perturbation model, it is a nonlinear one, so this is not guaranteed. The model is somewhat more regular than nature; the high-frequency fluctuations present only in the real atmosphere and ocean may account for the broader natural spectrum.

Figure 9 depicts the evolution of SST during the El Niño event of model

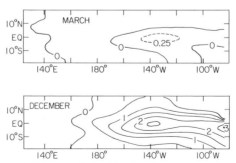

Figure 9   Coupled model SST anomalies for March and December during the model El Niño event in model year 31. (Note that the contour interval is 0.25°C for March and 0.5°C for December.) Compare with Figures 3 and 7.

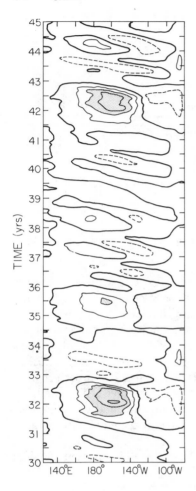

Figure 10 Time-longitude sections for model years 30–45 of the coupled model integration showing the forcing for the gravest mode oceanic Kelvin wave, a measure of zonal wind anomalies along the equator (cf Figure 5).

year 31. In December of the preceding year there was no discernible anomaly; by March of year 31 there is a small but systematic warming in the eastern Pacific; by December the anomaly extends to the date line, with a maximum at about 135°W. The model patterns and amplitudes are fairly realistic (cf Figure 3) except near the South American coast, where the model's coarse resolution precludes an accurate simulation of coastal upwelling processes.

Figure 10 shows the evolution of zonal wind along the equator. The prominent feature is the band of westerly anomalies in the central Pacific. The spatial and temporal patterns are realistic until the year following the event (year 32). The model westerly anomalies persist several months longer than is typical of El Niño events. The same is true for SST and other fields and is characteristic of model events. A possible cause is the model's inability to produce the easterly anomalies in the far western Pacific, which appear during the termination phase of observed events. As is the case even when observed SST anomalies are specified, the model winds are poorest in the Asian monsoon region and in the far eastern Pacific. As with observed events, El Niño anomalies disappear quite rapidly, to be replaced by cold SST in the eastern Pacific and stronger-than-normal easterlies along the equator.

In summary, a coupled ocean-atmosphere model incorporating the physical processes discussed in the preceding sections proved able to reproduce the spatial and temporal evolution of El Niño events. Moreover, without external forcing it was able to terminate these events and generate new ones in a never-ending ENSO cycle.

CONCLUSION

The coupled model results presented in the previous section have a number of implications for the real ENSO cycle. Since there is a leap from a model to reality, these are best regarded as hypotheses about the nature of the ENSO cycle. Nothing of what is suggested here is in conflict with observations, but alternative interpretations of the data are possible.

ENSO is an oscillation internal to the coupled ocean-atmosphere system. All of the interactions essential to creating and maintaining this cycle take place within the tropical Pacific region: No extratropical or extra-Pacific influences are required to account for the initiation or termination of El Niño events. The interactions required are deterministic: Random fluctuations in the atmosphere or ocean are not needed to initiate events or to account for the aperiodicity of ENSO.

Warm events result from a positive feedback between anomalies in the atmosphere and ocean. Warmer-than-normal SST in the east leads to increased atmospheric heating in the central Pacific. The anomalous inflow into this heating region includes westerly surface winds along the equator. The associated surface wind stress changes reduce upwelling, drive eastward currents, and deepen the thermocline in the east. The first two of these responses depend primarily on surface-layer dynamics, while the last involves the upper ocean down to the thermocline. All reinforce the warm anomaly.

Thus, the model calculations support Bjerknes' "chain reaction" hypothesis for the growth of ENSO anomalies. However, the hypothesis does not explain why there is a switch to an El Niño state (i.e. why the anomalies start). If this is explained by assuming that the coupled system is unstable, then the fact that El Niño does not occur all the time must be accounted for.

The answer must be that there is a necessary condition for the instability of the coupled system, which is not met every year: From time to time the normal seasonal cycle combines with the interannual variations to precondition it for an El Niño. In the real climate system the atmosphere has little year-to-year memory compared with the ocean. In the model the atmosphere has no memory at all: It is completely determined by the present state of the ocean. The necessary condition must reside in the ocean. The thermal damping time of the ocean surface mixed layer is only a few months, so its memory is short compared with ENSO time scales. Inspections of model SST fields before events show no discernible precursor pattern. The same may be said of the many diagnostic studies of the real SST data. In the model this leaves only the upper-layer depth (i.e. the oceanic heat content) as the possible locale for the necessary condition; this leads naturally to the hypothesis that the same is true for the real system. Preliminary analysis of model results suggests that the appropriate condition is that the heat content of the Pacific within the equatorial waveguide (5°S to 5°N) be above its average value. This applies to the zonal integral across the basin, and it is likely that further analysis will lead to refinements of this criterion. [On the basis of sea-level measurements, Wyrtki (1985) has recently proposed a similar criterion involving a broader meridional expanse of ocean.] The condition is physically plausible: Even given an initial westerly wind anomaly, the chain reaction will sputter if not enough warm water is readily available to maintain the warming of the eastern ocean.

If conditions are favorable, an event may be triggered by a variety of initial perturbations. As in other instability problems, the necessary and sufficient conditions for instability are a more crucial issue than the precise nature of the growing normal mode. From this point of view, one would expect great differences in the initial stages of El Niño from event to event. This is as observed. In nature the most readily available favorable perturbations are the bursts of westerly wind that occur with great frequency in the western equatorial Pacific (e.g. Luther & Harrison 1984). However, other fluctuations, such as intrusions of midlatitude westerlies, may also serve as triggers. The model does not simulate the westerly bursts and generally understates the variability of the atmosphere over the western Pacific. As a result, more of the model events begin in the central Pacific. (Note that the model wind anomalies lack the eastward propagation often observed in real events; compare Figure 10 with Figure 5.)

The variations of the atmosphere and ocean over their annual cycle strongly influence the susceptibility of the coupled system to instability. Conditions in the northern summer and fall are most favorable, so once begun, ENSO anomalies are best able to grow to large amplitude during these seasons. In the following spring, the normal seasonal changes (reductions in trade winds, upwelling, and zonal temperature gradient) weaken the coupling between the atmosphere and ocean sufficiently so that the warm event can no longer be sustained. With the coupling strength reduced the warm state of the ocean overbalances the wind, and the ocean retreats toward its normal, non–El Niño state. Modulation by the seasonal cycle is not the only way to generate oscillations (e.g. the models reviewed in McCreary 1985), but it appears to be the principal cause in nature. For example, a calculation with perpetual May conditions (Zebiak 1984) also gives long-period oscillations, but its evolution and amplitude are quite unrealistic.

As the system relaxes, it overshoots its mean state, resulting in the cold eastern Pacific SST and stronger-than-normal equatorial easterlies characteristic of the year following an El Niño event (cf Figures 2, 3, 5). Sea level slopes upward to the west more steeply than normal, while the overall heat content of the equatorial ocean is lower than normal: The El Niño event results in a heat loss from the equatorial Pacific. This is in part a loss of heat to the atmosphere, which provides the substantial anomalous atmospheric heating needed to power the global changes associated with the Southern Oscillation. However, the ocean's thermal capacity is very great, and this diabatic loss is much smaller than the dynamical export of heat through the ocean to higher latitudes.

Over the years following an event, the equatorial Pacific heat reservoir is refilled until the ocean is once again prepared for an El Niño event. It is not obvious that this must happen, that the coupled system cannot remain in a colder equilibrium with strong easterlies and a large zonal temperature contrast. A possible explanation lies in the nature of linear equatorial ocean dynamics (Cane & Sarachik 1981): The response to an equatorially confined easterly wind anomaly includes a positive zonally integrated heat anomaly in low latitudes. Thus, the strong easterlies that go with the enhanced zonal SST gradient following El Niño necessarily restore the condition needed for the next event.

In both the model and nature, ENSO has the character of a relaxation oscillation of the coupled system: The slow buildup to the necessary condition for instability during the cold phase leads to a rapid warming during the event itself; an even more rapid change to cold conditions follows. In both model and nature the ENSO cycle is aperiodic. The fact that the model system is deterministic supports the idea that the aperiodicity of the natural system also results from the deterministic physics discussed above and not from extraneous random fluctuations. This idea should not be surprising: There is an extensive literature documenting chaotic behavior in far simpler deterministic dynamical systems (e.g. Guckenheimer & Holmes 1983). In such systems, behavior can change in complicated ways as parameters are varied, so it is not likely to prove a simple task to delineate the parameter dependencies important for ENSO.

Results on the predictability of dynamical systems indicate that it will be impossible to predict ahead several events, so it is fortunate that only the next El Niño is of paramount practical interest. Even for the next one, there may be circumstances where the coupled system is so close to a bifurcation point that unobservable differences decide between its occurrence or nonoccurrence. This does not rule out forecasting procedures with useful (if imperfect) skill, and it has already been shown that the 1982/83 El Niño could have been forecast several months ahead by statistical methods if appropriate data had been available in real time (Barnett 1984).

Pezet had only Peru in mind when he asserted that the El Niño question is important, "not only from the oceanographic point of view, but also from the climatic," and therefore "calls for the serious attention of the men of science of the whole world." Some 70 years later, Bjerknes established a link between El Niño and the Southern Oscillation pattern of global climate effects. Advances since then have motivated the creation in 1985 of an international program, TOGA (Tropical Ocean Global Atmosphere), to study the question. At the time of this program's birth, it appears that a satisfactory theory for the ENSO cycle is within reach and that successful prediction of El Niño will follow within the decade.

ACKNOWLEDGMENTS

Special thanks to Philippe Hisard for bringing the Pezet article to my attention, and to Gene Rasmusson for sharing his understanding of the history of the subject. Comments by Steve Zebiak, Tony Busalacchi, Eli Katz, and Gilles Reverdin on an earlier draft greatly improved the manuscript. Thanks also to Karen Streech for her assistance in preparing the manuscript. Support for the author's work on El Niño under grants NAGW-582 from the National Aeronautics and Space Administration, OCE84-44718 from the National Science Foundation, and NA-84-RAD-05082 from the EPOCS program of the National Oceanic and Atmospheric Administration are gratefully acknowledged. Contribution Number 3883 of the Lamont-Doherty Geological Observatory.

Literature Cited

Barber, R. T., Chavez, F. P. 1983. Biological consequences of El Niño. *Science* 222: 1203–10

Barnett, T. P. 1977. An attempt to verify some theories of El Niño. *J. Phys. Oceanogr.* 7:633–47

Barnett, T. P. 1984. Prediction of El Niño of 1982–83. *Mon. Weather Rev.* 112:1403–7

Bjerknes, J. 1966. A possible response of the atmospheric Hadley circulation to equatorial anomalies of ocean temperature. *Tellus* 18:820–29

Bjerknes, J. 1969. Atmospheric teleconnections from the equatorial Pacific. *Mon. Weather Rev.* 97:163–72

Bjerknes, J. 1972. Large-scale atmospheric response to the 1964–65 Pacific equatorial warming. *J. Phys. Oceanogr.* 2:212–17

Bryden, H. L., Brady, E. C. 1985. Diagnostic model of circulation in the equatorial Pacific ocean. *J. Phys. Oceanogr.* 15:1255–73

Busalacchi, A. J., Cane, M. A. 1985. Hindcasts of sea level variations during the 1982–83 El Niño. *J. Phys. Oceanogr.* 15:213–21

Busalacchi, A. J., O'Brien, J. J. 1981. Interannual variability of the equatorial Pacific in the 1960's. *J. Geophys. Res.* 86:10901–7

Busalacchi, A. J., Takeuchi, K., O'Brien, J. J. 1983. Interannual variability of the equatorial Pacific—revisited. *J. Geophys. Res.* 88:7551–62

Canby, T. Y. 1984. El Niño's ill wind. *Natl. Geogr.* 165:144–83

Cane, M. A. 1979. The response of an equatorial ocean to simple wind stress patterns II: model formulation and analytic results. *J. Mar. Res.* 37:253–99

Cane, M. A. 1983. Oceanographic events during El Niño. *Science* 222:1189–94

Cane, M. A. 1984. Modeling sea level during El Niño. *J. Phys. Oceanogr.* 14:1864–74

Cane, M. A., Sarachik, E. S. 1981. The response of a linear baroclinic equatorial ocean to periodic forcing. *J. Mar. Res.* 39:651–93

Cane, M. A., Sarachik, E. S. 1983. Equatorial oceanography. *Rev. Geophys. Space Phys.* 21:1137–48

Enfield, D. B. 1981. Annual and nonseasonal variability of monthly low-level wind fields over the southeastern tropical Pacific. *Mon. Weather Rev.* 109:2177–90

Enfield, D. B., Allen, J. S. 1980. On the structure and dynamics of monthly mean sea level anomalies along the Pacific coast of North and South America. *J. Phys. Oceanogr.* 10:557–88

Geisler, J. E., Stevens, D. E. 1982. On the vertical structure of damped steady circulation in the tropics. *Q. J. R. Meteorol. Soc.* 108:87–93

Gill, A. E. 1980. Some simple solutions for heat-induced tropical circulation. *Q. J. R. Meteorol. Soc.* 106:447–62

Gill, A. E., Rasmusson, E. 1983. The 1982–83 climate anomaly in the equatorial Pacific. *Nature* 306:229–34

Glantz, M. H. 1984. Floods, fires and famine: Is El Niño to blame? *Oceanus* 27:14–19

Guckenheimer, J., Holmes, P. 1983. *Nonlinear Oscillations, Dynamical Systems, and Bifurcation of Vector Fields.* New York: Springer-Verlag, 453 pp.

Horel, J. D., Wallace, J. M. 1981. Planetary scale atmospheric phenomena associated with the Southern Oscillation. *Mon. Weather Rev.* 109:813–29

Hurlburt, H., Kindle, J., O'Brien, J. J. 1976. A numerical simulation of the onset of El Niño. *J. Phys. Oceanogr.* 6:621–31

Julian, P. R., Chervin, R. M. 1978. A study of the Southern Oscillation and Walker Circulation phenomena. *Mon. Weather Rev.* 106:1433–51

Lau, N. C. 1985. Modeling the seasonal dependence of the atmospheric response to observed El Niños in 1962–1976. *Mon. Weather Rev.* In press

Lau, N. C., Oort, A. H. 1985. Reponse of a GFDL general circulation model to SST fluctuations observed in the tropical Pacific Ocean during the period 1962–1976. In *Hydrodynamics of the Equatorial Ocean,* ed. J. C. J. Nihoul, pp. 289–302. New York: Elsevier

Leetmaa, A. 1983. The role of local heating in producing temperature variations in the offshore waters of the eastern tropical Pacific. *J. Phys. Oceanogr.* 13:467–73

Luther, D. S., Harrison, D. E. 1984. Observing long-period fluctuations of surface winds in the tropical Pacific: initial results from island data. *Mon. Weather Rev.* 112:285–302

Matsuno, T. 1966. Quasi-geostrophic motions in the equatorial area. *J. Meteorol. Soc. Jpn.* 44:25–42

McCreary, J. P. 1976. Eastern tropical ocean response to changing wind systems: with application to El Niño. *J. Phys. Oceanogr.* 6:632–45

McCreary, J. P. Jr. 1985. Modeling equatorial ocean circulation. *Ann. Rev. Fluid Mech.* 17:359–409

McWilliams, J. C., Gent, P. R. 1978. A coupled air-sea model for the tropical Pacific. *J. Atmos. Sci.* 35:962–89

Meyers, G. 1982. Interannual variation in sea level near Truk Island—a bimodal seasonal cycle. *J. Phys. Oceanogr.* 12:1161–68

Moore, D. W., Philander, S. G. H. 1977. Modeling of the tropical ocean circulation. In *The Sea*, ed. E. D. Goldberg, I. N. Cave, J. J. O'Brien, J. H. Steele, 6:319–61. New York: Interscience. 1048 pp.

Pezet, F. A. 1896. The counter-current "El Niño" on the coast of northern Peru. *Geogr. J. (London)* 7:603–6

Philander, S. G. H. 1979. Nonlinear coastal and equatorial jets. *J. Phys. Oceanogr.* 9: 739–47

Philander, S. G. H. 1983. El Niño–Southern Oscillation phenomena. *Nature* 302:295–301

Philander, S. G. H., Seigel, A. D. 1985. Simulation of El Niño of 1982–83. In *Hydrodynamics of the Equatorial Ocean*, ed J. C. J. Nihoul, pp. 517–42. New York: Elsevier

Quinn, W. H., Dopf, D. O., Short, K. S., Kuo-Yang, R. T. W. 1978. Historical trends and statistics of the Southern Oscillation, El Niño, and Indonesian droughts. *Fish. Bull.* 76:663–78

Ramage, C. S., Hori, A. M. 1981. Meteorological aspects of El Niño. *Mon. Weather Rev.* 109:1827–35

Rasmusson, E. M. 1985. El Niño and variations in climate. *Am. Sci.* 73:168–77

Rasmusson, E. M., Carpenter, T. H. 1982. Variations in tropical sea surface temperature and surface wind fields associated with the Southern Oscillation/El Niño. *Mon. Weather Rev.* 110:354–84

Rasmusson, E. M., Wallace, J. M. 1983. Meteorological aspects of the El Niño/Southern Oscillation. *Science* 222:1195–1202

Shukla, J., Wallace, J. M. 1983. Numerical simulation of the atmospheric response to equatorial Pacific sea surface temperature anomalies. *J. Atmos. Sci.* 40:1613–30

Walker, G. T. 1924. Correlation in seasonal variations of weather, IX: a further study of world weather. *Mem. India Meteorol. Dep.* 24 (Part 9): 275–332

Walker, G. T., Bliss, E. W. 1932. World Weather V. *Mem. R. Meteorol. Soc.* 4(36):53–84

Weare, B. C. 1983. Interannual variation of net heating at the surface of the tropical Pacific Ocean. *J. Phys. Oceanogr.* 13:873–85

Webster, P. S. 1981. Mechanisms determining the atmospheric response to sea surface temperature anomalies. *J. Atmos. Sci.* 38:554–71

Wyrtki, K. 1975. El Niño, the dynamic response of the equatorial Pacific Ocean to atmospheric forcing. *J. Phys. Oceanogr.* 5:572–84

Wyrtki, K. 1979. The response of sea surface topography to the 1976 El Niño. *J. Phys. Oceanogr.* 9:1223–31

Wyrtki, K. 1981. An estimate of equatorial upwelling in the Pacific. *J. Phys. Oceanogr.* 11:1205–14

Wyrtki, K. 1985. Water displacements in the Pacific and the genesis of El Niño cycles. *J. Geophys. Res.* 90:7129–32

Zebiak, S. E. 1982. A simple atmospheric model of relevance to El Niño. *J. Atmos. Sci.* 39:2017–27

Zebiak, S. E. 1984. *Tropical atmosphere-ocean interaction and the El Niño/Southern Oscillation phenomenon.* PhD Thesis. Mass. Inst. Technol., Cambridge. 261 pp.

Zebiak, S. E., 1985. Atmospheric convergence feedback in a simple model for El Niño. *Mon. Weather Rev.* In press

Zebiak, S. E., Cane, M. A. 1985. A simulation of sea surface temperature anomalies during El Niño. Submitted for publication

Ann. Rev. Earth Planet. Sci. 1986. 14 : 71–83

MOLECULAR PHYLOGENETICS

Jerold M. Lowenstein

Department of Medicine, University of California, San Francisco,
California 94143

INTRODUCTION

The task of constructing a family tree of all living organisms has challenged
science ever since Charles Darwin proposed more than a century ago that
life evolved from simpler to more complex forms. Traditionally, comparative anatomy and embryology have provided the main tools for estimating
evolutionary relationships, and the fossil record has provided the time scale
and sequence for evolutionary successions. During the past three decades,
however, following the breakthroughs in understanding the chemistry of
the gene and the genetic code, molecular evidence for the timing and
topology of evolutionary relationships has been accumulating at a rapid
rate (Ayala 1976, Dayhoff 1978, Doolittle 1979, Dover & Flavell 1982,
Goodman 1982, Nei & Koehn 1983, Sigman & Brazier 1980, Wilson et al
1977).

Early in this century, Nuttall (1904) used immunological cross-reactions
between the sera of many animals to estimate their relative genetic
distances. In the 1960s Goodman (1963) employed the more sensitive
immunodiffusion technique to investigate evolutionary relationships
among primates, and Sarich & Wilson (1967) began applying the even more
quantitative microcomplement fixation method to the task of constructing
immunological family trees for specific proteins like albumin and transfer-
rin in primates, other mammals, and other vertebrates such as frogs (Wilson
et al 1977). Not long afterward, new methods for the sequencing of protein
amino acids made it possible to compare sequence data for proteins like
cytochrome c (Margoliash 1980) and globins that are common to plants
and animals and so have evolutionary histories going back more than a
billion years. Immunological and sequence data are highly concordant
(Prager et al 1978). During the past few years, new rapid methods for
sequencing DNA have made possible the most direct comparisons of the

71

0084–6597/86/0515–0071$02.00

complex genetic messages that determine each species' biochemical, morphological, and behavioral characteristics (Dover & Flavell 1982).

At present, many techniques are being applied to developing molecular phylogenetics of living organisms such as bacteria, fungi, plants, invertebrates, and vertebrates. In this review, I focus on advances in mammalian systematics and the recent application of molecular techniques to the taxonomy of extinct as well as living species.

THE ENCYCLOPEDIC GENOME

The eukaryotic genome includes nuclear DNA and mitochondrial DNA. Exploration of the genome on the molecular level has already proved to be as exciting and surprising as the exploration of a newly discovered continent—or perhaps a more apt metaphor would be the discovery of a huge encyclopedia describing everything about that new continent, written in a newly deciphered language (DNA). The chromosomes are the volumes of the encyclopedia. The message is written in only four letters, A, C, G, and T (four nucleotide bases), and each combination of three letters (called a codon) encodes an amino acid. The amino acids constitute another alphabet of 20 letters, whose long words are proteins. Since there are 64 possible codons and only 20 amino acids, the genetic code is "redundant"—on average, three different codons can determine the same amino acid. In the living cell, the DNA encyclopedia is translated by messenger RNA into the smaller protein encyclopedia, and many bits of genetic (and evolutionary) information are lost in the translation.

Much more is lost, as it turns out, than anyone would have guessed a few years ago. One of the great surprises from exploration of the eukaryotic genome has been how little of the nuclear DNA message is translated into proteins—about 1% or even less. The other 99% consists of noncoding sequences variously labeled "junk," "parasitic," or "selfish" DNA (Doolittle & Sapienza 1980, Orgel & Crick 1980). Nucleotide substitutions, or mutations, which change the letter order in this part of the genetic message, have little or no effect on the amino acid sequence of the proteins that determine the phenotype—the shape and form of the organism (Jukes 1980, Kimura 1983).

Another surprise has been the observation, contrary to long-held evolutionary theory, that the vast majority of amino acid changes in proteins, when they do occur, make little or no difference to the function of the proteins or the survival of the organism. Most of us have been taught that nearly all such mutations, except for the rare "favorable" one, would be deleterious to the organism and eliminated by natural selection. But now it is clear from the comparative analysis of many different proteins in many different organisms that most (but not all) mutations are selectively "neutral" or nearly so (Kimura 1983), and that the rate of protein evolution is many times more rapid than anticipated.

Depending on its structural constraints, each protein has differing proportions that are selectively conservative. In addition, different proteins evolve at different rates. For instance, every third amino acid in collagen (the main structural protein of skin and bone) is glycine, necessary for its triple-helical conformation and tensile strength; thus collagen will accept no amino acid substitutes in the one third of its sequence represented by glycine (Fietzek & Kuhn 1976). Fibrinopeptides, on the other hand, act mainly as "spacers" between other proteins in blood clotting and will accept almost any amino acid substitution; they also evolve much more rapidly than collagen (Doolittle 1979).

MOLECULAR CLOCKS

If rates of protein or DNA change were constant or approximately so, they could be used as clocks (much as radioactive decay rates are) to time evolutionary events. Zuckerkandl & Pauling (1965) first noted that a particular protein tends to evolve at a constant rate over long stretches of time. If the hemoglobins of mammals, amphibians, and fish are compared, the amino acid sequence differences are proportional to the time, as estimated from the fossil record, since these animals diverged from their common ancestors (Jeffreys 1982). From the mitochondrial protein cyto-chrome c, one can derive a family tree that includes plants, yeasts, insects, and vertebrates, with branch lengths approximately proportional to the elapsed time since evolutionary divergence (Figure 1; Margoliash 1980). A family tree that includes yeasts, plants, and animals could hardly be derived from the anatomical characteristics of these organisms!

It seems then that proteins and DNA do behave like molecular clocks, providing not only genetic but also temporal information about the evolving life-forms on this planet. The molecular clock controversy, centering on the question of whether molecular changes keep good time or not, has spawned a vast literature. The great preponderance of it supports the general validity of the concept that homologous DNA or protein sequence differences are roughly proportional to species divergence times (Wilson et al 1977, Doolittle 1979, Kimura 1983).

Goodman (1982, Goodman et al 1982), however, has for many years argued for accelerating and decelerating rates of protein evolution at different time periods. Kimura (1983) has rebutted this contention in great detail. There are, as indicated, many different clocks, depending on the degree to which particular proteins and particular sites within those

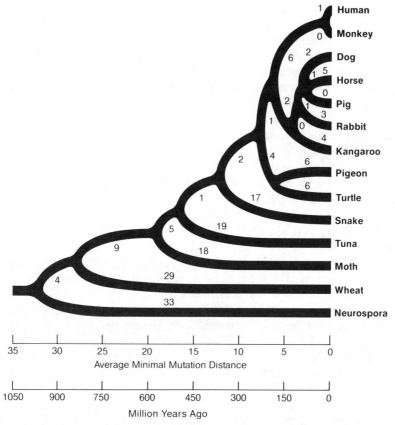

Figure 1 Cytochrome c phylogeny (after Margoliash 1980) that includes several mammals, a bird, two reptiles, a fish, an insect, a plant, and a yeast. On average, this protein of about 100 amino acids has undergone one mutation every 15 Myr, based on divergence times from the fossil record. The more ancient divergence times, such as that for the separation of plants and animals, are underestimated by this linear model because of the increased probability of multiple substitutions at a single amino acid site.

proteins are constrained by natural selection. Substitutions in the "silent codons" of DNA, those that do not lead to amino acid changes, take place at about five times the rate of those codon substitutions that do change amino acids. The relatively unconstrained substitutions in "introns" or "junk DNA" seem to have gone on at the remarkably constant rate of about 5 per site per billion years (Kimura 1983).

Mitochondrial DNA

The fastest molecular clock of all is mitochondrial DNA (mtDNA), which evolves at several times the average rate of nuclear DNA. There are thousands of mitochondria in the cytoplasm of each somatic cell. These tiny organelles, which probably originated as prokaryotic invaders of the eukaryotic cell and which contain only about 15,000 base pairs (as opposed to 3 billion in the nucleus of eukaryotes), are inherited maternally and have proved most useful as a kind of "second hand" for timing evolutionary change among closely related groups such as hominoids (Brown et al 1982, Brown 1983).

In terms of taxonomic information content, there is an increase by orders of magnitude in going from simpler to more complex organisms: Viruses consist of thousands of DNA or RNA bases, whereas prokaryotes (bacteria and blue-green algae) contain millions of DNA base pairs, and vertebrate nuclei contain billions of DNA base pairs. The reduction in information content also proceeds by orders of magnitude as we go from nuclear DNA (billions of base "letters") to messenger RNA (tens of millions) to proteins (millions) to gross morphology (where it is possible to analyze dozens of character traits, but their relation to the genetic DNA "message" remains uncertain). And in the fossil record, on which so much of the reconstruction of organic evolution has relied, we often have no more than a few shells or bones or teeth, so our information is reduced to tens or less. Given these enormous differences in information density, it is not surprising that morphological taxonomists so often disagree in their interpretation of the same material, whereas molecular phylogenies generally do agree (at least in branching orders), no matter which molecule is studied or in which laboratory.

FAMILY TREES OF MAMMALS

Mammalian taxonomy provides many examples of the complementary efforts of paleontologists, morphological systematists, and practitioners of molecular evolution. As Novacek (1982) has written:

> Although there is nearly unanimous agreement on the tripartite division of mammals into monotremes, marsupials, and eutherians, agreement generally stops at this point. The majority of living mammals, the eutherians or "placentals," are often presented as a series of some 15 ordinal-level taxa without reference to a hierarchic classification. Similarly, the phylogenetic history of eutherians is commonly depicted as a bushlike radiation sprouting from mysterious roots at the end of the Mesozoic.

Teeth are often the only parts of an animal remaining in the fossil record. The great influence of paleontologists such as Gregory (1910) and Simpson (1945) may help to account for the heavy reliance on dental comparisons for deducing systematic relationships. Depending on their eating habits, unrelated animals may have very similar teeth, and conversely, teeth may vary widely in related species. Functional features like teeth subject to strong selection are perhaps the least reliable indicators of evolutionary

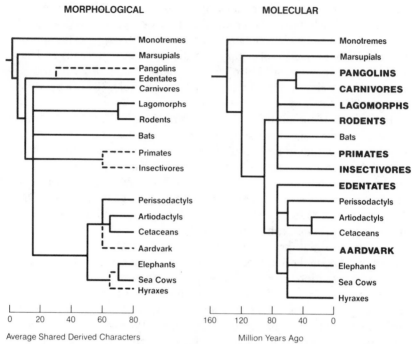

Figure 2   Comparison of recent morphologically derived and molecularly derived phylogenies of mammalian orders. The morphological tree (after Novacek 1982) is based on a cladistic analysis of shared derived character traits. Dashed lines indicate tentative assignments. The molecular tree (after Goodman et al 1982) is a consensus of sequence data from several different proteins. Boldface indicates significantly different placement in the molecular tree as compared with the morphological tree. For instance, morphologists have long disagreed on whether the African pangolins are related to the South American edentates, the lagomorphs (rabbits, hares, and pikas) to the rodents, and the primates to the insectivores. Molecular data do not support any of these pairings. Affinities of the aardvark have also been much debated. Proteins place it with the elephant–sea cow–hyrax group.

kinship, but Szalay (1977) has built a mammalian classification on a single ankle bone, the tarsus. Comparison of a morphological and a molecular view of higher-level mammalian taxonomy is given in Figure 2.

Questions in Mammalian Taxonomy

There are major unresolved questions in higher-level mammalian taxonomy. Are eutherian (placental) mammals monophyletic or polyphyletic? The most convincing morphological support for monophyly is an embryonic tissue, the trophoblast, which is not present in marsupials and monotremes; but obviously there can be no evidence for its presence or absence in fossil mammals. Do the lagomorphs (rabbits) belong with the rodents? Is the armor-plated African pangolin related to the South American edentates (sloths, armadillos, and anteaters), which it resembles, or is this an example of convergent evolution? Do the carnivores sort with the ungulates or nonungulates? Do the hyraxes and aardvarks associate with the elephants and sea cows? From which group of land mammals are the cetaceans (whales and dolphins) derived?

All these questions have been debated by morphological systematists for many decades without a consensus being reached. All have been resolved within the past 20 years by molecular techniques, but in many cases morphological taxonomists have been slow to accept the biochemical results. From the molecular standpoint, eutherian mammals are monophyletic relative to marsupials. Analysis of DNA and many different proteins reveals fewer sequence differences between eutherian orders than between the Eutheria and Marsupialia. The lagomorphs do not, in most molecular comparisons, form a close association with the rodents or with any other order; instead, they comprise a separate mammalian unit.

The pangolin is not related to the edentates, even though it eats ants and looks like an armadillo. The carnivores group with the nonungulates. The hyraxes and aardvarks do form a "paenungulate" group with the elephants and sea cows (de Jong et al 1981, Rainey et al 1984). Whales and dolphins are most closely related to the artiodactyls (even-toed ungulates).

BOVID EVOLUTION

Proceeding from the higher (ordinal) level to the lower (tribes, genera, and species) levels, molecular approaches have provided details of systematic relationships not always derivable from morphological data. The bovids, for example, which include cattle, sheep and goats, and gazelles, consist of 11 tribes, dozens of genera, and hundreds of species with a worldwide distribution and adaptation to many different climates and habitats. Their abundant fossil record, especially in Africa, has been extensively used for faunal dating of fossil and archeological sites and for reconstruction of paleoclimates. Yet the taxonomy of this large and important group of animals, based mainly on tooth and horn morphology, has remained rather uncertain, as indicated by the broken and unconnected lines of a phylogeny published a few years ago (Vrba 1980). In contrast, an immune phylogeny of this same group, done by radioimmunoassay (RIA), shows clear genetic and temporal relations among the tribes (Lowenstein 1985b,c).

Comparison of the bovid family trees derived morphologically and immunologically (Figure 3) reveals that the major subdivision into Boodontia and Aegodontia (derived from dental features) is not supported serologically. Instead, the Bovini, Boselaphini, and Tragelaphini form one

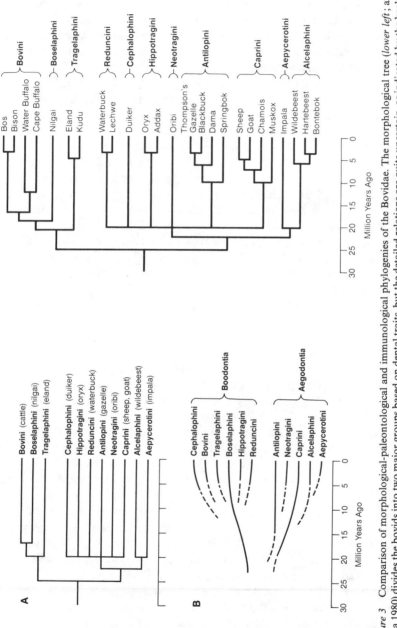

Figure 3   Comparison of morphological-paleontological and immunological phylogenies of the Bovidae. The morphological tree (*lower left*; after Vrba 1980) divides the bovids into two major groups based on dental traits, but the detailed relations are quite uncertain, as indicated by the broken, unconnecting lines. In contrast, the immunological tree (*upper left*; Lowenstein 1985b,c) depicts definite relationships, indicates a different dichotomy than the dentally derived one, and shows the origin of all modern bovid tribes by 15 Myr ago. A more detailed bovid phylogeny (*right*) resolves long-standing controversies about branching orders within the Bovini, the Caprini, and the Alcelaphini.

cluster and the eight other tribes another. Immunological distances place bovid origins about 25 Myr ago, and all the tribes seem to have originated by 15 Myr ago. This corresponds to a period of time when world climates became progressively cooler and more seasonal, and when the habitat of eastern and southern Africa shifted from closed forest to patchy open forest, woodland, and grassland (Laporte & Zihlman 1983). The profuse bovid radiation probably reflects the variety of ecological niches opening up at that time.

Debates About Bovid Relations

There have been numerous points of debate about bovids among systematists. Are the Cape buffalo (African) and water buffalo (Asian) more closely related to each other than either is to domestic cattle, or are water buffalo and cattle closer? What is the branching order within the Alcelaphini (hartebeest, wildebeest, and bontebok) group? How are sheep, goats, chamois, and muskox related? Do they form a single "clade" relative to the other bovids?

The answers to these questions, at least as provided by the immunological technique of radioimmunoassay, are shown in Figure 3. Water buffalo and Cape buffalo are closer to each other than to domestic cattle. Hartebeest and bontebok are closer to each other than to wildebeest. Sheep, goats, chamois, and muskox are monophyletic relative to the other bovids. Muskox (*Ovibos*) diverged from the common caprid lineage about 10 Myr ago, chamois (*Rupicapra*) about 7 Myr ago, and sheep and goats about 5 Myr ago.

The bovid immune phylogeny provides an example of the way in which molecular data can complement the work of morphological systematists, archeologists, paleontologists, and geologists in attempting to retrieve information about paleoclimates, divergence times, and the tempo of evolution in a major mammalian group.

MOLECULAR SYSTEMATICS OF EXTINCT SPECIES

Though molecular approaches have been extremely helpful in addressing systematic questions about living taxa, analysis of extinct forms (which far outnumber the extant) has until recently continued to rely exclusively on comparative morphology. It would obviously be useful to obtain molecular genetic information from fossils, if that were available. Many studies have revealed the presence of amino acids in fossils, often in a pattern characteristic of collagen (that is, one-third glycine and about 10% hydroxyproline), and electron microscopy has shown collagenlike fibrils in fossils millions of years old (Wyckoff 1972, Hare et al 1980). Though these

observations were intriguing, they did not yield species-specific information. All fossil collagens look alike and have similar amino acid profiles.

During the past few years, I have applied sensitive solid-phase radioimmunoassay (RIA) to obtaining molecular genetic information from fossils
(Lowenstein 1980, 1981, 1985a). Small amounts of protein do survive in
fossils thousands or even millions of years old, and RIA results show that
they may retain their species specificity, even though we know that proteins
break down and undergo chemical changes with the passage of time (Hare
et al 1980).

Mammoths, Mastodons, and Sea Cows

A frozen Siberian mammoth offered the ideal test for the RIA method
(Lowenstein et al 1981). A baby mammoth known as Dima (^{14}C age of
40,000 yr) was retrieved from the permafrost near Magadan, USSR, in 1977
and refrigerated at once, so that the soft tissues were preserved. By RIA,
both albumin and collagen, nearly identical to those of living elephants,
were detected. Though most taxonomists have believed, on the basis of
tooth shape, that the extinct mammoth (*Mammuthus primigenius*) was more
closely related to the Indian elephant (*Elephas maximus*) than to the African
elephant (*Loxodonta africana*), mammoth albumin is equally (more than
99%) similar to the albumins of both extant elephants.

Proof that bones also can yield molecular data was forthcoming when
the skeleton of a mastodon (*Mammut americanum*; ^{14}C age of 10,000 yr) was
unearthed during excavation for a Michigan housing project. Proteins
extracted from the bone dust that was drilled out while mounting the
skeleton were also shown by RIA to contain albumin and collagen closely
related to those of the living species of elephants (Shoshani et al 1985).

As noted above, sea cows (Sirenia) are the closest living relatives of the
elephants (Proboscidea). There are only four extant species of sea cows:
three Atlantic manatees and the Pacific dugong. A second dugongid,
Steller's sea cow (*Hydrodamalis gigas*), was discovered on the icy
Commander Islands near Kamchatka in 1741, but by 1768 the species had
been exterminated (Haley 1980). A few skeletons remain in museums, and
RIA of albumin from one of these confirmed the affinity of Steller's sea cow
with the dugong (Rainey et al 1984). However, the recency of their
divergence (about 5 Myr ago based on the albumin "molecular clock") and
the recency of the divergence of the dugongs and the manatees (about 20
Myr ago) are in conflict with paleontological opinion, which from the
sirenian fossil record estimated these divergence times as being at least twice
as old. Paleontologists like to find fossil "ancestors" of living groups, but
physical resemblance does not always indicate close genetic relationship, as
the many known cases of convergence and parallelism testify.

The immune phylogeny shown in Figure 4 is remarkable for its inclusion of three extinct species: the woolly mammoth, the mastodon, and Steller's sea cow. A total of only six species of Proboscidea and Sirenia remain extant, so that proteins from the three extinct species have expanded this molecular phylogeny by 50%!

The Extinct Quagga: Horse or Zebra?

Recently, DNA as well as proteins was extracted from the skin of a quagga, a South African equid extinct for 100 years (Higuchi et al 1984, Lowenstein & Ryder 1985). This mitochondrial DNA, which represented about 0.5% of the original total, was cloned in bacteria and sequenced, an achievement that excited speculation in the popular media that the reconstruction of extinct forms, including dinosaurs, was imminent. Of more practical interest was the molecular information that both the DNA and proteins provided for resolving the long-standing controversy over quagga systematics. Striped on its forward half like a zebra, unstriped and chestnut-hued on its hindquarters like a horse, the quagga has been claimed by various taxonomists to have been a close relative of the horse (Bennett 1980), a fourth species of African zebra (the others being the Plains, Mountain, and Grevy's zebras), or merely a subspecies of the Plains zebra.

At a recent conference on fossil biochemistry, Russell Higuchi presented his data derived from DNA and I presented mine derived from skin proteins. Neither of us knew in advance the other's results, but we drew

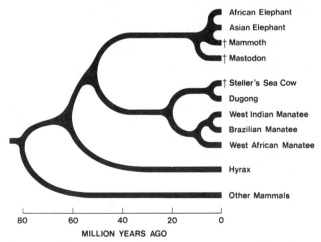

Figure 4   An immunological phylogeny of elephants and sea cows that includes three extinct species—the woolly mammoth, the American mastodon, and Steller's sea cow. By radioimmunoassay (RIA), skin and bone proteins of the extinct specimens were compared with serum, skin, and bone proteins of the extant species.

identical conclusions—namely, that the quagga is much more closely related to the Plains zebra than to any other member of the horse family (Lowenstein & Ryder 1985).

CONCLUSION

In summary, then, our expanding discoveries about the chemistry and evolution of the genome and of the species-specific proteins it encodes have revealed that molecular and morphological evolution may proceed at quite different rates. This knowledge has added a new dimension to our understanding of the evolution of life on this planet and our ability to deduce genetic relations among living and extinct species.

ACKNOWLEDGMENTS

My work has been supported by National Science Foundation Grant No. BSR-8400252, by the Wenner-Gren Foundation for Anthropological Research, and by the University of California, San Francisco Academic Senate. I thank A. L. Zihlman for critical comments on the manuscript and Gary Scheuenstuhl for technical assistance.

Literature Cited

Ayala, F. J., ed. 1976. *Molecular Evolution*. Sunderland, Mass: Sinauer Assoc. 277 pp.

Bennett, D. K. 1980. Stripes do not a zebra make, part I: a cladistic analysis of *Equus. Syst. Zool.* 29:272–87

Brown, W. M. 1983. Evolution of animal mitochondrial DNA. See Nei & Koehn 1983, pp. 62–88

Brown, W. M., Prager, E. M., Wang, A., Wilson, A. C. 1982. Mitochondrial DNA sequences of primates: the tempo and mode of evolution. *J. Mol. Evol.* 18:225–39

Dayhoff, M. O. 1978. *Atlas of Protein Sequences and Structure*, Vol. 5, Suppl. 3. Washington, DC: Natl. Biomed. Res. Found. 345 pp.

de Jong, W. W., Zweers, A., Goodman, M. 1981. Relationship of aardvark to elephants, hyraxes and sea cows from α-crystallin sequences. *Nature* 292:538–40

Doolittle, R. F. 1979. Protein evolution. In *The Proteins*, 4:1–118. New York: Academic

Doolittle, W. F., Sapienza, C. 1980. Selfish genes, the phenotype paradigm and genome evolution. *Nature* 284:601–3

Dover, G. A., Flavell, R. B., eds. 1982. *Genome Evolution*. London: Academic. 382 pp.

Fietzek, P. P., Kuhn K. 1976. The primary structure of collagen. *Int. Rev. Connect. Tissue Res.* 7:61–101

Goodman, M. 1963. Man's place in the phylogeny of the primates as reflected in serum proteins. In *Classification and Human Evolution*, ed. S. L. Washburn, pp. 204–34. Chicago: Aldine

Goodman, M., ed. 1982. *Macromolecular Sequences in Systematic and Evolutionary Biology*. New York: Plenum. 418 pp.

Goodman, M., Olson, C. B., Beeber, J. E., Czelusniak, J. 1982. New perspectives in the molecular biological analysis of mammalian phylogeny. *Acta Zool. Fenn.* 169:19–35

Gregory, W. K. 1910. The orders of mammals. *Bull. Am. Mus. Nat. Hist.* 27:1–524

Haley, D. 1980. The great northern sea cow: Steller's gentle siren. *Oceans* 13(5):7

Hare, P. E., Hoering, T. C., King, K., eds. 1980. *Biogeochemistry of Amino Acids*. New York: Wiley. 558 pp.

Higuchi, R., Bowman, B., Freiberger, M., Ryder, O. A., Wilson, A. C. 1984. DNA sequences from the quagga, an extinct member of the horse family. *Nature* 312:282–84

Jeffreys, A. J. 1982. Evolution of globin genes. See Dover & Flavell 1982, pp. 157–76

Jukes, T. H. 1980. Silent nucleotide substitutions and the molecular evolutionary clock. *Science* 210:973–78

Kimura, M. 1983. *The Neutral Theory of Molecular Evolution.* Cambridge: Cambridge Univ. Press. 367 pp.

Laporte, L. F., Zihlman, A. Z. 1983. Plates, climates, and hominoid evolution. *S. Afr. J. Sci.* 79:96–110

Lowenstein, J. M. 1980. Immunospecificity of fossil collagens. See Hare et al 1980, pp. 41–51

Lowenstein, J. M. 1981. Immunological reactions from fossil material. *Philos. Trans. R. Soc. London Ser. B* 292:143–49

Lowenstein, J. M. 1985a. Radioimmunoassay and molecular phylogeny. *BioEssays* 2:60–62

Lowenstein, J. M. 1985b. Bovid relations based on serum immunology. *S. Afr. J. Sci.* In press

Lowenstein, J. M. 1985c. Radioimmunoassay of extinct and extant species. In *The Past, Present and Future of Hominoid Evolution*, ed. P. V. Tobias. New York: Alan R. Liss. In press

Lowenstein, J. M., Ryder, O. A. 1985. Immunological systematics of the extinct quagga (Equidae). *Experientia* 41:1192–93.

Lowenstein, J. M., Sarich, V. M., Richardson, B. J. 1981. Albumin systematics of the extinct mammoth and Tasmanian wolf. *Nature* 291:409–11

Margoliash, E. 1980. Evolutionary adaptation of mitochondrial cytochrome c to its functional milieu. See Sigman & Brazier 1980, pp. 299–321

Nei, M., Koehn, R. K., eds. 1983. *Evolution of Genes and Proteins.* Sunderland, Mass: Sinauer Assoc. 331 pp.

Novacek, M. J. 1982. Information for molecular studies from anatomical and fossil evidence on higher eutherian phylogeny. See Goodman 1982, pp. 3–41

Nuttall, G. H. F. 1904. *Blood Immunity and Blood Relationships.* Cambridge: Cambridge Univ. Press. 444 pp.

Orgel, L. E., Crick, F. H. C. 1980. Selfish DNA: the ultimate parasite. *Nature* 284:604–7

Prager, E. M., Welling, G. W., Wilson, A. C. 1978. Comparison of various immunological methods for distinguishing among mammalian pancreatic ribonucleases of known amino acid sequence. *J. Mol. Evol.* 10:293–307

Rainey, W. E., Lowenstein, J. M., Sarich, V. M., Magor, D. 1984. Sirenian molecular systematics—including the extinct Steller's sea cow (*Hydrodamalis gigas*). *Naturwissenschaften* 71:586–88

Sarich, V. M., Wilson, A. C. 1967. Immunological time scale for hominid evolution. *Science* 158:1200–3

Shoshani, J., Walz, D. A., Goodman, M., Lowenstein, J. M., Prychodko, W. 1985. Protein and anatomical evidence of the phylogenetic position of *Mammuthus primigenius* within the Elephantinae. *Acta Zool. Fenn.* 170:237–40

Sigman, D. S., Brazier, M. A. B., eds. 1980. *The Evolution of Protein Structure and Function.* London: Academic. 350 pp.

Simpson, G. G. 1945. The principles of classification and a classification of mammals. *Bull. Am. Mus. Nat. Hist.* 85:1–350

Szalay, F. S. 1977. Phylogenetic relationships and a classification of the eutherian Mammalia. In *Major Patterns in Vertebrate Evolution*, ed. M. K. Hecht, P. C. Guody, B. M. Hecht, pp. 315–74. New York: Plenum

Vrba, E. S. 1980. The significance of bovid remains as indicators of environment and predation patterns. In *Fossils in the Making: Vertebrate Taphonomy and Paleoecology*, eds. A. K. Behrensmeyer, A. P. Hill, pp. 247–71. Chicago: Univ. Chicago Press

Wilson, A. C., Carlson, S. S., White, T. J. 1977. Biochemical evolution. *Ann. Rev. Biochem.* 46:573–639

Wyckoff, R. W. G. 1972. *The Biochemistry of Animal Fossils.* Baltimore: Williams & Wilkins. 144 pp.

Zuckerkandl, E., Pauling, L. 1965. Evolutionary divergence and convergence in proteins. In *Evolving Genes and Proteins*, ed. V. Bryson, H. J. Vogel, pp. 97–166. New York: Academic

Ann. Rev. Earth Planet. Sci. 1986. 14: 85–112

CONODONTS AND BIOSTRATIGRAPHIC CORRELATION

Walter C. Sweet and Stig M. Bergström

Department of Geology and Mineralogy, Ohio State University, Columbus, Ohio 43210

INTRODUCTION

At its 1969 meeting, The Pander Society convened a symposium on conodont biostratigraphy that resulted in a volume (Sweet & Bergström 1971a) in which specialists systematically summarized the status of conodont-based biostratigraphy up to 1969. One function of such a volume is to direct attention to areas that need more work and to focus on conclusions in need of more robust verification. The Pander Society's biostratigraphy volume provides this focus and has also been a very useful guide to the biostratigraphic uses of conodonts. In the 14 years since the volume appeared, additional effort has been expended on then-tentative biostratigraphic schemes, and revised or alternative biostratigraphic scales have been stimulated for several intervals.

We cannot hope in this review to bring the Pander Society's biostratigraphy volume fully up to date, or even to review all major lines of progress in conodont-based biostratigraphy of the past 14 years. What we can do is note major changes of fairly recent vintage, with emphasis on areas in which the application of different correlation algorithms has confirmed systems that were apparently well established in 1971, or in which use of different correlation methodology has resulted in a different view of these systems. Our discussion deals largely with Ordovician and Triassic biostratigraphy, our areas of expertise, and with the contrast in those areas between the results of traditional biozonal practices and those of less conventional graphic procedures.

85

0084–6597/86/0515–0085$02.00

CONODONTS AS BIOSTRATIGRAPHIC INDICES

Conodonts are an extinct phylum of Paleozoic and Triassic marine animals. They are represented in sedimentary rocks by a morphologically diverse array of tiny, phosphatic skeletal elements, which were parts of a cephalic apparatus that may have been used to grasp and aid in the ingestion of food. We suspect that conodonts were never numerous components of Paleozoic or Triassic marine faunas, but they are common as fossils because individual cephalic apparatuses may have included more than 20 discrete elements, which are remarkably resistant to chemical, physical, and biologic degradation. Hence, they have accumulated preferentially in sedimentary deposits, from which we now remove them by use of techniques that would destroy most other fossils. Because conodont elements are mostly less than a millimeter in size, they may be collected intact from even fine well cuttings. The widespread geographic distribution of many species and their common occurrence in black shales suggest that most conodonts were pelagic; and well-documented morphologic series indicate that they evolved rapidly along many different lines through their nearly 300 Myr of existence.

The qualities just mentioned are those commonly listed as attributes of ideal "index fossils." This combination of qualities—wide geographic distribution of stocks of rapidly evolving species; presence in, and collectability from, almost any type of marine sedimentary rock; and high preservation potential—has, in recent years, given conodonts "pride of place" in Paleozoic and Triassic biostratigraphy.

Conodonts also owe their eminence as biostratigraphic guides to another feature. As microfossils they cannot be seen in the field, and collectors commonly take bulk rock samples at regular intervals in the hope that conodonts will turn up in them after processing in the laboratory. Also, samples are routinely keyed to a specific position in a measured section. Consequently, students of conodonts are generally able to state with precision the actual, scaled range of the species represented in the sections sampled. This is not commonly the case with macrofossils, but it is an attribute that makes conodonts and other microfossils exceptionally useful in biostratigraphic procedures that use measured first and last occurrences, such as the graphic correlation program of Shaw (1964).

BIOSTRATIGRAPHIC CORRELATION

We favor a definition of biostratigraphy that includes all the activities involved in assembling and interpreting the fossil record, but we are concerned here just with biostratigraphic correlation, which involves a determination that rocks encountered in several places are similar or identical in terms of the species represented in them by fossils. Biostratigraphic correlation is important in developing the chronostratigraphic framework within which interpretations of much of Phanerozoic history are made, because close similarity or identity in the species assemblages of stratified rocks is commonly taken to indicate similarity or identity in the age of those rocks.

Biostratigraphic correlation is commonly preceded by a phase in which the interval of interest is divided into units termed biozones, which are bodies of rock distinguished from other such bodies by the fossils they contain. These then become the units between which similarity or identity is established by the several procedures used to match them. More attention has been paid to the means of establishing and naming the correlation units, or biozones, than to the means of recognizing them in a second section and thereby establishing the similarity or identity in age implied by that correlation.

Students of conodonts have divided the interval from the uppermost Cambrian through the Triassic into about 150 biozones. These are most commonly recognized in the various places they have been identified by the occurrence of distinctive guide species that serve as the index, or as indices, to the zone. Segments of this scale with the highest resolution are those in which zonal boundaries are at levels of evolutionary change in well-marked phylomorphogenetic series.

Shaw's (1964) method of graphic correlation bypasses establishment of biozones. Relationships between pairs of sections are stated in the form of equations for lines fitted to the arrays of first- and last-occurrence events determined from biaxial plots of the events common to the sections compared. In Shaw's method, correlation precedes, rather than follows, establishment of correlation units. That is, division of a network of correlated sections into such units (Shaw's "standard time units," or STUs) cannot be effected until a unique correlation of component sections has been accomplished. This means of correlation is not yet common biostratigraphic practice, but it has yielded interesting results in those parts of the geologic column in which it has been attempted. In the upper Middle and Upper Ordovician of the North American Midcontinent, for example, graphic correlation of 70 sections forms a network that is divisible into 80 STUs, each with a probable temporal extent of less than 500,000 yr; in contrast, traditional biozonal procedures lead to recognition of only three biozones in the same interval.

In what follows we cite the principal developments in conodont-based biostratigraphy since 1971, primarily as a guide for those interested in updating their copies of the Pander Society biostratigraphy volume (Sweet

& Bergström 1971a). At appropriate places we compare the results of biozonal and graphic procedures.

PROGRESS IN CONODONT-BASED BIOSTRATIGRAPHY

Cambrian

Cambrian conodonts are less well known than those of any other system, and they are widely used as index fossils only in the uppermost part of the system. Also, there are questions about the affinities of some of the Cambrian fossils commonly identified as conodonts. That is, one group (the protoconodonts) may include forms that are chaetognaths rather than conodonts (Szaniawski 1982).

Protoconodonts, which range from late Precambrian to at least middle Arenigian, have not been shown to have biostratigraphic significance, and no conodont-based biostratigraphy has been proposed for pre–Middle Cambrian strata. Paraconodonts, which may have evolved from proto-conodonts (Bengtson 1983), appear near the base of the Middle Cambrian (Miller 1984). In China, An (1982) distinguished two paraconodont zones in the upper Middle Cambrian and four in the Upper Cambrian, but these have not yet been identified in other areas.

Euconodonts, or "true conodonts," appear in the upper Upper Cambrian. Müller (1973) defined four assemblage zones based mainly on euconodonts in the Upper Cambrian of Iran. A more detailed scheme is summarized by Miller (1984) and Miller et al (1983) for the Upper Cambrian of North America, and units in it appear to have wide regional utility. The level of the Cambrian-Ordovician boundary has not yet been settled by international agreement, but it is clear that conodonts will be important in defining and correlating it. Middle and Upper Cambrian conodont-zonal schemes are compared in Figure 1.

Ordovician

In the Ordovician, North America was astride the equator, and Middle and Upper Ordovician conodont faunas from its broad midsection have little in common with those from the Appalachians southeast of the Saltville-Helena fault system or from most of Europe, which was evidently at a much higher latitude. Thus, in the Pander Society biostratigraphy volume, two separate biostratigraphic schemes were included, one (Lindström 1971, Bergström 1971) for areas with high-latitude conodont faunas and another (Ethington & Clark 1971, Sweet et al 1971a) for regions with low-latitude faunas.

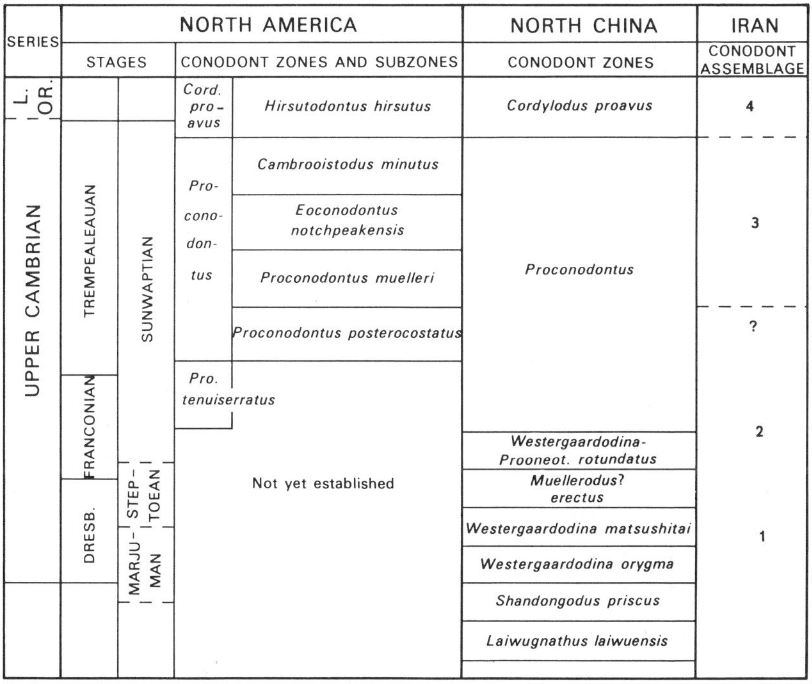

SERIES	NORTH AMERICA			NORTH CHINA	IRAN
	STAGES	CONODONT ZONES AND SUBZONES		CONODONT ZONES	CONODONT ASSEMBLAGE
L. OR.		Cord. pro-avus	Hirsutodontus hirsutus	Cordylodus proavus	4
UPPER CAMBRIAN	TREMPEALEAUAN — SUNWAPTIAN	Pro-cono-don-tus	Cambrooistodus minutus	Proconodontus	3
			Eoconodontus notchpeakensis		
			Proconodontus muelleri		
			Proconodontus posterocostatus		?
	FRANCONIAN	Pro. tenuiserratus			2
	DRESB. — MARJU-MAN STEP-TOEAN		Not yet established	Westergaardodina-Prooneot. rotundatus	
				Muellerodus? erectus	
				Westergaardodina matsushitai	1
				Westergaardodina orygma	
				Shandongodus priscus	
				Laiwugnathus laiwuensis	

Figure 1   Correlation of Cambrian conodont biostratigraphic schemes for the United States (Miller 1984), North China (An 1982), and Iran (Müller 1973).

BIOSTRATIGRAPHY OF THE LOW-LATITUDE FAUNAL REALM   Ordovician rocks are widespread in the North American Midcontinent and in other parts of the world that were then at low latitudes. In 1971, Ethington & Clark and Sweet et al recognized in these strata a succession of 5 Lower Ordovician and 12 Middle and Upper Ordovician faunas, which most subsequent authors have treated as hallmarks of a sequence of assemblage zones. Ethington & Clark (1982) have since increased the number of conodont faunas they recognize in the North American Lower Ordovician from 5 to 8 and have determined that the youngest of their 1971 faunas and the oldest of Sweet et al's Middle and Upper Ordovician faunas represent the same stratigraphic interval. Work since 1971 has also shown that stratigraphic overlap is the rule between Sweet et al's Middle and Upper Ordovician faunas, which have now grown in number to 13. Consequently, neither of the low-latitude schemes of 1971 has grown into a formal zonal biostratigraphy. On the contrary, intractability of Middle and Upper Ordovician rocks to zonation and correlation by traditional means has promoted development of an alternative system, assembled by graphic correlation and relative-abundance analysis. This system, which now resolves Middle and Upper Ordovician units about 0.5 Myr long, contrasts with the 1971 scheme, in which unit resolution is about 4.4 Myr.

Development of a nontraditional biostratigraphy for the Middle and Upper Ordovician of the North American Midcontinent began when Sweet (1979b) used measured ranges of 43 conodont species to effect a graphic correlation of Upper Ordovician strata at 10 localities in the western Midcontinent. By 1984 this network had been extended, and it now includes 70 stratigraphic sections in both the eastern and western Midcontinent. Means of assembling the network are discussed by Sweet (1984), who also reports on progress through 1983. Sweet (1984) provides graphic summaries of data that document each step in the development of the high-resolution biostratigraphic system for Middle and Upper Ordovician rocks of the North American Midcontinent. We summarize major aspects in the development of this system by reference to the example in Figure 2. In the graph on the left side of Figure 2, first and last occurrences of species identified by Rust (1968) in the Martinsburg Formation of the Narrows section, Virginia, are plotted against those of the same species in a composite section (CS) that includes maximized range information from sections at 69 other localities and that is based upon a standard section (SRS) in the Cincinnati Region of Kentucky. The array formed by the lowest of the first occurrences (numbered dots) and the highest of the last occurrences (numbered crosses) is rectilinear and may thus be expressed by the equation of a straight line fitted to it (CS = 0.405 MN + 977). This is read to mean that the base (or "0" level) in the Narrows section is equivalent to a point 977 m above the base of the CS, and that each meter of rock above the base of the Narrows section represents the same time interval recorded by units 0.405 m thick in the SRS, which serves as base for the CS of Figure 2. In other words, almost 2.5 times as great a thickness of rock accumulated in the Narrows region of Virginia during this part of the Ordovician as in north-central Kentucky, where the SRS is situated.

On the right side of Figure 2, a relative-abundance log of conodonts for the Narrows section is compared with one for that part of the Cincinnati Region SRS shown by graphic means to be equivalent. The vertical scale of the Narrows log has been reduced to that indicated by graphic correlation, so time-equivalent events in the two logs are now side by side. There are differences in the two logs, but many inflection points are at the same levels. If these points were plotted in the graph of Figure 2, they would join the array of first and last occurrences to which is fit the line whose equation we have used to determine correlation of the Narrows and composite sections.

In short, events recorded by log-inflection points have the same temporal significance as first and last occurrences.

Shaw (1964) used the standard error of estimate as a guide to dividing the SRS into correlation units that might be recognized at a given level of confidence in all component sections of a graphically assembled network. He termed the time intervals represented by these correlation units "standard time units" (or STUs). They are conceptually the same as the "chrons" of the *International Stratigraphic Guide* (Hedberg 1976). We use an index termed W in the same manner. Here W is the distance between the X-intercepts of parallel lines of the indicated slope through the points in the array farthest from its axis. In the graph of Figure 2, W has a value of 6 m, which means that events whose records in the SRS are less than 6 m apart cannot be resolved in the Narrows section. The maximum value of W in all the sections of the network determines the minimum value of divisions of

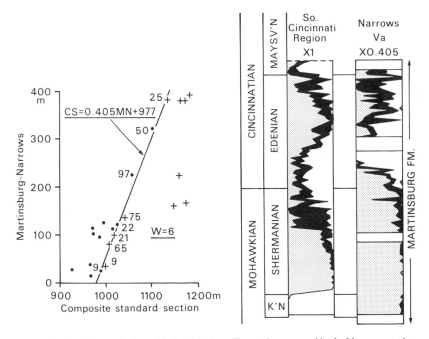

Figure 2   Graphic correlation of the Martinsburg Formation exposed in the Narrows section of New River, Virginia, and the composite standard section of Sweet (1984). On the right, the relative-abundance log of conodonts for Martinsburg is compared with a composite log for sections in the southern Cincinnati Region. In relative-abundance logs, the symbols are as follows: coarse-dot pattern = *Phragmodus*; black = *Plectodina*; white = *Aphelognathus* + *Oulodus*; fine-dot pattern = *Rhipidognathus*. Data from Rust (1968).

the SRS that may be resolved throughout the network. In the upper Middle and Upper Ordovician system of Sweet (1984), W has a maximum value of 6; hence the 477-m SRS is divisible into 79.5 chronozones that may be resolved in all sections that are parts of the network. In Figure 3 we show the relationship between the 79.5 chronozones of the Mohawkian-Cincinnatian biostratigraphic scheme developed by Sweet (1984) and divisions of the same stratigraphic interval determined in other ways. Note that it is now possible to determine the extent of the series and stages into which this part of the Ordovician is divided in North America, for reference sections of all these units are now parts of the graphically assembled network. It is also possible to evaluate the ranges of conodont faunas 7 through 12 of Sweet et al (1971a) and to relate these and other units to the three biozones established in this interval by Bergström (1971). Most importantly, the Mohawkian and Cincinnatian are now divisible into 79.5

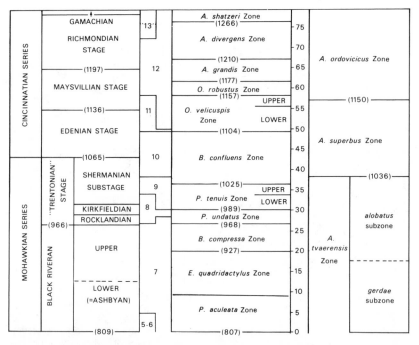

Figure 3 Chronostratigraphic and biostratigraphic divisions of the late Middle (Mohawkian) and Late Ordovician (Cincinnatian). Numbers in third column from left indicate conodont faunas of Sweet et al (1971a), numbers in second column from right are scale of standard time units of Sweet (1984), and numbers on horizontal lines indicate levels (in meters) of boundaries in Middle and Upper Ordovician composite standard section of Sweet (1984).

chrons, each (at least conceptually) of the same length. This represents an increase in resolution of almost 27 times.

In the central column of Figure 3 we indicate that Sweet (1984) also suggested retrospective division of the composite section into 11 zones, all with type sections in the Cincinnati Region SRS and bases marked in the SRS by the level (or projected level) of first occurrence of the species for which the zone is named. Like the 79.5 divisions marked off in the second column from the right side of Figure 3, these zones are chronozones, not the biozones of the *International Stratigraphic Guide* (Hedberg 1976) or the North American Stratigraphic Code (North American Commission on Stratigraphic Nomenclature 1983). That is, they are named bodies of rock that can be shown to be equivalent to segments of the SRS defined as just noted. Their boundaries are synchronous by definition, and they may be recognized in additional sections only by the same means used in establishing the synthetic composite section of which they are parts. Establishment of these zones may seem superfluous, considering that a greater degree of resolution is resident in the system. It is convenient, however, to have several levels of resolution in any such system, and the chronozones may facilitate reference to parts of the time interval in a manner more intelligible than, say, "STUs 50–55."

BIOSTRATIGRAPHY OF THE HIGH-LATITUDE FAUNAL REALM   Baltoscandia, where conodonts were first discovered (Pander 1856), is also the birthplace of Ordovician conodont zonal biostratigraphy. Lindström's (1955) study of upper Tremadocian and lower Arenigian conodonts from Sweden pioneered modern conodont research in Baltoscandia. Studies to 1970 were summarized by Lindström (1971), and most of the numerous contributions of the last 15 years are cited by Dzik (1983) and Bergström (1983). The Baltoscandic conodont succession is probably the best known Ordovician sequence in the world, and zonal units established in it have also been identified in many other regions. As Sweet & Bergström (1984) indicate, it is likely that typical Baltoscandian conodonts inhabited relatively cold waters. Baltoscandia itself was at high latitude for much of the Ordovician, but many Baltoscandian species apparently could migrate widely, and these forms also occur in low-latitude regions, mostly in relatively deep-water deposits.

The scheme of about 14 conodont zones and 13 subzones currently used in Baltoscandia was introduced by Lindström (1971) and Bergström (1971). Post-1971 modifications are minor, and we note only a few of the more important ones here. Biostratigraphically, it is interesting that the basis of the zonal units is not uniform through the system; that is, upper Tremadocian and Arenigian zones are mostly assemblage zones, whereas

Llanvirnian through Ashgillian ones are based on the vertical ranges of distinctive morphotypes in rapidly evolving lineages (Bergström 1973) and hence are phylozones in the sense of Hedberg (1976).

Modifications of the Lower Ordovician scheme proposed in the Pander Society biostratigraphy memoir (1971) include the one by Van Wamel (1974), who distinguished 20 assemblage zones in the Tremadocian and lower Arenigian of southeastern Sweden. Most of these units are unrecognizable outside his study area, appear to be of only local significance, and have not been accepted by subsequent workers. However, his study is carefully documented and is a useful source of information.

The conodont succession in the Baltoscandian Tremadocian is still incompletely known, largely because lithology in most sections is unfavorable for conodont extraction, and there is no established zonal scheme. This gap may soon be filled because V. Viira (personal communication) has made significant collections from the Tremadocian of Estonia, in which a succession of species of *Cordylodus* similar to that in North America may permit recognition of a series of widely traceable zones.

The Arenigian zonal succession outlined by Lindström (1971) has been widely used in Baltoscandia and in many other parts of the world where the diagnostic species are present. No formal reference sections for these zones have yet been proposed, but ranges of individual species and compositions of zonal assemblages are now well known, largely through the work of Löfgren (1978, 1985). Löfgren (1978) and Dzik (1983) found it difficult to separate the *Baltoniodus triangularis* and the *B. navis* zones, and (following these authors and others) we here recognize a combined *B. triangularis–B. navis* Zone. Dzik's (1983) rejection of the upper Arenigian *Microzarkodina flabellum parva* Zone on the grounds that it was defined as an assemblage zone seems unjustified, since most of the other Baltoscandian Lower Ordovician zones are also assemblage zones (Löfgren 1985).

As indicated by Bergström (1983) zonal classification of the Baltoscandian lower Llanvirnian is still provisional, and Lindström (1971), Dzik (1976, 1978), and Löfgren (1978) have each presented different schemes. Bergström (1983) reviewed the taxonomic and stratigraphic problems involved, and we follow his recommendation to adopt Löfgren's (1978) scheme pending further study. Zones and subzones established by Bergström (1971) in the upper Llanvirnian through Caradocian interval have been widely used in Baltoscandia and elsewhere and have required only minor modification (Bergström 1983). Dzik (1978) proposed several changes in this zonal scheme, based on a succession in the Holy Cross Mountains, Poland, but they do not appear to be real improvements, especially for long-distance correlation; hence, we do not adopt them. The same is the case with his suggestion (Dzik 1983) to abandon the *B. alobatus*

Subzone. Through the courtesy of Dr. Dzik, one of us (SMB) had the opportunity recently to study his collection, and the correlation between his units and those in common use in Baltoscandia (shown in Figure 4) is based largely on that study.

Baltoscandic conodont zones have been recognized in marginal areas of North America, such as the Appalachians (Barnes & Poplawski 1973, Bergström 1973, Bergström & Carnes 1976, Bergström et al 1974, Landing 1976, Fåhraeus & Nowlan 1978, Nowlan 1981, Stouge 1984, etc), the Ouachitas of Arkansas (Repetski & Ethington 1977), the Marathon area of western Texas (Bergström & Cooper 1973, Bergström 1978), and the Great Basin (Harris et al 1979). Sweet & Bergström (1984, Figure 1) summarize the occurrence of such faunas. Some Baltoscandic zones have also been recognized in the Midcontinent (e.g. Sweet & Bergström 1971b, 1976, Sweet 1979b, 1984). However, the profound provincialism of Ordovician conodont faunas necessitates use of other zonal schemes in central North America. We have commented on Sweet's (1984) scheme for upper Middle and Upper Ordovician rocks in the Midcontinent and show general relations to coeval Baltoscandic zones in Figure 4. We also note in Figure 4 that Stouge (1984) recognized two phylozones and four assemblage zones in the lower Llanvirnian of Newfoundland, and that Sweet (1984) informally distinguished four chronozones in about the same interval.

DEVELOPMENTS IN ASIA AND AUSTRALIA   Our regional review would be incomplete without consideration of the results of recent biostratigraphic work in Asia. Two regions are significant, the Siberian Platform and China, where distinctive conodont-based schemes reflect the remarkable biogeographic differentiation of Ordovician conodont faunas.

T. Moskalenko, G. Abaimova, and their co-workers have described the succession of conodonts in the widespread Ordovician deposits of the Siberian Platform. The faunas described have several features in common with those of the North American Midcontinent, but the numerous endemic species give these faunas such a distinct flavor that the Siberian Platform may well be recognized as a separate conodont-biogeographic unit. Moskalenko (1983) recently outlined the conodont zonal succession currently used and listed 12 assemblage zones (Figure 4), which were correlated with Baltoscandian zones and North American faunal intervals. Only a limited number of species are in common between these areas, so correlations are tentative, especially in the upper Middle and Upper Ordovician.

In China, extensive work in the last decade has produced a wealth of information, much of which is not readily available to workers unfamiliar with Chinese. Data at hand suggest that Ordovician conodont faunas from

CONODONT ZONES AND SUBZONES

SERIES	BALTOSCANDIA	POLAND	N. AMERICA	SIBERIA	S. CHINA	N.
ASHGILLIAN	Amorphognathus ordovicicus	Amorphognathus ordovicicus	*(CINCINNATIAN)* A. shatzeri / A. divergens / A. grandis / O. robustus	Aphelognathus pyramidalis	Proto-prioniodus insculptus	
CARADOCIAN	Amorphognathus superbus	Amorphognathus superbus	O. velicuspis / B. confluens	Acanth. nobilis / "S." dolboricus	Hamarodus europaeus	
CARADOCIAN	Amorphognathus tvaerensis *(B. variabilis / B. gerdae / B. alobatus)*	B. gerdae	*(MOHAWKIAN)* P. tenuis / Phr. undatus / B. compressa / E. quadridactylus / P. aculeata	Acanth. festus / Bel. compressa / C. mangazeica / C. sweeti - Ph. inflexus		
LLAN. / LLANVIRNIAN	P. anserinus *(in, ki, li)*	A. inaequalis / P. anserinus / E. lindstroemi	C. sweeti	P. anomalis - / B. lenaica	C. friendsvill-ensis / E. reclinatus / E. foliaceus / E. pseudo-planus	B. compressa - M. symmetricus / Au. serratus / Pl. onycho-donta / E. suecicus
LLANVIRNIAN	P. serra *(ro, re, fo)*	E. robustus / E. reclinatus / E. foliaceus	C. friendsvillensis	P. flexuosus / Card.- Polyplac. / C. mirabilis		
LLANVIRNIAN	E. suecicus *(su, gr)*	E. suecicus / E. pseudo-planus	Phr. "pre-flexuosus" / H. holodontata			
LLANVIRNIAN	E. variabilis *(oz, fl)*	E. variabilis / B. variabilis	*(WHITEROCKIAN)* H. sinuosa / H. altifrons / M. flabellum - T. laevis	A. antivariabilis		
ARENIGIAN	M. flab. parva	P. originalis	Prot. aranda - J. jaanussoni	S. quadra-plicatus - H. angulata	B. aff. B. navis / P. originalis	Tangshanodus tangshanensis
ARENIGIAN	P. originalis	B. navis	J. gananda - R. andinus		O. multicorr.- P. flab.	Aur. lep.-L.dis. / Sc. sunanensis
ARENIGIAN	B. nav. - B. triang.	O. evae	Oe. com.-"M." mar.		O. evae	P. paltodifor.
ARENIGIAN	O. evae	P. elegans	Acodus deltatus / Macerodus dianae		B. communis	S. extensus
ARENIGIAN	P. elegans / P. proteus	P. proteus	"S." quadraplicatus aff. S. rex		Serr. div.	S. bilobatus / Sc. tersus
TREMADOCIAN	P. deltifer	P. deltifer	*(IBEXIAN)* Loxodus bransoni	Ac. lineatus	P. deltifer / T. proteus / S. pauc. - S. barbatus / S. quadraplic.	S.quadraplic. - S. optimus / C. rotundatus - "A." oneotensis
TREMADOCIAN	Cordylodus ssp.		C. intermedius		A. costatus - "A." oneot. / D. simplex	U. beimadao-ensis - Monoc. sevierensis
TREMADOCIAN			C. proavus *(hin, sim, elo, ino, hir)*			C. proavus

northern China, which are similar to those of Korea, are strikingly different from most of those in south-central China, especially those from the carefully studied sections in the eastern Yangtze Gorges Region that serve as a stratotype of the Chinese Ordovician.

An's (1981) and An et al's (1983) useful summaries show that faunas from northern China share some species with coeval North American Mid-continent faunas, but also that they contain many never recorded in the latter area. Northern China may also have been a separate conodont province or subprovince during the Ordovician, but additional study would be needed to justify such a suggestion. From the lower Tremadocian to the lower Caradocian in this region, An (1981) recognizes nine zones, two of which (*Eoplacognathus reclinatus* and *E. foliaceus* zones) are interpreted by him to be the same as Baltoscandian units with the same names.

Tremadocian conodont faunas of south-central China (An 1981) are also more like those of the North American Midcontinent than those of Baltoscandia. However, in the lower Arenigian (basal Dawan Formation), there is a striking change in the biogeographic character of the faunas, which suddenly become almost identical with those of Baltoscandia. This is well shown in the Yangtze Gorges region, where Arenigian and Llanvirnian rocks are also similar to those of central Baltoscandia. With minor modifications the Baltoscandian zonal succession can be applied to this succession (An 1981), in which the distribution of other major faunal elements has also been studied in great detail (Zeng et al 1983). This produces an unparalleled correlation between zones based on conodonts, graptolites, brachiopods, cephalopods, and trilobites. Doubtless, this is one of the finest and most thoroughly studied Ordovician successions in the world. In Figure 4 we summarize the succession of zones and their correlation.

In general, Ordovician conodont faunas of Australia are less well known than those of any other continent with extensive Ordovician deposits except for South America. Webby et al (1981) give pre-1980 references, and more recent studies include those by Cooper (1981) and Burrett et al (1983). Most published work deals with local rather than regional biostratigraphy. Biogeographically, there is a complex pattern of faunas of diverse affinities, many of which appear to differ, especially at the species level, from coeval ones in other parts of the world. This may support

←

Figure 4 Tentative correlation of conodont biostratigraphic schemes for Baltoscandia (Bergström 1983, Lindström 1971, Löfgren 1978), Poland (Dzik 1978, 1983), North America (Ethington & Clark 1982, Miller 1984, Sweet 1984), Siberia (Moskalenko 1983), and China (An 1981, An et al 1983, Zeng et al 1983). Ethington & Clark's post–*C. proavus* Ibexian units in North America were defined as faunal intervals rather than as formal biozones.

Bergström's (1971) suggestion that at least part of Australia may represent a separate Ordovician conodont province.

Silurian

The conodont-based zonal scheme for the Silurian introduced by Walliser (1962, 1964, 1971) is widely applicable and still serves, with minor regional modifications, as an international standard. An informative review of worldwide Silurian conodont biostratigraphy was presented by Cooper (1980), who noted that Walliser's zones are either assemblage zones or range zones, the bases of which are defined as the level of appearance of the zonal indices. Studies by Nowlan (1983), Jeppsson (1983), Lin (1983), and others show that, apart from recognition of some more or less local units, the Wenlockian through Pridolian portion of Walliser's zonal scheme has remained relatively unchanged, but that the Llandoverian classification has been refined through recognition of new zones in the basal part of the system (Figure 5).

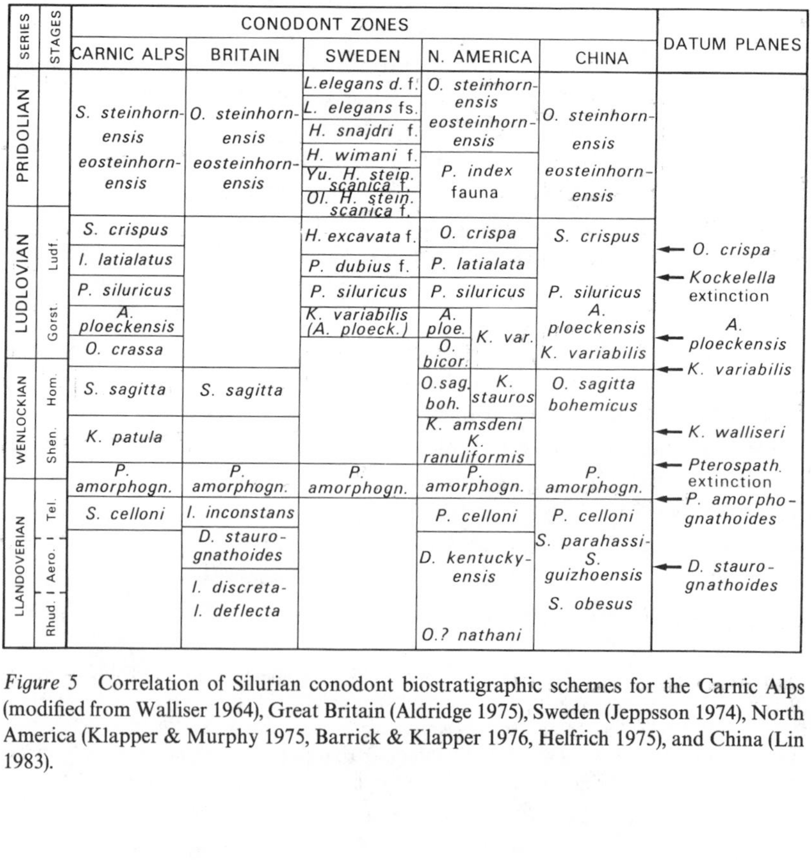

SERIES	STAGES	CONODONT ZONES					DATUM PLANES
		CARNIC ALPS	BRITAIN	SWEDEN	N. AMERICA	CHINA	
PRIDOLIAN		S. steinhornensis eosteinhornensis	O. steinhornensis eosteinhornensis	L.elegans d. f. / L. elegans fs. / H. snajdri f. / H. wimani f. / Yu. H. stein. scanica f. / Ol. H. stein. scanica f.	O. steinhornensis eosteinhornensis / P. index fauna	O. steinhornensis eosteinhornensis	
LUDLOVIAN	Ludf.	S. crispus / I. latialatus		H. excavata f. / P. dubius f.	O. crispa / P. latialata	S. crispus	← O. crispa / ← Kockelella extinction
	Gorst.	P. siluricus / A. ploeckensis / O. crassa		P. siluricus / K. variabilis (A. ploeck.)	P. siluricus / A. ploe. O. bicor. | K. var.	P. siluricus / A. ploeckensis / K. variabilis	← A. ploeckensis / ← K. variabilis
WENLOCKIAN	Hom.	S. sagitta	S. sagitta		O.sag. boh. | K. stauros	O. sagitta bohemicus	
	Shen.	K. patula			K. amsdeni K. ranuliformis		← K. walliseri / ← Pterospath. extinction
		P. amorphogn.	P. amorphogn.	P. amorphogn.	P. amorphogn.	P. amorphogn.	→ P. amorpho- gnathoides
LLANDOVERIAN	Tel.	S. celloni	I. inconstans		P. celloni	P. celloni	
	Aero.		D. staurognathoides		D. kentuckyensis	S. parahassi- S. guizhoensis	← D. stauro- gnathoides
	Rhud.		I. discreta- / I. deflecta		O.? nathani	S. obesus	

Figure 5 Correlation of Silurian conodont biostratigraphic schemes for the Carnic Alps (modified from Walliser 1964), Great Britain (Aldridge 1975), Sweden (Jeppsson 1974), North America (Klapper & Murphy 1975, Barrick & Klapper 1976, Helfrich 1975), and China (Lin 1983).

Cooper (1980) introduced a concept not previously used in Silurian conodont biostratigraphy: that of the datum or datum plane. Although such conceptual surfaces are in common use in the Mesozoic and Cenozoic, we are aware of only one previous application in Lower Paleozoic conodont biostratigraphy, namely in the Middle Ordovician of Virginia (Tillman in Markello et al 1979). In the Llandoverian through Ludlovian, Cooper (1980) defined eight datum planes based on the level of appearance or disappearance of key conodont genera or species. Five coincide with zonal bases previously defined on the appearance of the same species, and another coincides with the top of a zone; therefore, stratigraphic resolution was not greatly increased. However, some of the planes are useful in correlating between provincial zonal schemes, particularly because Cooper provided a precise definition of each one. Silurian conodont biostratigraphic schemes, including datum planes, are shown in Figure 5.

Devonian

Conodont zonal biostratigraphy has probably reached the pinnacle of its development in the Devonian, in which 50 conodont-based biozones are now recognized (Johnson et al 1985, and references therein). The development of these zones has been thoroughly summarized in recent reports (Klapper 1977, Klapper & Ziegler 1979, Johnson et al 1985), so we note here only that it is claimed that each of the 28 Late Devonian biozones represents about 0.5 Myr (Sandberg & Poole 1977) and that Johnson et al (1985) hypothesize that boundaries between these zones "identified on two or more conodont lineages may have a probable error of no more than a tenth of the average resolution of the enclosing zones." Johnson et al have also shown how such a biostratigraphic scheme can serve as the high-resolution framework for determining the synchroneity of Devonian eustatic fluctuations in Euramerica and for evaluating the rate and timing of faunal turnover in the interval of the famous Frasnian "extinction event."

Carboniferous and Permian

The Carboniferous and Permian, which together represent about 115 Myr (or a bit more than a third of the Paleozoic), have been divided by traditional biozonal procedures into 50 units, termed zones or faunal complexes. A majority of these have not yet been shown to be applicable outside the region in which they were developed, and hence we neither list nor discuss them further here. The eight conodont-based biozones of the lowermost Carboniferous (or Tournaisian) (Sandberg et al 1978, Lane et al 1980) are exceptions, however, for they form a system that rivals in resolution and wide recognizability the finely tuned biostratigraphic scheme of the Upper Devonian. Boundaries of the lower six Tournaisian biozones are based on events in the evolutionary history of *Siphonodella* that may be recognized worldwide. Boundaries of the upper two Tournaisian zones (Typicus, Anchoralis-Latus) are less securely established, are based on events in unrelated lineages (*Gnathodus, Scaliognathus*), and can be correlated from one continent to another with reduced resolution.

Triassic

In the Pander Society's biostratigraphy volume, Sweet et al (1971b) divided the Triassic into 22 biozones, which were compared in extent with the 31 ammonoid-based ones of Silberling & Tozer (1968). Detailed work had just begun on Triassic conodont faunas, so it could not be claimed that conodonts would ultimately provide the basis for a higher-resolution biostratigraphy than one based on ammonoids, the traditional guides to the Triassic. But it did appear in 1971 that conodonts might eventually displace ammonoids as primary guides to Lower Triassic (or Scythian) rocks, in which Sweet et al recognized 13 conodont-based biozones.

Space limitations preclude a review here of all post-1971 progress in conodont biozonation of Middle and Upper Triassic rocks. It is sufficient to note that studies have been both intensive and numerous and have resulted in division of these rocks into 21 biozones in one recent publication (Kozur 1980). This is an increase of 12 over the number recognized in 1971 and represents a doubling in biostratigraphic resolution.

TRADITIONAL VERSUS NONTRADITIONAL BIOSTRATIGRAPHY OF THE LOWER TRIASSIC   We are more familiar with conodonts and conodont-based biostratigraphy of Scythian rocks. Since 1971 there has been a great increase in our knowledge of conodont distribution in these strata. Most of this new information has been used to modify the zonal scheme of 1971, but a substantial part has been presented in a form amenable to graphic analysis. Thus we use this interval, as we did the upper Middle and Upper Ordovician, to contrast the results of traditional and nontraditional correlation algorithms in building an Early Triassic chronostratigraphy.

In Figure 6 we compare conodont-based biostratigraphic schemes for the Lower Triassic. All are the results of traditional zonal procedures, and differences between them are minor. The zonal scheme of Sweet (1970b), the basis for the Scythian zonation of Sweet et al (1971b), reflected data on the distribution of conodonts in 11 sections in Pakistan. Information was best near the base and top of the Scythian sequence, and much thinner in between. Data lacking or poorly represented in the Pakistan sequence were supplied later from sections in the western United States, which also provided additional insight into the upper Scythian conodont distribution.

The two data sets were then tied together by Sweet et al (1971b) at about the base of the Waageni Zone. The Lower Triassic zonal scheme of 1971 was thus a butt-spliced synthesis that could not be verified at any locality then known.

Since 1971, much new information has accumulated on the ranges of Lower Triassic conodonts in Kashmir (Matsuda 1981a,b, 1982, 1983, 1984), Pakistan (Pakistani-Japanese Research Group 1981), the western United States (Solien 1979, Paull 1982), and the far-eastern USSR (Buryi 1979). Sweet has assembled this information in a form amenable to graphic analysis, which, with reconsideration of other data (e.g. Staesche 1964), helps bridge the gap between the Pakistan data and that from the western United States used to flesh out the 1971 zonal scheme. Sweet's study will be

SWEET, 1970b	Stage	SWEET et al 1971	KOZUR and MOSTLER 1972	1973	SOLIEN 1979 CARR & PAULL 1983
Timorensis	SPATHIAN STAGE	Timorensis	Timorensis	Timorensis	8
Jubata	SPATHIAN STAGE	Jubata	Homeri	Homeri	7
Jubata	SPATHIAN STAGE	Neosp. n.sp.G	Homeri	Homeri	6
Jubata	SPATHIAN STAGE	Platyvillosus			5 B / A
Waageni	SMITHIAN STAGE	Milleri	Waageni	Elongata	4 D
Waageni	SMITHIAN STAGE	Conservativus	– – – – –	– – – – – G.aff. Milleri	4 C
Waageni	SMITHIAN STAGE	Parachirogn. Furnishius	Milleri	– – – – Eotriassica	B
	SMITHIAN STAGE				A
Pakistanensis	SMITHIAN STAGE	Pakistanensis	N.gen. n.sp.	Gondolella n.sp. B	
Cristagalli	DIENERIAN STAGE	Cristagalli		(Not discussed)	3 Dieneri
Dieneri	DIENERIAN STAGE	Dieneri	Dieneri	(Not discussed)	3 Dieneri
Kummeli	DIENERIAN STAGE	Kummeli			2 Kummeli
Carinata	GRIESBACHIAN	Carinata	Carinata	Carinata	1 Typicalis
	GRIESBACHIAN		Typicalis	Parvus	1 Typicalis
Typicalis	GRIESBACHIAN / PERMIAN	Typicalis	Typicalis	Isarcica	

Figure 6   Five biostratigraphic schemes for the Lower Triassic (Scythian). All zones are based on conodonts. Stages are based on ammonoids, and their boundaries are only approximately located with respect to those of conodont zones.

published in full elsewhere; here, we use preliminary results as the basis for evaluating the zonal schemes compared in Figure 6.

From the sources just listed, information was assembled on the measured positions of first and last occurrences of 45 conodont species in sections in Kashmir, Pakistan, northern Italy, Utah, and Primor'ye. This information was then used graphically (Figures 7, 8) to determine relationships between these sections and the one in Kashmir selected as a standard reference section (SRS). The main function of the graphic step in the correlation procedure is to aid identification of the array of biologic events that best

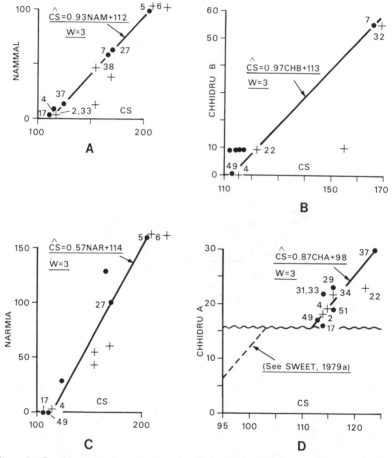

Figure 7   Graphic correlation of four sections from Pakistan, with composite section based on SRS in Kashmir. Equations are for lines fitted to arrays composed of numbered first (dots) and last (crosses) occurrences of conodont species. Scale is in meters.

expresses the temporal relationship between the sections compared. In Figure 7*A*, for example, first-occurrence events 5, 7, 17, 27, 33, and 37 join last-occurrence events 2, 4, 6, and 38 to form a narrow, rectilinear array to which is fit a line with the equation CS = 0.93 NAM + 112. Geologically, this equation says that the base of the Nammal section in Pakistan is equivalent to a level 112 m above the base of the SRS in Kashmir, and that 1.00 m of rock formed at Nammal in the same time interval required for 0.93 m to accumulate in Kashmir. The equation is then used to project positions of first- and last-occurrence events from the Nammal section into the SRS. Following each phase of graphic comparison, a new composite section is assembled from the lowest value for each first occurrence and the highest value of each last occurrence. Thus, each data set shown in the graphs of Figures 7 and 8 is compared with the best information available anywhere in the system.

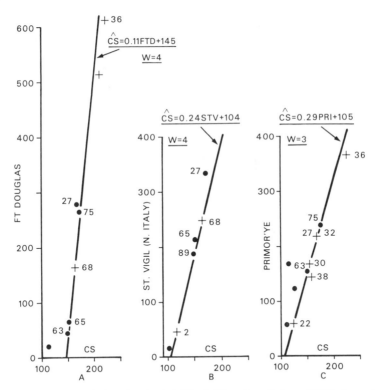

Figure 8   Graphic correlation of section in (*A*) Utah, (*B*) northern Italy, and (*C*) the far-eastern USSR (Primor'ye), with composite section based on SRS in Kashmir. Equations are for lines fitted to arrays composed of numbered first (dots) and last (crosses) occurrences of conodont species. Scale is in meters.

The exercise is complete when a full round of graphic recorrelations produces results exactly like those of the preceding round. The final composite-range table (still in terms of the SRS) then becomes the best available statement of ranges of the species represented in the sections correlated. The composite section derived in Sweet's restudy of Lower Triassic conodont distribution is depicted here in Figure 9. Note that the

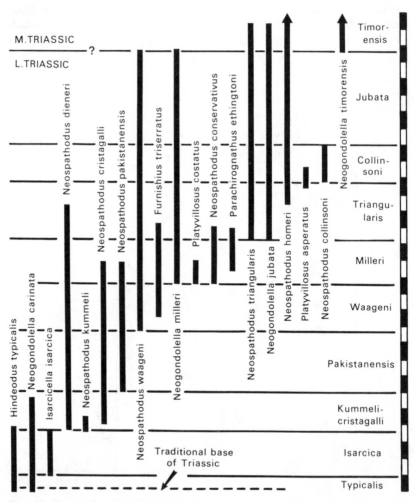

Figure 9   Ranges of stratigraphically significant conodont species in composite section of the Lower Triassic created by graphic correlation of sections in Kashmir, Pakistan, Utah, northern Italy, and Primor'ye. The vertical bar scale at right is divided into standard time units, each of the same length. Chronozones bounded by solid horizontal lines and named at right are the ones preferred in this report.

vertical dimension of this chart is that of the SRS and that the right margin is marked off in 4-m units, the length of which is determined by the maximum value of the index W shown in Figures 7 and 8. W, it may be recalled, is a measure of the width on the X-axis of the array to which the equation applies and has a maximum value of 4 m in both the Fort Douglas, Utah, section (Figure 8A) and the St. Virgil, Italy, section (Figure 8B). In Figure 9 the interval between the base of the Triassic and that of the Timorensis Zone (here regarded as earliest Middle Triassic) is thus divisible into 26.5 STUs or chrons. If the Scythian were only 5 Myr long, as was suggested in a recent report (Salvador 1985), each STU represents about 189,000 yr.

EVALUATION OF TRADITIONAL BIOZONAL SCHEMES   The composite section (Figure 9) may be used to evaluate the schemes of Figure 6, all originally delineated and currently correlated by biozonal procedures. Sweet's current preference in Scythian conodont zones is indicated along the right side of Figure 9.

Scythian strata with *Hindeodus typicalis* (Sweet) are clearly not everywhere separated from ones above with *Neospathodus kummeli* Sweet by an interval dominated by *Neogondolella carinata* (Clark). Graphic assembly of data indicates that, regionally, rocks with *H. typicalis* are directly overlain by ones with *N. kummeli*, and that there is no justification for a Carinata Zone above the Typicalis Zone. This is not a novel conclusion. Regional utility of a Carinata Zone was questioned by Sweet (1970b) when he established it for Pakistan; more serious doubts were raised by study of isolated collections from India (Sweet 1973); and the zone was abandoned by Collinson & Hasenmueller (1978), Paull (1982), and Carr & Paull (1983).

Kozur & Mostler (1973) divided lower Scythian strata with *Isarcicella isarcica* (Huckriede) into Isarcica and Parvus zones (Figure 6). Sweet (1970b) noted that "*Anchignathodus*" *isarcicus* is represented in the Triassic part of the Typicalis Zone; Staesche (1964) and Assereto et al (1973) confirmed that the species characterizes an interval in the Lower Triassic Seis beds of northern Italy; Sweet (1979a) and Matsuda (1981a) summarized occurrences in Iran and Kashmir; and Paull (1982) reported specimens from the lower 30 m of the Triassic in the Terrace Mountains, Utah. *Isarcicella isarcica* is thus widespread and marks a distinctive interval, so it makes sense to recognize an Isarcica Zone in the Lower Triassic, with a base at the first occurrence of *I. isarcica* in the composite section (CS) and a top at the first occurrence in that section of *Neospathodus kummeli*. But there is no resolvable interval between the highest known occurrence of *I. isarcica* and the lowest known occurrence of *N. kummeli*. Hence, there is no room for an additional biozone between the two. Furthermore, Sweet (in Ziegler 1977) concludes that *Anchignathodus parvus*

Kozur & Pjatakova is a synonym of *Isarcicella isarcica* (Huckriede). There are thus no grounds for a Parvus Zone, either above or below the Isarcica Zone. In Figure 9 the Isarcica Zone is shown above a much restricted Typicalis Zone, which continues to straddle the Permian-Triassic boundary but needs redefinition that will involve Permian rocks.

Sweet (1970a,b) divided strata in Pakistan and Kashmir between the first-occurrence levels of *Neospathodus kummeli* Sweet and *N. pakistanensis* Sweet into Kummeli, Dieneri, and Cristagalli biozones (Figure 6), but graphic assembly (Figure 9) suggests that, regionally, this interval is not so divisible. That is, *N. dieneri, N. cristagalli,* and *N. kummeli* seem to have appeared within the same STU, and the stratigraphic interval formerly divided into three zones represents only 2.5 STUs. These are thus combined into a single Kummeli-Cristagalli Zone, defined basally by the first appearance of *N. kummeli* and at the top by the first occurrence of *N. pakistanensis.*

The Pakistanensis Zone (Sweet 1970b) has been recognized only in Pakistan and Kashmir, but its guide species, *Neospathodus pakistanensis* Sweet, is now known from areas as far apart as Spitzbergen, Utah, and Primor'ye. In the CS (Figure 9), the compiled range of *N. pakistanensis* defines an interval below the first occurrence of *N. waageni* Sweet, so the Pakistanensis Zone is retained. Kozur & Mostler (1972, 1973) used different names for this interval because they confused *N. pakistanensis* with *N. homeri* (Bender), a younger species.

The interval above the Pakistanensis Zone has an involved history and in most recent studies (e.g. Carr & Paull 1983) is either lumped into one biostratigraphic unit (the Waageni Zone) or divided into as many as four, based primarily on various interpretations of the taxonomy and distribution of *Neogondolella milleri* (Müller) and species of *Parachirognathus, Furnishius,* and *Platyvillosus.* Relations summarized in Figure 9 suggest a different scheme.

In the CS (Figure 9), the appearance of *Neospathodus waageni* Sweet is followed, less than one STU later, by that of *Furnishius triserratus* Clark. These two events were probably not synchronous, but they were surely closely spaced and mark a distinctive mid-Scythian level. For stability of nomenclature, the first occurrence of *N. waageni* in the CS is used to define the base of a Waageni Zone, and the first CS occurrence of *N. milleri* (Müller) its top. The Waageni Zone thus includes the lower part of Sweet et al's (1971b) *Parachirognathus-Furnishius* Zone, Collinson & Hasenmueller's (1978) *Furnishius triserratus* Zone, Solien's (1979) *Furnishius* Zone, and at least subzone A of Carr & Paull's (1983) Zone 4.

Sweet et al (1971b) divided the upper part of Sweet's (1970b) Waageni Zone into Conservativus and Milleri zones, but that division is not supported by data from the western United States, is abandoned by Collinson & Hasenmueller (1978), Solien (1979), and Carr & Paull (1983), and is shown in Figure 9 to be indefensible, for although *Neospathodus conservativus* (Müller) appeared 3.5 STUs earlier than *Neogondolella milleri* (Müller) in western North America, the two appeared simultaneously in Primor'ye (Buryi 1979). In a zonal scheme for worldwise use, rocks in the SRS between the projected first-occurrence levels of *N. milleri* and *Neospathodus triangularis* (Bender) might be included in a Milleri Zone, which would thus embrace the Conservativus and Milleri zones of Sweet et al (1971b). Note, however, that the Milleri Zone of Figure 9 also includes the complete range of *Platyvillosus costatus* (Staesche), which was thought by Sweet et al (1971b), Collinson & Hasenmueller (1978), and Carr & Paull (1983) to characterize an interval younger than the Milleri Zone.

The simultaneous first appearances in the CS of *Neospathodus triangularis* (Bender) and *Neogondolella jubata* Sweet mark the base of Sweet's (1970b) Jubata Zone and a level close to that at which Kozur & Mostler (1972, 1973) based their Homeri Zone. It is also apparently the level at which Sweet et al (1971b), Collinson & Hasenmueller (1978), and Carr & Paull (1983) placed the base of their *Platyvillosus* Zone. It has been noted, however, that *P. costatus* first appears in the Milleri Zone in Kashmir and northern Italy, and that *P. asperatus* Clark, Sincavage & Stone appears about 3 STUs later than the beginning of Sweet's (1970b) Jubata interval, although most of its distribution is in the division above. There is undoubtedly much to learn about the stratigraphic ranges of *P. costatus* and *P. asperatus*, and perhaps about the taxonomy of this group of conodonts. However, it was evidently premature to use their ranges in 1971 to outline units in a formal zonal biostratigraphy. Consequently, in Figure 9 the level of simultaneous first occurrence of *N. triangularis* and *N. jubata* marks the base of a Triangularis Zone, which continues to the level at which *N. collinsoni* Solien makes its debut.

Sweet et al (1971b) and later authors have noted that *Neospathodus collinsoni* Solien (= *N*. n. sp. G of Sweet et al) defines a distinctive interval in the upper Scythian that merits recognition as a separate biozone. That distinctiveness carries over to the CS (Figure 9), and the name Collinsoni Zone is retained. Between the highest known occurrence of *N. collinsoni* and the earliest occurrence of *N.* (or *Neogondolella*) *timorensis* (Nogami), however, is a segment of the Scythian, representing about 6 STUs, which it appears impossible to subdivide by the procedures of zonal biostratigraphy. Sweet et al (1971b) included this interval in a restricted Jubata Zone. Experience with the earlier Scythian Carinata and Dieneri zones suggests, however, that such interval zones have little utility and rather short shelf lives.

The Timorensis Zone (top of Figure 6) has its base at the first occurrence of *Neospathodus* (or *Neogondolella*) *timorensis* (Nogami). Collinson & Hasenmueller (1978) report *N. timorensis* in samples from the Haugi ammonoid zone, which Silberling & Tozer (1968) regard as highest Scythian, but Nicora (1977) writes that all specimens of *N. timorensis* known to her are from rocks above those with definite Scythian ammonoids or are from strata that contain early Anisian forms. Thus, the Timorensis Zone may span the Scythian-Anisian boundary. In Figure 9, this boundary is drawn at the base of the Timorensis Zone, but with no particular conviction.

In summary, graphic assembly of a Scythian composite section makes it possible to suggest that the zonal scheme along the right side of Figure 9 is more representative of the distributional data now available than any of those in Figure 6. But it should also be pointed out that the composite scale provides the means for resolving 26.5 equal-length STUs in component sections of the network, and that it might thus become the basis for a considerably more refined chronostratigraphy than any of the zonal schemes reviewed, including the one proposed in Figure 9. In short, work completed thus far suggests that nontraditional procedures may provide a very high-resolution chronostratigraphic scale for the Scythian, but that eight or nine biozones may be about the limit of resolution possible in this interval by biozonal procedures.

SUMMARY

Conodonts, with all the requisites of ideal index fossils, are now the basis for recognition and correlation of a succession of more than 150 biozones in marine rocks of Late Cambrian through Triassic age. The extent, resolution, and regional recognizability of these units varies considerably from one level to another in the column. Intervals in which biozonal boundaries are based on evolutionary events in well-controlled phylomorphogenetic sequences, such as the early Middle Ordovician and Devonian, may be divisible into biostratigraphic units with unit-resolution values of about 0.5 Myr. In parts of the column in which such evolutionary events are difficult to recognize, such as the upper Middle and Upper Ordovician, or the Lower Triassic, graphic procedures may provide not only the best means of correlation, but also a means of recognizing correlation units, or chronozones, of potentially higher resolution than is available even in intervals controlled by events in well-controlled phyletic sequences. A combination of the results of both procedures should produce a chronostratigraphic framework of exceptionally high resolution.

Literature Cited

Aldridge, R. J. 1975. The stratigraphic distribution of conodonts in the British Silurian. *J. Geol. Soc. London* 131 : 607–18

An, T.-X. 1981. Recent progress in Cambrian and Ordovician conodont biostratigraphy. *Geol. Soc. Am. Spec. Pap.* 187 : 209–26

An, T.-X. 1982. Study on the Cambrian conodonts from north and northeast China. *Sci. Rept. Inst. Geosci. Univ. Tsukuba, Sect. B, Geol. Sci.* 3 : 113–59

An, T.-X., Zhang, F., Xiang, W., Zhang, Y., Xu, W., et al. *The Conodonts of North China and the Adjacent Regions.* Beijing: Science Press of China. 223 pp. (In Chinese)

Assereto, R., Bosellini, A., Fantini Sestini, N., Sweet, W. C. 1973. The Permian-Triassic boundary in the Southern Alps (Italy). *Can. Soc. Pet. Geol. Mem.* 2 : 176–99

Barnes, C. R., Poplawski, M. L. S. 1973. Lower and Middle Ordovician conodonts from the Mystic Formation, Quebec, Canada. *J. Paleontol.* 47 : 760–90

Barrick, J. E., Klapper, G. 1976. Multielement Silurian (late Llandoverian-Wenlockian) conodonts of the Clarita Formation, Arbuckle Mountains, Oklahoma, and the phylogeny of Kockelella. *Geol. Palaeontol.* 10 : 59–98

Bengtson, S. 1983. The early history of the Conodonta. *Fossils and Strata* 15 : 5–19

Bergström, S. M. 1971. Conodont biostratigraphy of the Middle and Upper Ordovician of Europe and eastern North America. See Sweet & Bergström 1971a, pp. 83–161

Bergström, S. M. 1973. Biostratigraphy and facies relations in the lower Middle Ordovician of easternmost Tennessee. *Am. J. Sci.* 172-A : 261–93

Bergström, S. M. 1978. Middle and Upper Ordovician conodont and graptolite biostratigraphy of the Marathon, Texas graptolite zone reference standard. *Palaeontology* 21 : 723–58

Bergström, S. M. 1983. Biogeography, evolutionary relationships, and biostratigraphic significance of Ordovician platform conodonts. *Fossils and Strata* 15 : 35–58

Bergström, S. M., Carnes, J. B. 1976. Conodont biostratigraphy and paleoecology of the Holston Formation (Middle Ordovician) and associated strata in eastern Tennessee. *Geol. Assoc. Can. Spec. Pap.* 15 : 27–57

Bergström, S. M., Cooper, R. A. 1973. Didymograptus bifidus and the trans-

Atlantic correlation of the Lower Ordovician. *Lethaia* 6 : 313–40

Bergström, S. M., Riva, J., Kay, M. 1974. Significance of conodonts, graptolites, and shelly faunas from the Ordovician of western and north-central Newfoundland. *Can. J. Earth Sci.* 11 : 1625–60

Burrett, C., Stait, B., Laurie, J. 1983. Trilobites and microfossils from the Middle Ordovician of Surprise Bay, southern Tasmania, Australia. *Mem. Assoc. Australas. Palaeontol.* 1 : 177–93

Buryi, G. 1979. *Lower Triassic Condonts, South Primor'ye.* Moscow : Nauka. (Acad. Sci. USSR, Siberian Div., Inst. Geol. Geophys.) 144 pp. (In Russian)

Carr, T. R., Paull, R. K. 1983. Early Triassic stratigraphy and paleogeography of the Cordilleran miogeocline. In *Mesozoic Paleogeography of the West-Central United States,* ed. M. W. Reynolds, E. D. Dolly, pp. 39–55. Denver : Rocky Mt. Sect., Soc. Econ. Paleontol. Mineral.

Collinson, J. W., Hasenmueller, W. A. 1978. Early Triassic paleogeography and biostratigraphy of the Cordilleran miogeosyncline. In *Mesozoic Paleogeography of the Western United States,* ed. D. G. Howell, K. A. McDougall, pp. 175–87. Los Angeles : Pacific Sect., Soc. Econ. Paleontol. Mineral.

Cooper, B. J. 1980. Toward an improved Silurian conodont biostratigraphy. *Lethaia* 13 : 209–27

Cooper, B. J. 1981. Early Ordovician conodonts from the Horn Valley Siltstone, central Australia. *Palaeontology* 24 : 147–83

Dzik, J. 1976. Remarks on the evolution of Ordovician conodonts. *Acta Palaeontol. Pol.* 21 : 395–455

Dzik, J. 1978. Conodont biostratigraphy and paleogeographical relations of the Ordovician Mojcza Limestone (Holy Cross Mts. Poland). *Acta Palaeontol. Pol.* 23 : 51–72

Dzik, J. 1983. Relationships between Ordovician Baltic and North American Midcontinent conodont faunas. *Fossils and Strata* 15 : 59–85

Ethington, R. L., Clark, D. L. 1971. Lower Ordovician conodonts in North America. See Sweet & Bergström 1971a, pp. 63–82

Ethington, R. L., Clark, D. L. 1982. Lower and Middle Ordovician conodonts from the Ibex area, western Millard County, Utah. *Brigham Young Univ. Geol. Stud.* 28(2) : 1–160

Fåhraeus, L. E., Nowlan, G. S. 1978. Fran-

110   SWEET & BERGSTRÖM

conian (Late Cambrian) to Early Champlainian (Middle Ordovician) conodonts from the Cow Head Group, western Newfoundland. *J. Paleontol.* 52 : 444–71

Harris, A. G., Bergström, S. M., Ethington, R. L., Ross, R. J. Jr. 1979. Aspects of Middle and Upper Ordovician conodont biostratigraphy of carbonate facies in Nevada and southeast California and comparison with some Appalachian successions. *Brigham Young Univ. Geol. Stud.* 26(3): 7–43

Hedberg, H., ed. 1976. *International Stratigraphic Guide.* New York: Wiley. 200 pp.

Helfrich, C. T. 1975. Silurian conodonts from the Wills Mountain anticline, Virginia, West Virginia, and Maryland. *Geol. Soc. Am. Spec. Pap.* 161 : 1–82

Jeppsson, L. 1974. Aspects of Late Silurian conodonts. *Fossils and Strata* 6 : 1–54

Jeppsson, L. 1983. Silurian conodont faunas from Gotland. *Fossils and Strata* 15 : 121–44

Johnson, J. G., Klapper, G., Sandberg, C. A. 1985. Devonian eustatic fluctuations in Euramerica. *Geol. Soc. Am. Bull.* 96 : 567–87

Klapper, G. 1977. Lower and Middle Devonian conodont sequence in central Nevada. *Univ. Calif. Riverside Mus. Contrib.* 4 : 33–54

Klapper, G., Murphy, M. A. 1975. Silurian–Lower Devonian conodont sequence in the Roberts Mountains Formation of central Nevada. *Univ. Calif. Publ. Geol. Sci.* 111 : 1–62

Klapper, G., Ziegler, W. 1979. Devonian conodont biostratigraphy. *Spec. Pap. Palaeontol.* 23 : 199–224

Kozur, H. 1980. Revision der Conodontenzonierung der Mittel- und Obertrias des tethyalen Faunenreichs. *Geol. Paläontol. Mitt. Innsbruck* 10 : 79–172

Kozur, H., Mostler, H. 1972. Die Bedeutung der Conodonten für stratigraphische und paläogeographische Untersuchungen in der Trias. *Mitt. Ges. Geol. Bergbaustud. Wien* 32 : 777–810

Kozur, H., Mostler, H. 1973. Beiträge zur Mikrofauna permotriadischer Schichtfolgen. Teil I: Conodonten aus der Tibetzone des niederen Himalaya (Dolpogebiet, Westnepal). *Geol. Paläontol. Mitt. Innsbruck* 3 : 1–23

Landing, E. 1976. Early Ordovician (Arenigian) conodont and graptolite biostratigraphy of the Taconic allochthon, eastern New York. *J. Paleontol.* 50 : 614–46

Lane, H. R., Sandberg, C. A., Ziegler, W. 1980. Taxonomy and phylogeny of some Lower Carboniferous conodonts and preliminary standard post-*Siphonodella* zonation. *Geol. Palaeontol.* 14 : 117–64

Lin, B.-Y. 1983. New developments in conodont biostratigraphy of the Silurian of China. *Fossils and Strata* 15 : 145–47

Lindström, M. 1955. Conodonts from the lowermost Ordovician strata of south-central Sweden. *Geol. Fören. Stockholm Förh.* 76 : 517–614

Lindström, M. 1971. Lower Ordovician conodonts of Europe. See Sweet & Bergström 1971a, pp. 21–82

Löfgren, A. 1978. Arenigian and Llanvirnian conodonts from Jämtland, northern Sweden. *Fossils and Strata* 13 : 1–129

Löfgren, A. 1985. Early Ordovician conodont biozonation at Finngrundet, south Bothnian Bay, Sweden. *Bull. Geol. Inst. Univ. Uppsala* (new ser.) 10 : 115–28

Markello, J. R., Tillman, C. G., Read, J. F. 1979. Field Trip No. 3. Lithofacies and biostratigraphy of Cambrian and Ordovician platform and basin facies carbonates and clastics, southwestern Virginia. *Geol. Soc. Am., Southeastern Sect. Field Trip Guideb., April 27–29, 1979,* pp. 41–85

Matsuda, T. 1981a. Early Triassic conodonts from Kashmir, India. Part I: *Hindeodus* and *Isarcicella. J. Geosci. Osaka City Univ.* 24 : 75–108

Matsuda, T. 1981b. Appendix to conodonts of Guryul Ravine. *Palaeontol. Indica* (new ser.) 46 : 187–88

Matsuda, T. 1982. Early Triassic conodonts from Kashmir, India. Part 2: *Neospathodus* I. *J. Geosci. Osaka City Univ.* 25 : 87–103

Matsuda, T. 1983. Early Triassic conodonts from Kashmir, India. Part 3: *Neospathodus* 2. *J. Geosci. Osaka City Univ.* 26 : 87–110

Matsuda, T. 1984. Early Triassic conodonts from Kashmir, India. Part 4: *Gondolella* and *Platyvillosus. J. Geosci. Osaka City Univ.* 27 : 119–41

Miller, J. F. 1984. Cambrian and earliest Ordovician conodont evolution, biofacies and provincialism. *Geol. Soc. Am. Spec. Pap.* 196 : 43–68

Miller, J. F., Taylor, M. E., Stitt, J. H., Ethington, R. L., Hintze, L. F., Taylor, J. F. 1983. Potential Cambrian-Ordovician boundary stratotype sections in the western United States. In *The Cambrian-Ordovician Boundary: Sections, Fossil Distributions, and Correlations,* ed. M. G. Bassett, W. T. Dean, pp. 155–180. Cardiff: Natl. Mus. Wales, Geol. Ser. 3

Moskalenko, T. 1983. Conodonts and biostratigraphy in the Ordovician of the Siberian Platform. *Fossils and Strata* 15 : 87–94

Müller, K. J. 1973. Late Cambrian and Early

Ordovician conodonts from northern Iran. *Iran Geol. Surv. Rep. 30.* 77 pp.

Nicora, A. 1977. Lower Anisian platform-conodonts from the Tethys and Nevada: Taxonomic and stratigraphic revision. *Palaeontogr. A* 157:88–107

North American Commission on Stratigraphic Nomenclature. 1983. North American stratigraphic code. *Am. Assoc. Pet. Geol. Bull.* 67:841–75

Nowlan, G. 1981. Late Ordovician–Early Silurian conodont biostratigraphy of the Gaspe Peninsula—a preliminary report. In *Subcommission on Silurian Stratigraphy, Ordovician-Silurian Boundary Working Group. Field Meeting, Anticosti-Gaspé, Quebec 1981, Vol. II: Stratigraphy and Paleontology,* ed. P. J. Lespérance, pp. 257–91.

Nowlan, G. 1983. Early Silurian conodonts of eastern Canada. *Fossils and Strata* 15:95–110

Pakistani-Japanese Research Group. 1981. Stratigraphy and correlation of the marine Permian–Lower Triassic in the Surghar Range and the Salt Range, Pakistan. *Kyoto Univ. Rep.* 25 pp.

Pander, C. H. 1856. Monographie der fossilen Fische des silurischen Systems der russisch-baltischen Gouvernements. *K. Akad. Wiss. St. Petersburg.* 91 pp.

Paull, R. K. 1982. Conodont biostratigraphy of Lower Triassic rocks, Terrace Mountains, northwestern Utah. In *Overthrust Belt of Utah,* ed. D. B. Nielsen, pp. 235–49. Salt Lake City: Utah Geol. Assoc. Publ. 10

Repetski, J. E., Ethington, R. L. 1977. Conodonts from graptolite facies in the Ouachita Mountains, Arkansas and Oklahoma. *Ark. Geol. Comm. Symp. Geol. Ouachita Mts.,* 1:92–106

Rust, C. C. 1968. *Conodonts of the Martinsburg Formation (Ordovician) of southwestern Virginia.* PhD dissertation. Ohio State Univ., Columbus. 189 pp.

Salvador, A. 1985. Chronostratigraphic and geochronometric scales in COSUNA stratigraphic correlation charts of the United States. *Am. Assoc. Pet. Geol. Bull.* 68:181–89

Sandberg, C. A., Poole, F. G. 1977. Conodont biostratigraphy and depositional complexes of Upper Devonian cratonic-platform and continental-shelf rocks in the western United States. *Univ. Calif. Riverside Mus. Contrib.* 4:144–82

Sandberg, C. A., Ziegler, W., Leuteritz, K., Brill, S. M. 1978. Phylogeny, speciation, and zonation of Siphonodella (Conodonta, Upper Devonian and Lower Carboniferous). *Newsl. Stratigr.* 7:102–20

Shaw, A. B. 1964. *Time in Stratigraphy.* New York: McGraw-Hill. 365 pp.

Silberling, N. J., Tozer, E. T. 1968. Biostratigraphic classification of the marine Triassic in North America. *Geol. Soc. Am. Spec. Pap. 110.* 63 pp.

Solien, M. A. 1979. Conodont biostratigraphy of the Lower Triassic Thaynes Formation, Utah. *J. Paleontol.* 53:276–306

Staesche, U. 1964. Conodonten aus dem Skyth von Südtirol. *Neues Jahrb. Geol. Paläontol. Abh.* 119:247–306

Stouge, S. 1984. Conodonts of the Middle Ordovician Table Head Formation, western Newfoundland. *Fossils and Strata* 16:1–145

Sweet, W. C. 1970a. Permian and Triassic conodonts from a section at Guryul Ravine, Vihi District, Kashmir. *Univ. Kans. Paleontol. Contrib.* 49:1–10

Sweet, W. C. 1970b. Uppermost Permian and Lower Triassic conodonts of the Salt Range and Trans-Indus ranges, West Pakistan. *Univ. Kans., Dep. Geol. Spec. Publ.* 4:207–75

Sweet, W. C. 1973. Late Permian and Early Triassic conodont faunas. *Can. Soc. Pet. Geol. Spec. Publ.* 2:630–46

Sweet, W. C. 1979a. Graphic correlation of Permo-Triassic rocks in Kashmir, Pakistan and Iran. *Geol. Palaeontol.* 13:239–48

Sweet, W. C. 1979b. Late Ordovician conodonts and biostratigraphy of the western Midcontinent Province. *Brigham Young Univ. Geol. Stud.* 26(3):45–85

Sweet, W. C. 1984. Graphic correlation of upper Middle and Upper Ordovician rocks, North American Midcontinent Province, U.S.A. In *Aspects of the Ordovician System,* ed. D. L. Bruton, pp. 23–35. Oslo: Universitetsforlaget

Sweet, W. C., Bergström, S. M., eds. 1971a. *Symposium on Conodont Biostratigraphy. Geol. Soc. Am. Mem.* 127. 499 pp.

Sweet, W. C., Bergström, S. M. 1971b. The American Upper Ordovician Standard. XIII. A revised time-stratigraphic classification of North American upper Middle and Upper Ordovician rocks. *Geol. Soc. Am. Bull.* 82:613–28

Sweet, W. C., Bergström, S. M. 1976. Conodont biostratigraphy of the Middle and Upper Ordovician of the United States Midcontinent. In *The Ordovician System, Proc. Palaeontol. Assoc. Symp., Birmingham, Sept. 1974,* ed. M. G. Bassett, pp. 121–151. Cardiff: Univ. Wales Press and Natl. Mus. Wales

Sweet, W. C., Bergström, S. M. 1984. Conodont provinces and biofacies of the Late Ordovician. *Geol. Soc. Am. Spec. Pap.* 196:69–87

Sweet, W. C., Ethington, R. L., Barnes, C. R.

1971a. North American Middle and Upper Ordovician conodont faunas. See Sweet & Bergström 1971a, pp. 163–93

Sweet, W. C., Mosher, L. C., Clark, D. L., Collinson, J. W., Hasenmueller, W. A. 1971b. Conodont biostratigraphy of the Triassic. See Sweet & Bergström 1971a, pp. 441–65

Szaniawski, H. 1982. Chaetognath grasping spines recognized among Cambrian proto-conodonts. J. Paleontol. 56:806–10

Van Wamel, W. A. 1974. Conodont biostratigraphy of the Upper Cambrian and Lower Ordovician of north-western Öland, south-eastern Sweden. Utrecht Micropaleontol. Bull. 10:1–119

Walliser, O. H. 1962. Conodontenchronologie des Silurs (Gotlandiums) und des tieferen Devons mit besonderer Berücksichtigung der Formationsgrenze. In Svmp.-Band Int. Arbeitstag. Silur/Devon-Grenze und die Stratigraphie von Silur und Devon, 2nd, Bonn, Bruxelles, 1960, ed.

H. K. Erben, pp. 281–87. Stuttgart: Schweizerbart'sche Verlagsbuchhandlung

Walliser, O. H. 1964. Conodonten des Silurs. Hess. Landesamt. Bodenforsch. Abh. 41. 106 pp.

Walliser, O. H. 1971. Conodont biostratigraphy of the Silurian of Europe. See Sweet & Bergström 1971a, pp. 195–206

Webby, B. D., Vandenberg, A. H. M., Cooper, R. A., Banks, M. R., Burrett, C. F., et al., eds. 1981. The Ordovician System in Australia, New Zealand and Antarctica, Correlation Chart and Explanatory Notes, Int. Union Geol. Sci. Publ. 6:1–64

Zeng, Q. L., Ni, S. Z., Xu, G. H., Zhou, T. M., Wang, Z. F., et al. 1983. Subdivision and correlation of the Ordovician in the eastern Yangtze Gorges, China. Bull. Yichang Inst. Geol. Min. Res., Chin. Acad. Sci. 6:21–68

Ziegler, W., ed. 1977. Catalogue of Conodonts, Vol. 3. Stuttgart: Schweizerbart'sche Verlagsbuchhandlung. 574 pp.

Ann. Rev. Earth Planet. Sci. 1986. 14 : 113–47

OCCURRENCE AND FORMATION OF WATER-LAID PLACERS

Rudy Slingerland

Department of Geosciences, The Pennsylvania State University, University Park, Pennsylvania 16802

Norman D. Smith

Department of Geological Sciences, University of Illinois at Chicago, Chicago, Illinois 60680

INTRODUCTION

A placer is a deposit of residual or detrital mineral grains in which a valuable mineral has been concentrated by a mechanical agent. The agent is usually running water, and the valuable mineral is usually denser than quartz (for example, gold, diamond, cassiterite, ilmenite, or chromite). It is difficult to overemphasize the importance of placers as sources of mineral wealth; the Witwatersrand (South Africa) paleoplacers, discovered in 1886, alone have provided over half of all the gold ever mined in the world (Pretorius 1976), and alluvial placer deposits of cassiterite in southeast Asia presently are the world's major source of tin (Toh 1978). The earliest evidence of placer mining comes from stone carvings in Egypt dated at 2500 B.C. (MacDonald 1983), and as early as 7 B.C., Strabo pointed out that in Turdetania (Spain) more gold was procured from washing sand in the rivers than by digging in the mines.

Given the economic importance of water-laid placers and the long history of their exploitation, it is remarkable that they have received so little attention in sedimentary research. Mining and exploration practices for the most part have been guided by empirical rules of thumb that are locally useful but lack any firm physical basis of understanding [see Bateman

113

0084–6597/86/0515–0113$02.00

(1950) and MacDonald (1983) for summaries]. For example, MacDonald (1983, p. 145) states that "on normally stable beaches the highest concentrations [of heavy minerals] are found where the pounding of waves has been most intensive . . . ," although the causes of this association, even if true, remain obscure. It is significant, then, that in the last decade a body of knowledge on the occurrence and genesis of placers has grown out of both the physical principles of hydrodynamics and sediment transport and an improved understanding of the sedimentary environments of deposition. To be sure, whether or not an economic placer deposit forms may depend ultimately on such external factors as the stage of stream evolution (Tuck 1968, Kartashov 1971, Schumm 1977, Adams et al 1978), tectonic history (Henley & Adams 1979, Sigov et al 1972), local geology and physiography (Jenkins 1964), or climate (Krook 1968). But common to the origins of all water-laid placers are concentrating mechanisms that involve interactions among the fluid, sediment bed, and transported particles, and it is these mechanisms for sorting dense sediment grains from light grains that we emphasize in this review. Unfortunately, our coverage is not exhaustive. Russian papers are numerous in this field, and only a few translations are considered here.

OCCURRENCES OF WATER-LAID PLACERS

Water-laid placers occur in a variety of geomorphological sites (observed or interpreted) and over a wide range of physical scales. Undoubtedly the most common and important group of economic placer deposits are those formed by streams, with beaches and shallow nearshore areas probably second in importance. Other settings, by comparison, are relatively minor, either in modern environments or as ancient paleoplacers.

Table 1 lists sites of heavy-mineral concentrations described from both modern and ancient examples and classifies them according to their spatial scales (Smith & Minter 1980, p. 1; Slingerland 1984, p. 138). Large-scale concentrations (order 10^4 m) occur on regional or system-wide scales as products of long-term interactions among time-averaged flow variables, available heavy minerals, and substrate characteristics. The intermediate scale (order 10^2 m) refers to concentrations associated with major depositional or erosional topography within the sediment-transporting system. In fluvial settings, these are commonly bars, short channel segments, riffles, and the like. Small-scale concentrations (order 10^0 m) occur at the sediment-bed scale and are commonly manifested as heavy-mineral-rich laminations in sequences of stratification. Sorting associated with the formation and migration of bedforms is a dominant cause of small-scale segregations. These scales are clearly hierarchical in that smaller scales

Table 1 Observed sites of water-laid placers

Sites	References
Large scale (10^4 m)	
Bands parallel to depositional strike	Minter 1970, 1978, Sestini 1973, McGowen & Groat 1971
Heads of wet alluvial fans	Schumm 1977
Points of abrupt valley widening	Kuzvart & Bohmer 1978, Crampton 1937
Points of exit of highland rivers onto a plain	Toh 1978
Regional unconformities	Minter 1976, 1978
Strand-line deposits	Nelson & Hopkins 1972, Komar & Wang 1984, Eliseev 1981
Incised channelways	Minter 1978, Yeend 1974, Buck 1983
Pediment mantles	Krapez 1985
Intermediate scale (10^2 m)	
Concave sides of channel bends	Kuzvart & Bohmer 1978, Crampton 1937
Convex banks of channel bends	Kuzvart & Bohmer 1978
Heads of midchannel bars	Toh 1978, Smith & Minter 1980, Kartashov 1971, Boggs & Baldwin 1970
Point bars with suction eddies	Toh 1978, Bateman 1950
Scour holes, especially at tributary confluences	Kuzvart & Bohmer 1978, Mosley & Schumm 1977
Inner bedrock channels and false bedrock	Schumm 1977, Kuzvart & Bohmer 1978, Adams et al 1978
Bedrock riffles	Cheny & Patton 1967, Toh 1978
Constricted channels between banks and bankward-migrating bars	Smith & Minter 1980, Smith & Beukes 1983
Beach swash zones	Stapor 1973, Reimnitz & Plafker 1976, Kogan et al 1975
Small scale (10^0 m)	
Scoured bases of trough cross-strata sets	Toh 1978, McGowen & Groat 1971, Smith & Minter 1980, Buck 1983
Winnowed tops of gravel bars	Toh 1978, McGowen & Groat 1971
Thin ripple-form accumulations	Brady & Jobson 1973
Dune crests	Brady & Jobson 1973
Dune foresets	Brady & Jobson 1973, McGowen & Groat 1971, Buck 1983
Plane parallel laminae	Slingerland 1977, Clifton 1969, Buck 1983, Stavrakis 1980
Leeward side of obstacles	Lindgren 1911
Beach berms	Stapor 1973

are superimposed on larger ones; thus, a few heavy minerals might concentrate on a small scale over a very short time in response to some sorting event, but increasingly larger scales of concentrations require successively greater areas over which the sorting mechanism is applied. For example, a strand-line placer (large scale) may be dominated by patchy concentrations in the beach swash zone (intermediate scale); these concentrations in turn are composed of segregated heavy-mineral-rich laminations (small scale). A basic understanding of placer formation, therefore, requires knowledge of sorting arising from the finer scales of fluid-sediment interactions; such knowledge is only partial at this time.

CONDITIONS FOR PLACER DEVELOPMENT

The various placer sites listed in Table 1 contain five factors common to all members. First, in the parlance of mining technology, each site is a natural geomorphological "dressing mill." The mill may be the toe of a dune avalanche face or a berm crest on a beach, but in each case it receives "pulp" (that is, granule- to silt-sized, discrete heavy and light mineral grains mixed with water) and remains relatively fixed in time and space as it sizes and separates the pulp. Second, heavy minerals are present in the pulp and have a size distribution that is proper for the natural dressing mill. Most alluvium or natural pulp contains some heavy minerals, but usually in concentrations of less than 1%. Gemstones and gold are much rarer, of course, and are linked to particular plate tectonic settings (Henley & Adams 1979). The initial size distributions of heavy and light minerals in various source terranes are still not known with any accuracy. Given the crystal sizes of minerals in igneous and metamorphic rocks (Feniak 1944) and the increased susceptibility of some heavy minerals to comminution, the global average grain size of heavy minerals in alluvium is probably less than that of light minerals, even before hydraulic sorting has occurred (Rittenhouse 1943, Van Andel 1959, Briggs 1965, Stapor 1973). This assumption is made in the subsequent discussions. Third, in each case the pulp is fed at a proper rate for the mill size. Fourth, and especially critical, each dressing mill contains the proper combination and sequence of hydraulic sorting mechanisms for separating the pulp by size and density. Fifth, the geometry of the mill circuits is such that tailings are properly disposed (transported away) and ore concentrate is quasi-permanently stored.

A tributary junction of a stream is a good example of an intermediate-scale natural dressing mill (Schumm 1977). It processes pulp from a tributary, stores the ore concentrate in deep scour holes, and disposes of tailings by sediment transport due to flows of the main channel. The pulp rate is governed by a feedback loop from the mill through a base-level control on the amount of upstream erosion. The sorting mechanisms created by the local flow geometry are the sediment entrainment and transport processes discussed in what follows.

HYDRAULIC SORTING MECHANISMS—
THE SINE QUA NON

However favorable other factors may be, without mechanisms for sorting dense sediment grains from light grains, there will be no water-laid placer deposit. Sorting of a heterogeneous size-density pulp is accomplished in natural mill circuits by an alternation of at least four mechanisms, some sorting more by size and some by density. The sorting may be subdivided by scale into two types: local and progressive (Brush 1965, Rana et al 1973, Deigaard & Fredsøe 1978). Progressive sorting occurs by the cumulative effects of local sorting and along-flow changes in competency and capacity.

The mechanisms of local sorting are the following (Slingerland 1984): (a) the free or hindered settling of grains, usually in turbulent water, (b) the entrainment of grains from a granular bed by flowing water, (c) the transport of grains by flowing water, and (d) the shearing of grains in a moving granular dispersion. The categories are not mutually exclusive— shear sorting may play a role in transport sorting, for example—but we find this organization useful at our present level of knowledge.

Before discussing each mechanism in detail, we briefly highlight some pertinent principles of fluid mechanics and sediment entrainment. Middleton & Southard (1977) and Yalin (1977) provide excellent, more in-depth reviews. Flows in natural stream channels are either predominantly or wholly turbulent (i.e. they are characterized by random velocity fluctuations and flow pathlines that are strongly three dimensional). The gross character of the flow can be visualized by considering a vertical cross section in the downstream direction with a planar channel bottom (Figure 1). Three intergrading flow zones can be identified for turbulent flows above a smooth boundary (Hinze 1975): (a) Immediately above the channel bottom is the *viscous sublayer*, which contains turbulent fluctuations but is dominated by viscous rather than turbulent transfer of momentum across fluid layers; it rarely exceeds a few millimeters in thickness. (b) A thin *turbulence-generation layer* (or "buffer layer") is located just above the viscous sublayer. Here the shear stresses are very high, and small but strong turbulent eddies are generated and carried outward above the boundary or downward into the viscous sublayer. (c) The *outer* (or *core*) *region* occupies the remainder of the flow boundary layer; in most natural streams, this zone comprises most of the flow depth and the highest mean velocities.

In hydrodynamically "rough" bottoms, coarse sediment particles project upward sufficiently to disrupt flow structure near the bed; the viscous sublayer is destroyed, and the turbulence-generation layer extends to the bottom. This situation occurs when the height of the roughness elements, K [taken by Einstein (1950) as the 65th percentile of the bottom grain size distribution] exceeds the "potential" thickness of the viscous sublayer [i.e. its thickness if the boundary were smooth under existing flow conditions (Figure 1)].

The curved profile of upward-increasing velocity (Figure 1), typical of open-channel flow, is caused by bottom frictional drag, which is transferred upward by viscous (in laminar flow) and inertial (turbulent eddies) momentum exchanges between successive fluid layers. If we assume a straight channel segment in which neither depth nor average velocity changes downstream (i.e. "uniform flow"), the force exerted by the gravity-driven flow on the channel boundary (bottom and sides) is balanced by the frictional resistance of the boundary, because the assumed flow neither accelerates nor decelerates. The temporal mean boundary fluid force, called the tractive or boundary shear stress $\bar{\tau}_0$, is given by

$$\bar{\tau}_0 = \rho_f gRS, \tag{1}$$

in which ρ_f is the fluid density, g the acceleration due to gravity, S the slope of the stream bed and water surface (equal in uniform flow), and R the hydraulic radius. For natural streams in which the width greatly exceeds

TURBULENT BOUNDARY LAYER

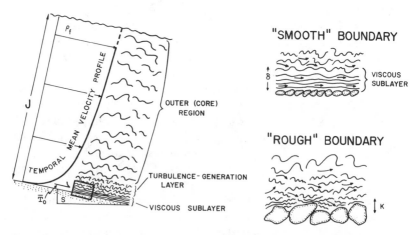

Figure 1   Internal structure of natural turbulent flows, where J is the flow depth, ρ_f the fluid density, $\bar{\tau}_0$ the temporal mean boundary shear stress, S the bed surface slope, δ the thickness of the viscous sublayer, and K the height of roughness elements. See text for discussion.

the depth J, it follows that $R \sim J$ and thus that

$$\bar{\tau}_0 = \rho_f g J S. \tag{2}$$

The instantaneous shear stress at the bed follows a positive skewed distribution, with a coefficient of variation equal to 0.4 (Grass 1983); thus, grains may experience shear stresses up to twice the temporal mean given in Equations (1) and (2).

For certain problems in fluid mechanics, a parameter defined as

$$U_* = \sqrt{\frac{\bar{\tau}_0}{\rho_f}} = \sqrt{gJS} \tag{3}$$

is used, in which U_* is termed the shear velocity, or friction velocity. Although U_* has the dimension of velocity, it need be considered only as a surrogate or "convenience" variable for tractive shear stress. The shear velocity cannot be measured directly (such as with a current meter) and is always much smaller than the average velocity.

The presence, and thickness, of the viscous sublayer is important to bottom sediment movement because this sublayer affects the nature and distribution of fluid forces (viscous and pressure) acting on the grains. For smooth boundaries, the thickness of the sublayer, δ, depends on shear velocity and viscosity as

$$\delta = \frac{cv}{U_*}, \tag{4}$$

where c is a constant. Since protruding grains of sufficient size will destroy the viscous sublayer, it is reasonable to think that the ratio of bottom roughness size to sublayer thickness, K/δ, may serve to define hydro-dynamically rough and smooth boundaries. Substituting from (4), we have

$$\frac{K}{\delta} = \frac{U_* K}{cv} = \text{constant} = R_*, \tag{5}$$

where the dimensionless quantity $U_* K/v$ is termed the boundary Reynolds number R_*. Earlier workers (see Inman 1949) considered that the value $R_* = 3.5$ distinguished smooth from transitional boundaries, but $R_* = 5$ is the more commonly accepted value today. The value of R_*, then, reflects the degree to which grains on the bed project into the turbulent zone of the boundary layer, and we should expect the distribution of fluid forces acting on grains to be a function of that value. For $R_* > 70$, the wall is considered to be fully rough, with the range $5 < R_* < 70$ representing a transition in which turbulence only periodically disrupts the viscous sublayer and impinges directly on the grains.

Another Reynolds number, the grain Reynolds number Re, is also important because it determines the degree of turbulence in a boundary layer around a grain undergoing relative motion in a fluid. Its velocity term is the relative velocity, its length term is the grain diameter, and its viscosity term is the kinematic viscosity. The coefficient of drag C_d, a proportionality factor between the mean relative velocity and the total drag force on a grain, is a direct function of Re.

Settling of Grains

Grains of differing sizes, densities, and shapes fall through a fluid at differing velocities, thereby sorting themselves (Figure 2). If the grains are spherical and fall at very low concentrations through still water, their constant terminal settling velocity w_∞ can be obtained by equating expressions for gravity and fluid drag forces on a sphere; the resulting calculation yields

$$w_\infty = \left[\frac{4}{3}\frac{(\rho_p - \rho_f)gD}{\rho_f C_d}\right]^{1/2}, \tag{6}$$

where ρ_p is the density of the particle, D is its diameter, and the other variables are as defined previously. For grain Reynolds numbers Re less than 0.5 (e.g. for quartz sphere diameters less than 0.1 mm), accurate settling velocities can be computed theoretically from Stokes' law, where

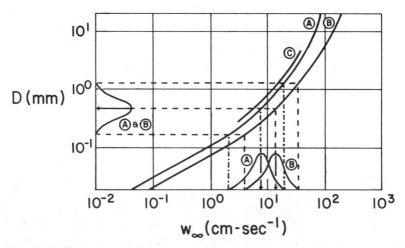

Figure 2   Unhindered terminal settling velocities w_∞ of spheres in still water versus their diameter D. Curve A is for quartz (density = 2.65 g cm^{-3}), B is for magnetite (density = 5.10 g cm^{-3}) (both calculated from Warg 1973), and C is for natural quartz grains from Oregon beaches (Baba & Komar 1981). A log-normal size distribution of quartz and magnetite spheres produces two overlapping, skewed settling velocity distributions.

$C_d = 24/Re$ (see, for example, Graf 1971). Settling velocities of grains in this range vary as the square of the diameter and the first power of the buoyant density. For larger or denser grains, velocities must be computed from semitheoretical, empirically calibrated formulae; the more recent and useful of these are presented by Gibbs et al (1971), Warg (1973), Lerman et al (1974), Komar & Reimers (1978), Baba & Komar (1981), Komar (1981), and Doyle et al (1983). Settling velocities of grains in this range vary as the square root of diameter and buoyant density. Thus, smaller particles are more effectively sorted by settling than larger particles because doubling their density, for example, quadruples their settling velocity.

Without doubt, shape effects on settling behavior can also be considerable (Briggs et al 1962), although these are often overlooked in studies of heavy-mineral sorting. Data by Doyle et al (1983), for example, show that flat biotite grains may settle from 4 to 12 times slower than quartz spheres of equal diameter. Even natural quartz grains from beaches are nonspherical enough to settle on average at three quarters the velocity of an equivalent-sized sphere (Baba & Komar 1981; see Figure 2). As these studies have demonstrated, however, the effects of shape on settling velocities can be estimated and, in principle, accounted for. Tourtelot & Riley (1973), Kolesov (1975a,b), and Saks (1976) stress the importance of particle shape (especially flattening) in hydraulic sorting of gold.

An additional and more severe complication is that in most placer-forming environments, grains settle in turbulent rather than still water. Numerous studies have attempted to define the effects of turbulence on grain settling (for example, Torobin & Gauvin 1961, Murray 1970, Ludwick & Domurat 1982), but the results are contradictory. Murray (1970) found that quartz spheres whose still-water grain Reynolds numbers were less than 70 settled up to 40% slower in quasi-isotropic turbulence. Over the range of quartz sphere diameters from ~ 0.2 to 1 mm, the amount of reduction was proportional to the intensity of turbulence, and a maximum reduction was predicted to occur at a diameter of ~ 0.3 mm. At higher grain Reynolds numbers ($\sim 10^3$), Torobin & Gauvin (1961) found exactly the opposite effect. In their wind tunnel experiments, quasi-isotropic turbulence increased the settling velocities of spheres, again as a function of the intensity of turbulence and the sphere diameter. Neither of these two experiments can be considered definitive, however, and it must be emphasized that we presently cannot predict what sizes of heavy and light minerals will settle together to the bed of a turbulent flow. Until this glaring gap in our knowledge is rectified, one can only assume that size ratios of heavy and light minerals are predicted, albeit roughly, by the still-water settling laws.

If the volumetric concentration of the settling grains becomes appreciable, say greater than 5%, their fall is hindered by grain-grain interactions and an upward counterflow of the suspending fluid. In monodisperse systems, it is generally agreed that a grain's constant terminal settling velocity w_∞ is retarded according to the expression $w/w_\infty = (1 - C)^n$, where w is the hindered settling velocity, C the volumetric grain concentration, and n an exponent equal to $4.4\,\mathrm{Re}^{-0.1}$ in the range $200 < \mathrm{Re} < 500$ (Richardson & Zaki 1954). In polydisperse systems of mixed sizes and densities, no general theory yet exists, but the experimental studies of Davies (1968), Lockett & Al-Habbooby (1974), Mirza & Richardson (1979), and especially Richardson & Meikle (1961) have shown that the above equation applies to individual species in a mixture if C is defined as the total concentration of all species present.

How does free settling of grains aid in the formation of placers? If the settling velocity distributions of the heavy and light minerals in a pulp are equal [i.e. hydraulically equivalent in Rubey's (1933, 1938) sense], then free settling by itself will not sort the grains. Many heuristic explanations of placer development and heavy-mineral occurrences low in an alluvial fill assume that the heavy minerals settle faster than the light minerals (MacDonald 1983); these explanations do not consider that the starting heavy-mineral size distributions probably have smaller means and therefore are potentially hydraulically equivalent to the light minerals. Free settling of hydraulically equivalent distributions is, however (as is discussed later), important, because it produces deposits that can be subsequently sorted by size. If the pulp contains heavy and light minerals with unequal settling velocity distributions, then its free settling in a stationary fluid will sort the grains and produce a deposit vertically layered by density. If the free settling occurs in a unidirectional flow, then a deposit laterally segregated by density will be formed. Finally, if the pulp is the suspended load of a turbulent flow, grains will be fractionated into different elevations above the bed (Rouse 1950, Brush 1965, Middleton & Southard 1977, p. 6.27) such that at any specified elevation, the ratio of concentration C to the concentration at some reference level C_a for heavy (h) versus light (l) mineral grains is

$$\left(\frac{C}{C_a}\right)_h = \left(\frac{C}{C_a}\right)_l^{w_h/w_l}, \tag{7}$$

where w_h and w_l are the settling velocities of the heavy and light grains, respectively (Slingerland 1984). If the mean settling velocities of the heavy grains are smaller than those of the light grains, perhaps as a result of local size deficiencies, then higher levels of the flow will be relatively enriched in suspended heavy grains, which can then be carried to higher elevations of the floodplain. Such a mechanism is suggested by the recent work of Nami

(1983), who shows in a small section of the Witwatersrand Carbon Leader Reef that interchannel highs contain greater concentrations of placer gold than thicker contiguous channel deposits.

How does hindered settling of grains aid in the formation of placers? As an example, consider a mixture used by Richardson & Meikle (1961) consisting of equal volumes of two species of sand-sized spheres, one of density 2.9 g cm^{-3} and diameter 0.071 mm and the other of density 1.04 g cm^{-3} and diameter 0.382 mm such that they have equal free settling velocities. A grain density of 1.04 is unrealistic for natural minerals, but the results should still be applicable to common mixtures of light and heavy grains. If the total concentration C is less than 8%, the two species settle with equal (but reduced) velocities and produce a single mixed layer. If C is between 8 and 10%, settling produces three layers of sediment, the lower consisting solely of the denser grains, the upper consisting solely of the lighter grains, and the intermediate consisting of a mixture. At concentrations above 10%, settling produces two layers, each completely segregated by size and density. The results were explained by Richardson & Meikle in terms of the buoyant forces on the large light grains created by the mixture of fluid and small dense grains. This phenomenon may be the principal source of mineral sorting in jigs and hydraulic classifiers, and it deserves more study. In natural dressing mills, the necessary conditions for its operation might be realized in decelerating overwash flows on beaches or points of flow expansion in streams where sediment drops quickly out of suspension. The resulting deposits would be laminae or beds with heavy-mineral-enriched bases.

Entrainment of Grains

Entrainment is the dislocation of grains from a granular bed and their initial movement by a superimposed fluid flow. In realistic situations, the grains possess differing sizes, densities, and shapes, and the flow is unidirectional, nonuniform, unsteady, and turbulent. It is generally accepted that under these conditions the probability of a grain's entrainment increases as the mean values and variances of the fluid forces increase over the gravity and frictional forces holding the grain in place [see Yalin (1977) for a review]. Sorting occurs and placers may be formed because the forces depend upon a grain's size, shape, and density. The phenomenon is a stochastic one because (a) the magnitudes of the fluid forces in a turbulent fluid comprise a Gaussian or positively skewed distribution in time (Grass 1983) and (b) the detaining forces vary with the local grain geometries of the bed. For these reasons it is convenient to define a dimensionless parameter N equal to nD/U_*, where n is the number of grains of diameter D in motion per unit area per unit time, and U_* is the fluid shear velocity (Yalin 1977).

124   SLINGERLAND & SMITH

Then the "critical" fluid, grain, and bed values at the threshold of motion may be evaluated for any arbitrarily small N; Yalin suggests $N = 10^{-6}$.

The theoretical prediction of critical conditions proceeds from a torque balance on a grain in the bed (White 1940, Everts 1973, Yalin 1977, Middleton & Southard 1977, Slingerland 1977; see Figure 3). Most researchers (see Graf 1971) simplify the situation by assuming that the bed is horizontal, planar, and cohesionless, that no grains are yet moving in the flow, and that the grains are equant. For a grain to rotate about a pivot point A (Figure 3), a moment balance shows that the ratio of the fluid forces F to the gravity forces G must be

$$\frac{F}{G} \geq \frac{a_g \sin \alpha}{a_f \cos (\alpha - \zeta)},$$

(8)

where a_g and a_f are the moment arms about A, α is the reactive angle, ζ is the angle between the fluid force vector and the horizontal, and the point of application of the fluid force is assumed to be along the normal to the pivot

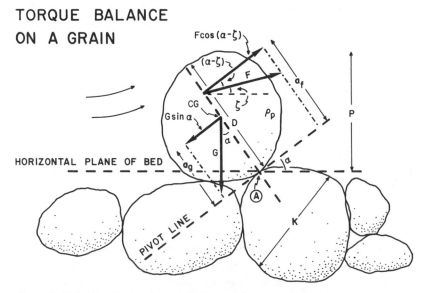

Figure 3   Definition diagram for calculating the initiation of grain motion on a horizontal bed. A subspherical grain of diameter D and density ρ_p, protruding above the bed a distance P, must pivot about point A on a downflow grain of diameter K. A fluid force vector F, representing both drag and lift forces, acts at a distance a_f away from a pivot line through A and at an angle ζ from the horizontal. A grain weight vector G acts through the grain center of gravity CG at a distance a_g away from the pivot line. At the moment of entrainment, the fluid torque must be greater than the resisting torque.

line. The magnitude of the fluid force vector F and its orientation ζ are determined by the magnitudes of the drag and lift forces and their points of application on the grain, all inadequately known. Some derivations (Slingerland 1977, Sundborg 1956) relate the drag and lift forces to the square of the local mean flow velocity near the top of the grain using lift and drag force equations, whereas others (Yalin 1977) relate them to the temporal mean boundary shear stress $\bar{\tau}_0$ or its surrogate, the shear velocity U_*. Because the lift and drag forces may be related to U_* and the boundary Reynolds number R_* (Einstein & El-Samni 1949, Coleman & Ellis 1976), and also because of the difficulty of defining the velocity at the top of the grain, it is probably preferable to express the resultant fluid force magnitude in the latter manner, such that

$$F = f(bD^2, \bar{\tau}_0, R_*), \tag{9}$$

where b is a proportionality factor relating D^2 to the true area of the grain on which the fluid forces act, D is the diameter of the grain, and $\bar{\tau}_0$ and R_* are defined in Equations (1) and (5), respectively. The gravity force G may be written as the grain's submerged weight,

$$G = cD^3(\rho_p - \rho_f)g, \tag{10}$$

where c accounts for the grain's nonspherical shape. The moment arm a_g is equal to the grain radius $D/2$ if the grain is a sphere, or more generally to cD. The moment arm a_f, proportionality factor b, and angle ζ vary with the structure of the fluid boundary layer (measured by R_*) and the protrusion of the grain above the mean bed elevation (measured by P/D; Abbott 1974) and must be evaluated experimentally. The reactive angle α is the angle of repose of a single grain on a fixed bed and decreases as D/K, D, sphericity, and roundness increase (Miller & Byrne 1966, Carrigy 1970, Luque 1974, Z. Li & P. D. Komar; see Refs. Added in Proof).

Defining $\bar{\tau}_c$ as the mean critical boundary shear stress at the threshold of motion and substituting Equations (9) and (10) into (8) while considering the above yields

$$\frac{\bar{\tau}_c}{(\rho_p - \rho_f)gD} \geq f(D/K, P/D, R_*). \tag{11}$$

The left-hand side is called θ_c, the critical dimensionless Shields parameter (Shields 1936), and is the ratio of the shear stress exerted by the fluid on the bed to the weight of a potentially entrainable grain layer over a unit area.

In the simplest case, where well-sorted sediments are entrained and transported over a bed of equal-sized particles, θ_c is solely a function of the boundary Reynolds number. A graph of θ_c versus R_* is difficult to interpret, however, because both parameters contain the fluid shear stress. To circumvent this difficulty, Yalin (1977) divided R_*^2 by θ_c and obtained a dimensionless parameter Ξ, whose square root is the Yalin parameter. At Yalin numbers above 40, experimental data from numerous studies using differing grain and fluid densities define the relationship quite well (Figure 4; Miller et al 1977, Yalin & Karahan 1979). The variance of data at any one boundary Reynolds number is in part due to the various definitions of N (that is, of how many grains must be moving at the threshold of motion). At Yalin numbers between 4 and 40, Collins & Rigler (1982) found that θ_c was overestimated for grains of density from 4 to 7 g cm^{-3}. In fact, for grain settling velocities w_∞ less than 10 cm s^{-1} and a wide range of densities, the critical boundary shear stress in their experiments was solely a function of grain settling velocity and independent of R_*, such that

$$\bar{\tau}_c = 1.24 \, w_\infty^{1/3}. \tag{12}$$

This is not a theoretically predictable relationship (P. D. Komar & K. E. Clemens; see Refs. Added in Proof), and until more experiments are completed, it remains a troublesome complication.

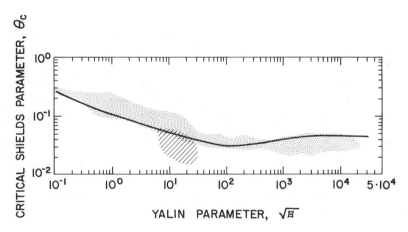

Figure 4 Standard threshold conditions for entrainment of grains of uniform size (redrawn from Miller et al 1977). The parameters θ_c and $\sqrt{\Xi}$ are defined by

$$\theta_c = \bar{\tau}_{0c}/[(\rho_p - \rho_f)gD],$$

$$\sqrt{\Xi} = [(\rho_p - \rho_f)gD^3/(\rho_f v^2)]^{1/2},$$

where $\bar{\tau}_{0c}$ is the critical boundary shear stress, ρ_p is the grain density, ρ_f is the fluid density, g is the gravitational acceleration, D is the grain diameter, and v is the kinematic viscosity of the fluid. The stippled pattern denotes the data field used by Miller et al in defining their preferred curve (shown as a black line). The diagonal pattern denotes the data field of Collins & Rigler (1982).

How does entrainment of well sized-sorted sediments lead to the separation of grains into populations of differing densities? Recasting the curve in Figure 4 as a plot of critical shear stress against grain diameter (Figure 5) yields simple monotonically rising curves; curves for different densities are of similar form but progressively displaced toward higher values of shear stress. As intuitively expected, less dense grains can be selectively winnowed from a population of equal-sized but denser grains, leaving a denser lag. Ljunggren & Sundborg (1968), McQuivey & Keefer (1969), and Grigg & Rathbun (1969) were among the first to analyze heavy-mineral entrainment using Shields criteria of this type. It must be remembered, however, that this analysis applies only to well-sorted sediments on a plane bed. It is not possible to predict from Figure 5 which sizes of different density minerals will be entrained off a bed together, because the presence of different sizes violates the experimental conditions under which the curve was obtained. Certain thin, planar, heavy-mineral-rich laminae in fluvial deposits such as described by Lucchitta & Suneson (1981) and Stavrakis (1980) might be explained in this manner.

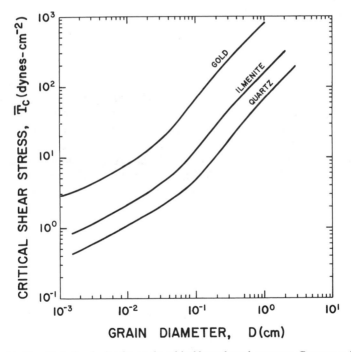

Figure 5   The effect of grain density on the critical boundary shear stress. Curves are derived from Figure 4 using densities for quartz, ilmenite, and gold of 2.65, 4.70, and 19.3 g cm^{-3}, respectively.

In the more general case, where grains of varying sizes and densities compose the bed, all three variables on the right side of Equation (11) are important. Larger grains protruding higher into the flow present a greater surface area for drag and lift forces, experience greater instantaneous turbulent shear stresses, and have smaller reactive angles α. Smaller grains are either sheltered from the flow or experience increased turbulence from the wakes of larger grains. A curious result is that coarser grains may be entrained at shear stresses lower than those for finer grains (Gilbert 1914, Meland & Norrman 1966, 1969, Brady & Jobson 1973, Saks & Gavshina 1975, Slingerland 1977, Day 1980, Raudkivi & Ettema 1982); thus a sorting mechanism based more on size exists to complement the sorting mechanisms based more on density.

In the simplified case of a binary mixture where particles of diameter D rest in a bed of particles of diameter K and protrude to varying heights P above the mean bed level, the appropriate Shields parameter in Equation (11) for the superjacent particle (labeled θ_{ci} to separate it from its value for a level, uniformly sized bed) can be calculated from an equation due to Slingerland (1977, Equation 12). Written in terms of shear stress for a horizontal bed, the equation is

$$\theta_{ci} = \frac{4}{3} \frac{(\beta_1 \beta_2)^2 \tan \alpha}{C_d}. \qquad (13)$$

Here, β_1 is a turbulent velocity fluctuation coefficient, β_2 is a coefficient accounting for the point of application of fluid forces and angle ζ in Figure 3, and C_d is the coefficient of drag appropriate for the grain as it rests in the bed. All are functions of the boundary Reynolds number R_* and the protrusion distance P/D.

The reactive angle α is determined by the protrusion distance P/D and the size of the grain relative to the underlying grains, D/K (Figure 3). In the case where a grain protrudes above an adjacent downflow grain of equal size, the geometry is such that

$$\alpha = \arccos P/D. \qquad (14)$$

In the case where a grain of size D rests on grains of size K, Z. Li & P. D. Komar (see Refs. Added in Proof) have shown that

$$\alpha = e(D/K)^{-f}, \qquad (15)$$

where e varies from 35 to 70 and f varies from 0.30 to 0.75. For nearshore marine sands, Miller & Byrne (1966) give $e = 61.5$ and $f = 0.3$; for gravel-sized spheres undergoing grain-top rotation, Z. Li & P. D. Komar (see Refs. Added in Proof) give $e = 36.3$ and $f = 0.72$. For river gravels where K is taken as equal to the mean grain size $\bar{D}$, P. D. Komar & Z. Li (see

Refs. Added in Proof) give $e = 35$ and $f = 0.9$ for $D_i/\bar{D} > 1$ and $e = 35$ and $f = 0.6$ for $D_i/\bar{D} < 1$.

The influence of protrusion distance on θ_{ci} is seen in Figure 6, in which are plotted the data of Fenton & Abbott (1977) collected in a fully turbulent boundary layer. In their experiment, particles were slowly pushed up into the flow through adjacent fixed grains until swept away. At the moment of entrainment, the relative distance of protrusion above the local bed level was P/D. Also shown in Figure 6 is the critical Shields parameter $\theta_c = 0.045$ for a level, uniformly sized, hydrodynamically rough bed. It can be seen from the figure that particles that protrude above the top of the bed a distance less than one third to one quarter their diameter possess θ_{ci} larger than 0.045, whereas particles protruding farther into the flow possess lesser θ_{ci}. The causes of this deviation are not well understood, but most certainly

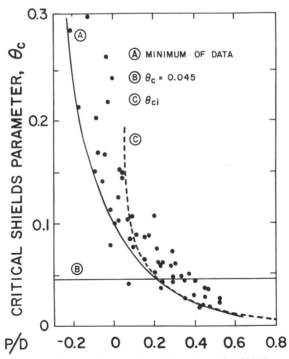

Figure 6   The effects of relative grain protrusion P/D on the critical Shields parameter. Data points are from flume experiments by Fenton & Abbott (1977, experiment B) at $\sqrt{\Xi} = 500$ and define a minimum curve A (excluding outliers). Line B is the conventional critical Shields criterion from Figure 4 for these conditions. Curve C is a plot of Equation (13) in the text. Grains can be less or more easily entrained depending upon their relative protrusion above the bed.

they include a reduction in the reactive angle α, an increase in the surface area of the grain upon which the fluid shear stress acts, and an increase in the intensity of turbulence as grains protrude higher into the flow. To explain the relative contributions of each, it may be assumed as a first approximation that for a fully rough bed the variation due to turbulence and surface area is much less than that of tan α. Then Equation (13) can be plotted in Figure 6 by substituting Equation (14) for α and evaluating the remaining terms as a constant, calculated to yield $\theta_{ci} = 0.045$ when $P/D = 0.2$. Although the data reflect a unique grain geometry and Equation (13) fits only moderately well, the trends and relative magnitudes of θ_{ci} seem indisputable. Much of the decrease in the critical Shields parameter with increasing protrusion can be ascribed to a decrease in the reactive angle.

Similar theoretical results have been obtained by Komar & Wang (1984) and P. D. Komar & Z. Li (see Refs. Added in Proof) for the variation in entrainment threshold due to variation in $\bar{D}/K$ (or more generally, in $D_i/\bar{D}$, where the subscript i refers to the ith grain size in a distribution of mean size). P. D. Komar & Z. Li combined Equations (13) and (15) into the form

$$\frac{\theta_{ci}}{\theta_{cr}} = \left(\frac{D_i}{\bar{D}}\right)^{-h}, \tag{16}$$

where θ_{cr} is the reference Shields parameter at $D_i/\bar{D} = 1$. The exponent h and θ_{cr} are functions of R_*, but most workers assume that they are constants for fully rough boundaries. The coefficients have been evaluated in flume experiments by Day (1980) using quartz sand-gravel mixtures and in field measurements in quartz gravel-bed streams by Parker et al (1982) and Andrews (1983) (Table 2). Day found that θ_{cr} equaled the conventional Shields parameter, whereas Parker et al and Andrews obtained a constant value almost double that. The disparity among h values is also important and deserves further study. Day's mixtures were finer grained and possibly yielded different coefficients in Equation (15) as a result of grain size effects or different coefficients in Equation (16) as a result of varying boundary Reynolds numbers. Regardless of these differences among studies, the trends in Equation (16) are clear. Critical Shields parameters of sand-gravel

Table 2 Representative values of coefficients in Equation (16) for sand-gravel mixtures

θ_{cr}	h	References
$\theta_{c\bar{D}}$	0.53	Day 1980
0.0876	0.982	Parker et al 1982
0.0834	0.872	Andrews 1983

mixtures are increased over their conventional values for fine sizes and decreased for coarse sizes; thus all sizes become more nearly equally mobile.

There must be some asymptotic limit to the decrease in θ_{ci} with increasing D/K or $D_i/\bar{D}$. Fenton & Abbott's (1977) data suggest a value of θ_{ci} equal to 0.01; Ramette & Heuzel (1962) and Andrews (1983) suggest 0.02. Also, with ever-increasing D/K values, the flume experiments by Raudkivi & Ettema (1982) demonstrate that increasingly more vigorous wakes are shed off the larger particles. Scour holes are created on their downflow sides and the particles become embedded, which thereby decreases their protrusion and increases their θ_{ci}. In experiments using binary mixtures, D-sized particles embed almost immediately if $D/K > 17$.

It is possible to explain entrainment sorting of size-density mixtures qualitatively by combining these ideas in two diagrams, one representing pebbly sand (Figure 7) and the other representing sandy gravel (Figure 8). In each figure the conventional Shields entrainment curve for quartz is presented for comparison with the entrainment curves for mixtures. The latter are calculated by combining Equations (13) and (15) and ignoring the variations in β_1, β_2, and C_d with R_*. As expected, the critical Shields parameters for mixtures are higher for grains finer than the mean and lower for grains coarser than the mean when compared with the conventional Shields values (Figures 7, 8). The shapes of the two entrainment curves for mixtures are different because the coefficients in Equation (15) depend upon mean grain size, as shown by Z. Li & P. D. Komar (see Refs. Added in Proof). The most easily entrainable size (minimum with respect to the boundary shear stress isograms) in the pebbly sand mixture is about 0.7 times the mean grain size. Flume data due to Day (1980) are in reasonable agreement with this shape. In the sandy gravel mixture, a wide range of sizes $(0.35 < D_i/\bar{D} < 2)$ possess the same minimum boundary shear stress. The upper bound of this zone could be as high as 5 depending upon the true minimum θ_{ci} for coarse fractions of mixtures. These results are consistent with those of Everts (1973), who found in flume experiments using binary mixtures that quartz spheres either less than 1/2 or greater than 3 times the bed size would not stay in transport at a θ_{ci} just less than that of the bed.

Five different grain behaviors important to entrainment sorting are noted in Figures 7 and 8. Trapping contributes to entrainment sorting if grains are already in motion and a flow's boundary shear stress plots just below one of the θ_{ci} curves. In this situation, denser particles may be trapped or selected out of the waning bed load in preference to less dense particles because the denser particles test the bed more often and once in place are not reentrained. The efficiency of this process is probably low, although it may explain some of the gold concentrations in thick, massive,

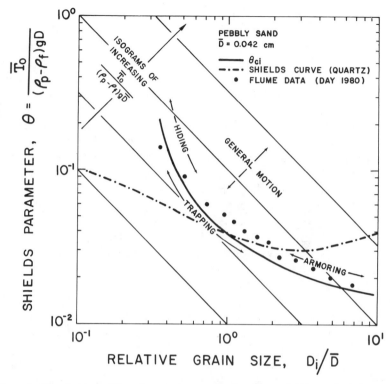

Figure 7   Threshold conditions for entrainment of grains from a pebbly sand mixture. The ordinate is the Shields parameter θ, in which the grain size is D for the conventional Shields curve and D_i for the θ_{ci} curve; the abscissa is the relative grain size $D_i/\bar{D}$. Isograms of constant dimensionless boundary shear stress appear as lines of slope equal to -1 and possess values equal to θ at their intersection with the $D_i/\bar{D} = 1$ line. Here $\bar{D}$ equals 0.042 cm, the mean size of the mixture from which Day's (1980) flume data were obtained. The entrainment curve for mixtures is $\theta_{ci} = 0.026 \tan{[61.5(D_i/\bar{D})^{-0.3}]}$, derived from Equations (13) and (15) in the text. Important grain behaviors in entrainment sorting are (*a*) hiding, in which grains smaller than the mean size require larger boundary shear stresses because of larger reactive angles; (*b*) armoring, in which grains coarser than the mean size experience higher boundary shear stresses because of larger grain weights; and (*c*) trapping, in which grains of size $D_i/\bar{D} \sim 1$ already in motion at a θ just less than the θ_{ci} curve are captured in sites on the bed.

crudely horizontally stratified conglomerates of the Ventersdorp Contact placer (Krapez 1985).

Overpassing or winnowing, corresponding to Case 1 of Slingerland (1977), contributes to heavy-mineral enrichment because mode-sized and less dense particles are preferentially entrained at lower boundary shear stresses than are finer- and coarser-sized and denser particles. If the pulp consists of finer heavy and intermediate-sized light minerals—for example,

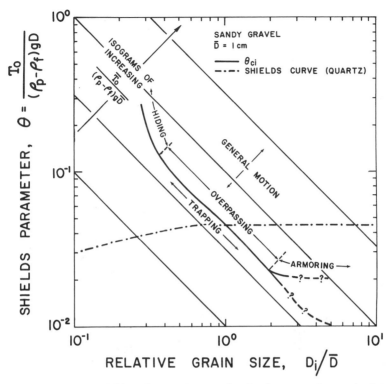

Figure 8   Threshold conditions for entrainment of grains from a sandy gravel mixture. See Figure 7 for details. Here $\bar{D} = 1$ cm and $\theta_{ci} = 0.0643 \tan [35(D_i/\bar{D})^{-j}]$, where $j = 0.9$ if $D_i/\bar{D} > 1$ and $j = 0.6$ if $D_i/\bar{D} < 1$. In gravel mixtures, grains from one third to twice the mean size possess roughly the same critical boundary shear stress and will overpass, whereas finer and coarser sizes will not.

settling-equivalent size distributions where the light-mineral distribution defines $\bar{D}$—then the bed should become enriched in heavy minerals, and thus the settling velocity ratios of the grains, w_h/w_l, should become greater than 1, a condition often observed at the top of beach swash zones (Slingerland 1977, Hand 1967, McIntyre 1959, Stapor 1973, Komar & Wang 1984) and suggested by the data of Stavrakis (1980; his "mixed lamina") from an ancient fluvial deposit. Brady & Jobson (1973) invoke a similar mechanism for heavy-mineral segregation on dune crests.

Armoring (corresponding to Case 3 of Slingerland 1977) is important in entrainment sorting because the finer grains in a mixture can be winnowed away, leaving the coarse tail. As mentioned previously, large clasts of size D embed themselves rapidly in a bed of size K when $D/K \geq 17$ or, for a mixture of sizes, when the geometric standard deviation of the mixture is greater than $\simeq 1.3$ (Little & Mayer 1976). The amount of enrichment occurring by this process depends upon the sizes in the pulp—specifically, the spread between the means of the heavy- and light-mineral distributions. If the mean size of the heavy minerals is similar to $\bar{D}$, the armor may become impoverished and the sediment load enriched because turbulent wakes around large immobile grains may entrain and suspend all the finer (and consequently also denser) particles. The increased θ_{ci} resulting from the increased density of the particles is less than the decreased θ_{ci} resulting from their lesser $D_i/\bar{D}$. As the mean size of the particles increases, the increased θ_{ci} due to increased density may come to predominate, and the armor may become enriched. Concentrations immediately underneath armor layers on top of gravel bars and pebble stringers in ancient deposits [described by Toh (1978), Smith & Minter (1980), and Krapez (1985)] can be explained in this manner.

In hiding, both smaller sizes and greater densities act to increase a grain's resistance to motion. Fine heavy and light grains come to rest among fixed-roughness elements in a greater ratio than in the transported load. Hiding can explain the increased concentrations of heavy minerals, especially gold, in open framework gravels (Smith & Minter 1980).

General motion does not contribute to entrainment sorting per se, because all particles are moving. The rates of motion are unequal, however, and lead to transport sorting by size and density. This process is considered next.

Differential Transport of Grains

Transport sorting results when one size or density fraction of a sediment mix is transported at a different rate from another and so may come to rest at a different location. It involves not only differences in grain transport velocities but also entrainment sorting, since bed-load particles tend to frequently "stop and go" while moving downstream. (Each "stop" requires reentrainment.) Size sorting of quartz-density grains due to differential bed-load transport has been well studied by Gilbert (1914), Einstein & Chien (1953), Egiazaroff (1965), Meland & Norrman (1966), Gessler (1971), Bridge (1981), and Parker et al (1982). Few researchers, however, have studied transport sorting as a function of density, although this is an important mechanism for generating heavy-mineral enrichments (Meland & Norrman 1969).

As with entrainment, key variables in transport sorting are the means and variances of the light- and heavy-mineral size distributions, the bed roughness, and some measure of the fluid force. The phenomenon has been treated theoretically by Slingerland (1984) by using H. A. Einstein's bed-load function, which allows for grain shielding effects. Slingerland calculated transport rates for the different size fractions in a 90% quartz, 10% magnetite sand mix under different combinations of bed roughness and shear velocity. The starting size distributions of the quartz and magnetite were constructed to yield nearly equal settling velocity distributions. Flow strength was calculated by using Equation (2) for various hypothetical depths and slopes. He found that for a given shear velocity, transport rates for all sizes and both densities decrease with increasing roughness, and that the relative proportion of magnetite in the moving bed load increases with increasing U_* for a given roughness and decreases with increasing roughness for a given U_* (Figure 9). These results are corroborated by experimental data (Meland & Norrman 1966, Steidtmann 1982). Meland & Norrman (1966) conducted flume experiments in which transport velocities of glass spheres ($\rho_p \sim 2.54$) were measured over fixed beds of ideally packed, perfectly sorted spheres. Diameters of both the transported and fixed spheres were varied from 0.21 to 0.78 mm to test the effects of relative roughness on transport rates. Results showed that for any particle size and roughness within the experimental range, transport velocity increased with increasing shear velocity, but that for any roughness size, larger grains moved faster than smaller ones (Figure 10). For a given shear velocity, the fastest transport rates were shown by the largest grains (0.78 mm) moving

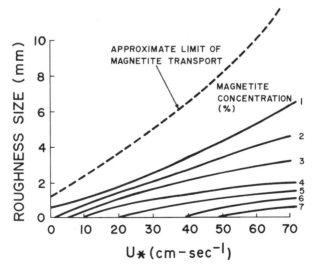

Figure 9   Effects of bed roughness and shear velocity U_* on the magnetite concentration in the sediment load of a stream as predicted by the Einstein bed-load function (after Slingerland 1984). The bed is composed of 10% magnetite and 90% quartz, with mean sizes of 0.02 cm for the magnetite and 0.042 cm for the total distribution. Maximum bed enrichment of magnetite occurs for higher roughnesses and lower U_*.

over the smallest roughness (0.21 mm). As discussed in the section on entrainment sorting, this is a commonly observed phenomenon that is caused by the larger particles projecting farther upward into the velocity profile and rolling more easily over the relatively smooth (small roughness) surface. Conversely, the lowest transport rates were shown by small particles (0.21 mm) moving over the roughest bed (0.78 mm), a result of shielding and higher reactive angles. Because Meland & Norrman did not vary grain density, the significance of this result to heavy-mineral sorting is indirect (e.g. for cases where initial heavy and light grain size distributions are not alike).

In somewhat similar experiments using glass spheres and fixed bed roughness, Steidtmann (1982) investigated the effects of density on particle transport rates (Figure 11). He observed that grains less than 0.2 mm (D/K

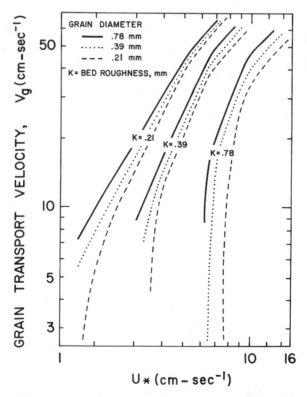

Figure 10   Relationships between grain transport velocity and shear velocity (U_*) for equal-density spheres over fixed beds of perfectly sorted spheres. For a given shear velocity and roughness size, larger grains are transported faster than are smaller grains. Also, note that all particles move slower with increased roughness size (after Meland & Norrman 1966).

< 0.6) are not transported at the shear velocities used in his experiments, a finding that corroborates the results of Everts (1973). This is a consequence of the high reactive angle in Equation (13) and therefore is dependent more on size than density. Grains of larger relative size have transport velocities V_g that increase with both increasing D/K and increasing shear velocity. Transport velocities of the light and heavy minerals are nearly equal for smaller D/K, diverge for intermediate D/K, and probably converge again at the largest D/K. This is because V_g is a measure of a grain's velocity while moving (proportional to its average elevation off the bed), the number of times it tests the bed for a stable resting place, its duration there, and its entrainability. All four factors depend upon a grain's free settling velocity. At the values of U_*/w_∞ used in Steidtmann's study, small heavy and light grains probably traveled high enough in the flow to seldom test the bed and therefore experienced almost continuous motion. Intermediate-size heavy grains, because of their smaller U_*/w_∞ values, probably traveled in closer contact to the bed in the zone of lower fluid velocity (Figure 1), tested the

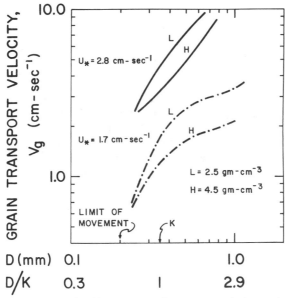

Figure 11 Generalized relationships among grain transport velocity, grain size D, shear velocity U_*, and roughness size K for light (L) and heavy (H) spheres moving over a bed of fixed roughness. Curves are drawn through the flume data of Steidtmann (1982). For particles significantly smaller than the bed roughness size, there is little difference in transport velocity between equal-sized light and heavy grains because both types travel in intermittent suspension. For particles near the roughness size, light grains move significantly faster than heavy grains of the same size because heavy grains must be repeatedly reentrained.

bed more times for stable sites, and possessed higher θ_{ci} compared with equal-sized light grains. One would expect that heavy and light grains with very large D/K would again possess similar transport velocities, because once these grains are accelerated, the forces needed to keep them in motion are equivalent. Steidtmann extended his experiments to include bulk transport rates in which the moving sediment defined its own roughness (i.e. mobile instead of fixed bed). He obtained qualitatively similar results for a plane-bed condition—heavy grains were found in decreased concentrations relative to light grains in downstream deposits. For a rippled bed, however, no systematic downstream differences in transport rates were observed, presumably because of sorting and transport complexities associated with irregular bedforms (Brady & Jobson 1973), but possibly also because of the restricted flume length and limited sampling intervals used by Steidtmann. Certain enigmatic results of an earlier flume study by Minter & Toens (1970) are now more understandable in light of the above. Using a sediment composed of quartz and magnetite grains, with median diameters of 0.57 and 0.083 mm, respectively, Minter & Toens observed that when the sediment was transported over a porous gravel bed, the sediment trapped by the gravel contained a lower magnetite concentration than was contained in the moving bed load. Of course this result contradicts intuition. Apparently, the much smaller sizes of the magnetite fraction, although denser, moved over the very rough gravel bed higher in the flow and therefore tested it less than the larger quartz grains.

How does differential transport form placers? An aggrading bed can be enriched or depleted in heavy minerals relative to the bed load, depending upon U_* and the sizes of the heavy- and light-mineral grains relative to the roughness-determining size. Enrichment of the bed is most intense when $\bar{D}_h$ and $\bar{D}_l$ are near the roughness size and U_* is such that the heavy grains travel with more bed contact than the light grains. Defining the U_*/w_∞ values that effect this must await further study. If the heavy and light minerals possess fall-equivalent diameters, the heavy minerals will always lag behind the light minerals and thus will produce enriched deposits in the upflow portions of the transport path.

Shearing of Grains

Theoretical and experimental studies by Bagnold (1954, 1956) showed that when a concentrated flow of cohesionless particles is sheared by gravity or fluid forces, grain interactions create a force perpendicular to the plane of shearing such that the granular mass expands toward the free surface (i.e. away from the bed). Bagnold called this force the "dispersive pressure." He showed that dispersive pressures are greater on larger and denser grains than on smaller or less dense grains in the same horizon of a grain flow; this result suggests that larger or denser grains would migrate upward toward the surface of a nonuniform sediment mix. "Shear sorting" (Inman et al 1966, p. 800) refers to this vertical fractionation of particles caused by dispersive pressures in a moving granular dispersion.

Sallenger (1979), assuming that the magnitudes of the dispersive pressures predicted by Bagnold also hold for sediment populations of mixed sizes and densities, proposed that the diameters of heavy and light grains in the same horizon of a sheared granular mass would be governed by their densities as

$$d_\mathrm{h} \approx d_\mathrm{l}(\rho_\mathrm{l}/\rho_\mathrm{h})^{1/2}. \tag{17}$$

Thus, the diameters would be independent of fluid density, which implies that the same relation would hold in air or water. Equality is approximated because several variables in Bagnold's equation were only assumed, but not demonstrated, to be correct for nonuniform sediment. Sallenger showed that dispersion-equivalent sizes could be expected in such deposits as grain flows (e.g. on avalanche faces of dunes) and beach swash zones, and furthermore that both heavy-mineral laminations and inversely graded beds could be formed by this process. Clifton (1969) had earlier invoked a similar interpretation for inversely graded heavy-mineral-rich beach laminae.

APPLICATIONS

Our discussion of hydraulic sorting mechanisms has been at the sediment grain scale (millimeter to centimeter), whereas economic placers require enrichments on much larger spatial scales. These enrichments come about by many means, but principally by numerous geomorphological dressing mills operating simultaneously over a large area and migrating slowly in space. The following examples show the types of enriched zones left behind.

Small-scale enrichments frequently are associated with asymmetrical bedforms; commonly, heavy-mineral segregations occur on the stoss sides, crests, and slip faces of active ripples and dunes (Brady & Jobson 1973, McQuivey & Keefer 1969) and in foresets and trough surfaces of dune-formed cross-beds (McGowen & Groat 1971, Smith & Minter 1980, Buck 1983). Such enrichments by bedform activity may combine to attain economic grades at regional scales (Theis 1979, Buck 1983). It is possible that all four sorting mechanisms described above interact to produce heavy-mineral segregations in dunes and ripples (Slingerland 1984, Brady & Jobson 1973). Over the upstream-dipping stoss sides, bed shear stress (and U_*) is greatest and the intensity of turbulence is lowest. Larger light grains are preferentially transported to the dune crest, leaving the smaller, slower-moving heavy grains behind either by entrainment transport or by shear sorting. Foreset segregation may result from shear sorting of grain avalanches down the slip face of the migrating bedform. Concentrations lining scour surfaces of trough cross-beds probably result from a combination of entrainment and settling sorting as grains brought to this area by avalanching and suspension settling are reworked by high turbulence and backflow caused by flow separation. [For reviews of bedform mechanics, see Middleton & Southard (1977).]

Planar or horizontal stratification may make up major portions of fluvial deposits, and heavy-mineral-rich laminations are frequently observed in such deposits (Ljunggren & Sundborg 1968, Brady & Jobson 1973, Stavrakis 1980, Lucchitta & Suneson 1981, Buck 1983, Cheel 1984). As is the case for ripples and dunes, no single sorting mechanism produces all plane-bed segregations. Slingerland (1980) suggested that as the boundary Reynolds number is increased over a plane sand bed, the dominant sorting process progresses from entrainment to settling to dispersive (shear) sorting. Recent flume observations by Cheel (1984) indicate that transport sorting is important as well. The grain-size data of Stavrakis (1980, his Figure 6) suggest at least two sorting mechanisms for the heavy-mineral laminae he describes.

Much of the past and present placer mining has been aimed at intermediate scales of concentration. Representative examples are given by Crampton (1937), Boggs & Baldwin (1970), Cobb (1973), Komar & Wang (1984), and Smith & Minter (1980). Commonly, economic placer concentrations occur in the interstices of well-packed gravels, which in modern streams often define specific topographic features such as bars, channel junctions, bends, and riffles. Such deposits probably derive from entrainment and transport sorting of fine bed load as it passes over a rough bed, as suggested by Figures 8 and 9.

A type of intermediate-scale segregation in fluvial sandy sediment is described by Smith & Beukes (1983) and is illustrated in Figure 12. Narrow channelways confined by a bank on one side and a bankward-moving bar on the other were found to be enriched in heavy minerals when the channel was constricted to the narrowest extent allowed by the converging flows. High concentrations of magnetite and chromite, up to five times the background concentrations, appear to result from transport sorting. Transport-equivalent heavy and light grains supplied to the channelway by the bar are subjected to increased shear stresses and strong turbulence in the combined flow of the channelway; the largest light grains are rolled out quickly, and light grains of nearly equal size to the heavy grains are momentarily suspended by turbulent eddies and then transported away, leaving behind a slower moving mass of heavy grains only slightly smaller than the remaining light grains. The situation described by Smith & Beukes [and earlier by Smith & Minter (1980) in a case study of Witwatersrand gold/uranium concentration in conglomerate] may be only a special case for converging flow features in general. In flume experiments, both Wertz (1949) and Mosley & Schumm (1977) observed magnetite enrichments in the downstream ends of channel scours; in the latter study, the scours were formed by flows converging at channel junctions. The scours are probably similar to the bar-to-bank channelways in that turbulence is high and local shear stresses are at or above threshold values for most of the bed load.

Beach placers are another example of intermediate-scale segregation (Rao 1957, Nelson & Hopkins 1972, Stapor 1973). The dressing mill is the inner surf zone (Slingerland 1977, Komar & Wang 1984), and the sorting process is a combination of (a) free settling of heavy and light grains out of landward flows and (b) entrainment and transport of the larger light grains by seaward flows. Seasonal erosion-deposition cycles enhance the process by providing repeated deposits of near settling-equivalent sizes that can be beneficiated during erosive phases.

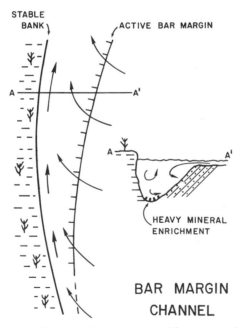

Figure 12 Sketch showing the generalized geometry and flow patterns formed by an active bar migrating toward a stable bank. Heavy minerals concentrate in the constricted channelway, which gradually widens in the downcurrent direction to accommodate increased flow from over the bar surface. Figure summarized from Smith & Minter (1980) and Smith & Beukes (1983).

Large-scale (regional) fluvial concentrations as described by Minter (1976, 1978) and McGowen & Groat (1971) must ultimately be attributed to optimal combinations of available grain sizes of heavy and light grains, regional slope [as it governs mean shear stress and shear velocity; see Equations (2) and (3)], and average bed roughness (Slingerland 1984). In any aggrading fluvial system, slope and bed roughness size can be expected to decrease downstream; thus, heavy grains entering the system will tend to be transported intermittently until they reach sites of permanent deposition dictated by local roughness, mean shear stress, and initial densities and size distributions of the transported grains. In degrading systems, areal patterns of heavy-mineral concentrations may be more confused because downstream changes in slope and especially roughness are less predictable. In such cases, roughness may be controlled by bedrock substrate or (more likely) by thin but coarse lag deposits, and downstream size decreases of bed material would be minimal along the erosional surface. Because of their large roughness (lag gravel) and relative stability (slowly degrading surface uncomplicated by deposition), such sites provide excellent settings for placers because large volumes of sediment can be processed by a relatively consistent set of slope-roughness-flow conditions. Many of the most important Witwatersrand orebodies occur along unconformities, either regional where degradation has been general and widespread (e.g. the Basal, Vaal, Ventersdorp Contact Reefs) or more restricted where placer bodies follow distinct channelways dissected into the substrate (e.g. the Saaiplaas and "B" Reefs) (Pretorius 1976, Minter 1978, Buck 1983, Krapez 1985). On smaller scales, segregations along local gravel-lined scour surfaces within alluvial sequences are well known in both modern and ancient placers.

Regional nearshore or beach concentrations appear to result from the slow shore-normal migration of the inner surf zone either landward (greater enrichment) or seaward (lesser enrichment). Alongshore variation is controlled by differential transport in the net littoral drift direction. Such variations have been shown by Trask & Hand (1985) for a Lake Ontario beach and by Peterson et al (1985) for the Oregon coast, where placers are best developed south of headlands at points of shoreline inflection.

CONCLUDING REMARKS

Presently we know that heavy-mineral segregations in water-laid deposits occur on hierarchically different scales (from millimeters to kilometers). The many geomorphological sites where segregation occurs act as natural dressing mills, remaining relatively fixed in space as a selective sorting circuit processes a pulp and stores the concentrate where it will not be diluted. The sorting circuits rely upon different responses of heavy and light grains to local fluid forces acting on a sediment bed. Sorting by density may occur during entrainment, transport, settling from suspension, or dispersion by grain-to-grain interactions, and natural heavy-mineral enrichments often involve combinations of these processes. Pure settling behavior, embodied in the classical idea of "hydraulic equivalence," probably does not play as important a role in placer origins as is believed by some workers.

Several processes in placer formation remain imperfectly known. For example, we still cannot quantify how fluid turbulence in polydispersed concentrations of grains modifies the free settling velocities of grains. Nor can we predict the transport rates and ultimate fates of size-density fractions in a mixture undergoing unsteady, nonuniform flow and therefore aggrading or degrading bed conditions. At larger scales, additional work is needed to determine the optimal regional sediment and water yields and fluid energy slopes for placer development. These goals will be best met by flume studies using heterogeneous size-density mixtures, analysis of well-exposed modern and ancient depositional systems, and two- or three-dimensional sediment-water routing calculations.

ACKNOWLEDGMENTS

RS thanks numerous colleagues who through the years have collaborated in or otherwise encouraged his research on placers: E. Williams, P. Holbrook, J. Warg, S. Reed, K. Gerety, R. Sauermann, H. Azuola, H. Hanson, M. Nami, E. Wecker, and P. Komar. He also wishes to acknowledge the Breen Foundation (Princeton, New Jersey), Narex Ore Search, Inc., the Chamber of Mines of South Africa, and the National Science Foundation (NSF Grant EAR-8418355) for support. NS thanks numerous colleagues and friends in South Africa who introduced him to those most fascinating of placer deposits: the Witwatersrand paleoplacers. W. E. L. Minter and N. J. Beukes were particularly helpful. Financial support from the Anglo-American Corporation and the Council for Scientific and Industrial Research are also acknowledged.

Literature Cited

Abbott, J. E. 1974. *The dynamics of a single grain in a stream.* PhD thesis. Imperial Coll., London, Engl.

Adams, J., Zimpfer, G. L., McLane, C. F. 1978. Basin dynamics, channel processes, and placer formation: a model study. *Econ. Geol.* 73:416–26

Andrews, E. D. 1983. Entrainment of gravel from a naturally sorted riverbed material. *Geol. Soc. Am. Bull.* 94:1225–31

Baba, J., Komar, P. D. 1981. Measurements and analysis of settling velocities of natural quartz sand grains. *J. Sediment. Petrol.* 51:631–40

Bagnold, R. A. 1954. Experiments on a gravity-free dispersion of large solid spheres in a Newtonian fluid under shear. *Proc. R. Soc. London Ser. A* 225:49–63

144 SLINGERLAND & SMITH

Bagnold, R. A. 1956. The flow of cohesionless grains in fluids. *Proc. R. Soc. London Ser. A* 265: 315–19

Bateman, A. M. 1950. *Economic Mineral Deposits.* New York: Wiley. 916 pp. 2nd ed.

Boggs, S. Jr., Baldwin, E. M. 1970. Distribution of placer gold in the Sixes River, southwestern Oregon. *US Geol. Surv. Bull. 1312-I*

Brady, L. L., Jobson, H. E. 1973. An experimental study of heavy-mineral segregation under alluvial-flow conditions. *US Geol Surv. Prof. Pap. 562-K.* 38 pp.

Bridge, J. S. 1981. Hydraulic interpretation of grain-size distributions using a physical model for bedload transport. *J. Sediment. Petrol.* 51: 1109–24

Briggs, L. I. 1965. Heavy mineral correlations and provenances. *J. Sediment. Petrol.* 35: 939–55

Briggs, L. I., McCulloch, D. S., Moser, F. 1962. The hydraulic shape of sand particles. *J. Sediment. Petrol.* 32: 645–56

Brush, L. M. Jr. 1965. Sediment sorting in alluvial channels. In *Primary Sedimentary Structures and Their Hydrodynamic Interpretation,* ed. G. V. Middleton, pp. 125–33. *Soc. Econ. Paleontol. Mineral. Spec. Publ. No. 12.* 285 pp.

Buck, S. G. 1983. The Saaiplaas Quartzite Member: a braided system of gold- and uranium-bearing channel placers within the Proterozoic Witwatersrand Supergroup of South Africa. In *Modern and Ancient Fluvial Systems,* ed. J. D. Collinson, J. Lewin, pp. 549–62. *Int. Assoc. Sedimentol. Spec. Publ. 6*

Carrigy, M. A. 1970. Experiments on the angles of repose of granular materials. *Sedimentology* 14: 147–58

Cheel, R. J. 1984. Heavy mineral shadows, a new sedimentary structure formed under upper flow regime conditions: its directional and hydraulic significance. *J. Sediment. Petrol.* 54: 1173–82

Cheny, E. S., Patton, T. C. 1967. Origin of the bedrock values of placer deposits. *Econ. Geol.* 62: 852–53

Clifton, H. E. 1969. Beach lamination: nature and origin. *Mar. Geol.* 7: 553–59

Cobb, E. H. 1973. Placer deposits of Alaska. *US Geol. Surv. Bull. 1374.* 213 pp.

Coleman, N. L., Ellis, W. M. 1976. Model study of the drag coefficients of a stream-bed particle. *Proc. Fed. Inter-Agency Sediment. Conf., 3rd, Denver, Colo. Preprints of Papers,* pp. 4.1–4.11. *Rep. SEDCOM-03,* Water Resour. Counc., Sediment. Comm., Washington, DC

Collins, M. B., Rigler, J. K. 1982. The use of settling velocity in defining the initiation of motion of heavy mineral grains under unidirectional flow. *Sedimentology* 29: 419–26

Crampton, F. A. 1937. Occurrence of gold in stream placers. *Min. J.* 20: 3–4, 33–34

Davies, K. 1968. The experimental study of the differential settling of particles in suspension at high concentrations. *Powder Technol.* 2: 43–51

Day, T. J. 1980. A study of initial motion characteristics of particles in graded bed material. In *Current Research, Part A,* pp. 281–86. *Geol. Surv. Can. Pap. 80-1A*

Deigaard, R., Fredsøe, J. 1978. Longitudinal grain sorting by current in alluvial streams. *Nord. Hydrol.* 9: 7–16

Doyle, L. J., Carder, K. L., Steward, R. G. 1983. The hydraulic equivalence of mica. *J. Sediment. Petrol.* 53: 643–48

Egiazaroff, I. U. 1965. Calculation of nonuniform sediment concentrations. *J. Hydraul. Div. ASCE* 91: 225–47

Einstein, H. A. 1950. The bedload function for sediment transportation in open channel flows. *US Dep. Agric. Tech. Bull. 1026.* 78 pp.

Einstein, H. A., Chien, N. 1953. Transport of sediment mixtures with large ranges of grain sizes. *US Army Corps Eng., Missouri River Div. Sediments Ser. No. 2.* 73 pp.

Einstein, H. A., El-Samni, E. A. 1949. Hydrodynamic forces on a rough wall. *Rev. Mod. Phys.* 21: 520–24

Eliseev, V. I. 1981. Placers of the coastal areas outside the USSR and their genetic types. *Lithol. Miner. Resour. (USSR)* 15: 324–32

Everts, C. A. 1973. Particle overpassing on flat granular boundaries. *J. Waterw. Harbors Coastal Eng. Div. ASCE* 99: 425–538

Feniak, M. W. 1944. Grain sizes and shapes of various minerals in igneous rocks. *Am. Mineral.* 29: 415–21

Fenton, J. D., Abbott, J. E. 1977. Initial movement of grains on a stream bed: the effects of relative protrusion. *Proc. R. Soc. London Ser. A* 352: 523–37

Gessler, J. 1971. Beginning and ceasing of sediment motion. In *River Mechanics,* ed. H. W. Shen, 1: 7-1–7-22. Fort Collins, Colo: Water Resour. Publ.

Gibbs, R. J., Matthews, M. D., Link, D. A. 1971. The relationship between sphere size and settling velocity. *J. Sediment. Petrol.* 41: 7–18

Gilbert, G. K. 1914. Transportation of debris by running water. *US Geol. Surv. Prof. Pap. 86.* 263 pp.

Graf, W. H. 1971. *Hydraulics of Sediment Transport.* New York: McGraw-Hill. 513 pp.

Grass, A. J. 1983. The influence of boundary layer turbulence on the mechanics of sediment transport. In *Mechanics of Sediment Transport,* ed. B. Mutlu Sumer, A. Muller, pp. 3–17. *Proc. Euromech 156, Istanbul, 1982.* Rotterdam: A. A. Balkema

Grigg, N. S., Rathbun, R. E. 1969. Hydraulic equivalence of minerals with consideration of the reentrainment process. *US Geol. Surv. Prof. Pap. 650-B*, pp. B77–B80

Hand, B. M. 1967. Differentiation of beach and dune sands, using settling velocities of light and heavy minerals. *J. Sediment. Petrol.* 37:514–20

Henley, R. W., Adams, J. 1979. On the evolution of giant gold placers. *Inst. Min. Metall. Trans.* 88:B41–B50

Hinze, J. O. 1975. *Turbulence.* New York: McGraw-Hill. 790 pp. 2nd ed.

Inman, D. L. 1949. Sorting of sediments in the light of fluid mechanics. *J. Sediment. Petrol.* 19:51–70

Inman, D. L., Ewing, G. C., Corliss, J. B. 1966. Coastal sand dunes of Guerno Negro, Baja California, Mexico. *Geol. Soc. Am. Bull.* 77:787–802

Jenkins, O. P. 1964. Geology of placer deposits. *Calif. Div. Mines Bull.* 135:147–216

Kartashov, I. P. 1971. Geological features of placers. *Econ. Geol.* 66:879–85

Kogan, B. S., Naprasnikova, L. A., Ryabtseva, G. I. 1975. Distribution and origin of local beach concentrates of gold as in one of the bays in South Primor'ye. *Int. Geol. Rev.* 17:945–49

Kolesov, S. V. 1975a. Flattening and hydrodynamic sorting of placer gold. *Int. Geol. Rev.* 17:940–44

Kolesov, S. V. 1975b. Subsidence of gold grains in allochthonous alluvial and coastal-marine placers. *Int. Geol. Rev.* 17:951–53

Komar, P. D. 1981. The applicability of the Gibbs equation for settling velocities to conditions other than quartz grains in water. *J. Sediment Petrol.* 51:1125–32

Komar, P. D., Reimers, C. E. 1978. Grain shape effects on settling rates. *J. Geol.* 86:193–209

Komar, P. D., Wang, C. 1984. Processes of selective grain transport and the formation of placers on beaches. *J. Geol.* 92:637–55

Krapez, B. 1985. The Ventersdorp Contact placer: a gold-pyrite placer of stream and debris-flow origins from the Archean Witwatersrand Basin of South Africa. *Sedimentology* 25:223–34

Krook, L. 1968. Origin of bedrock values of placer deposits. *Econ. Geol.* 63:844–46

Kuzvart, M., Bohmer, M. 1978. *Prospecting and Exploration of Mineral Deposits,* Vol. 8. *Developments in Economic Geology.* Amsterdam: Elsevier. 431 pp.

Lerman, A., Lal, D., Dacey, M. F. 1974. Stokes settling and chemical reactivity of suspended particles in natural waters. In *Suspended Solids in Water,* ed. R. J. Gibbs,

pp. 17–47. New York: Plenum

Lindgren, W. 1911. The Tertiary gravels of the Sierra Nevada of California. *US Geol. Surv. Prof. Pap. 73.* 226 pp.

Little, W. C., Mayer, P. G. 1976. Stability of channel beds by armoring. *J. Hydraul. Div. ASCE* 102:1647–61

Ljunggren, P., Sundborg, A. 1968. Some aspects of fluvial sediments and fluvial morphology. II. A study of some heavy mineral deposits in the valley of the river Lule Alv. *Geogr. Ann.* 50A:121–35

Lockett, M. J., Al-Habbooby, H. M. 1974. Relative particle velocities in two-species settling. *Powder Technol.* 10:67–71

Lucchitta, I., Suneson, N. 1981. Flash flood in Arizona—observations and their application to the identification of flash-flood deposits in the geologic record. *Geology* 9:414–18

Ludwick, J. C., Domurat, G. W. 1982. A deterministic model of the vertical component of sediment motion in a turbulent fluid. *Mar. Geol.* 45:1–15

Luque, R. F. 1974. *Erosion and transport of bed-load sediment.* PhD thesis. Tech. Hogesch. Delft, Neth. 65 pp.

MacDonald, E. H. 1983. *Alluvial Mining.* New York: Chapman & Hall. 508 pp.

McGowen, J. H., Groat, C. G. 1971. Van Horn Sandstone, West Texas: an alluvial fan model for mineral exploration. *Bur. Econ. Geol., Univ. Tex. at Austin Rep. Invest. No. 72.* 57 pp.

McIntyre, D. D. 1959. The hydraulic equivalence and size distributions of some heavy mineral grains from a beach. *J. Geol.* 67:278–301

McQuivey, R. S., Keefer, T. N. 1969. The relation of turbulence to deposition of magnetite over ripples. *US Geol. Surv. Prof. Pap. 650-D*, pp. D244–D247

Meland, N., Norrman, J. O. 1966. Transport velocities of single particles in bed-load motion. *Geogr. Ann.* 48:165–82

Meland, N., Norrman, J. O. 1969. Transport velocities of individual size fractions in heterogeneous bed load. *Geogr. Ann.* 51:127–43

Middleton, G. V., Southard, J. B. 1977. *Mechanics of Sediment Movement.* Soc. Econ. Paleontol. Mineral. Short Course Notes 3. 241 pp.

Miller, M. C., Byrne, R. J. 1966. The angle of repose for a single grain on a fixed rough bed. *Sedimentology* 6:303–14

Miller, M. C., McCave, I. N., Komar, P. D. 1977. Threshold of sediment motion under unidirectional currents. *Sedimentology* 24:507–27

Minter, W. E. L. 1970. Gold distribution related to the sedimentology of a Precambrian Witwatersrand conglomerate, South

Africa, as outlined by moving-average analysis. *Econ. Geol.* 65:963–69

Minter, W. E. L. 1976. Detrital gold, uranium, and pyrite concentration related to sedimentology in the Precambrian Vaal Reef placer, Witwatersrand, South Africa. *Econ. Geol.* 71:157–76

Minter, W. E. L. 1978. A sedimentological synthesis of placer gold, uranium, and pyrite concentrations in Proterozoic Witwatersrand sediments. In *Fluvial Sedimentology*, ed. A. D. Miall, pp. 801–29. *Can. Soc. Pet. Geol. Mem. 5*

Minter, W. E. L., Toens, P. D. 1970. Experimental simulation of gold deposition in gravel beds. *Trans. Geol. Soc. S. Afr.* 73:89–98

Mirza, S., Richardson, J. F. 1979. Sedimentation of suspensions of particles of two or more sizes. *Chem. Eng. Sci.* 35:447–54

Mosley, M. P., Schumm, S. A. 1977. Stream junctions—a probable location for bedrock placers. *Econ. Geol.* 72:691–94

Murray, S. P. 1970. Settling velocities and vertical diffusion of particles in turbulent water. *J. Geophys. Res.* 75:1647–54

Nami, M. 1983. Gold distribution in relation to depositional processes in the Proterozoic Carbon Leader placer, Witwatersrand, South Africa. In *Modern and Ancient Fluvial Systems*, ed. J. D. Collison, J. Lewin, pp. 563–75. *Int. Assoc. Sedimentol. Spec. Publ. 6*

Nelson, C. H., Hopkins, D. M. 1972. Sedimentary processes and distribution of particulate gold in the northern Bering Sea. *US Geol. Surv. Prof. Pap. 689.* 27 pp.

Parker, G., Klingman, P. C., McLean, D. L. 1982. Bedload and size distribution in paved-gravel streams. *J. Hydraul. Div. ASCE* 108:544–71

Peterson, C. D., Komar, P. D., Scheidegger, K. F. 1985. Distribution, geometry, and origin of heavy mineral placer deposits on Oregon beaches. *J. Sediment. Petrol.* In press

Pretorius, D. A. 1976. The nature of the Witwatersrand gold-uranium deposits. In *Handbook of Strata-Bound and Stratiform Ore Deposits*, ed. K. H. Wolf, pp. 29–32. Amsterdam: Elsevier

Ramette, M., Heuzel, N. F. N. 1962. Le Rhone à Lyon Étude de l'entraînement des galets à l'aide de traceurs radioactifs. *Houille Blanche No. Spéc. A*, pp. 389–99

Rana, S. A., Simons, D. B., Mahmood, K. 1973. Analysis of sediment sorting in alluvial channels. *J. Hydraul. Div. ASCE* 99:1967–80

Rao, C. B. 1957. Beach erosion and concentration of heavy mineral sands. *J. Sediment. Petrol.* 27:143–47

Raudkivi, A. J., Ettema, R. 1982. Stability of armour layers in rivers. *J. Hydraul. Div. ASCE* 108:1047–57

Reimnitz, E., Plafker, G. 1976. Marine gold placers along the Gulf of Alaska margin. *US Geol. Surv. Bull. 1415.* 16 pp.

Richardson, J. F., Meikle, R. A. 1961. Sedimentation and fluidization. Part III. The sedimentation of uniform fine particles of two-component mixtures of solids. *Trans. Inst. Chem. Eng.* 39:348–56

Richardson, J. F., Zaki, W. N. 1954. Sedimentation and fluidization. Part I. *Trans. Inst. Chem. Eng.* 32:35–53

Rittenhouse, G. 1943. Transportation and deposition of heavy minerals. *Geol. Soc. Am. Bull.* 54:725–80

Rouse, H., ed. 1950. *Engineering Hydraulics.* New York: Wiley. 139 pp.

Rubey, W. W. 1933. The size distribution of heavy minerals within a water-laid sandstone. *J. Sediment. Petrol.* 3:3–29

Rubey, W. W. 1938. The force required to move particles on a stream bed. *US Geol. Surv. Prof. Pap. 189-E*, pp. 121–41

Saks, S. Ye. 1976. Principle of hydrodynamic equivalence of clastic particles. *Int. Geol. Rev.* 18:553–62

Saks, S. E., Gavshina, A. N. 1975. Stream transport of cassiterite. *Litol. Polezn. Iskop.* 2:129–34

Sallenger, A. H. 1979. Inverse grading and hydraulic equivalence in grain-flow deposits. *J. Sediment. Petrol.* 49:553–62

Schumm, S. A. 1977. *The Fluvial System.* New York: Wiley. 338 pp.

Sestini, G. 1973. Sedimentology of a paleoplacer: the gold-bearing Tarkwaian of Ghana. In *Ores in Sediments*, ed. G. L. Amstutz, A. J. Bernard, pp. 275–306. *Int. Union Geol. Sci. Ser. A3.* New York: Springer-Verlag

Shields, A. 1936. Application of similarity principles and turbulence research to bedload movement. Transl. W. P. Ott, J. C. Uchelen as *Rep. No. 167*, Calif. Inst. Technol., Pasadena. 43 pp. (From German)

Sigov, A. P., Lomayev, A. V., Sigor, V. A., Storozhenko, L. Ye., Khypov, V. N., et al. 1972. Placers of the Urals, their formation, distribution and elements of geomorphic prediction. *Sov. Geogr.* 13:375–87

Slingerland, R. L. 1977. The effects of entrainment on the hydraulic equivalence relationships of light and heavy minerals in sands. *J. Sediment. Petrol.* 47:753–70

Slingerland, R. L. 1980. Origin of composition-size sorting in some planar laminae. *Geol. Soc. Am. Abstr. with Programs* 12:83

Slingerland, R. L. 1984. Role of hydraulic sorting in the origin of fluvial placers. *J. Sediment. Petrol.* 54:37–50

Smith, N. D., Beukes, N. J. 1983. Bar to bank convergence zones: a contribution to the origin of alluvial placers. *Econ. Geol.* 78: 1342–49

Smith, N. D., Minter, W. E. 1980. Sedimentological controls of gold and uranium in two Witwatersrand paleoplacers. *Econ. Geol.* 75: 1–14

Stapor, F. W. 1973. Heavy mineral concentrating processes and density/shape/size equilibria in the marine and coastal dune sands of the Apalachicola, Florida, region. *J. Sediment. Petrol.* 43: 396–407

Steidtmann, J. R. 1982. Size-density sorting of sand-size spheres during deposition from bedload transport and implications concerning hydraulic equivalence. *Sedimentology* 29: 877–83

Stavrakis, N. 1980. Opaque heavy minerals of the Katberg sandstone, South Africa. *Trans. Geol. Soc. S. Afr.* 83: 17–21

Sundborg, A. 1956. The river Klaralven, a study of fluvial processes. *Geogr. Ann.* 38: 125–316

Theis, N. J. 1979. Uranium-bearing and associated minerals in their geochemical and sedimentological context, Elliot Lake, Ontario. *Can. Geol. Surv. Bull. 304.* 50 pp.

Toh, E. S. C. 1978. Comparison of exploration for alluvial tin and gold. *Proc. 11th Commonwealth Min. Metallurgy Congr., 11th, Hong Kong,* ed. M. J. Jones, pp. 269–78. *Trans. Inst. Min. Metall.*

Torobin, L. B., Gauvin, W. H. 1961. The drag coefficients of single spheres moving in steady and accelerated motion in a turbulent fluid. *Am. Inst. Chem. Eng. J.* 7: 615–19

Tourtelot, H. A., Riley, L. B. 1973. Size and shape of gold and platinum grains. In *Ores in Sediments,* ed. G. L. Amstutz, A. J. Bernard, pp. 307–19. *Int. Union. Geol. Sci. Ser. A3.* New York: Springer-Verlag

Trask, C. B., Hand, B. M. 1985. Differential transport of fall-equivalent sand grains,

Lake Ontario, New York. *J. Sediment. Petrol.* 55: 226–33

Tuck, K. 1968. Origin of the bedrock values of placer deposits. *Econ. Geol.* 63: 191–93

Van Andel, T. H. 1959. Reflections on the interpretation of heavy mineral analysis. *J. Sediment. Petrol.* 29: 153–63

Warg, J. B. 1973. An analysis of methods for calculating constant terminal-settling velocities of spheres in liquids. *Math. Geol.* 5: 59–72

Wertz, J. B. 1949. Logarithmic pattern in river placer deposits. *Econ. Geol.* 44: 193–209

White, C. M. 1940. The equilibrium of grains on the bed of a stream. *Proc. R. Soc. London Ser. A* 174: 332–38

Yalin, M. S. 1977. *Mechanics of Sediment Transport.* Oxford: Pergamon. 298 pp.

Yalin, M. S., Karahan, E. 1979. Inception of sediment transport. *J. Hydraul. Div. ASCE* 105: 1433–43

Yeend, W. E. 1974. Gold-bearing gravel of the ancestral Yuba River, Sierra Nevada, California. *US Geol. Surv. Prof. Pap. 772,* pp. 1–39

References Added in Proof

Komar, P. D., Clemens, K. E. 1986. The relationship between a grain settling velocity and threshold of motion under unidirectional currents. *J. Sediment. Petrol.* In press

Komar, P. D., Li, Z. 1986. Pivoting analyses of the selective entrainment of sediments by shape and size with application to gravel threshold. *Sedimentology.* In press

Li, Z., Komar, P. D. 1986. Laboratory measurements of pivoting angles for applications to selective entrainment of gravel in a current. *Sedimentology.* In press

Ann. Rev. Earth Planet. Sci. 1986. 14:149–75

EARTHQUAKES AND ROCK DEFORMATION IN CRUSTAL FAULT ZONES

Richard H. Sibson

Department of Geological Sciences, University of California, Santa Barbara, California 93106

INTRODUCTION

The physical origin of earthquakes lies ultimately in the geological structure of fault zones and the deformation processes that occur therein in response to tectonic stress. Although the possibility now exists that the nucleation regions for large earthquakes at depths of ∼ 10 km may shortly become directly accessible by deep drilling, our present knowledge of fault structure and the shallow earthquake source is derived largely from a variety of *indirect* sources. These include seismological studies, surface studies of fault zones and earthquake ruptures, geodetic information on modes of fault slip, geophysical constraints on fault zone structure and rheology, and information garnered from materials science and experimental rock deformation.

Studies of fault zone structure and the rock products of faulting provide complementary information on deformation processes at depth in fault zones. Although descriptions of fault rocks are widespread in the geological literature (e.g. Spry 1969, Higgins 1971), it is only in the past decade that they have begun to be interpreted in the context of the physical conditions and processes prevalent in seismically active fault zones at different crustal depths (Sibson 1977, Watts & Williams 1979, Anderson et al 1983, Wise et al 1984). Such interpretations are still at an early stage, but the deformation textures and structural associations of fault rocks have already been shown to have the potential to yield information on such diverse topics as shear stress levels, power dissipation, and seismic efficiency during earthquake faulting; on fluid pressure levels and episodic fluid flow accompanying

149

faulting; on physical mechanisms for earthquake rupture arrest; and on factors governing the depth of the seismogenic zone in continental crust.

This review seeks to demonstrate the extent to which field data from fault rocks and associated structures can be integrated with information from the other sources to build *conceptual fault zone models* accounting for a wide range of observed fault behavior. It is not intended to be a comprehensive review of such studies but, rather, a demonstration of our present level of understanding, focusing in particular on the areas of uncertainty and the problems in interpretation.

SAMPLING CONSIDERATIONS

Methodology

In developing conceptual fault zone models, the general aim is to establish for appropriate host rocks the changes in macroscopic faulting style (e.g. slip on discrete planes, homogeneous or heterogeneous shear across zones of finite width) and the associated mineral deformation mechanisms that occur with depth for both seismic and aseismic slip modes. Information of this kind may be gathered from ancient fault zones exposed at different erosion levels or from dip-slip fault zones, perhaps still active, which have associated with them rock products of faulting generated over a range of depths. Note that fault rocks derived from deeper crustal levels should generally occur on the hanging walls of reverse-slip faults (e.g. Sibson et al 1979) and on the footwalls of normal-slip faults (e.g. Davis 1983). The distribution, metamorphic state, and textural-microstructural characteristics of different fault rocks may then be correlated with the associated style of faulting to build up fault models (Figure 1).

Fault Zone Evolution

Conceptual models relating to long-established fault systems are of greatest interest because the largest earthquake ruptures and most intense subsidiary seismicity are localized within tabular fault zones, commonly 10–10^3 m in thickness, that have presumably undergone considerable strain

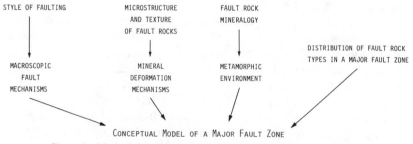

Figure 1   Methodology for evolving conceptual models of fault zones.

softening with respect to the surrounding crust. Our prime concern, then, is with the *residual* infrastructure of mature fault zones, which have arisen from the progressive evolution and coalescence of faults, fractures, and ductile shear zones that originally nucleated in more-or-less intact crust (Segall & Pollard 1983). Major fault zones may also undergo substantial geochemical evolution with time through their role as conduits for fluid flow at all crustal levels (Beach 1980, Kerrich et al 1984). For example, progressive hydrothermal alteration of fault rocks to clay-rich assemblages within the seismogenic regime may drastically change the mechanical properties of fault zones with time (Wu 1978, Wang 1984). Unfortunately, the residual infrastructure of mature fault zones is inherently less easy to study than that of low-displacement faults and shear zones, where comparatively undamaged country rock may, for example, allow correlation of finite displacement with style and intensity of deformation (e.g. Ramsay & Graham 1970, Sibson 1975, Aydin & Johnson 1978, Gay & Ortlepp 1979, Mitra 1979, Muraoka & Kamata 1983, Simpson 1983).

Further complications arise in connection with the evolution of faults with a major component of dip-slip and with fault zones that have undergone progressive unroofing throughout their history. In such situations, fault rocks generated at one crustal level may be transported to become texturally overprinted in different deformation environments (Grocott 1977, Watts & Williams 1979). Perturbation of the thermal environment by sustained dip-slip, by shear heating, or by some combination of these effects (Brewer 1981) may cause additional problems in interpretation.

Range of Slip and Strain Rates

A major consideration in the interpretation of fault-related deformation is the enormous variety of slip and strain rates prevalent within active fault zones, which has the important implication that rates of energy dissipation within fault zones likewise vary over a broad range (Sibson 1977). Measured slip rates across active fault zones span ten orders of magnitude, from steady aseismic rates of $1-30$ mm yr^{-1} to transitory seismic slip rates that may reach $0.1-2$ m s^{-1} for periods of up to a few seconds (Brune 1976), repeating at intervals commonly between 10^2 and 10^4 yr. Slip at intermediate rates may also occur. Aseismic shear strain rates inferred for mylonite belts at depth range from perhaps 10^{-9} to 10^{-12} s^{-1}, but much higher rates may pertain locally where strain continuity is maintained during seismic slip.

Deformation and the Earthquake Stress Cycle

Within the seismogenic regime, which generally occupies at least the top third of continental crust (Chen & Molnar 1983, Sibson 1983), it is now generally accepted that the bulk of fault displacements occur by large-scale earthquake rupturing (Thatcher et al 1975, Hyndman & Weichert 1983), though aseismic slip may predominate over local fault segments, such as along the San Andreas fault in central California (Burford & Harsh 1980). Thus, as a general rule one cannot think in terms of steady progressive deformation at constant stress within the upper levels of fault zones, nor of a simple succession of deformational events. Rather, one must consider deformation in relation to multiple cycles of stress and elastic strain energy accumulation, with intermittent, abrupt stress and energy release accompanying each major rupture event.

A range of stress-time and displacement-time regimes may be envisaged for seismically active fault zones (Figure 2). Crude sawtooth stress

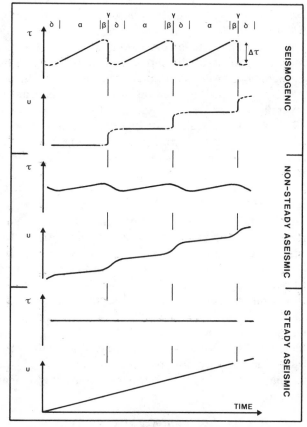

Figure 2 Hypothetical stress-time and displacement-time relationships at different levels in crustal fault zones.

oscillations should prevail throughout the seismogenic regime and presumably to some depth beneath it, with the oscillations progressively becoming damped and smoothed. At still greater depths, aseismic shearing should proceed at near-constant stress. The average amplitude of the stress oscillation within the seismogenic regime, corresponding to the static stress drop of moderate to large earthquakes ($\Delta\tau$) is well constrained to 1–10 MPa (Kanamori & Anderson 1975, Hanks 1977). This may be equated with a characteristic release of elastic shear strain ($\Delta\gamma \sim 10^{-4}$–$10^{-5}$). However, it is important to note that at present there is no consensus on the average absolute level of tectonic shear stress (τ) within crustal fault zones; current estimates range from ~ 10 MPa to ~ 100 MPa (see discussion by Hanks & Raleigh 1980).

The stress cycle for a large shallow earthquake is conveniently partitioned into four phases: an α-phase of secular, mainly elastic strain accumulation; a β-phase of preseismic anelastic deformation perhaps involving dilatant cracking and terminating in foreshock activity and accelerating precursory slip; a coseismic γ-phase of mainshock slip, rupture propagation, and energy release; and a postseismic δ-phase of decelerating afterslip and aftershock activity decaying inversely with time in accordance with Omori's law (Figure 2). With a view to understanding details of the faulting process, it is clearly desirable that deformation arising from different phases of the stress cycle should be distinguished. For example, β-phase deformation might provide an explanation for observed precursory phenomena, while γ-phase deformation could potentially yield information on power dissipation during seismic slip and on earthquake efficiency (Sibson 1980a).

Unfortunately, the cyclical nature of fault deformation leads to major interpretative difficulties: Structural features developed in one particular phase of a stress cycle are liable to be overprinted or obliterated by deformation during a later phase of the same cycle or during subsequent cycles. The seriousness of this problem becomes apparent when one considers that a finite displacement of, say, 10 km might have developed by $\sim 10^4$ 1-m seismic slip increments. Clearly, features related to particular phases of the earthquake stress cycle may be easier to distinguish around faults with small total displacements, but such faults then may not have attained the mature residual infrastructure determining long-term behavior.

Foreshock-Mainshock-Aftershock Sequences

Within the seismogenic regime, the primary stress cycle for large earthquakes may be locally perturbed by subsidiary stress release accompanying foreshocks and, more particularly, by the extensive aftershock sequences that invariably follow large crustal earthquakes. While the cumulative slip produced by such sequences is generally subordinate to that of the main earthquake, subsidiary deformation may be distributed over a broad region surrounding the mainshock rupture. Figure 3 illustrates concentrations of aftershock activity following two moderate strike-slip ruptures that terminated in different structural configurations. It is apparent that substantial deformation may be induced well outside the causative fault zone by a large earthquake.

DEFORMATION PROCESSES IN FAULT ZONES

Crustal fault processes encompass diverse rock and mineral deformation mechanisms as a consequence of the wide range of slip and strain rates involved and of the different physical environments encountered. Much experimental rock deformation has been devoted to establishing conditions for uniform flow of rocks and constitutive flow laws (see reviews by Tullis 1979, Schmid 1982, Kirby 1983) and to studies of frictional sliding with the particular aim of understanding the instabilities leading to stick-slip behavior (Byerlee 1978, Dieterich 1981, Rice 1983, Okubo & Dieterich 1984, Shimamoto & Logan 1986). However, it is important to appreciate

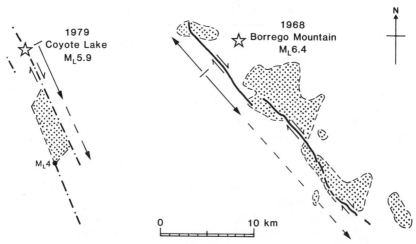

Figure 3 Seismotectonic maps illustrating aftershock concentrations associated with right-lateral strike-slip rupturing within the San Andreas fault system, California (after Sibson 1985). (*Left*) The 1979 $M_L = 5.9$ Coyote Lake earthquake rupture, which terminated in a dilational fault jog; (*Right*) The 1968 $M_L = 6.4$ Borrego Mountain rupture, which was at least partly arrested at an antidilational jog. (Epicenters represented by stars, propagation direction and extent of mainshock ruptures by arrows, surface breaks by broad lines, microearthquake lineaments by dash-dot lines, areas of intense aftershock activity by stippling.)

that adequate laboratory simulation has yet to be achieved for many natural deformation processes for which there is abundant field evidence. Pressure solution (water-assisted diffusive mass transfer) is an important example of one such mechanism (Rutter 1983). Furthermore, laboratory friction experiments have yet to simulate the displacements and anticipated levels of frictional energy dissipation likely to accompany moderate or greater earthquake ruptures. Some physical appreciation of these fast dissipation processes can be gained by considering that slip at seismic rates of ~ 1 m s^{-1} against an assumed constant shear resistance of 10 MPa (the lowest likely value) leads to power dissipation at 10 MW m^{-2} on the fault, albeit only for a few seconds at a time (see Sibson 1980a).

Thus, in considering fault-related deformation, we emphasize those processes for which there is abundant field evidence in the form of textures and microstructures. It is convenient to partition the deformation mechanisms into those capable of allowing steady aseismic shearing, those associated with the intermittent bursts of energy dissipation accompanying the δ slip phase, and those that may be associated with the β and γ phases of the earthquake stress cycle, with the recognition that varieties of inter-mediate behavior may also occur. Note also that individual mechanisms rarely operate in isolation.

Mechanisms Allowing Steady Shearing

In truly steady-state deformation, shearing would occur at a uniform rate under constant stress across a tabular zone of fixed thickness containing a statistically invariant microstructure. These requirements are unlikely to be met in most natural fault environments; we therefore relax the definition to include here all mechanisms that allow steady shearing under more-or-less constant stress. It is useful to distinguish further between mechanisms for stable sliding on more-or-less discrete surfaces and those involving volumetric shearing flow in tabular zones, although intermediate behavior is again possible and transitions may occur with progressive deformation.

PLANAR SLIDING   Stable frictional sliding is well documented from experi-mental studies (Engelder et al 1975, Dieterich 1978) and leads to the progressive accumulation of gouge by processes of frictional wear involving ploughing of asperities into opposing walls, sidewall cracking and plucking, and shearing of asperities (Engelder 1978). Clearly, with time this may lead to shear across a cataclastically flowing gouge layer, though there are usually indications that deformation continues to be concentrated near the gouge-rock interface, with the layer growing laterally with increasing displacement. In natural, fluid-saturated fault zones it seems inevitable that subcritical crack growth involving stress corrosion (Atkinson 1982, 1984) becomes an integral part of frictional wear processes accompanying slow stable sliding.

Evidence in the form of fibrous growth "slickensides" coating some natural fault surfaces, and the presence of related features such as stylolites, suggests the existence of a nonfrictional mechanism for stable sliding involving dissolution of asperities, diffusive mass transfer in an aqueous phase, and precipitation into fibrous accretion steps (Durney & Ramsay 1973); this mechanism has been termed pressure solution slip by Elliot (1976). Rutter & Mainprice (1979) predict linear viscous behavior with an inverse dependence of sliding rate on grain size. They also point out that the process may lead to substantial local fluctuations in porosity, possibly affecting fluid pressures and fault stability. The frequent association of these fiber-coated fault surfaces with arrays of extension veins, interpretable as natural hydraulic fractures, suggests that this mechanism is commonly operative under conditions of very low effective confining pressure (Sibson 1980a).

SHEARING FLOW   Schmid (1982) has reviewed deformation mechanisms allowing steady flow of crustal rocks. Cataclastic flow, intracrystalline dislocation creep, and different variants of grain boundary diffusion creep may contribute to aseismic shearing flow within crustal fault zones. Intracrystalline lattice diffusion (Nabarro-Herring creep) is unlikely to be significant. Deformation maps have been constructed for quartz and calcite illustrating the dominant mechanisms operating under different grain size, stress, temperature, and strain rate conditions (Rutter 1976, White 1976, McClay 1977, Etheridge & Wilkie 1979, Schmid 1982). However, there is at present no adequate means for representing fields of cataclastic flow on such maps. Moreover, the boundaries must still be regarded as preliminary, and their application to deformation of polymineralic rocks remains uncertain. In most crustal environments, flow of polymineralic rocks probably involves a mixture of mechanisms. On a further cautionary note, one should also consider inferences drawn by Etheridge et al (1984) to the effect that under metamorphic conditions, high-strain deformation invariably occurs at near-zero effective confining pressure in the presence of a mobile, high-pressure fluid. If true, this has profound implications for the rheology and strength of fault zones in the mid-to-deep crust.

Cataclastic flow   This mechanism may involve both brittle fragmentation of rocks (brecciation) and comminution of mineral grains, so that grain size distribution and microstructure evolve with time (Engelder 1974). Micromechanisms involve intragranular and transgranular cracking, frictional grain boundary sliding, and grain rotation. No adequate constitutive law has yet been devised, but the mechanism is highly sensitive to effective confining pressure (high pressures inhibit cracking and frictional grain sliding) and is relatively unaffected by variations in temperature and strain rate. Subcritical crack growth by stress corrosion is again likely to play an integral role in natural cataclastic flow. Possible examples of cataclastic flow related to faulting have been described by Engelder (1974), Brock & Engelder (1977), House & Gray (1982), Mitra (1984), and Blenkinsop & Rutter (1986). Comparison of natural gouge infrastructure with that developed during experimental stable shearing of gouge layers has revealed similarities in the form of systematic arrays of Riedel shears and subsidiary fractures (Logan et al 1979).

Dislocation creep   With the exception of calcite-rich rocks, where large strains may develop by twinning, the dominant intracrystalline deformation mechanism in crustal rocks is dislocation creep. Steady-state flow occurs by a combination of dislocation glide leading to work hardening and of processes of recovery involving dislocation climb and cross slip (see Nicolas & Poirier 1976). The recovery processes are thermally activated, so that in general steady flow is only achieved for absolute temperatures $T > 0.5\ T_m$, where T_m is the melting temperature. Flow to large strains by dislocation creep generally involves dynamic recrystallization and the development of a strong crystallographic preferred orientation, possibly with a statistically invariant microstructure (White 1976). Dislocation creep in silicates is facilitated by hydrolytic weakening through the incorporation of structural "water" into the lattice (Griggs 1967), though the details of the weakening process are not yet fully understood (Kirby 1984). Constitutive flow laws for dislocation creep take the form of a power law

$$\dot{e} = A\ \exp(-H/RT)\sigma^n, \tag{1}$$

where e and σ are the flow strain rate and the differential stress, respectively, A is a material constant, R is the gas constant, H is an activation energy, and $3 < n < 5$. Thus, there is no explicit dependence on grain size nor on confining pressure, though recent experimental work suggests that the diffusion rate of "water" into silicates to allow hydrolytic weakening is affected by pressure and may also lead to a grain size dependence (Blacic & Christie 1984, Kronenberg & Tullis 1984).

   Textural evidence from naturally deformed rocks suggests that dynamic recrystallization accompanying dislocation creep becomes significant in quartz under greenschist facies metamorphic conditions ($T > \sim 300°C$) and in feldspars under high greenschist to amphibolite facies conditions ($T > \sim 450°C$) (Voll 1976, Tullis et al 1982, Hanmer 1982, Simpson 1985). Grain refinement accompanying dynamic recrystallization in quartz-bearing rocks appears to play a major role in promoting strain localization within mylonitic shear belts in the mid-to-deep crust (Bell & Etheridge 1973, White 1976, Etheridge & Wilkie 1979, White et al 1980, Tullis et al 1982).

Grain boundary diffusion Coble creep and pressure solution are two deformation mechanisms that both involve diffusive mass transfer along grain boundaries; in the latter, more geologically significant mechanism, diffusion is believed to be greatly enhanced by the presence of an aqueous intergranular film (Rutter 1976, 1983, McClay 1977). In the broad sense, pressure solution encompasses all processes of stress-controlled solution, diffusional mass transfer, and redeposition. Solute transport may occur on the scale of individual grains, or over distances of millimeters to meters with redeposition in extension veins (the crack-seal mechanism of Ramsay 1980), or the dissolved material may be flushed from the deforming system, which leads to volume losses as high as 50% (Etheridge et al 1984). For diffusive mass transfer on the grain scale, strain rate depends inversely on the cube of grain size. Situations may thus arise where grain boundary diffusion takes over as the dominant deformation mechanism once initial grain size has been reduced below some critical value, perhaps by cataclastic processes or by dynamic recrystallization. Field studies of regional (low strain rate) metamorphic terrains suggest that pressure solution is an important flow mechanism for fine-grained quartz and calcite-bearing rocks over the temperature interval 150–450°C (McClay 1977, Kerrich et al 1977). However, the role of these processes in comparatively fast strain rate shear zones is less clear. Textures characteristic of pressure solution have been recognized in the fine groundmass of some cataclastic shear zones (Brock & Engelder 1977, Mitra 1978, 1984, Rutter 1983, Blenkinsop & Rutter 1986) and also in some mylonites (White & White 1983).

Superplastic flow It has been postulated that intense grain refinement accompanying dynamic recrystallization in mylonitic shear zones may allow a transition to low-stress superplastic flow (Boullier & Gueguen 1975). In this mechanism, large strains develop dominantly by nonfrictional grain boundary sliding (involving diffusive mass transfer and/or dislocation processes) in a constant microstructure of fine equiaxed grains. Grain switching and rotation occur readily; superplastic tectonites should not therefore possess a strong crystallographic preferred orientation. A super-plastic flow regime has been attained experimentally for Solnhofen limestone (Schmid et al 1977), and on textural grounds the mechanism has been identified in the fine-grained (<10 μm) Lochseiten calc-mylonite along the Glarus overthrust (Schmid et al 1981). Kerrich et al (1980) have inferred superplastic behavior within a quartzo-feldspathic ductile shear zone. The possibility of low-temperature superplastic flow has also been raised for ultrafine-grained cataclastic shear zones in quartzo-feldspathic host rocks (Phillips 1982, Anderson et al 1983). However, in general the extent to which quartz may develop superplastic behavior, even under mylonitic conditions, remains controversial (Etheridge & Wilkie 1979, White et al 1980).

Preseismic (β-Phase) Deformation

Despite its potential importance in accounting for observed precursory behavior to large earthquakes, little has been done to distinguish permanent deformation around ancient fault zones that might be related to the preseismic phase of the earthquake stress cycle. Possibilities include high-stress microcrack dilatancy, intensified through repetitive stress cycling (Scholz & Kranz 1974); cyclic dilatancy involving the reactivation of old or the development of new macrofracture systems, perhaps by subcritical growth of extension cracks (Crampin et al 1984); and possibly some form of sand-grain dilatancy within the fault zone itself (Nur 1975). Arrays of macroscopic extension veins in the vicinity of faults with textures recording incremental development in a shared stress regime have been interpreted as resulting from cyclical preseismic hydrofracture dilatancy (Sibson 1981). This form of dilatancy is a low-stress phenomenon that can only develop at shallow depths or under conditions of abnormally high fluid pressure, but it seems plausible that much of the repetitive crack-seal deformation described by Ramsay (1980) may also be caused by similar fault-related stress cycling. Mawer & Williams (1985) have attributed periodic microfracturing and crack healing in crystalline rocks to seismic activity.

Coseismic (γ-Phase) Deformation

Through consideration of likely power dissipation effects, it appears that the character of deformation accompanying seismic slip is critically dependent on the thickness of the slip zone (Sibson 1980a). Field evidence from deeply exhumed fault zones (Flinn 1977, Aydin & Johnson 1978, Davis et al 1980, Segall & Pollard 1983, Grocott 1981) suggests the existence, at least locally, of concentrated slip zones commonly centimeters or less in thickness (and even discrete sliding surfaces) throughout the seismogenic zone to depths of 10–15 km. Often, these principal slip surfaces (PSS) are located at the margins of broader cataclastic shear zones.

Given the high strain rates and levels of power dissipation anticipated for seismic slip, one may expect related deformation to be primarily cataclastic, though minor crystal plastic strains may develop by dislocation glide and through twinning and kinking of appropriate minerals. Within the broader slip zones, frictional wear processes may include attrition brecciation of the wall rock and cataclastic grain comminution. Transitory fluid pressure differentials caused by seismic slip transfer across dilational fault jogs may also induce brecciation of wall rock by hydraulic implosion (Sibson 1985). High-dilation breccias resulting from this process generally possess a matrix of hydrothermal minerals with textures recording multiple episodes of brecciation and cementation.

Depending on the degree of slip localization, significant increases in temperature can be expected to develop locally as a consequence of the rapid energy dissipation (Cardwell et al 1978). Overpressuring of fluid inclusions from sudden temperature rises may induce crystal fragmentation, with thermally induced stresses also contributing to fracturing adjacent to slip surfaces (Moore & Sibson 1978). Provided that seismic slip is localized to within a few millimeters, sufficient heat may be generated under *dry* conditions for friction-melting to occur (McKenzie & Brune 1972, Sibson 1975, Allen 1979). This is borne out by the restricted occurrence of pseudotachylyte friction-melt to minor faults in crystalline rock (Sibson et al 1979, Grocott 1981, Maddock 1983, Macaudiere et al 1985). In mature fluid-saturated fault zones, high transient fluid pressures are likely to develop on faults as a consequence of initial power dissipation at the onset of slip, leading to dramatic reductions in shear resistance and preventing large temperature increases (Sibson 1980a, Lachenbruch 1980, Raleigh & Evernden 1981). Possible field evidence for these high transient fluid pressures is sometimes found in the form of gouge-laden clastic dikes leading off slip surfaces (Gretener 1977).

These alternative feedback processes for lowering shear resistance during seismic slip (friction melting and the development of high transient fluid pressures) should both lead to near-total stress relief and high seismic efficiency, provided that the fluids remain contained within the slip zones. Acoustic fluidization has been suggested as a further possible mechanism for drastically lowering shear resistance during earthquake slip (Melosh 1979).

Postseismic (δ-Phase) Deformation

As previously discussed, distributed aftershock activity must lead to substantial deformation, much of which may be well away from the mainshock rupture (Figure 3). However, significant postseismic deformation may also develop within the primary slip zone as a consequence of afterslip. Following some large earthquakes, afterslip has increased surface displacement significantly over periods of weeks or months (e.g. Bucknam et al 1978). Such behavior can be expected to impose substantial strains and fabrics on the rock products of the preceding coseismic slip phase. A possible example of this behavior could be the local imposition of strong fabrics within some pseudotachylyte fault veins (Sibson 1980b, Passchier 1982).

In the postseismic phase, displacements may also extend with time over a much greater area than the coseismic rupture. Thatcher (1975) has shown from geodetic data that in the decades following the great 1906 San Francisco earthquake, displacements matching the coseismic strike-slip of ~ 4 m in the top 10 km of the crust extended downward to depths of perhaps 30 km. Short-term aseismic deformation at unusually high but decaying strain rates may thus be induced in the surrounds of the mainshock rupture.

Interseismic Healing

It seems probable that earthquake rupture surfaces undergo some form of healing and strengthening between successive major events. Apart from the time-dependent frictional "healing" observed in laboratory experiments (Dieterich 1978), processes of recrystallization, pressure solution, and hydrothermal cementation from circulating fluids may contribute to this effect (Angevine et al 1982). The concept of interseismic healing accords with observations of a correlation between recurrence interval and magnitude of stress drop (Kanamori & Allen 1986).

CONTINENTAL FAULT ZONE MODEL

A general model for a mature active fault zone in quartzo-feldspathic crust has been developed on the basis of the deformation textures and the distribution of fault rocks in ancient fault zones (Sibson 1977, 1983, Anderson et al 1983), and has been used to account for the depth distribution of earthquakes in continental crust. In this model, a seismo-genic frictional regime dominated by pressure-sensitive deformation involving cataclasis and frictional sliding gives way with increasing depth and temperature to a quasi-plastic regime where largely aseismic and continuous shearing is localized within mylonite belts (Figure 4). The passage from frictional to quasi-plastic behavior is believed to be determined most commonly by the changing response of quartz to deviatoric stress with increasing temperature; dislocation creep becomes an important quartz deformation mechanism at $T > \sim 300°C$, corresponding to the onset of a greenschist facies metamorphic environment. Complex transitional behavior incorporating mixed continuous and discontinuous deformation over a large range of strain rates is inferred in the vicinity of the frictional/quasi-plastic transition.

These fault models have been crudely quantified using available

162   SIBSON

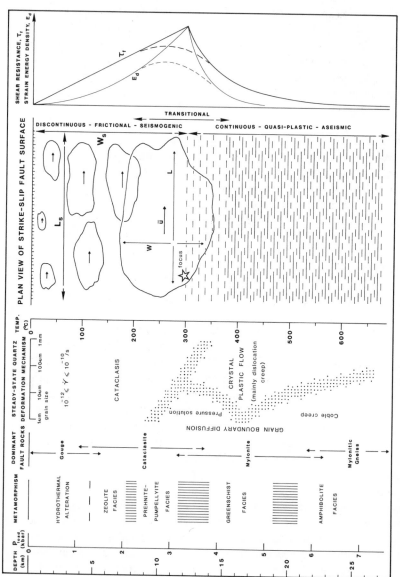

Figure 4 Conceptual model for a major strike-slip fault zone in continental crust (uniform density of 2.80 g cm⁻³, and geothermal gradient of 25°C km⁻¹), schematically relating different fault regimes to likely metamorphic environment, dominant quartz steady-state deformation mechanism, and associated fault rocks (from Sibson 1983). Rupture parameters (length L, width W, mean slip $\bar{u}$) shown for events within a seismogenic regime of total dimensions $L_s \times W_s$. Profiles of shear resistance and distortional strain energy density are in arbitrary units.

laboratory data on the frictional and rheological properties of crustal rocks to construct composite profiles illustrating the variation of shear resistance with depth for different faulting modes and heat flow provinces (Meissner & Strehlau 1982, Sibson 1982, 1983, 1984, Smith & Bruhn 1984). The peak shear resistance occurs at the frictional/quasi-plastic transition, though in reality its value is likely to be considerably reduced and smoothed out below the theoretical sharp intercept (Figures 4, 6). Observed depth distributions of microearthquakes defining the seismogenic zone can be correlated fairly satisfactorily with the modeled depths of frictional interaction in different heat flow provinces (Figure 5). Larger ($M_L > 5.5$) earthquakes tend to nucleate toward the base of this seismogenic zone in the region of inferred peak shear resistance, rupturing upward and laterally

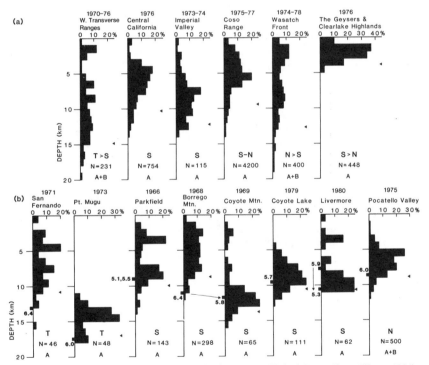

Figure 5 Earthquake depth distributions in the western United States (from Sibson 1984). Solid triangles indicate depths above which 90% of the activity occurs; the dominant faulting mode (T, thrust; S, strike-slip; N, normal), the sample size, and the data quality are listed at the base of each histogram: (a) background microearthquake activity (note that the Geysers–Clearlake Highlands region is an area of intense geothermal activity); (b) aftershock sequences following moderate earthquakes. Depth and local magnitude (M_L) of mainshock or mainshocks are indicated by solid squares. Related larger events are linked by arrows.

so that their aftershock sequences are also largely restricted to the same zone (Sibson 1982, Chen & Molnar 1983). Variations in conductive heat flow and crustal composition (especially the quartz-feldspar ratio) appear the most important of a range of factors affecting the depth of frictional interaction and seismic activity (Figure 6; Sibson 1984).

Given the heterogeneous nature of continental crust and the variety of deformation mechanisms involved in faulting, it is clear that these simple fault models can only represent bulk fault behavior to a first approximation. Many uncertainties remain in our understanding of the fine structure of fault zones and the changing physical conditions and deformation processes operative at different crustal levels. To illustrate these problems, we consider each of the main fault regimes in turn.

Frictional Regime

BRITTLE INFRASTRUCTURE   Detailed maps of the surface traces produced by individual and repeated earthquake ruptures (Vedder & Wallace 1970, Clark 1972, Tchalenko & Berberian 1975) and of the brittle infrastructure of exhumed fault zones (Flinn 1977, Segall & Pollard 1983), when coupled with high-precision aftershock studies (Reasenberg & Ellsworth 1982), suggest that throughout the seismogenic regime much of the displacement within mature fault zones is localized on principal slip surfaces (PSS), which are commonly discontinuous or curved. Segmentation may be systematically en echelon, or the PSS may step abruptly from one margin to another of major fault zones that range up to a kilometer or so in width. Geomorphic evidence indicates that the broad infrastructure of PSS may persist in a particular configuration through many episodes of rupturing over lengthy time periods, in some instances exceeding 10^4 yr. Detailed seismological analyses of rupture propagation, precision microearthquake studies, and theoretical consideration of the effects of fault segmentation (Bakun et al 1980, Segall & Pollard 1980, King & Nabelek 1985, Sibson 1985) suggest that the broad PSS infrastructure exerts major controls on rupture nucleation and arrest and on patterns of aftershock activity (see Figure 3).

FAULT ROCKS AND DEFORMATION   The dominant rock products of faulting in the frictional regime are incohesive gouge, breccias, and cohesive fault rocks of the cataclasite series (Sibson 1977). Cataclastic deformation may be localized to thin tabular zones associated with PSS or may be distributed through large volumes. In general, there is abundant evidence of hydration in mature fault zones; it seems probable that most seismogenic faulting takes place in the presence of an aqueous fluid that is at least at hydrostatic pressure. Arrays of hydrothermal extension veins in the vicinity of some faults indicate the local presence of suprahydrostatic fluid pressures exceeding the least principal compressive stress (Sibson 1981).

One of the main problems in interpreting fault rocks within the seismogenic regime arises from the lack of good depth control on most samples. A major controversy exists concerning the relative dominance of cataclasite series material or clay gouge, rich in montmorillonite. This has important implications for the strength of fault zones. Microbreccias and cataclasites may crudely be expected to follow Byerlee's (1978) general rock friction relationship, with a static shear resistance approximated by

$$\tau_f = \mu_s \sigma_n' = \mu_s(\sigma_n - P), \tag{2}$$

where σ_n is the normal stress across the sliding surface, P is the fluid pressure, and $\mu_s \sim 0.75$ is the static friction coefficient (Sibson 1983). In contrast, montmorillonite-rich gouge may have an effective friction coefficient ≤ 0.35 and a complex mechanical response under shear (Bombolakis et al 1978, Bird 1984). Moreover, the low permeability of clay gouges may contribute to the development and maintenance of suprahydrostatic fluid pressures in fault zones (Morrow et al 1984). It can be argued on thermodynamic and other grounds that given an appropriate geochemical environment, the montmorillonite-rich gouge frequently found at high levels in fault zones remains the dominant intrafault material to depths in excess of 10 km (Wu 1978, Bird 1984, Wang 1984). However, studies of ancient exhumed fault zones generally reveal a preponderance of microbreccias and cataclasites (Brock & Engelder 1977, Flinn 1977, Davis et al 1980, Anderson et al 1980, 1983), with the implication that clay-rich gouges are largely restricted to the top few kilometers. This view is supported by studies of clastic shales undergoing diagenesis during progressive burial; these studies demonstrate that the smectite-illite transition generally begins below temperatures of $\sim 90°C$ and is largely complete by $150°C$ (Ramseyer & Boles 1986). It is tempting to speculate that the comparatively low level of microseismic activity generally observed in the top few kilometers of active fault zones (e.g. Bakun et al 1980, Reasenberg & Ellsworth 1982) may correlate with the depth extent of weak montmorillonite-rich gouge.

For most of the rock products of faulting within the frictional regime, there are major problems in distinguishing deformation that results from seismic slip (γ-phase) from that induced during other phases of the earthquake stress cycle or during stable aseismic sliding and cataclastic flow. For example, while the development of shape fabrics in some cataclasites (House & Gray 1982, Chester et al 1985) seems likely to be a consequence of slow aseismic flow, it is unclear whether the flow was stable or was instead induced by δ-phase afterslip. High-dilation, hydrothermally recemented breccias associated with dilational fault jogs are inferred to have developed by hydraulic implosion induced by seismic slip (Sibson 1985). Gouge-filled clastic dikes sometimes observed leading off sliding surfaces (Gretener 1977) may be diagnostic of transient high fluid pressures developed as a consequence of temperature increases during earthquake rupturing (Sibson 1980a). Anomalous vitrinite reflectances recognized on some thrust faults may likewise result from seismic power dissipation (Bustin 1983). Pseudotachylyte friction-melt, the product of localized seismic slip under dry conditions, can be recognized only rarely in crystalline rocks that have preserved their integrity. Fibrous slickensides diagnostic of aseismic pressure solution slip (Elliot 1976) are often associated with extension vein arrays characterizing areas of anomalously high fluid pressure. Microstructural evidence for pressure solution has likewise been recognized in the groundmass of a range of gouge and cataclasite material (Brock & Engelder 1977, Mitra 1978, 1984, Rutter 1983, Blenkinsop & Rutter 1986).

Thus, given the heterogeneity of continental crust, the picture that emerges of fault deformation within the frictional regime is of extreme variability. In mature fault zones, fluid pressures may be expected to range upward from hydrostatic values; portions of faults probably obey the Byerlee friction relationship, but these may give way both laterally and vertically to clay-rich patches of lower strength or to sliding surfaces and zones where pressure solution mechanisms are operative.

Quasi-Plastic Regime

Deformation within this regime is largely aseismic and continuous. Strain rates range upward from $\sim 10^{-14}$ s^{-1} to the higher values of 10^{-12}–10^{-10} s^{-1} inferred for mylonitic shear zones. However, significant strain rate fluctuations, dying out with depth, must still occur as a consequence of the intermittent stress relief accompanying earthquake rupturing in the frictional regime. Shearing is mostly localized in mylonite belts, with widths that in the mid-crust typically range from tens to hundreds of meters but in the deep crust may broaden to as much as 10 km (Jegouzo 1980, Sorensen 1983). Deformation is heterogeneous over a wide range of scales; typically, an anastomosing mesh of high-strain shear belts encloses lozenges of comparatively undeformed material (Mitra 1979, Simpson 1983, Sorensen 1983). Characteristic rock products are well-ordered L-S tectonites of the mylonite series and, in the lower crust, mylonitic gneisses (Sibson 1977). High-shear strain gradients may give rise to considerable mesoscopic structural complexity (Bell & Hammond 1984).

Mylonitization of quartzo-feldspathic rocks under mid-crustal green-schist facies conditions at $T > \sim 300°C$ involves flow of quartz by dislocation creep around resistant feldspar grains; this gives rise to the characteristic fluxion texture of mylonites. This contrasting mechanical response becomes less marked as dynamic recrystallization of feldspars and other minerals sets in under amphibolite facies conditions in the deep crust (Voll 1976, Tullis et al 1982, Hanmer 1982, Simpson 1985). Microstructural characteristics of mylonitic rocks are reviewed by Simpson & Schmid (1983) and Lister & Snoke (1984). Although there is no question that dislocation creep plays a major role in mylonitic shear zones, especially during their initial development, the possibility of a transition to mechanisms involving grain boundary diffusion, and perhaps even to superplastic flow, has to be considered once grain size has been sufficiently reduced by dynamic recrystallization (Boullier & Gueguen 1975). If such a transition occurs and, more particularly, if it involves shearing flow under near-zero effective confining pressure, as Etheridge et al (1984) surmise, the shear resistance will drop well below values inferred from dislocation creep flow laws.

Transitional Regime

Deformation processes associated with the frictional/quasi-plastic transition are of particular interest because of the tendency for larger earthquake ruptures to nucleate in the inferred transition region, where shear resistance should be at a maximum (Sibson 1982). Problems of strain compatability in the vicinity of the transition, arising during time-dependent loading of the frictional regime from below by aseismic quasi-plastic shearing, are likely to play a key role in the nucleation process. Moreover, it has been suggested that behavior in the transitional regime may control whether or not moderate earthquakes develop into very large ruptures with lengths greatly exceeding their width (Das 1982). However, as pointed out by Carter & Kirby (1978), deformation within the transition region is likely to be highly varied, involving mixed continuous and discontinuous behavior over an enormous range of slip and strain rates. For a vertical extent of perhaps several kilometers, aseismic flow at varying strain rates may involve a mixture of cataclasis, pressure solution, and crystal plastic processes, intermittently punctuated by the sudden energy dissipation accompanying brittle earthquake rupturing. Much experimental work is now being directed at understanding this complex semibrittle behavior (Shimamoto & Logan 1986). Possible examples of the mixed deformation styles derived from this regime have been described by Sibson (1980b), Passchier (1982), Mitra (1978, 1984), and White & White (1983).

Shear Resistance Profiles

Composite profiles illustrating the variation of fault shear resistance with depth (e.g. Figure 6) are constructed on a number of simplifying assumptions. In the frictional regime, it is usually assumed that frictional resistance is independent of rock type, obeying Byerlee's (1978) simple friction laws with $\mu_s \sim 0.75$, and that fluid pressures are hydrostatic. Within the quasi-plastic regime, flow shear resistance is estimated for an appropriate conductive geotherm and a constant strain rate using a laboratory-determined flow law for an assumed representative rock type (Meissner & Strehlau 1982, Sibson 1982, 1984). More complex multilayer rheologies may also be modeled (Smith & Bruhn 1984). The power laws employed depend largely on the flow of a dominant constituent such as quartz by dislocation creep. Choice of the quasi-plastic strain rate depends on the assumption of a uniformly deforming lower crust ($\sim 10^{-14}$ s^{-1}) or on the more reasonable inference from geologic evidence that

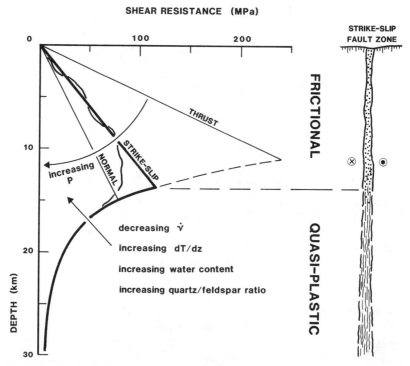

Figure 6 Composite shear resistance profile for a simple fault model (constructed for $\mu_s = 0.75$, hydrostatic fluid pressures P, a geothermal gradient $dT/dz = 25°$C km^{-1}, and a Westerly Granite flow law at a strain rate of 10^{-11} s^{-1}). Expected subsidiary roughness in the frictional regime and the smoothing out of the peak shear resistance are illustrated schematically for the strike-slip profile. Effects of perturbing factors on the general profiles are indicated (after Sibson 1984).

quasi-plastic shearing is concentrated in narrow mylonitic shear belts $(\sim 10^{-12}-10^{-10} \text{ s}^{-1})$.

It is instructive to consider the implications of the profiles with regard to the controversy over shear stress levels in crustal fault zones (see Hanks & Raleigh 1980). The strike-slip profile given in Figure 6 yields a peak shear resistance of ~ 120 MPa at the frictional/quasi-plastic transition and an average shear resistance of ~ 60 MPa over a 14-km-deep seismogenic zone. Within the quasi-plastic regime, values of flow shear resistance are broadly consistent with the shear stresses of 10–100 MPa derived from paleopiezometric studies of mylonites (Kohlstedt & Weathers 1980, Ord & Christie 1984). However, the average frictional shear stress is three times the 20-MPa maximum inferred from heat flow measurements around the San Andreas fault (Lachenbruch & Sass 1980).

If we take account of the simplifying input assumptions, there are, in fact, good grounds for believing that the modeled profiles generally represent upper bounds to possible stress values in quartz-bearing crust. Shear resistance in the frictional regime is likely to be lowered by concentrations of impermeable montmorillonite-rich gouge, by suprahydrostatic fluid pressures, and (especially in the vicinity of the frictional/quasi-plastic transition) by pressure solution processes. Within the quasi-plastic regime, the effect of shear zone broadening with depth will be to decrease strain rates and the flow shear resistance. Note, though, that diminishing quartz content and a change to a feldspar-dominated rheology would increase the flow resistance. A transition to diffusional mechanisms and superplastic flow consequent on grain size reduction would lead to linear viscous behavior and extreme softening (White et al 1980). This weakening would be carried to extremes if, as postulated by Etheridge et al (1984), quasi-plastic mylonite belts generally act as conduits for high-pressure aqueous fluids, so that shearing is occurring under near-zero effective confining pressure.

DISCUSSION: OUTSTANDING PROBLEMS

The conceptual models developed to date account for a wide variety of observed fault behavior and earthquake source characteristics, but they can only be regarded as preliminary. Refinement of existing models will come through the combination of new field, theoretical, and experimental studies. The following interrelated topics appear to be particularly fruitful avenues for future research.

Fault zone geometry Further investigations of the geometrical characteristics of fault systems at all crustal levels and on all scales are needed.

Particular attention should be paid to large-scale brittle infrastructure and to the continuation of fault zones beneath the seismogenic regime for different faulting modes; both may play critical roles in determining the character of earthquake rupturing.

Evolution of fault zones   Our understanding of fault maturation is still limited, and a number of questions remain unanswered. What factors control how a fault zone widens with time and with increasing displacement? What degree of strain softening is involved at different crustal levels, and how does it occur? How does a residual brittle infrastructure develop, and what controls its longevity in one particular configuration? What processes of interseismic healing occur within the frictional regime?

Pressure-temperature controls   Much more and improved quantitative information is needed on the *P-T* conditions and, in particular, the depths of formation of the different fault rock types, especially within the seismogenic regime.

Physical conditions and stress levels   The level of tectonic shear stress remains the outstanding problem in crustal fault mechanics. Fault rock studies on power dissipation in relation to seismic slip bear directly on this important question. It is intimately related to other poorly constrained physical controls on deformation, such as fluid pressure levels, and to whether heat flow in the vicinity of major faults is dominantly a conductive or a convective process. The relative dominance of impermeable clay gouge or cataclasite within the frictional regime and the effects of pressure solution and hydrothermal self-sealing on fault permeability all have implications for the development of suprahydrostatic fluid pressures requiring further assessment. Systematic evaluation of the commonness of hydrothermal extension veining, diagnostic of near-lithostatic fluid pressures, is needed within both the frictional and quasi-plastic regimes of ancient fault zones, particularly in view of the hypothesis that mylonitic deformation generally occurs in the presence of a high-pressure fluid phase.

Rheological modeling   As new laboratory-determined constitutive flow laws become available, it should be possible to develop more elaborate rheological models for fault zones that reflect continental heterogeneity. Prime requirements are a better appreciation of the clay gouge to cataclasite ratio throughout the frictional regime, its controls and mechanical effects, quantitative understanding of the role of pressure solution processes at all crustal levels in realistically "dirty" fault zone environments, and improved knowledge of the factors affecting hydrolytic weakening of silicates. Detailed modeling of the time-dependent loading of the frictional regime to failure and of the likely effects throughout the quasi-plastic regime of sudden stress drops in the seismogenic zone would be instructive. A rheological model for fault zones in oceanic crust should also be developed.

Rupture controls   Research should be directed toward improving our nascent understanding of the role of brittle infrastructure in controlling earthquake rupture nucleation and arrest within the frictional regime. An improved ability to distinguish the rock products of steady aseismic shearing from those associated with different phases of the earthquake stress cycle is desirable in this regard. The complex mixture of continuous and discontinuous deformation inferred to be associated with the nucleation of larger ruptures in the frictional/quasi-plastic transitional regime warrants special attention.

ACKNOWLEDGMENTS

I acknowledge with gratitude help afforded me over the past several years by my colleagues at Imperial College, London, and at the University of California, Santa Barbara, and by researchers in the Office of Earthquakes, Volcanoes and Engineering, US Geological Survey, Menlo Park. Figures 4, 5, and 6 are reproduced by kind permission of the Geological Society of London and the American Geophysical Union. Research leading to this review was supported by National Science Foundation grant #EAR83-05876.

Literature Cited

Allen, A. R. 1979. Mechanism of frictional fusion in fault zones. *J. Struct. Geol.* 1: 231–43

Anderson, J. L., Osborne, R. H., Palmer, D. F. 1980. Petrogenesis of cataclastic rocks within the San Andreas fault zone of southern California, U.S.A. *Tectonophysics* 67: 221–49

Anderson, J. L., Osborne, R. H., Palmer, D. F. 1983. Cataclastic rocks of the San Gabriel fault—an expression of deformation at deeper crustal levels in the San Andreas fault zone. *Tectonophysics* 98: 209–51

Angevine, C. L., Turcotte, D. L., Furnish, M. D. 1982. Pressure solution lithification as a mechanism for the stick-slip behavior of faults. *Tectonics* 1: 151–60

Atkinson, B. K. 1982. Subcritical crack propagation in rocks: theory, experimental results and applications. *J. Struct. Geol.* 4: 41–56

Atkinson, B. K. 1984. Subcritical crack growth in geological materials. *J. Geophys. Res.* 89: 4077–4114

Aydin, A., Johnson, A. M. 1978. Development of faults as zones of deformation bands and as slip surfaces in sandstone. *Pure Appl. Geophys.* 116: 931–42

Bakun, W. H., Stewart, R. M., Bufe, C. G., Marks, S. M. 1980. Implication of seismicity for failure of a section of the San Andreas Fault. *Bull. Seismol. Soc. Am.* 70: 185–201

Beach, A. 1980. Retrogressive metamorphic processes in shear zones with special reference to the Lewisian complex. *J. Struct. Geol.* 2: 257–64

Bell, T. H., Etheridge, M. A. 1973. Microstructures of mylonites and their descriptive terminology. *Lithos* 6: 337–48

Bell, T. H., Hammond, R. L. 1984. On the internal geometry of mylonite zones. *J. Geol.* 92: 667–86

Bird, P. 1984. Hydration-phase diagrams and friction of montmorillonite under laboratory and geologic conditions, with implications for shale compaction, slope stability and strength of fault gouge. *Tectonophysics* 107: 235–60

172   SIBSON

Blacic, J. D., Christie, J. M. 1984. Plasticity and hydrolytic weakening of quartz single crystals. *J. Geophys. Res.* 89:4223–39

Blenkinsop, T. G., Rutter, E. H. 1986. Cataclastic deformation in quartzites of the Moine Thrust zone. *J. Struct. Geol.* In press

Bombolakis, E. G., Hepburn, J. C., Roy, D. C. 1978. Fault creep and stress drops in saturated silt-clay gouge. *J. Geophys. Res.* 83:818–29

Boullier, A. M., Gueguen, Y. 1975. SP-mylonites: origin of some mylonites by superplastic flow. *Contrib. Mineral. Petrol.* 50:93–104

Brewer, J. 1981. Thermal effects of thrust faulting. *Earth Planet. Sci. Lett.* 56:233–44

Brock, W. G., Engelder, T. 1977. Deformation associated with the movement of the Muddy Mountain overthrust in the Buffington window, southeastern Nevada. *Geol. Soc. Am. Bull.* 88:1667–77

Brune, J. N. 1976. The physics of earthquake strong motion. In *Seismic Risk and Engineering Decision*, ed. C. Lomnitz, Rosenblueth, pp. 141–77. Amsterdam: Elsevier. 425 pp.

Bucknam, R. C., Plafker, G., Sharp, R. V. 1978. Fault movement (afterslip) following the Guatemala earthquake of February 4, 1976. *Geology* 6:170–73

Burford, R. O., Harsh, P. W. 1980. Slip on the San Andreas fault in central California from alinement array surveys. *Bull. Seismol. Soc. Am.* 70:1233–61

Bustin, R. M. 1983. Heating during thrust faulting in the Rocky Mountains: friction or fiction? *Tectonophysics* 95:309–28

Byerlee, J. D. 1978. Friction of rocks. *Pure Appl. Geophys.* 116:615–26

Cardwell, R. K., Chinn, D. S., Moore, G. F., Turcotte, D. L. 1978. Frictional heating on a fault zone with finite thickness. *Geophys. J. R. Astron. Soc.* 52:525–30

Carter, N. L., Kirby, S. H. 1978. Transient creep and semibrittle behavior of crystalline rocks. *Pure Appl. Geophys.* 116:807–39

Chen, W. P., Molnar, P. 1983. Focal depths of intracontinental and intraplate earthquakes and their implications for the thermal and mechanical properties of the lithosphere. *J. Geophys. Res.* 88:4183–4214

Chester, F. M., Friedman, M., Logan, J. M. 1985. Foliated cataclasites. *Tectonophysics* 111:134–46

Clark, M. M. 1972. Surface rupture along the Coyote Creek fault. *US Geol. Surv. Prof. Pap.* 787, pp. 55–86

Crampin, S., Evans, R., Atkinson, B. K. 1984. Earthquake prediction: a new physical basis. *Geophys. J. R. Astron. Soc.* 76:147–56

Das, S. 1982. Appropriate boundary conditions for modelling very long earthquakes and physical consequences. *Bull. Seismol. Soc. Am.* 72:1911–26

Davis, G. A., Anderson, J. L., Frost, E. G., Shackelford, T. J. 1980. Mylonitization and detachment faulting in the Whipple-Buckskin-Rawhide Mountains terrane, southeastern California and western Arizona. *Geol. Soc. Am. Mem.* 153:79–129

Davis, G. H. 1983. Shear-zone model for the origin of metamorphic core complexes. *Geology* 11:342–47

Dieterich, J. H. 1978. Time-dependent friction and the mechanics of stick-slip. *Pure Appl. Geophys.* 116:790–806

Dieterich, J. H. 1981. Constitutive properties of faults with simulated gouge. *Am. Geophys. Union Monogr.* 24:103–20

Durney, D. W., Ramsay, J. G. 1973. Incremental strains measured by syntectonic crystal growths. In *Gravity and Tectonics*, ed. K. A. de Jong, R. Scholten, pp. 67–96. New York: Wiley

Elliot, D. 1976. The energy balance and deformation mechanisms of thrust sheets. *Philos. Trans. R. Soc. London Ser. A* 283:289–312

Engelder, J. T. 1974. Cataclasis and the generation of fault gouge. *Geol. Soc. Am. Bull.* 85:1515–22

Engelder, J. T. 1978. Aspects of asperity-surface interaction and surface damage of rocks during experimental frictional sliding. *Pure Appl. Geophys.* 116:705–16

Engelder, J. T., Logan, J. M., Handin, J. 1975. The sliding characteristics of sandstone on quartz fault-gouge. *Pure Appl. Geophys.* 113:68–86

Etheridge, M. A., Wilkie, J. C. 1979. Grain-size reduction, grain boundary sliding and the flow strength of mylonites. *Tectonophysics* 58:159–78

Etheridge, M. A., Wall, V. J., Cox, S. F., Vernon, R. H. 1984. High fluid pressures during regional metamorphism and deformation: implications for mass transport and deformation mechanisms. *J. Geophys. Res.* 89:4344–58

Flinn, D. 1977. Transcurrent faults and associated cataclasis in Shetland. *J. Geol. Soc. London* 133:231–48

Gay, N. C., Ortlepp, W. D. 1979. Anatomy of a mining-induced fault zone. *Geol. Soc. Am. Bull.* 90:47–58

Gretener, P. E. 1977. On the character of thrust faults with particular reference to the basal tongues. *Bull. Can. Pet. Geol.* 25:110–22

Griggs, D. T. 1967. A model of hydrolytic weakening in quartz and other silicates. *Geophys. J. R. Astron. Soc.* 14:19–31

Grocott, J. 1977. The relationship between

Precambrian shear belts and modern fault systems. *J. Geol. Soc. London* 133: 257–62

Grocott, J. 1981. Fracture geometry of pseudotachylyte generation zones: a study of shear fractures formed during seismic events. *J. Struct. Geol.* 3: 169–78

Hanks, T. C. 1977. Earthquake stress drops, ambient tectonic stresses, and the stresses that drive plate motions. *Pure Appl. Geophys.* 115: 441–58

Hanks, T. C., Raleigh, C. B. 1980. The conference on magnitude of deviatoric stresses in the Earth's crust and upper mantle. *J. Geophys. Res.* 85: 6083–85

Hanmer, S. K. 1982. Microstructure and geochemistry of plagioclase and microcline in naturally deformed granite. *J. Struct. Geol.* 4: 197–213

Higgins, M. W. 1971. Cataclastic rocks. *US Geol. Surv. Prof. Pap. 687.* 97 pp.

House, W. M., Gray, D. R. 1982. Cataclasites along the Saltville thrust, U.S.A., and their implications for thrust-sheet emplacement. *J. Struct. Geol.* 4: 257–69

Hyndman, R. D., Weichert, D. H. 1983. Seismicity and rates of relative motion along the plate boundaries of western North America. *Geophys. J. R. Astron. Soc.* 72: 59–82

Jegouzo, P. 1980. The South Armorican shear zone. *J. Struct. Geol.* 2: 39–47

Kanamori, H., Allen, C. R. 1986. Earthquake repeat time and average stress drop. In *Earthquake Source Mechanics, Maurice Ewing Ser.*, ed. S. Das, Vol. 5. Washington, DC: Am. Geophys. Union. In press

Kanamori, H., Anderson, D. L. 1975. Theoretical basis of some empirical relations in seismology. *Bull. Seismol. Soc. Am.* 65: 1073–95

Kerrich, R., Beckinsdale, R. D., Durham, J. J. 1977. The transition between deformation regimes dominated by intercrystalline diffusion and intracrystalline creep evaluated by oxygen isotope geothermometry. *Tectonophysics* 38: 241–57

Kerrich, R., Allison, I., Barnett, R. L., Moss, S., Starkey, J. 1980. Microstructural and chemical transformations accompanying deformation of granite in a shear zone at Mieville, Switzerland; with implications for stress corrosion cracking and superplastic flow. *Contrib. Mineral. Petrol.* 73: 221–42

Kerrich, R., La Tour, T. E., Willmore, L. 1984. Fluid participation in deep fault zones: evidence from geological, geochemical and $^{18}O/^{16}O$ relations. *J. Geophys. Res.* 89: 4331–43

King, G. C. P., Nabelek, J. 1985. Role of fault bends in the initiation and termination of earthquake rupture. *Science* 228: 984–87

Kirby, S. H. 1983. Rheology of the litho-

sphere. *Rev. Geophys. Space Phys.* 21: 1458–87

Kirby, S. H. 1984. Introduction and digest to the special issue on chemical effects of water on the deformation and strengths of rocks. *J. Geophys. Res.* 89: 3991–95

Kohlstedt, D. L., Weathers, M. S. 1980. Deformation-induced microstructures, paleopiezometers and differential stresses in deeply eroded fault zones. *J. Geophys. Res.* 85: 6269–85

Kronenberg, A. K., Tullis, J. 1984. Flow strengths of quartz aggregates: grain size and pressure effects due to hydrolytic weakening. *J. Geophys. Res.* 89: 4281–97

Lachenbruch, A. H. 1980. Frictional heating, fluid pressure, and the resistance to fault motion. *J. Geophys. Res.* 85: 6097–6112

Lachenbruch, A. H., Sass, J. H. 1980. Heat flow and energetics of the San Andreas fault zone. *J. Geophys. Res.* 85: 6185–6222

Lister, G. S., Snoke, A. W. 1984. S-C mylonites. *J. Struct. Geol.* 6: 617–38

Logan, J. M., Friedman, M., Higgs, N., Dengo, C., Shimamoto, T. 1979. Experimental studies of simulated fault gouge and their application to studies of natural fault zones. *US Geol. Surv. Open-File Rep.* 79-1239, pp. 305–43

Macaudiere, J., Brown, W. L., Ohnenstetter, D. 1985. Microcrystalline textures resulting from rapid crystallization in a pseudotachylite melt in a meta-anorthosite. *Contrib. Mineral. Petrol.* 89: 39–51

Maddock, R. H. 1983. Melt origin of fault-generated pseudotachylytes demonstrated by textures. *Geology* 11: 105–8

Mawer, C. K., Williams, P. F. 1985. Crystalline rocks as possible paleoseismicity indicators. *Geology* 13: 100–2

McClay, K. R. 1977. Pressure solution and Coble creep in rocks and minerals: a review. *J. Geol. Soc. London* 134: 57–70

McKenzie, D., Brune, J. N. 1972. Melting on fault planes during large earthquakes. *Geophys. J. R. Astron. Soc.* 29: 65–78

Meissner, R., Strehlau, J. 1982. Limits of stresses in continental crust and their relation to the depth-frequency distribution of shallow earthquakes. *Tectonics* 1: 73–89

Melosh, H. J. 1979. Acoustic fluidization: a new geologic process? *J. Geophys. Res.* 84: 7513–20

Mitra, G. 1978. Ductile deformation zones and mylonites: the mechanical processes involved in the deformation of crystalline basement rocks. *Am. J. Sci.* 278: 1057–84

Mitra, G. 1979. Ductile deformation zones in Blue Ridge basement rocks and estimation of finite strains. *Geol. Soc. Am. Bull.* 90: 935–51

Mitra, G. 1984. Brittle to ductile transition

174 SIBSON

due to large strains along the White Rock thrust, Wind River Mountains, Wyoming. *J. Struct. Geol.* 6:51–61

Moore, H. E., Sibson, R. H. 1978. Experimental thermal fragmentation in relation to seismic faulting. *Tectonophysics* 49:T9–T17

Morrow, C. A., Shi, L. Q., Byerlee, J. D. 1984. Permeability of fault gouge under confining pressure and shear stress. *J. Geophys. Res.* 89:3193–3200

Muraoka, H., Kamata, H. 1983. Displacement distributions along minor fault traces. *J. Struct. Geol.* 5:483–95

Nicolas, A., Poirier, J. P. 1976. *Crystalline Plasticity and Solid State Flow in Metamorphic Rocks.* London: Wiley. 444 pp.

Nur, A. 1975. A note on the constitutive law for dilatancy. *Pure Appl. Geophys.* 133:197–206

Okubo, P. G., Dieterich, J. H. 1984. Effects of physical fault properties on frictional instabilities produced on simulated faults. *J. Geophys. Res.* 89:5817–27

Ord, A., Christie, J. M. 1984. Flow stresses from microstructures in mylonitic quartzites of the Moine Thrust Zone, Assynt area, Scotland. *J. Struct. Geol.* 6:639–54

Passchier, C. W. 1982. Pseudotachylyte and the development of ultramylonite bands in the Saint-Barthelemy Massif, French Pyrenees. *J. Struct. Geol.* 4:69–79

Phillips, J. C. 1982. Character and origin of cataclasite developed along the low-angle detachment fault, Whipple Mountains, California. In *Mesozoic-Cenozoic Tectonic Evolution of the Colorado River Region, California, Arizona and Nevada,* ed. E. G. Frost, D. L. Martin, pp. 109–16. San Diego: Cordilleran Publ. 608 pp.

Raleigh, C. B., Evernden, J. 1981. Case for low deviatoric stress in the lithosphere. *Am. Geophys. Union Monogr.* 24:173–86

Ramsay, J. G. 1980. The crack-seal mechanism of rock deformation. *Nature* 284:135–39

Ramsay, J. G., Graham, R. H. 1970. Strain variation in shear belts. *Can. J. Earth Sci.* 7:786–813

Ramseyer, K., Boles, J. R. 1986. I/S clay minerals in Tertiary sediments, San Joaquin Valley, California. *Clays Clay Miner.* In press

Reasenberg, P., Ellsworth, W. L. 1982. Aftershocks of the Coyote Lake, California, earthquake of August 6, 1979: a detailed study. *J. Geophys. Res.* 87:10,637–55

Rice, J. R. 1983. Constitutive relations for fault slip and earthquake instabilities. *Pure Appl. Geophys.* 121:443–75

Rutter, E. H. 1976. The kinetics of rock deformation by pressure solution. *Philos.*

Trans. R. Soc. London Ser. A 283:203–19

Rutter, E. H. 1983. Pressure solution in nature, theory and experiment. *J. Geol. Soc. London* 140:725–40

Rutter, E. H., Mainprice, D. H. 1979. On the possibility of slow fault slip controlled by a diffusive mass transfer process. *Gerlands Beitr. Geophys., Leipzig* 88:154–62

Schmid, S. M. 1982. Microfabric studies as indicators of deformation mechanisms and flow laws operative in mountain building. In *Mountain Building Processes,* ed. K. J. Hsü, pp. 95–110. London: Academic. 263 pp.

Schmid, S. M., Boland, J. N., Paterson, M. S. 1977. Superplastic flow in finegrained limestone. *Tectonophysics* 43:257–91

Schmid, S. M., Casey, M., Starkey, J. 1981. The microfabric of calcite tectonites from the Helvetic nappes (Swiss Alps). *Geol. Soc. London Spec. Publ.* 9:151–58

Scholz, C. H., Kranz, R. 1974. Notes on dilatancy recovery. *J. Geophys. Res.* 79:2132–35

Segall, P., Pollard, D. D. 1980. Mechanics of discontinuous faults. *J. Geophys. Res.* 85:4337–50

Segall, P., Pollard, D. D. 1983. Nucleation and growth of strike-slip faults in granite. *J. Geophys. Res.* 88:555–68

Shimamoto, T., Logan, J. M. 1986. Velocity-dependent behavior in halite-simulated fault gouge: an analog for silicates. In *Earthquake Source Mechanics, Maurice Ewing Ser.,* ed. S. Das, Vol. 5. Washington, DC: Am. Geophys. Union. In press

Sibson, R. H. 1975. Generation of pseudotachylyte by ancient seismic faulting. *Geophys. J. R. Astron. Soc.* 43:775–94

Sibson, R. H. 1977. Fault rocks and fault mechanisms. *J. Geol. Soc. London* 133:191–213

Sibson, R. H. 1980a. Power dissipation and stress levels on faults in the upper crust. *J. Geophys. Res.* 85:6239–47

Sibson, R. H. 1980b. Transient discontinuities in ductile shear zones. *J. Struct. Geol.* 2:165–71

Sibson, R. H. 1981. Fluid flow accompanying faulting: field evidence and models. In *Earthquake Prediction: An International Review,* ed. D. W. Simpson, P. G. Richards, *Maurice Ewing Ser.* 4:593–603. Washington, DC: Am. Geophys. Union. 680 pp.

Sibson, R. H. 1982. Fault zone models, heat flow, and the depth distribution of earthquakes in the continental crust of the United States. *Bull. Seismol. Soc. Am.* 72:151–63

Sibson, R. H. 1983. Continental fault structure and the shallow earthquake source. *J. Geol. Soc. London* 140:741–67

Sibson, R. H. 1984. Roughness at the base of the seismogenic zone: contributing factors. *J. Geophys. Res.* 89:5791–99

Sibson, R. H. 1985. Stopping of earthquake ruptures at dilational fault jogs. *Nature* 316:248–51

Sibson, R. H., White, S. H., Atkinson, B. K. 1979. Fault rock distribution and structure within the Alpine Fault Zone: a preliminary account. *R. Soc. N. Z. Bull.* 18:55–65

Simpson, C. 1983. Displacement and strain patterns from naturally occurring shear zone terminations. *J. Struct. Geol.* 5:497–506

Simpson, C. 1985. Deformation of granitic rocks across the brittle-ductile transition. *J. Struct. Geol.* 7:503–12

Simpson, C., Schmid, S. M. 1983. An evaluation of criteria to deduce the sense of movement in sheared rocks. *Geol. Soc. Am. Bull.* 94:1281–88

Smith, R. B., Bruhn, R. L. 1984. Intraplate extensional tectonics of the eastern Basin-Range: inferences on structural style from seismic reflection data, regional tectonics and thermo-mechanical models of brittle-ductile deformation. *J. Geophys. Res.* 89:5733–62

Sorensen, K. 1983. Growth and dynamics of the Nordre Stromfjord shear zone. *J. Geophys. Res.* 88:3419–37

Spry, A. 1969. *Metamorphic Textures.* Oxford: Pergamon. 350 pp.

Tchalenko, J. S., Berberian, M. 1975. Dasht-e-Bayaz Fault, Iran: earthquake and earlier related structures in bed rock. *Geol. Soc. Am. Bull.* 86:703–9

Thatcher, W. 1975. Strain accumulation and release mechanism of the 1906 San Francisco earthquake. *J. Geophys. Res.* 80:4862–72

Thatcher, W., Hileman, J. A., Hanks, T. C. 1975. Seismic slip distribution along the San Jacinto fault zone, southern Cali-fornia, and its implications. *Geol. Soc. Am. Bull.* 86:1140–46

Tullis, J. A. 1979. High temperature deformation of rocks and minerals. *Rev. Geophys. Space Phys.* 17:1137–54

Tullis, J. A., Snoke, A. W., Todd, V. R. 1982. Penrose Conference Report on significance and petrogenesis of mylonitic rocks. *Geology* 10:227–30

Vedder, J. G., Wallace, R. E. 1970. Map showing recently active breaks along the San Andreas and related faults between Cholame Valley and Tejon Pass, California. *US Geol. Surv. Misc. Geol. Invest. Map 1-574*, scale 1:24,000

Voll, G. 1976. Recrystallization of quartz, biotite and feldspars from Erstfeld to the Levantina Nappe, Swiss Alps, and its geological significance. *Schweiz. Mineral. Petrogr. Mitt.* 56:641–47

Wang, C.-Y. 1984. On the constitution of the San Andreas fault zone in central California. *J. Geophys. Res.* 89:5858–66

Watts, M. J., Williams, G. D. 1979. Fault rocks as indicators of progressive shear deformation in the Guingamp region, Brittany. *J. Struct. Geol.* 1:323–32

White, J. C., White, S. H. 1983. Semi-brittle deformation within the Alpine Fault Zone, New Zealand. *J. Struct. Geol.* 5:579–89

White, S. H. 1976. The effects of strain on the microstructures, fabrics and deformation mechanisms in quartzites. *Philos. Trans. R. Soc. London Ser. A* 283:69–86

White, S. H., Burrows, S. E., Carreras, J., Shaw, N. D., Humphreys, F. J. 1980. On mylonites in ductile shear zones. *J. Struct. Geol.* 2:175–87

Wise, D. U., Dunn, D. E., Engelder, J. T., Geiser, P. A., Hatcher, R. D., et al. 1984. Fault-related rocks: suggestions for terminology. *Geology* 12:391–94

Wu, F. T. 1978. Mineralogy and physical nature of clay gouge. *Pure Appl. Geophys.* 116:655–89

Ann. Rev. Earth Planet. Sci. 1986. 14:177–99

GENESIS OF MISSISSIPPI VALLEY–TYPE LEAD-ZINC DESPOSITS

Dimitri A. Sverjensky

Department of Earth and Planetary Sciences, The Johns Hopkins University, Baltimore, Maryland 21218

INTRODUCTION

The Paleozoic sedimentary rocks of the greater Mississippi Valley region are the hosts to major districts of sulfide mineralization, which have been important sources of sphalerite, galena, barite, and fluorite (Figure 1). Numerous smaller districts and many minor deposits are scattered between the major deposits, all of which possess so many geologic and geochemical similarities that they constitute a class of mineral deposits known as Mississippi Valley–type deposits. Deposits with similar characteristics occur in other geologic provinces of the United States, where they have been either deformed or metamorphosed, and in Proterozoic-to-Cenozoic sedimentary rocks on most of the other continents. The most important characteristics of Mississippi Valley–type deposits (Bastin 1939, Ohle 1959, 1980, Roedder 1967, 1977, Snyder 1968, White 1968, 1974) are the following:

1. They occur principally in limestone or dolostone that forms a thin cover over an igneous or highly metamorphosed Precambrian basement.
2. They consist of bedded replacements, vuggy ores, and veins, but the ore is strongly controlled by individual strata.
3. They contain galena, sphalerite, pyrite, marcasite, fluorite, barite, chalcopyrite, dolomite, calcite, and quartz.
4. They are not associated with igneous rocks, except in the case of the Kentucky-Illinois district.
5. They always occur in areas of mild deformation, expressed in brittle fracture, broad domes and basins, and gentle folds.

177

0084–6597/86/0515–0177$02.00

6. The ore is never in the basement rocks, but its distribution is often spatially related to basement highs, with the ore located within sandbanks, ridges, and reef structures that surround the basement highs.

7. The ore is at shallow depths, generally less than 600 m relative to the present surface, and was probably never at depths greater than about 1500 m.

8. There is always evidence of dissolution of the carbonate host rock, expressed by slumping, collapse, brecciation, or thinning of the host rock, that provides clear proof that the ores are epigenetic.

9. The carbon and oxygen isotopic compositions of the host rocks are normal for such rocks but are lowered adjacent to ore, which suggests that the host rocks were recrystallized in the presence of a fluid.

10. Fluid inclusions in sphalerite, fluorite, barite, and calcite always

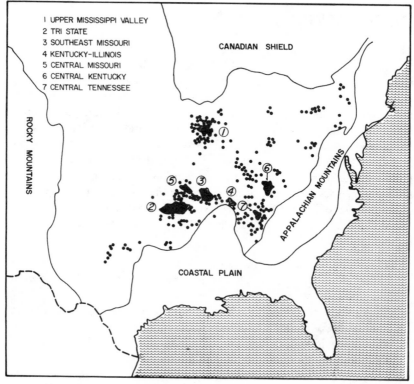

Figure 1   Major and minor Mississippi Valley–type deposits within the greater Mississippi Valley region (after Heyl 1968).

contain dense, saline, aqueous fluids and often oil and/or methane. The total dissolved salts range from 10 to 30 wt% and are predominantly chloride, sodium, and calcium, with much smaller amounts of potassium and magnesium. Homogenization temperatures are generally in the range 50–200°C.

11. Reconstruction of the total sediment thickness over the ore, together with normal geothermal gradients, suggests temperatures much lower than the fluid inclusion homgenization temperatures.

12. The hydrogen and oxygen isotopic compositions of the water in the fluid inclusions are similar to those of the pore fluids in sedimentary basins.

13. The ranges of sulfur isotopic values and the degree of approach to isotopic equilibrium between sulfides are different for each district. In some districts, the source of the sulfur could not have been magmatic and thus must have been sedimentary.

14. The isotopic composition of the lead in galena is extremely radiogenic and thus yields future model ages, which suggest sources in the upper crust. The lead isotopic values are often zoned across whole districts, within individual deposits, and even within single crystals of galena; such zoning suggests multiple sources of lead, a long period of mineralization, or both.

It is well established that Mississippi Valley–type deposits formed from hot, saline, aqueous solutions some time after the lithification of their host rocks. The major element compositions, high salinities, D/H and $^{18}O/^{16}O$ ratios, and temperatures of Mississippi Valley–type ore-forming fluids are remarkably similar to those of oil-field brines found in present-day sedimentary basins (Carothers & Kharaka 1978, 1980, Carpenter et al 1974, Hall & Friedman 1963, Hanor 1979, Kharaka et al 1980, Roedder 1967, 1977, White 1968, 1974, 1981). According to the basinal brine hypothesis of ore formation (e.g. White 1968, Ohle 1980), hot saline fluids similar to oil-field brines migrated out of sedimentary basins and along aquifers, eventually forming ore deposits in sedimentary host rocks at distances of the order of 100 km from the basins (Figure 2). This hypothesis raises a whole host of questions, summarized by Ohle (1980):

1. What were the mechanisms of fluid flow and the pathways during migration?

2. How long did the flow systems persist, and how much fluid passed through the site of ore deposition?

3. Did the brines become ore-forming fluids before, during, or after migration?

4. What were the sources of the ore-forming constituents, their mechanisms of transport, and their concentrations in the brines?
5. What chemical reactions were responsible for the precipitation of the sulfide ore minerals?

Over the last century, numerous models have been proposed to answer some of these questions. With regard to the transportation and precipitation questions, it is possible to group all the genetic models into three different categories (Sverjensky 1981), as summarized in Table 1. Until recently, it has not been possible to unambiguously demonstrate which of these three models is appropriate for any of the major Mississippi Valley–type districts.

According to the first model (the mixing model), it is suggested that base metals were transported in one fluid, that sulfide was transported separately in another fluid, and that precipitation took place where the two fluids mixed (e.g. Anderson 1975, Beales & Jackson 1966). Other theories included under the mixing category hypothesize that the source of reduced sulfur encountered was in the form of sour gas, diagenetic iron disulfides, or sulfur associated with petroliferous materials (e.g. Skinner 1967). In the second model in Table 1 (the sulfate reduction model), it is suggested that base metals were transported in fluids together with sulfate, rather than sulfide, and that precipitation occurred because of a reaction during which sulfate was reduced to sulfide, possibly by oxidation of organic matter or methane (e.g. Anderson 1983, Barton 1967, Macqueen & Powell 1983). According to the third model in Table 1 (the reduced-sulfur model), base metals and reduced sulfur were transported together in the same fluid, with the cause of precipitation being pH changes, cooling, dilution, or some combination of these (e.g. Anderson 1973, Helgeson 1970, Sverjensky 1981).

Since the review of the basinal brine model by Ohle (1980), substantial amounts of work have been reported, resulting in significant progress in identifying which of the three general models summarized above and in Table 1 applies to specific Mississippi Valley–type districts. In addition,

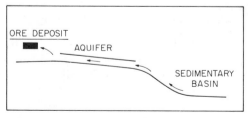

Figure 2   Schematic depiction of the migration of pore waters out of a sedimentary basin and along an aquifer to a site of ore formation (from Sverjensky 1984).

Table 1   Geochemical models for the transportation and precipitation of base metals and sulfur in Mississippi Valley–type deposits[a]

Transportation of ore-forming constituents		Precipitation mechanism	
I.	*Mixing models* Base metals transported by fluids without significant sulfur contents	(a)	Mixing with fluids containing H_2S
		(b)	Replacement of diagenetic iron disulfides
		(c)	Thermal degradation of organic compounds releasing sulfur
II.	*Sulfate reduction models* Base metals transported together with sulfate in the same solutions	(a)	Reduction of sulfate by reactions with organic matter or methane
III.	*Reduced-sulfur model* Base metals transported together with reduced sulfur in the same solutions	(a)	Changes of pH
		(b)	Dilution
		(c)	Decrease of temperature

[a] After Sverjensky (1981).

answers to some of the five questions summarized above have started to appear. In this review, I focus on work reported since the late 1970s that has helped to answer these questions and thereby define the applicability of the basinal brine hypothesis in more detail. During the last 10 years, the most intensively studied Mississippi Valley–type districts have been the Upper Mississippi Valley and southeast Missouri districts (see Figure 1) and the Pine Point district (Northwest Territories, Canada). Consequently, this paper is strongly oriented toward a comparison of the characteristics of these three districts and their genesis.

GEOLOGY AND GEOCHEMISTRY OF THE DEPOSITS

Petrology

Studies of parts of the Upper Mississippi Valley district (Figure 1) by McLimans & Barnes (1975), McLimans (1977), and McLimans et al (1980) have revealed important new features of the sulfide mineral paragenesis, the characteristics of the fluid inclusions, and additional sulfur isotopic data. Firstly, the deposition of sphalerite in the Upper Mississippi Valley has been divided into three stages (A, B, and C), each of which is composed of a number of distinctive color bands (McLimans et al 1980). The middle stage (B) contains a high proportion of dark, almost opaque sphalerite that contains fluid inclusions with homogenization temperatures as high as

220°C. McLimans & Barnes (1975) were the first to report such high fluid inclusion homogenization temperatures from a Mississippi Valley–type deposit. Similar temperatures had been calculated from the differences between the sulfur isotopic compositions of the sphalerite and coprecipitated galena found in the sulfur isotopic study by Pinckney & Rafter (1972) and in the reinterpretations of their data by Heyl et al (1974). In fact, the Pinckney & Rafter (1972) data for coprecipitated sphalerite and galena are consistent with precipitation from a solution with a constant $\delta^{34}S_{H_2S} \simeq +14°/_{oo}$ over a wide temperature range (D. M. Rye, personal communication). Secondly, McLimans et al (1980) demonstrated that individual bands in the sphalerite could be correlated over distances of hundreds of meters within an orebody, and that certain bands could be correlated many kilometers across the entire district. Finally, it was also demonstrated that significant dissolution of sphalerite took place during the three stages of overall deposition (i.e. sphalerite was precipitated and dissolved repetitively). The latter two features of the Upper Mississippi Valley district (the correlation of the color bands across the district, and the repetitive precipitation and dissolution of the sphalerite) have been interpreted by McLimans et al (1980) to be more consistent with a model for transport and precipitation involving a fluid containing both metals and sulfur (models II or III in Table 1) than with a model calling for mixing of metal-rich and sulfur-rich solutions (model I in Table 1).

Recent studies of the southeast Missouri lead district (Figure 1) have also clearly demonstrated that repetitive precipitation and dissolution of sulfides, particularly galena, occurred throughout the Viburnum Trend (Clendenin 1977, Grundmann 1977, Hagni & Trancynger 1977, Paarlberg & Evans 1977, Sverjensky 1981). In many respects, southeast Missouri is significantly different from the Upper Mississippi Valley district. Southeast Missouri is lead rich (with an atomic Zn/Pb ratio of about 0.2), the style of mineralization is dominated by massive replacement, and the sphalerite does not contain color bands that can be correlated across the district. In contrast, the Upper Mississippi Valley is zinc rich (with an atomic Zn/Pb ratio of about 40), the style of mineralization is dominated by open-space filling of joints, fractures, and faults, and the sphalerite contains the color banding described above. Important isotopic differences are reviewed below. However, in both districts the bulk of the sphalerite in the paragenesis was deposited before the bulk of the galena, and evidence of repetitive precipitation and dissolution of the most abundant base-metal sulfides is ubiquitous.

It can be argued that the observation of repetitive precipitation and dissolution of sulfides such as sphalerite or galena strongly supports a reduced-sulfur model (type III in Table 1) rather than either the sulfate reduction or mixing models (types I and II in Table 1) (Barnes 1983, Giordano & Barnes 1981, Sverjensky 1981). Neither the sulfate reduction model nor the mixing model can easily account for the dissolution of the sulfides at the site of ore deposition, whereas a reduced-sulfur model can (Sverjensky 1981). Consequently, it appears that some version of the reduced-sulfur model is most applicable to the southeast Missouri and the Upper Mississippi Valley districts. Specific mechanisms of transport and precipitation of the metals and the reduced sulfur are discussed below.

The choice of a reduced-sulfur model for two of the classic Mississippi Valley–type deposits does not imply that all Mississippi Valley–type deposits formed in this way. For example, in the Pine Point district there are no reports of textural evidence for repetitive precipitation and dissolution of the sphalerite and galena (Beales 1975, Kyle 1980, 1981, Roedder 1968a,b). The studies of Pine Point demonstrate that the typical ore textures are dominated by colloform banding of extremely fine-grained sphalerite and galena, quite different to the ore textures of southeast Missouri and the Upper Mississippi Valley. Either a mixing model or a sulfate reduction model (types I and II in Table 1) may be the most appropriate to explain the origin of the mineralization at Pine Point, as advocated by Anderson (1975, 1983), Beales & Jackson (1966), Kyle (1980, 1981), Macqueen & Powell (1983), Powell & Macqueen (1984), and Rhodes et al (1984).

Wall-Rock Alteration

Wall-rock alteration assemblages are useful in constraining the pH and other details of the aqueous solution chemistry of ore-forming fluids (e.g. Meyer & Hemley 1967). In the case of Mississippi Valley–type deposits, the pH is particularly controversial (Anderson 1973, 1975, 1977, Barnes 1967, 1979, Giordano 1985, Giordano & Barnes 1981, Helgeson 1970, Sverjensky 1981, 1984) because certain models for the transportation of metals (e.g. type III in Table 1) are only applicable under specific ranges of pH. It is therefore critical to gather all possible evidence of wall-rock alteration in Mississippi Valley–type deposits.

Dissolution, recrystallization, and brecciation of the host carbonate rocks in association with mineralization occur in all Mississippi Valley–type deposits (e.g. Snyder 1968, Ohle 1959, 1980) and, together with varied proportions of silicification and dolomitization, are the most obvious examples of wall-rock alteration in these deposits. Unfortunately, these observations cannot be used, at present, to place significant constraints on the chemistry of the ore-forming fluids. Reactions such as

$$CaCO_3 + 2H^+ \rightarrow Ca^{2+} + CO_{2(aq)} + H_2O$$

do not constrain the pH of the solution unless the CO_2 contents are also well known.

What is needed is evidence of the effect(s) of Mississippi Valley–type ore-forming solutions on rocks other than limestones or dolostones. To date, only a few studies have attempted to address this need. The most widely quoted is that by Heyl et al (1964), who prepared X-ray powder diffraction patterns of the acid-insoluble residue of the Quimbys Mill Shale Member in the Upper Mississippi Valley district. Samples away from ore zones showed evidence of Md illite and K-feldspar, whereas samples in the ore zones showed evidence of $2M_1$ illite and K-feldspar. These results have been interpreted as demonstrating that the fluids transporting ore-forming constituents in the veins of the Upper Mississippi Valley district had ratios of the activity of potassium ion to hydrogen ion (a_{K^+}/a_{H^+}) consistent with the assemblage K-feldspar–muscovite–quartz (Barnes 1979, 1983, Giordano & Barnes 1981, Giordano 1985). However, the results of Heyl et al's study may only indicate the a_{K^+}/a_{H^+} ratio of reacted fluid in the wall-rock of the veins in the Upper Mississippi Valley district (Sverjensky 1981). In a similar study to that of Heyl et al (1964), Daniel & Hood (1975) demonstrated a minor increase in the degree of crystallinity of 10-Å layer silicates in the clay-size fraction of wall-rock in the vicinity of ore in part of the Kentucky-Illinois fluorite district (Figure 1).

The only direct observations of silicate wall-rock alteration associated with Mississippi Valley–type deposits are from the southeast Missouri district. Ohle (1952), Snyder & Gerdemann (1968), and Tarr (1936) all noted that sulfide mineralization occurring in or immediately adjacent to detrital pebbles and boulders of Precambrian igneous basement in the Upper Cambrian Bonneterre Formation in the Old Lead Belt of southeast Missouri was associated with silicification, pyritization, and in some cases kaolinitization of the feldspar-bearing fragments of the basement. Recent studies by Stormo & Sverjensky (1983) of similar occurrences of sulfide mineralization rimming detrital fragments of Precambrian granite in the Viburnum Trend of southeast Missouri document alteration of K-feldspar and albite to kaolin, quartz, and minor sericite. The abundance of kaolin strongly suggests low a_{Na^+}/a_{H^+} and a_{K^+}/a_{H^+} ratios in the ore-forming fluids of the southeast Missouri district. Combination of the results of such studies with fluid inclusion data for Na^+ and K^+ concentrations should provide unambiguous evidence of the pH of the ore-forming solutions at the site of ore deposition.

Metal Ratios

The ratios of copper, lead, and zinc in Mississippi Valley–type deposits are remarkably restricted (Figure 3). With the exception of the southeast

Missouri district, Mississippi Valley–type deposits tend to be Zn rich with extremely low Cu contents (Bjorlykke & Sangster 1981, Gustafson & Williams 1981, Sangster 1983, Sverjensky 1984). The southeast Missouri district appears to be anomalously Pb rich. It is also the only district with significant amounts of mineralization in sandstone, and it is likely that the Upper Cambrian Lamotte Sandstone was an important aquifer during the formation of the district (Doe & Delevaux 1972). In these regards, the Southeast Missouri district shares characteristics with a separate class of stratabound base-metal sulfide ores shown in Figure 3—namely, the sandstone-hosted lead deposits (e.g. Laisvall, Sweden; and Largentière, France)—as noted by Bjorlykke & Sangster (1981), Sangster (1983), and Sverjensky (1984). The sandstone-hosted lead deposits are characterized by Zn/Pb ratios much lower than Mississippi Valley–type deposits and by low Cu contents similar to the Mississippi Valley–type deposits. The third important class of stratabound base-metal sulfide ore deposits shown in Figure 3, the red-bed copper deposits, have distinctive Cu/(Pb + Zn) ratios higher than those of either the Mississippi Valley–type deposits or the sandstone-hosted lead deposits.

Figure 3 suggests that there is a very strong association of lithologies with base-metal ratios for the three classes of stratabound sediment-hosted base-metal sulfide ore deposits depicted (Bjorlykke & Sangster 1981, Sangster 1983). Sverjensky (1984) has suggested that the association exists between the lithologies of the aquifer units and the metal ratios of the associated ores: Grey or white sandstone aquifers are associated with lead-rich deposits (the classic sandstone lead deposits), and carbonate aquifers are associated with zinc-rich deposits (the classic Mississippi Valley–type deposits). This suggestion can be extended to include the red-bed copper deposits because they have such distinctive Cu/(Pb + Zn) ratios (Figure 3) and distinctive aquifer lithologies. The copper mineralization in these deposits occurs in chemically reducing portions of the red-beds or in

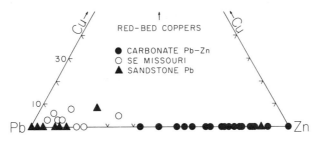

Figure 3 Relative abundances of lead, zinc, and copper in stratabound base-metal sulfide ores (after Sverjensky 1984).

adjacent organic-rich or pyrite-rich units, but the major aquifers associated with formation of the copper deposits appear to be the red-beds themselves (Bjorlykke & Sangster 1981, Rose 1976). According to this hypothesis, the southeast Missouri district is lead rich rather than zinc rich because of all the Mississippi Valley–type deposits, it is the most closely associated with a sandstone aquifer unit.

The association of the lithologies of the aquifer units and the metal ratios of the associated ores might be expected if the different aquifer lithologies have different effects on the chemistry of the ore-forming fluids that pass through them (Sverjensky 1984). For example, in carbonate aquifer units it would be expected that the fluid chemistry might easily be maintained near saturation with respect to sphalerite, galena, and chalcopyrite, because the availability of small quantities of metal and sulfide from such an aquifer and the possibility of the fluid reacting with trace quantities of detrital silicates will be enhanced by the retrograde solubility of the carbonates. Any cooling of the fluids in the pore spaces of the aquifer will tend to cause dissolution of carbonates and expose new sources of metals or sulfide and silicate minerals to the pore fluid. If, by these means, the state of saturation of the fluid is maintained near that for simultaneous saturation with respect to sphalerite, galena, and chalcopyrite, the relative abundances of base metals in the fluid will be $Zn > Pb \gg Cu$, and the overall Zn/Pb (atomic) ratio will be about 50 (Sverjensky 1984). From a fluid such as this, zinc-rich, copper-poor ores will be precipitated.

In contrast, in a quartz-rich sandstone that contains trace amounts of pyrite, without significant hematite, it may be possible for a potential ore-forming fluid to remain undersaturated with respect to sphalerite while saturated with respect to galena and chalcopyrite. The availability of detrital or diagenetic phases for reaction with the potential ore-forming fluid is likely to be minimal, compared with the case of the carbonate aquifer, because small amounts of cooling of the fluid in the aquifer will tend to cause precipitation of protective coatings of quartz or amorphous silica. From a fluid such as this (undersaturated with respect to sphalerite), lead-rich, copper-poor ores will be precipitated (Sverjensky 1984).

In a red-bed aquifer unit, it is likely that progressive reaction of a potential ore-forming fluid with hematite would result in gross undersaturation of the fluid with respect to sphalerite and galena, even though the fluid may be maintained near saturation with respect to chalcopyrite. Consequently, on encountering a reducing environment that causes precipitation of sulfides, only minute amounts of sphalerite and galena may be precipitated compared with copper-iron sulfides.

The relationships between aquifer lithologies, the states of saturation of

Table 2 Relationships between aquifer lithologies, states of saturation of migrating fluids, and metal abundances in resultant ores

| | Aquifer lithology | | |
	Carbonate	Quartz sandstone	Red-bed
Migrating fluid :[a]			
undersaturated	–	sp	sp, gn
saturated	sp, gn, cp, py	gn, cp, py	cp, py
Metal abundances in resultant ores	Zn > Pb ≫ Cu	Pb > Zn ≫ Cu	Cu > Pb, Zn
Stratabound ore examples	Most Mississippi Valley–type ores	Sandstone-Pb; southeast Missouri	Red-bed coppers

[a] State of saturation of fluid with respect to sphalerite (sp), galena (gn), chalcopyrite (cp), and pyrite (py).

the migrating fluids, and the base-metal ratios in the associated ores discussed above are summarized in Table 2. It should be emphasized that the relationships are sufficiently general that they are independent of specific models for metal transport and precipitation.

Lead and Sulfur Isotopic Geochemistry

The results of isotopic investigations of Mississippi Valley–type deposits carried out since the review by Heyl et al (1974) indicate that each district investigated in detail appears to have its own distinctive lead and sulfur isotopic characteristics.

Sulfur isotopic studies of the Upper Mississippi Valley district reported by Pinckney & Rafter (1972) and McLimans (1977) are consistent with precipitation of sphalerite and galena from fluids with a constant $\delta^{34}S_{H_2S}$ value of $+14°/_{oo}$ over a wide range of temperatures (as described above). Lead isotopic data for the same district vary widely, but they appear to show a spatial zonation pattern that becomes more radiogenic toward the Wisconsin Arch (see summaries in Heyl et al 1974).

At Pine Point, the average sulfur isotopic compositions of sphalerite and galena (Sasaki & Krouse 1969) and the fluid inclusion data (Roedder 1968b) suggest that the typical $\delta^{34}S_{H_2S}$ values of the fluids from which the sphalerite and galena were precipitated were approximately $22(\pm1)°/_{oo}$ and $23(\pm1)°/_{oo}$, respectively, if it is assumed that the equilibrium fractionations for sphalerite-H_2S and galena-H_2S pairs were $0(\pm1)$ and $-5(\pm1)°/_{oo}$ (Ohmoto 1972), respectively. Because of the similarity between the sulfur isotopic compositions of the sphalerite and galena and those of the anhydrite from Devonian evaporites in the adjacent Elk Point basin

188   SVERJENSKY

(19°/$_{oo}$; Sasaki & Krouse 1969), it has been suggested that reduction of sulfate derived from the evaporites provided the source of reduced sulfur for formation of the Pine Point orebodies (Macqueen & Powell 1983, Sasaki & Krouse 1969). If such a reduction process did take place, it must have been abiologic, because the distribution of isotopic compositions at Pine Point is unlike those resulting from any known biologically mediated sulfate reduction process. The details of the proposed reduction process are not yet well understood. Powell & Macqueen (1984) have suggested that abiologic reduction of evaporitic sulfate by H_2S produced native sulfur that reacted with sulfur-bearing bitumen to yield altered bitumens and additional H_2S, which caused precipitation of the sulfide ores at Pine Point. Such a model is suggested by the abundance of evaporites in adjacent portions of the stratigraphic section and by the occurrences of native sulfur and abundant bitumen and oil in the Pine Point orebodies. However, any proposed abiologic sulfate reduction process must be able to explain why most of the sulfur isotopic data for sulfides at Pine Point are significantly heavier than the evaporitic $\delta^{34}S$ value of 19°/$_{oo}$, as well as why the isotopic compositions of H_2S in equilibrium with sphalerite or galena (calculated above) appear to be about 22–23°/$_{oo}$.

The uniformity of the calculated values of $\delta^{34}S_{H_2S}$ for Pine Point is paralleled by the uniform lead isotopic compositions of galena from Pine Point. The latter display a very small range of values, which are not radiogenic (Cumming & Robertson 1969). Together, the lead and sulfur isotopic characteristics of the Pine Point ore-forming fluids appear to be consistent with derivation from single sources or well-homogenized multiple sources.

In marked contrast to the uniform variations of the $\delta^{34}S$ values of sulfide minerals in the Upper Mississippi Valley district and the essentially constant $\delta^{34}S$ values of sulfide minerals in the Pine Point deposit, the sulfur isotopic compositions of galena from the southeast Missouri district have a huge range of values (from about 0 to 25°/$_{oo}$), which correlate with the range of lead isotopic compositions measured (Sverjensky et al 1979, Sverjensky 1981). The correlation between the lead and sulfur isotopic compositions of galena from the southeast Missouri district strongly suggests that both the lead and sulfur isotopic compositions of the ore-forming fluids varied widely, which in turn suggests multiple sources of lead and sulfur. In addition, individual crystals of galena contain huge variations and complex lead isotopic zonation patterns (Hart et al 1981, 1983). These contribute to significant scatter in the overall correlation of the lead and sulfur isotopes and possibly reflect acquisition by the ore-forming solutions of small quantities of isotopically unusual lead from sources in or near the orebodies, such as detrital Precambrian feldspars or zircons.

TRANSPORT AND PRECIPITATION OF ORE-FORMING CONSTITUENTS

Table 1 shows that choosing one of the three models for transportation of metals and sulfur results in a limited number of choices for precipitation mechanisms. We have seen above that recent studies of the Upper Mississippi Valley and southeast Missouri districts support the choice of a reduced-sulfur model, whereas studies of Pine Point suggest a mixing model or sulfate reduction model. In the latter two cases, which involve fluids transporting base metals and negligible quantities of reduced sulfur, there is little doubt that the specific base metal–bearing species transported are metal-chloride complexes (e.g. Anderson 1975, 1977, 1983, White 1981). However, in the case of the reduced-sulfur model, the question of whether or not ore-forming concentrations of metal can be transported as chloride complexes in the presence of reduced sulfur is still controversial (Anderson 1973, 1975, 1977, 1983, Barnes 1979, Barrett & Anderson 1982, Giordano 1985, Giordano & Barnes 1981, Sverjensky 1981, 1984). Answers to this question for specific Mississippi Valley–type deposits are important, because otherwise it is unlikely that specific mechanisms of ore precipitation can be identified.

Metal-Chloride Complexing

At temperatures greater than about 200°C, chloride complexes of the base metals transport sufficient metal in the presence of dissolved H_2S and/or HS^- to form ore deposits (Barrett & Anderson 1982, Crerar & Barnes 1976, Helgeson 1964, 1969). The controversy refers to whether or not the solubilities of sphalerite and galena are sufficient at temperatures less than about 150°C and over specific pH ranges.

Experimental measurements of the solubilities of sphalerite and galena in 1–3 molal NaCl solutions at temperatures to 95°C under conditions of H_2S saturation were reported by Barrett & Anderson (1982). Calculated solubilities using stability constants of the metal-chloride complexes from Helgeson (1969) agree quite well with those measured. In 3 molal NaCl solutions at 100°C with pHs of 4.5, the solubilities of sphalerite and galena (calculated from Table 6 of Barrett & Anderson 1982) are $10^{-5.0}$ (0.5-ppm Zn) and $10^{-5.7}$ (0.4-ppm Pb), respectively, with equivalent concentrations of H_2S. At 150°C the corresponding values are 3-ppm Zn and 1.6-ppm Pb. In solutions with higher NaCl contents (e.g. 3–5 molal for the Upper Mississippi Valley district; McLimans 1977), significantly higher concentrations of metals and reduced sulfur could be transported (e.g. Sverjensky 1981). Clearly, under the conditions described above, it is possible to transport quantities of Zn or Pb of the order of about 1 ppm.

According to Barnes (1979), Giordano & Barnes (1981), and Giordano (1985), concentrations of Zn, Pb, and H_2S such as these are too low to constitute "ore-forming concentrations." However, there is no well-established lower limit to what constitutes an "ore-forming concentration" (Sverjensky 1981, 1984). Obviously, if smaller amounts of metal are transported, larger volumes of fluid and longer amounts of time for the formation of large districts are required. Recent quantitative modeling of fluid flow during the formation of Mississippi Valley–type ores (reviewed below) supports the possibility of very large quantities of fluid and geologically significant periods of time (of the order of 5 Myr or longer) for the formation of large districts (Cathles & Smith 1983, Garven & Freeze 1984a,b, Garven 1985). The latter values are consistent with concentrations of the order of 1-ppm metal (or less) arriving at the site of ore formation.

The most crucial issue involving the applicability of the chloride-complexing reduced-sulfur model to any Mississippi Valley–type district is that pH values significantly less than neutral are required for the ore-forming fluids. For example, a pH of 4.5 is approximately 1.5 units less than neutral pH in the temperature range 100–150°C. Sverjensky (1984) has demonstrated that fluids with similar pH occur in some oil fields, if it is assumed that the brines analyzed are saturated with respect to reservoir mineral assemblages such as kaolinite-muscovite-quartz. The adoption of acidic pH values for Mississippi Valley–type ore-forming fluids has been criticized by Anderson (1975, 1977, 1983), Barrett & Anderson (1982), and Beales (1975), who have maintained that such pHs are inconsistent with equilibrium between the ore-forming fluids and carbonate units because partial pressures of CO_2 greater than 1 atm would be required. However, the CO_2 contents of Mississippi Valley–type fluids are only known for one district. Total dissolved CO_2 contents of about 0.1 molal were reported from the Upper Mississippi Valley district by McLimans (1977), which are consistent with partial pressures of CO_2 of the order of 20 atm and with pH values significantly less than neutral (Sverjensky 1981). The CO_2 contents of fluid inclusions in the Pine Point and southeast Missouri districts are still unknown.

Values of pH less than neutral have also been rejected by Barnes (1979, 1983) and Giordano & Barnes (1981), who have interpreted observations by Heyl et al (1964) as indicating that the assemblage K-feldspar–muscovite–quartz controlled the a_{K^+}/a_{H^+} ratios of the ore-forming fluids in the Upper Mississippi Valley district. Combining this with fluid inclusion data indicates pH values for the ore-forming solutions that are neutral to alkaline. However, the observations in the paper by Heyl et al (1964) are of questionable relevance for the purpose of inferring a pH for the ore-forming fluids in the Upper Mississippi Valley district (as discussed above). There is no petrographic evidence that any Mississippi Valley–type ore-forming fluids were in equilibrium with the assemblage K-feldspar–muscovite–quartz. At present, the most promising district for obtaining evidence of silicate hydrothermal alteration assemblages through petrographic studies appears to be the southeast Missouri district (Stormo & Sverjensky 1983).

Clearly, unequivocal evidence of the pH of the ore-forming fluids in any single Mississippi Valley–type district is not yet available. The acidic pHs required by a reduced-sulfur model involving metal-chloride complexing appear to be consistent with equilibrium with respect to carbonates according to available fluid inclusion data, are representative of at least some present-day oil-field brines, and permit transport of the order of 1-ppm metals and H_2S to sites of ore formation.

Alternatives to Metal-Chloride Complexing

One alternative to metal-chloride complexing in Mississippi Valley–type ore-forming fluids that has been emphasized by Barnes (1967, 1979) and Barnes & Czamanske (1967) is metal-bisulfide complexing. Based on experimental studies of lead-bisulfide complexing, Giordano & Barnes (1979) and Hamann & Anderson (1978) concluded that ore-forming solutions at temperatures less than about 200°C and with total dissolved sulfur contents less than about 1 molal (32,000 ppm) cannot transport significant quantities of lead as bisulfide complexes. Similar conclusions were drawn by Barnes (1983) and Bourcier (1983) in the case of zinc-bisulfide complexes. Consequently, the metal-bisulfide complexing theory appears to be inadequate for Mississippi Valley–type ore-forming fluids.

A second alternative to metal-chloride complexing has recently been suggested, according to which metals such as zinc or lead are transported as organometallic complexes (Barnes 1979, 1983, Giordano & Barnes 1981, Giordano 1985). That organic ligands exist in concentrations up to about 0.1 molal in present-day oil-field brines has been demonstrated by Carothers & Kharaka (1978). However, the predominant organic ligand detected to date in brines with temperatures ranging from 100–150°C is acetate (with minor amounts of other short-chain aliphatic acid anions), which does not form strong enough complexes to compete effectively with the overwhelmingly abundant chloride ion. Giordano (1985) has demonstrated that extraordinarily strong organometallic complexes would be required because of the very low abundances of dissolved organic ligands in fluid inclusions and oil-field brines relative to the chloride ion; in addition, the propensity of dissolved organic ligands to form complexes with the abundant cations in solution, such as Na^+ and Ca^{2+}, effectively decreases even more the amount of organic ligand available to compete with chloride for trace elements such as Zn^{2+} and Pb^{2+}. As Giordano

(1985) has emphasized, there are at present no organic ligands known that could compete effectively with chloride under the conditions described above.

In summary, neither the metal-bisulfide nor the organometallic complexing theories appear to be viable alternatives to metal-chloride complexing for Mississippi Valley–type ore-forming fluids that transport both metals and reduced sulfur. Consequently, the metal-chloride complexing version of the reduced-sulfur model is presently the most appropriate for districts such as the Upper Mississippi Valley and southeast Missouri.

DYNAMICS OF FLUID FLOW

Mechanisms for fluid flow, the identities of the fluid pathways during the migration of fluids to sites of Mississippi Valley–type ore formation, the duration of ore formation, and the total quantities of fluid involved have all been modeled in a quantitative fashion recently. An important constraint on any model involving the migration of Mississippi Valley–type ore-forming fluids out of sedimentary basins is that the fluids are able to reach the sites of ore formation at relatively shallow depths and at temperatures consistent with those established by fluid inclusion studies. This constraint has enabled significant progress to be made in assessing specific mechanisms for fluid migration out of sedimentary basins.

Three general mechanisms for fluid migration out of sedimentary basins, in the absence of igneous activity (Hanor 1979), are (a) steady compaction of sediments containing fluids under essentially hydrostatic pressures, (b) episodic release of fluids at pressures approaching lithostatic pressures, and (c) gravity-driven fluid migration as a consequence of differences in hydrostatic head. In case (a), normal rates of basin subsidence and compaction will result in flow rates far too low to transport 100°C temperatures into a near-surface (less than 1 km depth) environment (Bethke 1983, Cathles & Smith 1983). However, mechanisms (b) and (c) appear to be capable of transporting elevated temperatures into near-surface environments. Although these mechanisms have been modeled separately, and are discussed separately below, they are by no means mutually exclusive alternatives for explaining fluid flow associated with Mississippi Valley–type ore deposits. Garven & Freeze (1984a,b) and Hanor (1979) have emphasized that development of fluids under pressures approaching lithostatic pressures is probably most important in the relatively early stages of the development of a deep sedimentary basin, whereas basin-wide gravity-driven fluid migration may predominate in more mature sedimentary basins with regional topographic gradients resulting from tectonic uplift and erosion.

Compaction of Sediments Containing Lithostatically Pressured Fluids

Sharp (1978) and Cathles & Smith (1983) have presented calculations of the extent of development of anomalously pressured fluids in subsiding sedimentary basins and the consequences of episodically rupturing the sediments containing these fluids. If we assume that the fluids drain into permeable aquifer units and that fluid migration after each episode of rupture is so rapid that significant heat is not lost from the aquifer units, a large number of individual pulses are postulated to have formed Mississippi Valley–type deposits. For example, Cathles & Smith (1983) suggested that about 10 to 50 dewatering pulses might occur, at a rate of about 1 Myr^{-1}, with each pulse lasting only about 10,000 yr. Consequently, they envision that the total period of time corresponding to expulsion of mineralizing fluids from a compacting basin might be of the order of tens of millions of years, even though individual periods of ore deposition lasted only a few thousandths of this total period.

Despite the advances made in quantitative modeling of the development of anomalously pressured fluids in sedimentary basins, it is still not well established as to how the sediments containing these fluids might be ruptured, or as to how the fluids drain down into the basal aquifers. Sharp (1978) and Cathles & Smith (1983) have suggested that these processes take place during faulting.

Gravity-Driven Fluid Migration

A model for the analysis of the flow patterns of potential ore-forming solutions driven by differences in hydraulic head across a sedimentary basin has been developed by Garven & Freeze (1984a,b) and applied to the Pine Point district by Garven (1985). Figure 4 (from Garven & Freeze 1984a) shows a schematic cross section through a basin in which the water table is assumed to approximately replicate the basin topography. It is also assumed that regional deformation and erosion of the basin have produced a gentle regional topographic gradient of about 2 m km^{-1}, which is sufficient to drive basin-wide fluid movement from recharge areas to updip discharge areas at the thin edge of the basin. Garven & Freeze (1984a,b) have investigated the effects of geologically reasonable salinity gradients in the basin, with varied amounts of heat flow from the basement, together with a range of basin geometries and sedimentary structures (including local anomalies in the configuration of the water table and position of the basement, and a variety of hydraulic conductivities of the sedimentary strata). In addition, mass transport of a single nonreactive chemical species was modeled by calculating the mechanical dispersion and diffusion of the

CONCEPTUAL MODEL

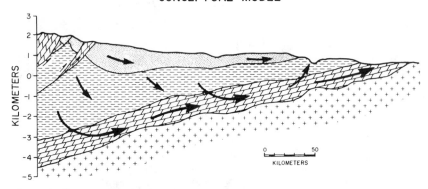

Figure 4   Conceptual model of gravity-driven fluid flow in sedimentary basins (from Garven & Freeze 1984a).

species in the flow system after it was introduced at a particular point in the recharge end of the basin. Garven & Freeze (1984a,b) have demonstrated that fluid flow in a basin with a basal highly permeable (aquifer) unit will be across formations at the elevated recharge end of the basin and will be strongly focused along the aquifer to the thin edge of the basin. Under these conditions, significant perturbations of the geothermal gradient elevate temperatures at the thin edge of the basin to values consistent with the fluid inclusion temperatures from Mississippi Valley–type ores. Because fluid is continuously recharged at one end of the basin and discharged at the other, enormous volumes of fluid could pass through the basin, provided that the gravity-driven flow system is not perturbed significantly. As a consequence, given stable tectonic conditions, these flow systems may persist for geologically significant lengths of time.

Application of such a gravity-driven fluid flow model to the Pine Point district (Garven 1985) suggests that a regional flow system could have developed across the Western Canada basin and along the Keg River Barrier Complex, with discharge in the vicinity of Pine Point at temperatures of 60–100°C at depths of 1–2 km. Calculated flow rates of 1–5 m yr^{-1} and the assumption that 5-ppm zinc sulfide precipitated from the ore-forming fluids imply that the ore deposits at Pine Point could have formed within a time interval of 0.5–5 Myr.

CONCLUDING REMARKS

Taking into account the studies summarized above, most of which were carried out within the last 10 years, it appears that the basinal brine hypothesis of ore formation can be applied to Mississippi Valley–type ores in detail. It is now possible to quantitatively model important aspects of the driving forces and the pathways of fluid migration out of sedimentary basins (e.g. at Pine Point; Garven 1985). The studies of the dynamics of fluid flow out of sedimentary basins summarized above also suggest that time scales of the order of several millions of years or longer may be involved in the formation of sediment-hosted base-metal sulfide ores, one or two orders of magnitude more than base-metal sulfide ores associated with igneous systems (e.g. Cathles 1981). In the gravity-driven flow model, the only limits on the duration of the flow system and the total amounts of fluid involved would appear to be those imposed by tectonic stability. Whether the basinal brines become ore-forming fluids before, during, or after migration is still not well understood, but their potential for chemical evolution during migration is now being addressed and may help to explain the strong association between the metal ratios of mineralized districts and the lithologies of the aquifer units and host rocks.

The regional nature of the fluid flow out of sedimentary basins suggests that multiple sources of the ore-forming constituents are to be expected in many districts. However, the ease with which the isotopic compositions of fluids can be modified during migration depends in part on the concentrations of the metals and sulfur in the fluids, which in turn depend on the mechanisms of transport of the ore-forming constituents. Complexing of base metals by bisulfide or organic species both appear unlikely mechanisms of transport in Mississippi Valley–type fluids. The long-standing hypothesis of metal-chloride complexing is the most likely means of metal transport in these fluids.

During the last 10 years, the most intensively studied Mississippi Valley–type districts have been the Upper Mississippi Valley, the southeast Missouri, and the Pine Point districts. The results of these studies, combined with earlier results, show that a number of significant differences exist between these districts (summarized in Table 3). The three districts are classic Mississippi Valley–type deposits, to which some form of the basinal brine hypothesis is applicable. However, the three districts may not have formed in exactly the same way. For example, as discussed earlier, at Pine Point the abundance of fine-grained sulfide mineralization (and associated oil, bitumen, and native sulfur), the lack of evidence for repetitive precipitation and dissolution of the sulfides, the uniformity of the lead and sulfur isotopic compositions of the fluids, and the low temperatures (typically about 80°C) are all consistent with a model for ore formation involving reduction of sulfate and precipitation of metal sulfides from basinal brines containing base metals transported as metal-chloride complexes. In the southeast Missouri district, however, the relatively low abundance of petroliferous materials or other organic matter, the absence of native sulfur, the abundant evidence of repetitive precipitation and

Table 3 Summary of the characteristics of three Mississippi Valley–type districts

	Pine Point[a]	Upper Mississippi Valley[b]	Southeast Missouri[c]
Typical temperatures (°C)	80	125	95 and 120
Sulfide textures	Abundant colloform	Abundant coarse banding of sphalerite	Minor colloform, abundant massive replacement
Evidence for precipitation/ dissolution of sulfides	None	Common in sphalerite	Common in both sphalerite and galena
Association of ore with organic matter	Abundant oil, bitumen	Some	Uncommon
Sulfur isotopic composition of H_2S in fluid (°/oo)	Constant (22–23)	Constant (14)	Highly varied (5–26)
Lead isotopic composition of fluid	Constant	Highly varied	Highly varied

[a] Kyle 1980, 1981, Roedder 1968a,b, Sasaki & Krouse 1969, Powell & Macqueen 1984.
[b] Barnes 1983, McLimans 1977, D. M. Rye, personal communication, Heyl et al 1974.
[c] Hagni 1983, Roedder 1977, Sverjensky 1981.

dissolution of sphalerite and galena, the highly varied nature of the lead and sulfur isotopic compositions of the fluids, and the higher temperatures are more consistent with a model for ore formation involving precipitation of small amounts of metal sulfide (about 1 ppm) from basinal brines that transported base metals as metal-chloride complexes together with H_2S under conditions of acidic pH and high total dissolved CO_2 concentrations.

ACKNOWLEDGMENTS

The preparation of this paper was supported in part by NSF Grant EAR-8419418. I wish to thank Grant Garven for reading the manuscript and offering helpful comments, and for many stimulating conversations about Mississippi Valley–type deposits.

Literature Cited

Anderson, G. M. 1973. The hydrothermal transport and deposition of galena and sphalerite near 100°C. Econ. Geol. 68:480–92

Anderson, G. M. 1975. Precipitation of Mississippi Valley–type ores. Econ. Geol. 70:937–42

Anderson, G. M. 1977. Thermodynamics and

sulfide solubilities. In *Short Course in Application of Thermodynamics to Petrology and Ore Deposits*, ed. H. J. Greenwood, pp. 136–50. Vancouver: Mineral. Assoc. Can. 231 pp.

Anderson, G. M. 1983. Some geochemical aspects of sulfide precipitation in carbonate rocks. See Kisvarsanyi et al 1983, pp. 61–76

Barnes, H. L. 1967. Sphalerite solubility in ore solutions of the Illinois-Wisconsin district. See Brown 1967, pp. 326–32

Barnes, H. L. 1979. Solubilities of ore minerals. In *Geochemistry of Hydrothermal Ore Deposits*, ed. H. L. Barnes, pp. 404–60. New York: Wiley-Interscience. 787 pp. 2nd ed.

Barnes, H. L. 1983. Ore-depositing reactions in Mississippi Valley–type deposits. See Kisvarsanyi et al 1983, pp. 77–85

Barnes, H. L., Czamanske, G. K. 1967. Solubilities and transport of ore minerals. In *Geochemistry of Hydrothermal Ore Deposits*, ed. H. L. Barnes, pp. 334–81. New York: Holt, Rinehart & Winston. 670 pp. 1st ed.

Barrett, T. J., Anderson, G. M. 1982. The solubility of sphalerite and galena in NaCl brines. *Econ. Geol.* 77:1923–33

Barton, P. B. Jr. 1967. Possible role of organic matter in the precipitation of the Mississippi Valley ores. See Brown 1967, pp. 371–78

Bastin, E. S., ed. 1939. Contributions to a knowledge of the lead and zinc deposits of the Mississippi Valley region. *Geol. Soc. Am. Spec. Pap. No. 24.* 156 pp.

Beales, F. W. 1975. Precipitation mechanisms for Mississippi Valley–type ore deposits. *Econ. Geol.* 70:943–48

Beales, F. W., Jackson, S. A. 1966. Precipitation of lead-zinc ores in carbonate reservoirs as illustrated by Pine Point ore field, Canada. *Inst. Min. Metall. Trans. Sect. B* 75:278–85

Bethke, C. M. 1983. Fluid flow and heat transport in compacting sedimentary basins. *Geol. Soc. Amer. Abstr. with Programs* 15:526

Bjorlykke, A., Sangster, D. F. 1981. An overview of sandstone lead deposits and their relationships to red-bed copper and carbonate-hosted lead-zinc deposits. See Skinner 1981, pp. 179–213

Bourcier, W. L. 1983. *Stabilities of chloride and bisulfide complexes of zinc in hydrothermal solutions.* PhD thesis. Pa. State Univ., University Park. 179 pp.

Brown, J. S., ed. 1967. *Genesis of Stratiform Lead-Zinc-Barite-Fluorite Deposits, Monograph 3.* Lancaster, Pa: Econ. Geol. Publ. Co./Lancaster Press. 443 pp.

Carothers, W. W., Kharaka, Y. K. 1978.

Aliphatic acid anions in oil-field waters—implications for origin of natural gas. *Am. Assoc. Pet. Geol. Bull.* 62:2441–53

Carothers, W. W., Kharaka, Y. K. 1980. Stable carbon isotopes of HCO_3^- in oil-field waters—implications for the origin of CO_2. *Geochim. Cosmochim. Acta* 44:323–32

Carpenter, A. B., Trout, M. L., Pickett, E. E. 1974. Preliminary report on the origin and chemical evolution of lead- and zinc-rich oil field brines in central Mississippi. *Econ. Geol.* 69:1191–1206

Cathles, L. M. 1981. Fluid flow and genesis of hydrothermal ore deposits. See Skinner 1981, pp. 424–57

Cathles, L. M., Smith, A. T. 1983. Thermal constraints on the formation of Mississippi Valley–type lead-zinc deposits and their implications for episodic basin dewatering and deposit genesis. *Econ. Geol.* 78:983–1002

Clendenin, C. W. 1977. Suggestions for interpreting Viburnum Trend mineralization based on field studies at Ozark Lead Company, southeast Missouri. *Econ. Geol.* 72:465–73

Crerar, D. A., Barnes, H. L. 1976. Ore solution chemistry V. Solubilities of chalcopyrite and chalcocite assemblages in hydrothermal solution at 200° to 350°C. *Econ. Geol.* 71:772–94

Cumming, G. L., Robertson, D. K. 1969. Isotopic composition of lead from the Pine Point deposit. *Econ. Geol.* 64:731–32

Daniel, M. E., Hood, W. C. 1975. Alteration of shale adjacent to the Knight ore-body, Rosiclare, Illinois. *Econ. Geol.* 70:1062–69

Doe, B. R., Delevaux, M. H. 1972. Source of lead in southeast Missouri galena ores. *Econ. Geol.* 67:409–25

Garven, G. 1985. The role of regional fluid flow in the genesis of the Pine Point deposit, Western Canada sedimentary basin. *Econ. Geol.* 80:307–24

Garven, G., Freeze, R. A. 1984a. Theoretical analysis of the role of groundwater flow in the genesis of stratabound ore deposits: 1. Mathematical and numerical model. *Am. J. Sci.* 284:1085–1124

Garven, G., Freeze, R. A. 1984b. Theoretical analysis of the role of groundwater flow in the genesis of stratabound ore deposits. 2. Quantitative results. *Am. J. Sci.* 284:1125–74

Giordano, T. H. 1985. A preliminary evaluation of organic ligands and metal-organic complexing in Mississippi Valley–type ore solutions. *Econ. Geol.* 80:96–106

Giordano, T. H., Barnes, H. L. 1979. Ore solution chemistry VI. PbS solubility in bisulfide solutions to 300°C. *Econ. Geol.* 74:1637–46

Giordano, T. H., Barnes, H. L. 1981. Lead

198 SVERJENSKY

transport in Mississippi Valley–type ore solutions. *Econ. Geol.* 76 : 2200–11

Grundmann, W. H. Jr. 1977. Geology of the Viburnum No. 27 mine, Viburnum Trend, southeast Missouri. *Econ. Geol.* 72 : 349–64

Gustafson, L. B., Williams, N. 1981. Sediment-hosted stratiform deposits of copper, lead and zinc. See Skinner 1981, pp. 139–78

Hagni, R. D. 1983. Ore microscopy, paragenetic sequence, trace element content, and fluid inclusion studies of the copper-lead-zinc deposits of the southeast Missouri Lead district. See Kisvarsanyi et al 1983, pp. 243–56

Hagni, R. D., Trancynger, T. C. 1977. Sequence of deposition of the ore minerals at the Magmont mine, Viburnum Trend, southeast Missouri. *Econ. Geol.* 72 : 451–64

Hall, W. E., Friedman, I. 1963. Composition of fluid inclusions, Cave-In-Rock fluorite district, Illinois, and Upper Mississippi Valley zinc-lead district. *Econ. Geol.* 58 : 886–911

Hamann, R. J., Anderson, G. M. 1978. Solubility of galena in sulfur-rich NaCl solutions. *Econ. Geol.* 73 : 96–100

Hanor, J. S. 1979. The sedimentary genesis of hydrothermal fluids. In *Geochemistry of Hydrothermal Ore Deposits*, ed. H. L. Barnes, pp. 137–42. New York : Wiley-Interscience. 787 pp. 2nd ed.

Hart, S. R., Shimizu, N., Sverjensky, D. A. 1981. Lead isotope zoning in galena : an ion microprobe study of a galena crystal from the Buick mine, southeast Missouri. *Econ. Geol.* 76 : 1873–78

Hart, S. R., Shimizu, N., Sverjensky, D. A. 1983. Toward an ore fluid lead isotope "stratigraphy" for galenas from the Vivurnum Trend, S.E. Missouri. See Kisvarsanyi et al 1983, pp. 257–70

Helgeson, H. C. 1964. *Complexing and Hydrothermal Ore Deposition.* New York : Macmillan. 128 pp.

Helgeson, H. C. 1969. Thermodynamics of hydrothermal systems at elevated temperatures and pressures. *Am. J. Sci.* 267 : 729–804

Helgeson, H. C. 1970. A chemical and thermodynamic model of ore deposition in hydrothermal systems. In *Mineral. Soc. Am. Spec. Pap., 50th Anniv. Symp.*, ed. B. A. Morgan, 3 : 155–86

Heyl, A. V. 1968. Minor epigenetic, diagenetic, and syngenetic sulfide, fluorite, and barite occurrences in the central United States. *Econ. Geol.* 63 : 585–94

Heyl, A. V., Hosterman, J. W., Brock, M. R. 1964. Clay mineral alteration in the Upper Mississippi Valley zinc-lead district. In *Clays and Clay Minerals, 12th Natl. Conf., Atlanta, 1963,* pp. 445–53. New York : Macmillan

Heyl, A. V., Landis, G. P., Zartman, R. E. 1974. Isotopic evidence for the origin of Mississippi Valley–type mineral deposits : a review. *Econ. Geol.* 69 : 992–1006

Kharaka, Y. K., Lico, M. S., Wright, V. A., Carothers, W. W. 1980. Geochemistry of formation waters from Pleasant Bayou No. 2 well and adjacent areas in coastal Texas. *Proc. Geopressured-Geotherm. Energy Conf., 4th, Austin,* pp. 168–93

Kisvarsanyi, G., Grant, S. K., Pratt, W. P., Koenig, J. W., eds. 1983. *International Conference on Mississippi Valley-Type Zinc-Lead Deposits, Rolla, Mo.* Rolla : Univ. Mo. 603 pp.

Kyle, J. R. 1980. Controls of lead-zinc mineralization, Pine Point district, Northwest Territories. *Min. Eng.* 32 : 1617–26

Kyle, J. R. 1981. Geology of the Pine Point lead-zinc district. In *Handbook of Stratabound and Stratiform Ore Deposits*, ed. K. H. Wolf, 9 : 643–741. New York : Elsevier

Macqueen, R. W., Powell, T. G. 1983. Organic geochemistry of the Pine Point lead-zinc ore field and region, Northwest Territories, Canada. *Econ. Geol.* 78 : 1–25

McLimans, R. K. 1977. *Geologic, fluid inclusion and stable isotope studies of the Upper Mississippi Valley zinc-lead district, southwest Wisconsin.* PhD thesis. Pa. State Univ., University Park. 175 pp.

McLimans, R. K., Barnes, H. L. 1975. Sphalerite stratigraphy in the Upper Mississippi Valley Pb-Zn deposits. *Geol. Soc. Am. Abstr. with Programs* 7 : 1197–98

McLimans, R. K., Barnes, H. L., Ohmoto, H. 1980. Sphalerite stratigraphy of the Upper Mississippi Valley, zinc-lead district, southwest Wisconsin. *Econ. Geol.* 75 : 351–61

Meyer, C., Hemley, J. J. 1967. Wall rock alteration. In *Geochemistry of Hydrothermal Ore Deposits*, ed. H. L. Barnes, pp. 166–235. New York : Holt, Rinehart & Winston. 670 pp. 1st ed.

Ohle, E. L. 1952. Geology of the Hayden Creek lead mine, southeast Missouri. *Am. Inst. Min. Metall. Trans.* 193 : 477–83

Ohle, E. L. 1959. Some considerations in determining the origin of ore deposits of the Mississippi Valley–type—Part I. *Econ. Geol.* 54 : 769–89

Ohle, E. L. 1980. Some considerations in determining the origin of ore deposits of the Mississippi Valley–type—Part II. *Econ. Geol.* 75 : 161–72

Ohmoto, H. 1972. Systematics of sulfur and carbon isotopes in hydrothermal ore deposits. *Econ. Geol.* 67 : 551–78

Paarlberg, N. L., Evans, L. L. 1977. Geology of the Fletcher Mine, Viburnum Trend, southeast Missouri. *Econ. Geol.* 72 : 391–97

Pinckney, D. M., Rafter, T. A. 1972. Fractionation of sulfur isotopes during ore deposition in the Upper Mississippi Valley zinc-lead district. *Econ. Geol.* 67:315–28

Powell, T. G., Macqueen, R. W. 1984. Precipitation of sulfide ores and organic matter: sulfate reactions at Pine Point, Canada. *Science* 224: 63–66

Rhodes, D., Lantos, E. A., Lantos, J. A., Webb, R. J., Owens, D. C. 1984. Pine Point orebodies and their relationship to the stratigraphy, structure, dolomitization, and karstification of the Middle Devonian barrier complex. *Econ. Geol.* 79:991–1055

Roedder, E. 1967. Environment of deposition of stratiform (Mississippi Valley–type) ore deposits, from studies of fluid inclusions. See Brown 1967, pp. 349–62

Roedder, E. 1968a. Temperature, salinity and origin of the ore-forming fluids at Pine Point, Northwest Territories, Canada, from fluid inclusion studies. *Econ. Geol.* 63:439–50

Roedder, E. 1968b. The noncolloidal origin of "colloform" textures in sphalerite ores. *Econ. Geol.* 63:461–71

Roedder, E. 1977. Fluid inclusion studies of ore deposits in the Viburnum Trend, southeast Missouri. *Econ. Geol.* 72:472–79

Rose, A. W. 1976. The effect of aqueous chloride complexes in the origin of red-bed copper and related deposits. *Econ. Geol.* 71:1036–48

Sangster, D. F. 1983. Mississippi Valley–type deposits: a geological mélange. See Kisvarsanyi et al 1983, pp. 7–19

Sasaki, A., Krouse, H. R. 1969. Sulfur isotopes and the Pine Point lead-zinc mineralization. *Econ. Geol.* 64:718–30

Sharp, J. M. Jr. 1978. Energy and momentum transport model of the Ouachita basin and its possible impact on formation of economic mineral deposits. *Econ. Geol.* 73:1057–68

Skinner, B. J. 1967. Precipitation of Mississippi Valley–type ores: a possible mecha-

nism. See Brown 1967, pp. 363–70

Skinner, B. J. ed. 1981. *Economic Geology, 75th Anniversary Volume*. Lancaster, Pa: Lancaster Press. 964 pp.

Snyder, F. G. 1968. Geology and mineral deposits, midcontinent United States. In *Ore Deposits of the United States, 1933–1967*, ed. J. D. Ridge, 1:257–86. New York: Am. Inst. Min. Metall. Pet. Eng. 1880 pp.

Snyder, F. G., Gerdemann, P. E. 1968. Geology of the southeast Missouri lead district. In *Ore Deposits of the United States, 1933–1967*, ed. J. D. Ridge, 1:326–58. New York: Am. Inst. Min. Metall. Pet. Eng. 1880 pp.

Stormo, S., Sverjensky, D. A. 1983. Silicate hydrothermal alteration in a Mississippi Valley–type deposit, Viburnum, southeast Missouri. *Geol. Soc. Am. Abstr. with Programs* 15:699

Sverjensky, D. A. 1981. The origin of a Mississippi Valley–type deposit in the Viburnum Trend, southeast Missouri. *Econ. Geol.* 76:1848–72

Sverjensky, D. A. 1984. Oil field brines as ore-forming solutions. *Econ. Geol.* 79:23–37

Sverjensky, D. A., Rye, D. M., Doe, B. R. 1979. The lead and sulfur isotopic compositions of galena from a Mississippi Valley–type deposit in the New Lead Belt, southeast Missouri. *Econ. Geol.* 74:149–53

Tarr, W. A. 1936. Origin of the southeastern Missouri lead deposit. Parts I and II. *Econ. Geol.* 31:712–54, 832–66

White, D. F. 1968. Environments of generation of some base-metal ore deposits. *Econ. Geol.* 63:301–35

White, D. F. 1974. Diverse origins of hydrothermal ore fluids. *Econ. Geol.* 69:954–73

White, D. F. 1981. Active geothermal systems and hydrothermal ore deposits. See Skinner 1981, pp. 392–423

Ann. Rev. Earth Planet. Sci. 1986. 14 : 201–35

CARBON DIOXIDE INCREASE IN THE ATMOSPHERE AND OCEANS AND POSSIBLE EFFECTS ON CLIMATE

Chen-Tung A. Chen and Ellen T. Drake

College of Oceanography, Oregon State University, Corvallis, Oregon 97331

INTRODUCTION

Although carbon dioxide constitutes only a small percentage (0.035%) of the atmospheric composition, it is an essential and beneficial compound for living, and changes in its concentration can have large effects on the global thermal regime and on life as we know it.

As early as 1827, the mathematician J. Fourier maintained that the atmosphere acts like "the glass of a hothouse, because it lets through the light rays of the sun but retains the dark rays from the ground" (cited in Arrhenius 1896). The British scientist Tyndall (1861) found that the elemental gases hydrogen, oxygen, and nitrogen mixed in the atmosphere have much less absorptive and radiative abilities than do the compound gases. He was thus convinced that air was a mixture and not a compound. His experiments showed that the compounds in the air such as "carbonic acid" (the nineteenth century term for carbon dioxide) or nitrous oxide have absorptive and radiative powers much greater than air or the elemental gases alone. He demonstrated that at certain pressures the absorption by "carbonic acid" was about 150 times that of oxygen alone.

Tyndall noted that various scientists, including H. De Saussure, J. Fourier, and C. Pouillet, regarded "the interception of the terrestrial rays as exercising the most important influence on climate." He believed that the compound gases augment "the differential action" between the heat coming from the Sun to the Earth and the heat radiating from the Earth into space.

201

0084–6597/86/0515–0201$02.00

He concluded:

It is not, therefore, necessary to assume alterations in the density and height of the atmosphere to account for different amounts of heat being preserved to the earth at different times; a slight change in its variable constituents would suffice for this. Such changes in fact may have produced all the mutations of climate which the researches of geologists reveal . . . the *extent* alone of the operation remaining doubtful.

In 1896, the distinguished Swedish chemist S. Arrhenius predicted that increases in CO_2 in the atmosphere would warm the Earth by as much as 9°C if the CO_2 level of his day could triple. He calculated that this 9°C warmer temperature is what prevailed in the balmy Tertiary Arctic regions. By the same token, for the Ice Age temperatures to prevail between the 40th and 50th parallels, the CO_2 level had to sink to 55–62% of the level of his day, which translates to a lowering of temperature by 4–5°C. T. C. Chamberlin, well-known American scientist and proponent of the principle of multiple working hypotheses, proposed similar ideas—that variations in climate, such as the advent of glaciation, could have been triggered by geologic processes that altered the carbon dioxide concentration in the atmosphere (Chamberlin 1899).

Since these early astute observations, much research has been conducted to investigate whether the CO_2 increase causes what is known as the "greenhouse effect." The radiating spectrum from the 6000 K Sun peaks in the visible range (about 0.55 μm), while radiation from the ~300 K Earth peaks at 11 μm, in the infrared. This difference in wavelengths is fundamental to the understanding of the greenhouse effect. CO_2 and other gases, such as methane, chlorofluorocarbons, nitrous oxide, and water vapor, allow the visible or near-visible radiation spectrum of the Sun to penetrate to the Earth to warm it. These same gases, however, block the reradiation of the infrared rays from the Earth back to space by absorbing them. The gases act, therefore, as a blanket for the Earth to keep its warmth. The imbalance between the incoming and outgoing radiation causes the Earth to warm to a higher average temperature.

The greenhouse effect has been a subject of deep concern for scientists, economists, politicians, and the lay public. This concern is partly generated from the knowledge that the increase in the atmospheric CO_2 content has been largely caused by the burning of fossil fuels and the large-scale clearing of forests, activities that are deeply rooted in human existence and are the very foundation of modern industrial society. The oceans act as a sink for excess CO_2 in the atmosphere, but at present it is unlikely that they can remove it at a rate that accords with the human time scale (i.e. in time so that the CO_2 increase does not affect life on Earth too drastically).

Aggravating the whole dilemma has been the constant rise in population in developing countries and the need to increase cultivation of farmland by deforestation in order to feed this growing population. At the same time, in spite of conservation efforts, people living in highly industrialized nations are not likely to reduce their standard of living substantially, so fuel consumption will not decrease.

Many symposia and conferences have been held to evaluate the CO$_2$ question. Experts in various fields have debated the issues at social, scientific, and even political gatherings, which has resulted in the publication of numerous books, articles, and reports. Space limitations do not allow us to cite all the important papers that have been published on this topic. By citing representative papers, however, we review the state of knowledge of the problem of increasing CO$_2$ in the atmosphere, discuss the carbonate system in the oceans, and report on recent investigations on the CO$_2$ increase in the oceans. We also relate the possible consequences of the rising atmospheric CO$_2$ content and explore the feasibility of tactics that might delay the "greenhouse" warming of the globe.

CARBON DIOXIDE IN THE ATMOSPHERE

As some early investigators observed, the carbon dioxide concentration has been increasing as a result of the Industrial Revolution from about the middle of the nineteenth century. Muntz & Aubin (1886) discovered through their measurements and observations that human activity, especially in the industrial belt of the Northern Hemisphere, has had a significant effect on the atmospheric CO$_2$ content. Their average value for the Northern Hemisphere was 282 ppmV,[1] which decreased as they made their measurements away from areas of dense population and industrial activity. Their average value for the Southern Hemisphere in 1882 was 266 ppmV, which is quite close to some modern estimates for the preindustrial atmospheric CO$_2$ level (Wigley 1983), although Siegenthaler (1984) questioned the reliability of the Muntz & Aubin results.

The Preindustrial CO$_2$ Level

In order to discover how much CO$_2$ has been added to the atmosphere as a result of human activity since the beginning of the Industrial Revolution, we must first ascertain the preindustrial CO$_2$ level. Various methods to estimate this value have been employed, and the range of disagreement

---

[1] Different authors use different units to express partial CO$_2$ pressure (PCO$_2$: ppmV, ppm, μatm). We cannot convert these units into a common unit without supportive data, such as temperature, pressure, humidity, etc. We have therefore reported in this article the original units used by the authors of the papers cited.

among the different techniques is narrowing. A few methods that have been used recently are described below.

SAMPLING OF AIR BUBBLES IN ICE CORES As in many branches of science, but especially in the Earth sciences, a useful principle to adopt is that the present holds the key to the past. A direct method of measuring the preindustrial CO_2 concentration is to analyze the composition of air trapped in pockets of permanent ice. The air bubbles are assumed to be actual air samples at the time the ice was formed.

Stauffer et al (1984) obtained a CO_2 concentration of 260–280 ppmV for samples from the period of A.D. 1800 to 1850, which represents the preindustrial CO_2 value. The results obtained by Barnola et al (1983) at their laboratories in Grenoble and Bern indicate that the mean CO_2 level in air bubbles in Antarctic ice for the period 800–2500 yr ago is about 260 ppmV. How closely do the bubbles match the original preindustrial atmospheric composition is one of the uncertainties in this method. Because of the high solubility of CO_2 in water, melting and freezing action of the ice may lead to higher-than-normal values. Sampling and drilling techniques may also affect the results, although careful selection of the ice usually eliminates such problems and yields results with less than 15 ppmV deviation from the atmospheric composition (Bojkov 1983).

Stauffer et al (1984) further discovered strong evidence supporting the relationships between atmospheric CO_2 concentration and climate (Broecker 1982). Low values of 180–220 ppmV were obtained for a very cold period during the end of the last glaciation, with higher values of 250–270 ppmV corresponding to a warming period.

FROM OCEANIC CO_2 DATA Since the onset of the Industrial Revolution, the increase of dissolved CO_2 in surface water has been about 1.8% (Chen & Millero 1979). Historical ocean CO_2 data are accurate to only a few percent and therefore cannot be compared directly with the present data (accurate to 0.2%) for the detection of such small secular variations. But the excess CO_2 signal is reflected in the carbonate chemistry of seawater. The farther away a deep water is from its source region, the older the water is and the less excess CO_2 it has been exposed to. As a result, by accurately measuring the dissolved CO_2 of seawaters of various ages, we obtain a trend that reveals the gradual increase of anthropogenic CO_2.

Further discussion on CO_2 in the oceans and possible uncertainties in this method is given in a later section. Using this direct sampling approach, Chen & Poisson (1984) have calculated that the preindustrial atmospheric CO_2 level was 268 ± 13 μatm, a value that agrees well with the 1882 Southern Hemisphere value of 266 ppmV obtained by Muntz & Aubin (1886).

ANALYSIS OF SPECTROSCOPIC DATA   Another direct CO$_2$ measurement is the analysis of spectroscopic data accumulated in the archives of the Smithsonian Institution. The spectrum of the Earth's atmosphere has been regularly recorded at various sites around the world for almost a century. Problems of calibration involve the determination of the relationships between the integrated column densities of atmospheric CO$_2$ and surface measurements and the level of performance of the original instrumentation used to obtain the spectra. Accordingly, the Smithsonian Table Mountain results are being compared with early measurements made by C. D. Keeling for the same period at Pasadena and San Diego, California. This effort gives some hope that if errors could be reduced, extrapolation of the data back to the mid-nineteenth century period could give an accurate preindustrial CO$_2$ value (Bojkov 1983, Stokes et al 1984).

TREE-RING ^{13}C RECORD   An indirect method of estimating atmospheric CO$_2$ variations is based on the ^{13}C record in tree rings. The assumption in this method is that such tree-ring ^{13}C records represent the ^{13}C/^{12}C ratio change for atmospheric CO$_2$ as a result of fossil fuel consumption and human activities in disturbing the terrestrial biosphere, such as large-scale deforestation and attendant decomposition in the soil. Plants prefer the lighter ^{12}C during photosynthesis, so that the ^{13}C/^{12}C ratios of organic material are lower than the carbon ratio in the atmosphere. Fossil fuel CO$_2$ also contains reduced amounts of the ^{13}C isotope, but the effect from this source can be separated out with the use of bomb-produced ^{14}C (Keeling 1979).

The ^{13}C isotopic method was described by Stuiver (1978), who used it to estimate the release of carbon to the atmosphere from plants, mainly trees (discussed later). He accepted the value of 268-ppm atmospheric CO$_2$ content for the preindustrial level.

Much is unknown about the significance of the ^{13}C/^{12}C ratio. For example, different substances in plant tissue, it seems, may show differences in relative concentrations of ^{12}C and ^{13}C, and the results may also be temperature dependent. In addition, pollution and sampling biases also affect Δ^{13}C records, so that the method is fraught with uncertainty; indeed, Stuiver's data still do not agree with those of Freyer (Stuiver 1978, and personal communication, 1983, Broecker et al 1979, Macdonald 1982, Francey & Farquhar 1982, Freyer & Belacy 1983, Stuiver et al 1984). Nevertheless, some investigators have also used this method to estimate the preindustrial CO$_2$ level. Peng et al (1984) estimated the pre-1850 atmospheric CO$_2$ value to be about 266 μatm. Stuiver et al (1984) obtained a value of 268 ppm for the year 1600 and an average of 276 ppm for the A.D. 235–1850 interval. There is fairly good agreement, therefore, on the approximate median value of the reported range of about 230–290 μatm for the reconstructed preindustrial CO_2 value.

Rise in CO_2 Level—The Mauna Loa Program

Now that we have some knowledge of the approximate preindustrial CO_2 level in the atmosphere, we need evidence that the concentration is increasing in a secular way. The first twentieth century direct and continuous measurement of atmospheric carbon dioxide over a period of more than two decades was initiated in 1957–1958 as part of the program of the International Geophysical Year. In these measurements, C. D. Keeling and his associates (Keeling et al 1976a,b, 1982) set up gas analyzers near the summit of Mauna Loa, Hawaii, and at the South Pole. Although the absolute values in parts per million were lower at the South Pole station, both records clearly show a steady rise in the CO_2 concentration since 1958.

The records at the 3400-m Mauna Loa site (Figure 1) also show the annual cycles of CO_2 uptake by the biosphere in the spring/summer and respiration in the fall/winter–early spring. The records may be complicated slightly by signals from CO_2 emission from nearby volcanic vents, or as a result of photosynthesis taking place in forests and sugar fields, or from man-made effects such as highway traffic as well as analytical problems from 1965–1969 (Keeling et al 1982). Great care, however, has been taken to eliminate the noise, and the general trend is unmistakable. The average of monthly average values for 1959 was 315.66 ppm, whereas that for 1980 was

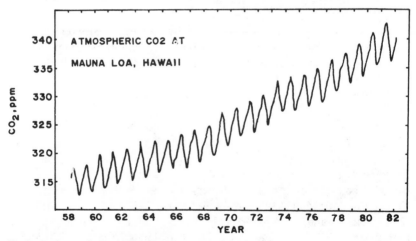

Figure 1   Atmospheric carbon dioxide concentration at Mauna Loa Observatory, Hawaii. The zigzag pattern reflects seasonal variations as a result of the removal of CO_2 by photosynthesis during the growing season followed by the return to the air of CO_2 by oxidation of plant tissues (after Keeling et al 1976a).

338.36 ppm, with an average yearly increment of 1.14 ppm. These data also show that the rate of increase has risen over the last 20 years, going from 1×10^{15} g yr^{-1} in the 1950s to 2.6×10^{15} g yr^{-1} by 1978 (Macdonald 1982).

Modeling Future CO$_2$ Increases

Approaching the problem from various directions, therefore, has yielded a range of preindustrial CO$_2$ values for which there is growing consensus among investigators. We also have excellent evidence that the atmospheric CO$_2$ level has been rising steadily over the last two decades. We must now address the difficult problem of making future predictions. Scientists have traditionally turned to modeling as a tool to generate information when variables and uncertainties within the variables approach an unmanageable level. Such efforts, however, are usually based on the best available estimate.

THE CARBON CYCLE   Models that simulate future changes in the carbon cycle assume an initial steady-state condition. Disturbances to the system through human activity or other variables are then evaluated. Accordingly, in order to attempt to predict future carbon levels in the atmosphere, one must have an understanding of the distribution of carbon among the various reservoirs. Various models have been proposed. Bolin (1981), for example, depicts a simple seven-reservoir steady-state model for the carbon cycle (Figure 2). The oceans are represented by a two-layer system

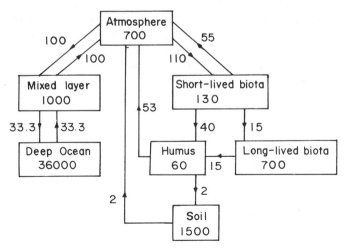

Figure 2   The carbon cycle depicted in a seven-reservoir steady-state model. The numbers represent units of 10^{15} g carbon (after Bolin 1981).

consisting of a mixed layer and a deep layer. The terrestrial ecosystem consists of four components.

As can be seen from such a model, the amount of carbon in the atmospheric reservoir is relatively small (7×10^{17} g). The exchanges among the reservoirs are rather large, so that small changes in the larger reservoirs (such as in the biosphere) could bring about significant changes in the atmospheric composition. Much disagreement, however, exists among various investigators as to how effective is the oceanic sink, or whether the biosphere is a net source or a net sink for carbon, or what fraction of the carbon released from fuel burning goes into the atmosphere and what fraction is absorbed by the oceans.

FOSSIL FUEL CONTRIBUTION   An estimated 40–60% of the carbon dioxide produced by human activities in the last 20 yr has remained in the air. Based on various statistics compiled by the United Nations on the annual fossil fuel consumption by industrial countries, on CO_2 production records from various sources published by the US Bureau of Mines, and on other records, Keeling (1973) estimates the cumulative increase of CO_2 production up to 1970 from fossil fuels plus that from kilning of limestone (another 1–2% added to annual totals) to be $4.1 \pm 0.5 \times 10^{17}$ g CO_2 ($1.12 \pm 0.14 \times 10^{17}$ g C). This amount is equivalent to about 18% of the amount of CO_2 in the atmosphere during the late nineteenth century. Rotty (1977, 1982) adds to this figure the CO_2 produced by flaring of natural gas, which increases the amount by another 2%.

Figure 3 shows the annual CO_2 production from fossil fuel combustion, expressed as carbon. From this curve a growth rate of 4.3% has been calculated, interrupted only by the world wars and economic slumps of period II (1914–1949). At this rate, the concentration would reach 600 ppm by about the year 2025. Another estimate, based on the drop in the annual growth rate (about 1.8% yr^{-1} for 1974–1980 after the oil-shortage scare) projects the 600-ppm concentration to occur in the middle of the twenty-first century (Rotty & Marland 1980).

Based on the overall record for the 120-yr period, Elliott (1983) calculates a growth rate of $3.5 \pm 0.18\%$. He argues that this figure is a more realistic rate, since the world's resources could not have sustained a growth rate of 4.3% for extended periods of time even without the disruptions in period II. W. P. Elliott (personal communication, 1984) believes that an even lower figure (2% or less) would be better for extrapolating. Other estimates range from 1 to 3.5% yr^{-1}, with most studies predicting between 2 and slightly above 3% yr^{-1} (National Research Council Commission on Physical Sciences, Mathematics and Resources 1983a). Nordhaus & Yohe (1983), based on their model of CO_2 emissions, provide a "best guess" of 1.6% yr^{-1} to the year 2025, with a reduction to slightly under 1% yr^{-1} after that.

At the present rate of fossil fuel use, most of the recoverable supply of petroleum and natural gas would be exhausted by the middle of the twenty-first century. Most of the world's coal supply, however, which is concentrated in the Soviet Union, China, and the United States, would still be recoverable. Future atmospheric CO$_2$ increases, as well as the increased emission of more undesirable and dangerous substances, would be largely associated with the burning of coal and shale oil. Nuclear power could help to alleviate the problem by replacing coal-burning plants, if nuclear energy generation could solve its safety problems and be freed from emotional barriers.

CONTRIBUTION FROM THE BIOSPHERE   What then is the role of the biosphere? CO$_2$, after all, is essential to life as the source of carbon fixed in photosynthesis by green plants, forming the basic trophic level that supports all plants and animals. According to Woodwell (1978), the world's forests hold almost 90% of all the carbon stored in the global ecosystem, with the largest net amount of photosynthesis taking place on land and not in the oceans, as had been previously believed. Woodwell and his colleagues estimate that the biotic contribution to CO$_2$ entering the atmosphere may have been as much if not larger than that from fossil fuel, which leads to a total estimated net carbon release since 1860 of 135–228 × 10^{15} g and an

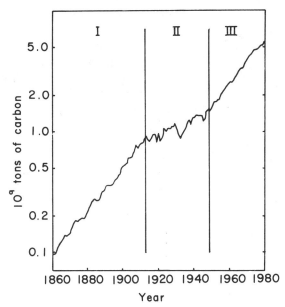

Figure 3   Annual CO$_2$ production from fossil-fuel combustion (expressed as carbon) for the period from 1860 to 1980 (after Elliott 1983).

annual estimated net release in 1980 of $1.8-4.7 \times 10^{15}$ g (Houghton et al 1983, Woodwell 1983). During the last 20 yr, however, the fossil fuel contribution has exceeded the carbon release from the terrestrial system.

Much uncertainty surrounds the mechanisms that drive the carbon cycle, but few would now deny that deforestation has taken place at an alarming rate, especially in the developing countries (Wong 1978). The reduction of forested areas has serious consequences other than contributing to the CO_2 problem. There is evidence that some regrowth in the United States has been slow and stunted by pollution (Baes & McLaughlin 1984, Shabecoff 1984).

Using carbon isotopic ratios in tree rings and comparing them with atmospheric ratios and those of fossil fuel, Stuiver et al (1984) calculated that for the period between 1600 and 1975, the biota released to the atmosphere as much carbon (149×10^{15} g) as did fossil fuel combustion (135×10^{15} g). Approximately half of this total amount has been transferred to the oceans. As was mentioned earlier, the difficulties in using carbon isotopic ratios in tree rings for such calculations causes a large uncertainty factor, and efforts to reforest the cleared land and the potential enhanced photosynthesis due to the excess CO_2 are all important factors that offset to a degree the effect of large-scale deforestation. Depending on the period of history under consideration, deforestation could contribute either more or less to the CO_2 production than does fossil fuel consumption, but the biota still appears to be a net source of CO_2 (Woodwell 1983).

The latest estimates of Broecker and his colleagues (Peng et al 1983, 1984), based on further study of the tree-ring ^{13}C record, show that the input of ^{13}C-depleted CO_2 from deforestation and soil decomposition increased from 1800 to 1860, and since then has been slowly declining with respect to the contribution from fossil fuel. During the two-decade period of the Mauna Loa record the forest-soil input has been about 20% that from fossil fuel, a level not considered by these authors as contributing significantly to the secular increase in atmospheric CO_2. On the other hand, in the opinion of Woodwell (1983), the management of the biotic pools could delay the reaching of any given concentration of atmospheric CO_2 by several decades, especially if the fossil fuel surge diminishes.

Even though controversy on this point still exists, agreement among scientists is nevertheless more complete that the biota appears to have been a net source of atmospheric CO_2. Little by little, we are reducing the uncertainties in this important area of human endeavor. Only the ocean now is left as the major CO_2 sink. We discuss its role in the next two sections.

CARBONATE SYSTEM IN THE OCEANS

The oceans contain an amount of dissolved carbon 60 times that present in the atmosphere and have acted as a major sink for excess CO_2. The surface wind-mixed layer (less than 200 m deep) of most of the oceans is more or less saturated with excess CO_2. The deep waters (from 200 m to an average depth of 3800 m), on the other hand, have not been affected by the excess CO_2 except near their source regions, such as the northern North Atlantic Ocean and the Weddell Sea. This slow penetration of excess CO_2 into the deep waters is due to the slow exchange of waters across the thermocline that separates the surface and deep waters. More and more excess CO_2, however, will gradually penetrate into the deep oceans. We examine this process first by discussing the role of the carbonate system.

Carbonate Chemistry

By far the most important system in the oceans, with the exception only of the chemistry of water itself, is the CO_2-carbonate system, a buffering system that helps to maintain the pH of seawater to within a narrow range (Horne 1969, Pytkowicz 1983).

CO_2 hydrates with water rather rapidly (in milliseconds) to form carbonic acid. Once CO_2 is hydrated, it is involved in a series of even more rapid proton transfer steps (submicroseconds; Patel et al 1973) to form bicarbonate (HCO_3^-) and carbonate (CO_3^{2-}): CO_2 (gas) $+ H_2O = H_2CO_3$, $H_2CO_3 = H^+ + HCO_3^-$, $HCO_3^- = H^+ + CO_3^{2-}$. The sum of the three aqueous carbonate species is defined as total CO_2 (TCO$_2$). A frequently used property is titration alkalinity (TA), the amount of acid needed to neutralize all bicarbonate and carbonate ions.

Because of the large buffer factor (or Revelle factor) of seawater, a 10% change in the partial CO_2 pressure results in only an approximately 1% change in TCO$_2$ (Baes 1981, Brewer 1983). At the current pH of seawater (about 8), the excess CO_2 added to the oceans reacts with the carbonate ion to form the bicarbonate ion without a change in pH: $CO_2 + CO_3^{2-} + H_2O = 2HCO_3^-$. The concentration of the carbonate ion is limited by the solubility of calcium carbonate as follows: $Ca^{2+} + CO_3^{2-} = CaCO_3$. In this reaction the solubility of $CaCO_3$ increases with pressure but decreases with temperature. The surface layer is usually saturated with $CaCO_3$, but except for isolated regions such as the Bahama Banks, spontaneous precipitation does not occur because organic material and magnesium in seawater retard $CaCO_3$ formation. Many marine organisms, however, utilize calcium and carbonate ions to form $CaCO_3$ skeletons and shells. This process occurs mainly in the surface layer. As the organisms die, their shells and skeletons sink to a level where the temperature becomes low enough or the pressure becomes high enough, and the water becomes undersaturated with respect to $CaCO_3$, at which point they start to dissolve (Pytkowicz 1983). Accumulation of undissolved $CaCO_3$ in the sediments over millions of years, however, makes the sediments a major sink of carbon.

Inorganic carbon is also used by marine organisms to grow soft tissue. Soft tissue decomposes quickly in the water column. Both soft and hard parts of the marine organisms act as a "biological pump," thus removing CO_2 from the surface ocean and transferring the gas into the deep ocean and ocean bottom.

Distribution of Carbon in the Oceans

For reasons not yet well understood, the distributions of pH, calcium ion, and the related properties [titration alkalinity (TA), total CO_2 (TCO_2), and partial CO_2 pressure (PCO_2)] in the surface waters seem to correlate linearly with temperature (Edmond 1974, Chen & Millero 1979, Chen et al 1982a), although temporal and spatial variations of these trends exist (Chen & Pytkowicz 1979, Chen 1982a, 1984, Takahashi et al 1983). An example for pH is shown in Figure 4 based on the data collected during the 1981 Joint US–USSR Weddell Polynya Expedition (WEPOLEX; Chen 1982c).

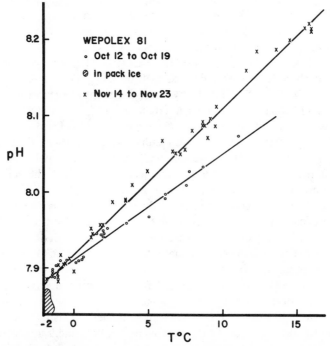

Figure 4   Surface pH values (measured at 25°C) vs surface temperature for samples collected during the Weddell Polynya Expedition (after Chen 1984).

The surface pH values (measured at 25°C) measured in October correlate linearly with surface temperature, with lower pH values (higher PCO$_2$ and TCO$_2$) corresponding to lower temperatures. The November data also show a linear trend but with a different slope.

We have mentioned the transfer of CO$_2$ into the deep waters by the "biological pump." The dissolution of CaCO$_3$ increases alkalinity, calcium, and total CO$_2$. The decomposition of organic matter also increases total CO$_2$ and PCO$_2$, but it decreases pH and perhaps also alkalinity (Brewer et al 1975). Overall, the dissolution of CaCO$_3$ contributes approximately 12% of the total CO$_2$ input into the deep waters (Chen et al 1982b, Bolin et al 1983). Takahashi et al (1981) have compiled the results of GEOSECS (Geochemical Ocean Section Studies) expeditions, and their total CO$_2$ results (adjusted to the same salinity) indicate that deep waters all have higher values than the surface, as explained above. The Pacific deep waters have higher concentrations than the Atlantic deep waters because of their older age (hence more CaCO$_3$ and organic carbon dissociation). The bottom water in the Pacific Ocean upwells to form the deep water. Consequently, the deep water, older than the bottom water, has higher CO$_2$ concentrations than the bottom water, again because more organic and inorganic carbon is dissociated.

TEMPORAL INCREASE OF CO$_2$ IN THE OCEANS

Although oceans are believed to act as an important sink for human-induced CO$_2$, very few studies have quantitatively shown the increase, and these only for surface seawaters (Postma 1964) or for deep waters coming from the same surface origin (Brewer 1978). As a result, most of our knowledge about the air-sea exchange and the distribution of excess CO$_2$ in the world oceans has come from the study of indirect data [namely, the radiotracers tritium (^{3}H), radiocarbon (^{14}C), and radon (^{226}Rn), and the stable freons]. Other tracers have seen very limited use.

The application of data from these tracers in various models has been quite successful. The three-dimensional distribution of excess CO$_2$, how-ever, cannot be generated from these models. Even the validity of some models is questioned; as Brewer (1983) states, "the models contain little that is recognizable about the physics of the real ocean."

We believe that the excess CO$_2$ signal should be retrievable from the oceanic carbonate data. Both the tracer studies and the direct approach are discussed below.

Tracer Studies

The injection into the atmosphere of massive amounts of ^{3}H and ^{14}C by the nuclear bomb tests of 1958–1962 provided oceanographers with two important tracers for estimating the air-sea exchange rate and the movement of deep-water masses.

AIR-SEA EXCHANGE   Revelle & Suess (1957) and Craig (1957) were among the first to use ^{14}C box models to evaluate the air-sea exchange of CO_2. The radon method, however, has been used more extensively recently (Peng et al 1979), although ^{3}H or stable chemicals have also been used.

The most comprehensive set of radon data is that of Peng et al (1979). Despite strong laboratory evidence (e.g. Merlivat & Memery 1983) that gas transfer is a function of wind speed, these data do not show a relationship between the gas exchange rate (κ) and the square of the wind speed measured at the time of the radon collection.

Peng et al (1979) pointed out that since the radon profile is dependent on the wind history of the sampling area, the wind as measured from the ship at the time the samples are taken is not necessarily relevant to the conditions before the station was occupied. C. T. A. Chen (unpublished, 1980) and Pytkowicz & Chen (1984) speculated that the wind speed history is reflected in the mixed-layer thickness (h) and the density gradient ($\partial\rho/\partial Z$) immediately below the mixed layer. A balance of changes in the stored potential energy in the mixed layer and the cumulative mechanical energy provided by wind leads to the relationship shown in Figure 5. The results are based on GEOSECS data north of 45°N and south of 45°S. As can be seen, a fair correlation ($1\sigma = 1.0$ m day^{-1}) exists.

The air-sea exchange rate across the ocean surface, although quite slow (meters per day) as compared with the time required for reaching equilibrium among the carbonate species, does not appear to be the rate-limiting step in preventing the deep oceans from taking up excess CO_2; this role is played instead by the even slower vertical mixing rate. Tritium and radiocarbon studies have provided important information regarding the vertical mixing rate.

VERTICAL MIXING   The vertical mixing rates have been studied extensively with the use of ^{3}H and ^{14}C as clocks (e.g. Broecker et al 1960, Fine et al 1981). A "snapshot" of tritium and radiocarbon distributions during the GEOSECS expeditions in 1972 in the western Atlantic Ocean is shown in Figure 6. This figure shows clearly that tritium has penetrated all the way to the seafloor in the northern North Atlantic in less than two decades after the bomb tests. Tritium also penetrates relatively deeper at 45°S near the Subtropical Front, but it is found to only a few hundred meters south of the Polar Front because of the upwelling of the old North Atlantic Deep Water around the Antarctic. It is somewhat surprising, however, that little or no tritium was found in the Antarctic Bottom Water, the major source of bottom waters in the oceans. An explanation is given in the next section.

The structure of the ^{14}C contours in Figure 6 is similar to that of tritium, with younger water (less negative values) penetrating deeper in the northern North Atlantic Ocean than near Antarctica. Freon data are now emerging (Smith & Cheek 1982, Cline et al 1985), and the distributions are similar to that of ^{3}H.

Figure 6 provides a vivid impression of the tritium and ^{14}C penetration in 1972. Recent tritium data collected during the Transient Tracers in the Oceans (TTO; Ostlund 1981) expeditions in the North Atlantic indicate that the tritium in the deep and bottom waters has progressed some 8° farther south than is shown in Figure 6. This increased penetration of tritium suggests that more excess CO$_2$ has also penetrated into the deep North Atlantic Ocean.

Many models have been devised based on tracer data to study the CO$_2$ exchange in the oceans. Just a few years ago, almost all calculations of uptake of excess CO$_2$ by the oceans were based on a one-dimensional box diffusion model (e.g. Oeschger et al 1975) that assumes a laterally

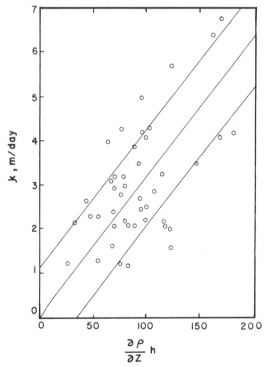

Figure 5   The air-sea gas exchange rate vs a measure of cumulative wind energy for areas north of 45°N and south of 45°S (after Pytkowicz & Chen 1984).

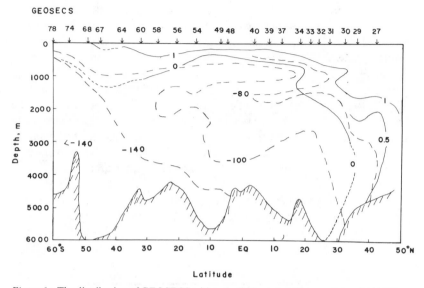

Figure 6   The distribution of GEOSECS tritium (solid lines, in tritium units) and $\Delta^{14}C$ (long dashed lines, in $\Delta^{14}C$ units) in the western Atlantic Ocean. Short dashed lines show regions where tritium data are scarce. Numbers on top scale are station numbers (from Chen 1982b).

homogeneous ocean. More complex models have since been developed (Broecker et al 1979, Killough & Emanuel 1981, Bolin et al 1983, and references therein).

The conversion of the tracer signal to that of CO_2, however, is still somewhat uncertain, because the results of the conversion depend on the model used. The uncertainty exists not only for tritium but also for all models that extract information indirectly. Such results, however, are valuable for comparison with results obtained by other techniques, e.g. the direct sampling method discussed in the next section. Recent model calculations also reveal that the earlier models perhaps underestimated the amount of excess CO_2 that has been absorbed by the oceans. It now appears that the oceans might have taken up the additional excess CO_2 released by the reduction of biomass. The global carbon budget, therefore, may not be "out of whack" after all (Kerr 1977, 1980).

Direct Carbonate Data

Consider a parcel of seawater originally at the surface and later conveyed by advection to some point in the deep ocean, where it is sampled and analyzed. The measured TCO_2 is likely higher than the original value because some organic material and $CaCO_3$ have decomposed in this water.

The amount of CaCO$_3$ dissolution is reflected in the calcium or TA concentrations. The amount of organic matter decomposition is reflected in the oxygen concentration. Since TA and oxygen are easy to measure, current work relies upon these two measurements to provide information on the alteration of a parcel of seawater after it leaves the surface (Brewer et al 1975, Chen 1978). On the other hand, calcium is more difficult to measure and has been used less frequently, although the application of the data is less ambiguous than that of the TA data (Brewer et al 1975, Shiller & Gieskes 1980, Chen et al 1982b, Chen 1983).

Various approaches have been applied to extract the CO$_2$ concentration for deep seawaters at the time they were originally at the surface (see references in Chen 1984). Figure 7 shows the distribution of anthropogenic CO$_2$ in the western Atlantic Ocean based on the GEOSECS data (Takahashi et al 1981). The results (Chen 1982b) indicate that the excess CO$_2$ has penetrated deeper than ^{3}H (Figure 6) and is below the thermocline everywhere in the western Atlantic. These results are to be expected, because excess CO$_2$ has been added over a century, whereas ^{3}H has only entered over the last 20 years. In general, Figure 7 looks similar to the distributions of ^{3}H and ^{14}C shown in Figure 6. In the North Atlantic Ocean, relatively older water (less excess CO$_2$, lower ^{3}H, more negative ^{14}C) at a depth of approximately 1000 m is sandwiched between younger

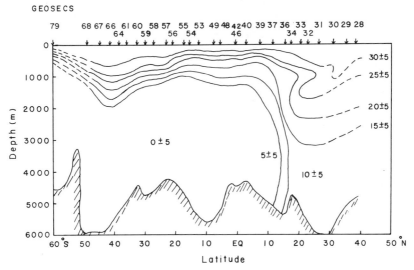

Figure 7   The distribution of excess CO$_2$ (μmol kg^{-1}) in the western Atlantic Ocean. Dashed lines show regions where results are subject to possible systematic errors. The uncertainties shown are random errors only (after Chen 1982b).

waters above and below. This old water coincides with minimum salinity, minimum oxygen, and maximum nutrients, and it is apparently influenced by the Antarctic Intermediate Water, which originated far away in the southern South Atlantic.

Little excess CO_2 was found in the Antarctic Bottom Water near Antarctica. This finding was confirmed by Chen (1984) and Chen & Poisson (1984) based on winter data collected during the Joint US–USSR Weddell Polynya Expedition. They found that the winter pack ice blocks the air-sea exchange of oxygen and CO_2, and hence the upwelled old deep water sinks back down as a result of cooling but does not pick up much excess CO_2. Several investigators (Weiss et al 1979, Minas 1980) had previously come to this conclusion based on summer oxygen data. Chen & Poisson report that only 6 μmol kg^{-1} excess CO_2 was found in the Weddell Sea Bottom Water, which is the main component of the Antarctic Bottom Water.

It should be cautioned, however, that like any other approaches, the excess CO_2 signal derived from oceanic CO_2 data contains known or unknown systematic errors (Shiller 1981, 1982, Baes 1981, Chen et al 1982a, Broecker et al 1982, 1983). Broecker and coworkers, for example, have pointed out that the method with which Chen and coworkers calculated the amount of organic matter decomposition is flawed. Chen has assumed that oxygen is at equilibrium at the surface when a water mass is formed. Lack of oxygen saturation, as correctly pointed out by Broecker et al, would indeed lead to an uncertainty in the direct carbonate approach. This problem can affect the results by 5 μmol kg^{-1} excess CO_2 (compared with a signal of 40 μmol kg^{-1}) and cannot be solved unless we obtain sufficient winter data at the deep-water formation regions.

Broecker and coworkers also evaluated the carbon vs oxygen ratio in the organic matter and proposed an average value of 0.80 ± 0.04. Chen and coworkers have been using a value of 0.78 ± 0.05. These reported values, therefore, do not differ very much, and the actual ratio (if it can be unequivocally obtained) will undoubtedly be somewhere around this figure when averaged over a large area. Chen et al (1982a) explained why they believe that their preferred value of 0.78 cannot be in large error.

Broecker et al (1983) believe that the major flaw in Chen's approach is that Chen did not address the question of "how much of the difference between the current atmospheric and the deep sea derived CO_2 partial pressure is anthropogenic in origin and how much is the result of differences between the CO_2 pressure in the atmosphere and that in winter surface water in the regions of formation of deep water." The only way to determine the magnitude of these errors is to collect winter data in the regions where subsurface waters are formed (Chen 1984). In the winter, the strong wind and the intensive cooling set up conditions most conducive to the air-sea

CO_2 exchange and the subsequent vertical CO_2 penetration. Yet because of logistic problems, most high-latitude CO_2 measurements have been conducted in the summer. Needless to say, modeling winter-formed deep waters using summer data adds more uncertainty.

Although Chen could not have evaluated the deep waters in the northern North Atlantic because there were no reliable data, he did evaluate extensively the deep waters in the Southern Ocean. As early as 1979 (Chen & Pytkowicz 1979) and in subsequent papers (Chen et al 1982a, Chen 1982b, Feely & Chen 1982), he raised the question about a possible systematic error (± 15 μmol kg^{-1}) due to a possible seasonal difference in the surface alkalinity and total CO_2 values. The lack of winter data (neither GEOSECS nor TTO produced winter data) is indeed one of the difficulties in using the available ocean carbonate data. Chen, therefore, collected winter data in the Weddell Sea and Bering Sea in 1981 and again in 1984 (Chen 1982c, 1985). His data, albeit collected from a limited region, indicate that the suspected possible error of ± 15 μmol kg^{-1} in the Southern Ocean did not exist. Prior to his collection of winter data, Chen used the ample data available for the Southern Hemisphere; the then-available North Atlantic carbonate data showed no differences in trend from the Antarctic. Chen, therefore, used only one equation to represent the carbonate data for the Atlantic Ocean, realizing that more sets of regional equations would be derived when more data became available (Chen 1982a).

The winter data in the Weddell Sea, which is the source of bottom waters, do not show much seasonality. These data, therefore, have established a good reference for comparison in the Southern Ocean. The North Atlantic, on the other hand, has been shown by Broecker et al (1983) to be extremely complex, with large spatial and seasonal variations. Furthermore, the North Atlantic deep waters, having been already contaminated by the excess CO_2, cannot be used as a baseline for comparison. Analyses of different data sets obtained from different regions, therefore, could have contributed to the large differences in opinion.

Recently we have received the first comprehensive winter carbonate data collected in the North Atlantic Ocean (SIO reference #84-14 1984). Analysis of these winter data and the TTO summer data should enable us to solve some of the problems pointed out by Broecker and coworkers and allow us to calculate the excess CO_2 values in the North Atlantic Ocean. Chen's earlier results in the South Atlantic and the Pacific Ocean, however, are unlikely to be significantly affected.

Science progresses by reducing uncertainties. In the case of assessing the increase of CO_2 in the oceans, both the indirect methods of using radiotracers and the direct sampling methods are valuable, as only when independent methods yield the same values can we have confidence with respect to the accuracy of the results. Many investigators around the world have planned, or are planning, to collect winter polar tracer and CO_2 data, and they are modeling the data in more realistic approaches (personal communications with C. F. Baes, B. Bolin, R. A. Feely, M. I. Hoffert, H. Oeschger, A. Poisson, T. Takahashi, and R. F. Weiss). What is needed at this juncture is a concerted effort, using different approaches, toward assessing the oceanic CO_2 problem.

POSSIBLE CONSEQUENCES OF THE RISING ATMOSPHERIC CO_2 LEVEL

Climate

The geologic record shows that during most of Earth's history, the atmosphere contained a much higher percentage of carbon dioxide than it does today. The carbonate-silicate computer model of Berner et al (1983), which takes into account plate tectonics and spreading rates, predicts that the atmospheric CO_2 level in the Cretaceous was severalfold greater than it is today. These authors are convinced that the higher CO_2 level then was at least partly responsible for the higher worldwide air temperatures that existed. Plass (1959) notes that even though plants are adapted to the spectral range and intensity of the light they receive, some grow much better in an environment that has 5 to 10 times the current CO_2 level, which indicates that they are "keyed to some much higher concentration in the atmosphere of the geologic past." Human interference with the natural processes by burning fossil fuels and cutting down forests for cultivation have made it necessary to study how the short-term effects could influence the existence of humankind.

Since Callendar's (1939) assessment of the influence of artificially produced CO_2 on temperature, various investigators have attempted to deduce the possible consequences of atmospheric carbon dioxide increase. Gates and his colleagues (Gates et al 1981, Schlesinger 1983) at Oregon State University have conducted a series of experiments on climatic effects of increased CO_2 with an atmospheric general circulation model (GCM) and a climatological ocean. Their simulations involved both doubling and quadrupling levels of CO_2. Their latest results show that the globally averaged tropospheric temperature would increase by 3°C for a doubled CO_2 level (W. L. Gates, personal communication, 1984).

Manabe and his colleagues (e.g. Manabe & Wetherald 1980) also constructed mathematical models to predict the responses of the atmosphere to doubled and quadrupled levels of CO_2. The warming of the model atmosphere was found to vary seasonally and latitudinally and resulted in an increase in moisture content of the air and in its transport poleward. The model predicted a much greater temperature response in the higher latitudes than in the tropical regions due to snow/ice albedo feedback and to generally greater atmospheric stability at the high latitudes. With a doubling of atmospheric CO_2 content, the average surficial temperature would rise by less than 2°C in the tropics; at 35°N, the temperature rise would be about 3°C; north of 50°N, the increase would be greater than 4°C; and above the Arctic Circle, it would be 7°C (Figure 8). These results agree well with the model predictions of Hansen et al (1981), who predict that a doubled CO_2 content in the air would create hot and dry conditions in the western two thirds of the United States, Canada, and much of Central Asia. They further predict that the hot, dry summer of 1980 may become the norm for the next century. Schlesinger (1983) provided detailed comparisons of various GCMs. These comparisons reveal similarities and differences among various simulations, and the authors present some possible explanations for the differences.

Because model predictions necessarily include some limitations (such as, for example, the simplified coupling between the atmosphere and the ocean, and incomplete information on the extent of sea ice and feedback mechanisms of ice sheets), past climates must be studied as possible analogues for modeling future climates. According to Flohn (1981), for instance, two paleoclimatic events may possibly recur in the not-too-distant future if the CO_2 concentration reaches levels above 600 ppm (along with corresponding increases of other greenhouse gases that would cause a 4–5°C increase in global average temperature). These events, which occurred 120,000 and 2,400,000 yr ago, were the disappearances of the West

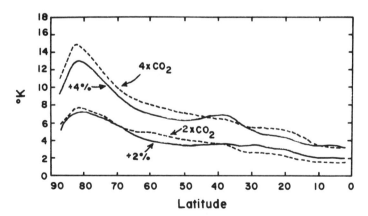

Figure 8 Latitudinal distribution of the mean (zonal) change of surface air temperature as a result of increases of 2 and 4% in the solar constant, and of doubled and quadrupled atmospheric CO_2 contents (after Manabe & Wetherald 1980).

Antarctic ice sheet and the Arctic Ocean drift ice, respectively. Flohn suggests that the thin Arctic sea ice, which is more sensitive to climatic change than large continental ice sheets, would require a shorter time to disappear than the Antarctic ice sheet and so would occur first. The unipolar-glaciation condition that would result would cause large shifts in climatic zones. Figure 9 shows the projected changes of annual surface temperature and precipitation with an ice-free Arctic. The northward displacement of the Intertropical Convergence Zone and the subtropical anticyclones by 2° or more would produce drying and warming in the regions around latitudes 40°N and 10°S. Precipitation would be greater between latitudes 10 to 20°N and north of 50°N. These results agree rather well with other model predictions (e.g. Kellogg & Bojkov 1982, and references cited).

Weller et al (1983) maintain, however, that temperature changes attributable to increasing CO_2 in the atmosphere cannot yet be confirmed because of inadequate data and large fluctuations in the last 100 yr. When other factors such as volcanic eruptions and solar variations that also influence climate are accounted for, Weller et al claim that "an apparent temperature trend consistent with the trend in CO_2 concentrations and simulations with climate models becomes more evident." These authors

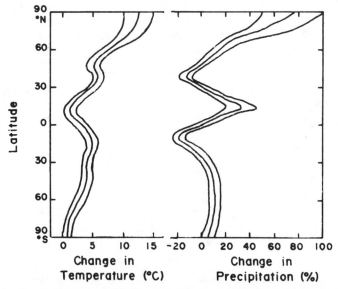

Figure 9   Changes in temperature (°C) and precipitation (%) by latitudes in the case of an ice-free Arctic. The range of uncertainty is shown by the three lines (after Flohn 1981).

warn, however, that because of large uncertainties, this conclusion can only be viewed as a tentative suggestion.

Sea Level

The other major paleoclimatic event that could happen again as a result of CO$_2$-induced warming (i.e. the disintegration of the West Antarctic ice sheet) could cause a sea-level rise of about 4–6 m (Kellogg 1979, Revelle 1983). According to Mercer (1978), this eventuality may occur after only a doubling of the atmospheric CO$_2$ content, a level that many investigators claim will be reached by about the middle of the next century. If the rate of fossil fuel consumption continues at the presently accelerated rate, this level could be reached within 50 yr. Mercer believes that the deglaciation of West Antarctica "would probably be the first disastrous result of continued fossil fuel consumption." A minimum time of 500 yr has been suggested for the disintegration of the West Antarctic ice sheet (Bentley 1983).

A 4–6 m rise in sea level would inundate many coastal cities and much of the Netherlands, Florida, Louisiana, and other low-lying regions around the world (Hoffman et al 1983, Revelle 1983). Some investigators (e.g. Revelle 1982) consider the disruption to human activity by these inundations not to be as potentially disastrous as might be imagined; they argue that depending on the rate of change, adjustment could very well keep pace with the change. On the other hand, other investigators (e.g. Hansen et al 1981) assert that the extent of our knowledge cannot provide the certainty that ice sheets would melt with CO$_2$ warming. If all the world ice were to melt in the time frame of several hundred years, the tens of meters or so rise in sea level would indeed severely disrupt human activities. Hansen et al claim, however, that with the warming of the ocean, the air above the world ice sheets might well remain below freezing, so that snow fall could increase, resulting in ice sheet growth and a lowering of sea level.

Hansen et al (1981) agree that the greatest danger of a rapid sea-level rise of up to around 5 or 6 m is with the melting of the West Antarctic ice sheet. This eventuality requires an average global warming of 2°C and about a 5°C warming of the West Antarctic ice sheet to initiate disintegration. The 2°C global warming may be reached and exceeded in the twenty-first century, even if we succeed in maintaining a no-growth energy policy and in banning coal use. The projected global warming of about 2.5°C for the next century "would approach the warmth of the Mesozoic, the age of dinosaurs" (Hansen et al 1981).

The complex interactions of the natural responses to CO$_2$ increase would also be reflected in huge changes in oceanic circulation, in oceanic biota, in the dissolution of calcium carbonate, and in many other possible processes about which we have very little information.

Social-Economic-Political Consequences

The climatic zonal changes associated with global warming could redistribute the world's water resources. The most devastating effect would be apparent in agricultural regions. Drought and excessive heat in the 40°N latitude band, for example, could cause the migration of the United States wheat belt north into Canada. The United States would then have to depend on Canada and other countries for its supply of this basic commodity, instead of the reverse situation as it is today. Adaptation to such economic changes would probably cause shifts in the global political structure (e.g. Kellogg & Schware 1982), but any discussion of the social, economic, and political consequences is obviously tentative.

Major river drainage systems around the world now supporting prime agricultural regions could decrease in flow, which would cause decreased food production; these areas include the Huanghe in China, the Amu Darya and Syr Darya in the USSR, the Tigris-Euphrates "fertile crescent," the Zambezi in Zimbabwe and Zambia, and the São Francisco in Brazil. At the same time the average flow of the Niger, Chari, Senegal, Volta, and Blue Nile rivers in northern Africa, of the Mekong in southeast Asia, and of the Brahmaputra in India could increase substantially because of increased precipitation; such an occurrence would lead to disastrous floods in these areas (Revelle 1982).

Underlying all the world crises is the unrelenting world population increase, especially in the developing countries. The increased production of food and consumption of fossil fuels needed to support this growth will insure that some of the predicted consequences of CO_2 increase will occur.

FEASIBILITY OF DELAYING GREENHOUSE WARMING

The economic disruption caused by the periodic occurrence of an El Niño–Southern Oscillation (ENSO) event is a well-documented fact. The 1982–1983 ENSO event is considered as possibly the largest natural perturbation of atmospheric CO_2 in the last 100 yr, with the observed fluctuation having been ± 1.5 ppm, or ± 3 Gtons C (Gammon et al 1984). Although disastrous for the individuals involved, for whom the recovery from its effects is often a slow process, the relatively short-term, cyclic nature of ENSO allows for adaptation without permanent, irreversible damage. Much more devastating could be the global, long-term effect of climatic zone changes caused by the CO_2 warming. For this reason, it is important to examine the ways in which to delay or adapt to such an eventuality. We should also be studying the natural feedback mechanisms to see which of the ongoing processes help us and which hinder us.

Natural Feedback Mechanisms

The effect of the CO_2 increase tends to be dampened somewhat by enhanced photosynthesis and increased productivity. Gorshkov (1982) maintains that the fixation of excess atmospheric carbon (without its subsequent oxidation) by ocean phytoplankton can result in an additional amount of carbon absorbed by the ocean equal to the amount of carbon ejected to the atmosphere from forest cutting. With this additional feedback at work, Gorshkov speculates, the concentration of atmospheric CO_2 should not exceed the preindustrial level by more than 40%. The quantities involved in such mechanisms, however, are difficult to ascertain.

Another example is the role of the world ice sheets. The Southern Ocean sea ice covers an area as large as 21,000,000 km² (Kukla & Gavin 1981). How much of this ice is permeable to CO_2 needs to be ascertained before we can determine how the ice affects an important link of the carbon cycle (Saltzman 1982). Current climatic models all consider the relationship between climate and sea ice. Because of the high albedo of ice, its presence or absence can change the surface solar energy balance in such a way that it tends to amplify any climatic change (Kellogg & Bojkov 1982). On the other hand, reduction of sea ice because of CO_2-induced global warming may enhance biological activities because of higher light intensity, and therefore it may reduce the CO_2 content in the atmosphere. More excess CO_2 could also penetrate into the polar oceans and reduce the greenhouse effect. These feedback mechanisms and their importance need to be evaluated.

Kellogg (1983) devised five feedback loops representing five different sets of conditions that govern the carbon dioxide concentration (under the assumption of our continued use of fossil fuels). The five loops are (a) CO_2–ocean circulation–upwelling, (b) CO_2–ocean stability–winter down-welling, (c) CO_2–Arctic sea ice–Arctic biomass, (d) CO_2–rainfall distribution–tropical biomass, and (e) CO_2–permafrost–tundra growth/decay and permafrost–methane release. Of these five, (a), (b), and (e) yield net results that amplify the rate of atmospheric CO_2 increase, whereas (c) and (d), which involve changes in biomass in the Arctic and the tropics, produce net weakly negative values. The physical processes involved in these five conditions are illustrated in Figure 10. Kellogg warns that the climate change due to CO_2 increase could be accelerated.

Possible Ways of Mitigating the Greenhouse Effect

REDUCTION OF FOSSIL FUEL USE   A US Environmental Protection Agency (EPA) report prepared by Seidel & Keyes (1983) analyzed various energy policies that could be adopted to delay or limit a greenhouse warming. Under the assumptions that a doubling of the preindustrial CO_2 level occurs by the year 2060 and that the average world temperature increases

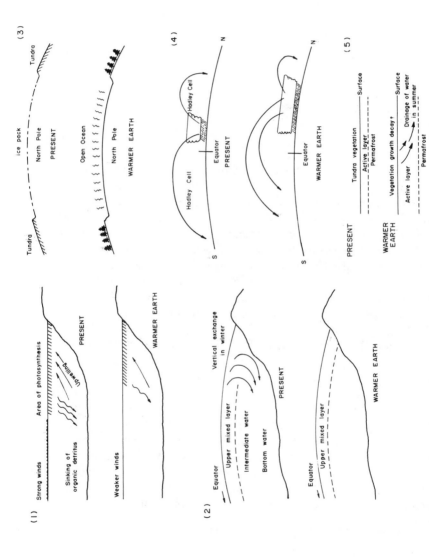

Figure 10 The physical processes involved in the five sets of conditions that govern the CO_2 concentration (assuming continued use of fossil fuels). Processes (1), (2), and (5) yield net results that amplify the rate of atmospheric CO_2 increase, whereas (3) and (4) produce weakly negative values (after Kellogg 1983).

by 2°C by around 2040, the report examines the effect of a 100% or 300% worldwide fossil fuel tax, a ban on synfuels and shale oil, a ban on coal, and a ban on coal and shale oil. The report concludes that a worldwide ban on coal instituted by the year 2000 would delay a 2°C change until 2055 and decrease the rate of temperature rise by the year 2100 from a 5°C warming to 3.5°C. A worldwide ban on both coal and shale oil would delay the 2°C change by another 10 years (or until the year 2065) and reduce the projected warming in the year 2100 from 5°C to 2.5°C. All other measures would affect the timing of the 2°C rise or reduce the warming effect by very little. The uncertainties involved in this assessment center on the unknown effect of the growth of other greenhouse gases, such as chlorofluoromethanes, nitrous oxide, and ozone. Also unknown is the temperature rise response to a given greenhouse gas increase once equilibrium is reached. These considerations are depicted in Figure 11.

The authors of the EPA report, however, recognize the infeasibility, both economic and political, of a worldwide ban on coal use. At the same time, they also dismiss nonenergy options such as scrubbing CO_2 emissions as technologically prohibitive, expensive, and of limited effectiveness.

OTHER ALTERNATIVES   While the EPA report gives a rather gloomy forecast of the world climate problem, the National Research Council Commission on Physical Sciences, Mathematics and Resources (1983a), in its Report of the Carbon Dioxide Assessment Committee, states that "overall, we find in the CO_2 issue reason for concern, but not panic." At the same time, the Commission (1983b) has proposed a global study program that, if implemented, would be unprecedented in its size and scope. According to Herbert Friedman, Chairman of the Commission, because changes in climate involve long and convoluted sets of interactions and "no single factor is clearly dominant," only a concerted, interdisciplinary approach on a global scale could hope to reveal "the physical, chemical and biological workings of the Sun-Earth system and the mysteries of the origin

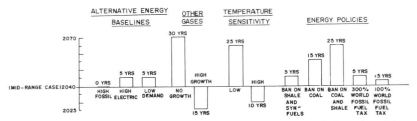

Figure 11   Changes in the date of a 2°C global warming. The year 2040 is the projected date for the midrange case. "Other gases" refers to greenhouse gases other than CO_2 (e.g. nitrous oxide, methane, chlorofluorocarbons) (after Seidel & Keyes 1983).

and survival of life in the biosphere." The program would coordinate the research activities of thousands of scientists around the world in various experiments and data-gathering schemes using space satellites, high-altitude balloons, and ocean research vessels equipped with the most sophisticated sensor devices. Clearly, the report seeks to reduce the uncertainties that plague the climate problem, giving the impression that a pessimistic outlook is premature.

Removal and disposal of CO_2   Until such a global fact-finding effort can be instituted and the results are known, individual investigators can only work on specific areas of the problem. Some (Baes et al 1980) have analyzed the feasibility of collecting and disposing of carbon dioxide from coal-fired power plants. These authors conclude that at present to scrub and strip carbon dioxide from the flue gas of a coal-fired power plant would require about 43% of the combustion energy of the coal. The coal could also be burned in pure oxygen (Marchetti 1977) in order to avoid the scrubbing and stripping method. The cost of supplying the oxygen would be about 30% of the output energy of the coal.

After recovery of the CO_2, Baes and his colleagues propose that the deep ocean is the most effective repository for disposal. Compressed CO_2 could be injected directly into seawater at a depth to prevent bubble formation (Marchetti 1977). This technique was analyzed by Hoffert et al (1979) using an atmosphere/mixed-layer/diffusive deep-ocean model for the carbon cycle, and they conclude that this method deserves further attention, especially if we continue to use fossil fuel at the present or an even more accelerated rate. This process also provides another buffering mechanism for the removal of CO_2 at the ocean bottom. The $CaCO_3$-bearing sediments on the seafloor are at present in a saturated or supersaturated water environment with respect to calcium carbonate. If fossil-fuel CO_2 is injected into the ocean bottom, the acidity could rise to the point at which the $CaCO_3$ is undersaturated and the sediments dissolve, which would further enhance the removal of CO_2 (Feely & Chen 1982). We do not know, however, in what time frame this enhanced removal would occur (Brewer 1983). The carbon dioxide could also be solidified into dry ice before disposal.

Baes et al (1980) further propose the development of floating coal-fired power plants that would circumvent the necessity for scrubbing/stripping by allowing for the direct disposal of CO_2 and a more efficient thermal cycle.

Carbon dioxide could also be disposed of on land. For example, it could be fixed onto a solid absorbent and stored in abandoned mines. The feasibility of such a technique needs to be studied (Steinberg 1983).

Recovered CO_2 could also be stored in depleted oil and gas wells, a process that would at the same time enhance the recovery of oil from depleted wells. Cavities in dissolved salt domes could serve as large storage spaces (Davis 1981, Horn & Steinberg 1982). Another possibility, albeit prohibitively expensive, is to package CO_2 as compressed gas (either solid or liquid) and dump it into outer space. Removal of other greenhouse gases such as N_2O and freons could also delay the global warming.

Direct measures to mitigate greenhouse warming   Measures that directly offset the greenhouse warming have been suggested. Such ideas as triggering volcanic explosions have been mentioned, since the dust emission has the effect of cooling the Earth by preventing part of the solar radiation from reaching the surface. Another idea is the injection of huge quantities of dust or sulfur dioxide into the stratosphere to reduce solar radiance by absorbing incoming visible sunlight. Aside from the cost factor, however, such measures have both predictable and unforeseen ramifications that could create situations as undesirable (or even more so) than CO_2 warming. For example, SO_2 in the atmosphere could affect chemical reactions that control the concentration of ozone and N_2O_2, both greenhouse gases. Furthermore, SO_2 would contribute to acid rain, itself a crisis-proportion problem. Because of the complexities involved and the compounded uncertainties of such measures, much more scrutiny and study would be needed before any of these options could be seriously considered.

The last alternative   The ultimate alternative for the survival of humankind is to study ways that we can adapt to the warming and other effects. In order to do so, much more information is needed.

CONCLUSION

Scientists have been aware of the potential consequences of changes in the atmospheric CO_2 level for many decades. Late nineteenth century investigators were especially impressed by the impact of the Industrial Revolution on the level of atmospheric CO_2. Modern scientists are alert to the accelerated effects of the same trends resulting from society's effort to meet the needs of an ever-growing world population. While the interaction of all the various factors may have either a compounded impact or a canceling effect, it is clear that two main factors emerge as the crux of the problem: the continued use of fossil fuels, and the large-scale deforestation necessary to provide land for cultivation of food.

Using various approaches, scientists have come to fair agreement over the preindustrial level of CO_2, and there is now good evidence that this level has been rising since the middle of the last century. If present trends continue, a global warming of a few degrees Celsius could result by the middle of the next century.

With the aid of mathematical models, specialists have presented scenarios of the future environment. While the models are based on the best estimates available, it is clear that there still remain all too many uncertainties in the responses of the systems and the feedback mechanisms. The role of the oceans in providing a sink for the gas is not fully understood with respect to the rates at which processes take place. Winter data in the polar regions that are critical to the understanding of CO_2 removal and of deep-water formation mechanisms are largely nonexistent. There is major disagreement on the effect of induced warming on the world ice sheets. The response of the biota to increased atmospheric CO_2 is complex and not well studied.

In order for us to know just what should be done about the CO_2 problem, we need much more information than we presently have. The exchange of knowledge that takes place at various national and international conferences and the studies that are now being carried out in individual laboratories throughout the world [notably those by the US Department of Energy (DOE) under its Carbon Dioxide Effects and Assessment Program (e.g. US Department of Energy 1979, 1980, 1982, 1983)] are all steps in the right direction. But a concerted global effort, as envisioned by DOE and proposed by the National Research Council, remains the hope for the eventual solution of the CO_2 problem, no matter whether such a solution lies in mitigating the effects of greenhouse gases or in adapting to the inevitable. Without the necessary knowledge upon which to base our decisions, any specific, problem-oriented, patchwork effort is just that— short-term, ineffectual, and potentially dangerous. The CO_2 issue illustrates once again the urgent need to support basic research.

ACKNOWLEDGMENTS

Our work was supported in part by grants from the Department of Energy's subcontract 19X-89608C with Oregon State University under Martin Marietta Energy Systems, Inc., contract DE-AC05-84OR 21400 with the US Department of Energy and the National Science Foundation (OCE 82-15053). We acknowledge with thanks the bibliographic assistance of Robert Lawrence and the Science-Technology staff of the Oregon State University (OSU) Kerr Library, Miriam Ludwig of the OSU College of Oceanography Pattullo Study, and the clerical assistance of Celeste Correia (OSU College of Oceanography). We thank the following colleagues who critically read a preliminary draft of this article and offered many helpful suggestions for improvement: Charles F. Baes, Jr. (Department of Chemistry, Oak Ridge National Laboratory), Charles W. Drake (Chairman, OSU Department of Physics), William P. Elliott (Air Resources Lab, National Oceanic and Atmospheric Administration), Mitchell Lyle and Nick Pisias (OSU College of Oceanography), and John Trabalka (Global Carbon Cycle Program, Oak Ridge National Laboratory). We also thank William W. Kellogg (National Center for Atmospheric Research) for his constructive comments, and two anonymous reviewers.

Literature Cited

Arrhenius, S. 1896. On the influence of carbonic acid in the air upon the temperature of the ground. *The London, Edinburgh, and Dublin Philos. Mag. and J. of Sci.*, 5th Ser. 41:237–76

Baes, C. F. Jr. 1981. The response of the oceans to increasing atmospheric carbon dioxide. *Rep. ORAU/IEA-816(M)*, Inst. Energy Anal., Oak Ridge, Tenn. 64 pp.

Baes, C. F. Jr., McLaughlin, S. B. 1984. Trace elements in tree rings: evidence of recent and historical air pollution. *Science* 224:494–97

Baes, C. F. Jr., Beal, S. E., Lee, D. W., Marland, G. 1980. Options for the collection and disposal of carbon dioxide. *Rep. ORNL-5657*, Oak Ridge Natl. Lab., Oak Ridge, Tenn. 35 pp.

Barnola, J. M., Raynaud, D., Neftel, A., Oeschger, H. 1983. Comparison of CO$_2$ measurements by two laboratories on air from bubbles in polar ice. *Nature* 303:410–13

Bentley, C. R. 1983. The west Antarctic ice sheet: diagnosis and prognosis. *Proc. Carbon Dioxide Res. Conf., Berkeley Springs, W. Va., 1982*, pp. IV.3–IV.50

Berner, R. A., Lasaga, A. C., Garrels, R. M. 1983. The carbonate-silicate geochemical cycle and its effect on atmospheric carbon dioxide over the past 100 million years. *Am. J. Sci.* 283:64–83

Bojkov, R. D., ed. 1983. *Report of the WMO (CAS) Meeting of Experts on the CO$_2$ Concentrations from Pre-Industrial Times to I.G.Y., Boulder, Colorado, June 22–25, 1983, WMO Project on Research and Monitoring of Atmospheric CO$_2$ Rep. No. 10*. 41 pp.

Bolin, B. 1981. Steady state and response characteristics of a simple model of the carbon cycle. In *Carbon Cycle Modelling*, ed. B. Bolin, pp. 315–31. *Sci. Comm. on Probl. Environ. (SCOPE) of the Int. Counc. Sci. Unions (ICSU)-SCOPE 16*. New York:

Wiley. 390 pp.

Bolin, B., Bjorkstrom, A., Holmen, K. 1983. The simultaneous use of tracers for ocean circulation studies. *Tellus* 35B:206–36

Brewer, P. G. 1978. Direct observation of the oceanic CO$_2$ increase. *Geophys. Res. Lett.* 5:997–1000

Brewer, P. G., Wong, G. T. F., Bacon, M. P., Spenser, D. W. 1975. An oceanic calcium problem? *Earth Planet. Sci. Lett.* 26:81–87

Brewer, P. G. 1983. The TTO North Atlantic Study—A progress report. *Proc. Carbon Dioxide Res. Conf., Berkeley Springs, W. Va., 1982*, II.93–II.122

Broecker, W. S. 1982. Ocean chemistry during glacial time. *Geochim. Cosmochim. Acta* 46:1689–1705

Broecker, W. S., Gerard, R. D., Ewing, M., Heezen, B. C. 1960. Natural radiocarbon in the Atlantic Ocean. *J. Geophys. Res.* 65:2903–31

Broecker, W. S., Takahashi, T., Simpson, H. J., Peng, T.-H. 1979. Fate of fossil fuel carbon dioxide and the global carbon budget. *Science* 206:409–18

Broecker, W. S., Peng, T. H., Brewer, P., Takahashi, T. 1982. Chenology challenged. *Eos, Trans. Am. Geophys. Union* 63:979 (Abstr.)

Broecker, W. S., Peng, T. H., Takahashi, T. 1983. Reconstructions of past atmospheric CO$_2$ contents from the chemistry of the contemporary ocean: an evaluation. Unpublished report

Callendar, G. 1939. The artificial production of carbon dioxide and its influence on temperature. *Q. J. R. Meteorol. Soc.* 64:223–37

Chamberlin, T. C. 1899. An attempt to frame a working hypothesis of the cause of glacial periods on an atmospheric basis. *J. Geol.* 7:545–84, 667–87, 751–87

Chen, C. T. 1978. Decomposition of calcium carbonate and organic carbon in the deep oceans. *Science* 201:735–36

232   CHEN & DRAKE

Chen, C. T. 1982a. Oceanic penetration of excess CO_2 in a cross-section between Alaska and Hawaii. *Geophys. Res. Lett.* 9:117–19

Chen, C. T. 1982b. On the distribution of anthropogenic CO_2 in the Atlantic and Southern Oceans. *Deep-Sea Res.* 29:563–80

Chen, C. T. 1982c. Carbonate chemistry during WEPOLEX-81. *Antarct. J. 1982 Rev.*, pp. 102–3

Chen, C. T. 1983. Distributions of dissolved calcium and alkalinity in the Weddell Sea in winter. *Antarct. J. 1983 Rev.*, pp. 136–37

Chen, C. T. A. 1984. Carbonate chemistry of the Weddell Sea. *US Dep. Energy Rep. DOE/EV/10611-4.* 118 pp.

Chen, C. T. A. 1985. Preliminary observations of oxygen and carbon dioxide of the wintertime Bering Sea marginal ice zone. *Cont. Shelf Res.* 4:465–83

Chen, C. T. A., Millero, F. J. 1979. Gradual increase of oceanic CO_2. *Nature* 277:305–7

Chen, C. T., Poisson, A. 1984. Excess carbon dioxide in the Weddell Sea. *Antarct. J. 1984 Rev.*, pp. 74–75

Chen, C. T., Pytkowicz, R. M. 1979. On the total CO_2-titration alkalinity-oxygen system in the Pacific Ocean. *Nature* 281:362–65

Chen, C. T., Millero, F. J., Pytkowicz, R. M. 1982a. Comment on calculating the oceanic CO_2 increase: a need for caution by A. M. Shiller. *J. Geophys. Res.* 87:2083–85

Chen, C. T., Pytkowicz, R. M., Olson, E. J. 1982b. Evaluation of the calcium problem in the South Pacific. *Geochem. J.* 16:1–10

Cline, J. D., Feely, R. A., Kelley-Hansen, K., Gendron, J. F., Wisegarver, D. P., Chen, C. T. 1985. Current inventory of anthropogenic carbon dioxide in the North Pacific Gyre. *NOAA Tech. Memo. ERL PMEL-60.* 46 pp.

Craig, H. 1957. The natural distribution of radiocarbon and the exchange time of carbon dioxide between atmosphere and sea. *Tellus* 9:1–17

Davis, R. M. 1981. National strategic petroleum reserve. *Science* 213:618–22

Edmond, J. M. 1974. On the dissolution of carbonate and silicate in the deep sea. *Deep-Sea Res.* 21:455–80

Elliott, W. P. 1983. A note on the historical industrial production of carbon dioxide. *Clim. Change* 5:141–44

Feely, R. A., Chen, C. T. 1982. The effect of excess CO_2 on the calculated calcite and aragonite saturation horizons in the northeast Pacific. *Geophys. Res. Lett.* 9:1294–97

Fine, R. A., Reid, J. L., Ostlund, H. G. 1981. Circulation of tritium in the Pacific Ocean.

J. Phys. Oceanogr. 11:3–14

Flohn, H. 1981. Major climatic events associated with a prolonged CO_2-induced warming. *Inst. Energy Anal. Rep.*, Oak Ridge, Tenn. 80 pp.

Francey, R. J., Farquhar, G. D. 1982. An explanation of $^{13}C/^{12}C$ variations in tree rings. *Nature* 297:28–31

Freyer, H. D., Belacy, N. 1983. $^{13}C/^{12}C$ records in Northern Hemispheric trees during the past 500 years—anthropogenic impact and climatic superpositions. *J. Geophys. Res.* 88:6844–52

Gammon, R. H., Komhyr, W. D., Waterman, L. S., Conway, T., Thoning, K. 1984. Estimating the natural variation in atmospheric CO_2 since 1860 from interannual changes in tropospheric temperature and the history of major El Niño events. *Abstr. AGU Chapman Conf. on Nat. Var. in Carbon Dioxide and the Carbon Cycle, Innisbrook, Tarpon Springs, Fla*

Gates, W. L., Cook, K. H., Schlesinger, M. E. 1981. Preliminary analysis of experiments on the climatic effects of increased CO_2 with an atmospheric general circulation model and a climatological ocean. *J. Geophys. Res.* 86:6385–93

Gorshkov, V. G. 1982. The possible global budget of carbon dioxide. *Nuovo Cimento C* 5:209–22

Hansen, J., Johnson, D., Lacis, A., Lebedeff, S., Lee, P., et al. 1981. Climate impact of increasing atmospheric carbon dioxide. *Science* 213:957–66

Hoffert, M. I., Wey, Y.-C., Callegari, A. J., Broecker, W. S. 1979. Atmospheric response to deep-sea injections of fossil-fuel carbon dioxide. *Clim. Change* 2:53–68

Hoffman, J. S., Keyes, D., Titus, J. G. 1983. Projecting future sea level rise—methodology, estimates to the year 2100, and research needs. *Off. Policy Resour. Manage. Rep.*, Environ. Prot. Agency Washington, DC. 121 pp. 2nd ed.

Horn, F. L., Steinberg, M. 1982. Possible sites for disposal and environmental control of atmospheric carbon dioxide. *Rep. BNL51597*, Brookhaven Natl. Lab., Upton, N.Y. 23 pp.

Horne, R. A. 1969. *Marine Chemistry.* New York: Wiley-Interscience. 568 pp.

Houghton, R. A., Hobbie, J. E., Melillo, J. M., Moore, B., Peterson, B. J., et al. 1983. Changes in the carbon content of terrestrial biota and soils between 1860 and 1980: a net release of CO_2 to the atmosphere. *Ecol. Monogr.* 53:235–62

Keeling, C. D. 1973. Industrial production of carbon dioxide from fossil fuels and limestone. *Tellus* 25:174–98

Keeling, C. D. 1979. The Suess effect. *Environ. Int.* 2:229–300

Keeling, C. D., Bacastow, R. B., Bainbridge, A. E., Ekdahl, C. A. Jr., Guenther, P. R., Waterman, L. S. 1976a. Atmospheric carbon dioxide variations at Mauna Loa Observatory, Hawaii. *Tellus* 28 : 538–51

Keeling, C. D., Adams, J. A. Jr., Ekdahl, C. A. Jr., Guenther, P. R. 1976b. Atmospheric carbon dioxide variations at the South Pole. *Tellus* 28 : 552–64

Keeling, C. D., Bacastow, R. B., Whorf, T. P. 1982. Measurements of the concentration of carbon dioxide at Mauna Loa Observatory, Hawaii. In *Carbon Dioxide Review 1982*, ed. W. C. Clark, pp. 377–85. Oxford/New York : Clarendon. 469 pp.

Kellogg, W. W. 1979. Influences of mankind on climate. *Ann. Rev. Earth Planet. Sci.* 7 : 63–92

Kellogg, W. W. 1983. Feedback mechanisms in the climate system affecting future levels of carbon dioxide. *J. Geophys. Res.* 88 : 1263–69

Kellogg, W. W., Schware, R. 1982. Society, science and climate change. *Foreign Aff.* 60 : 1076–1109

Kellogg, W. W., Bojkov, R. D. 1982. Report of the JSC/CAS meeting of experts on detection of possible climate change. *WMO Rep. WCP-29*. 43 pp.

Kerr, R. A. 1977. Carbon dioxide and climate : carbon budget still unbalanced. *Science* 197 : 1352–53

Kerr, R. A. 1980. Carbon budget not so out of whack. *Science* 208 : 1358–59

Killough, C. G., Emanuel, W. R. 1981. A comparison of several models of carbon turnover in the ocean with respect to their distributions of transit time and age, and responses to atmospheric CO_2 and ^{14}C. *Tellus* 33 : 274–90

Kukla, G., Gavin, J. 1981. Summer ice and carbon dioxide. *Science* 214 : 497–503

Macdonald, G. J., ed. 1982. *The Long-Term Impacts of Increasing Atmospheric Carbon Dioxide Levels*. Cambridge, Mass. Ballinger. 252 pp.

Manabe, S., Wetherald, R. T. 1980. On the distribution of climate change resulting from an increase in CO_2 since the atmosphere. *J. Atmos. Sci.* 37(1) : 99–118

Marchetti, C. 1977. On geoengineering the CO_2 problem. *Clim. Change* 1 : 59–68

Mercer, J. H. 1978. West Antarctic ice sheet and CO_2 greenhouse effect : a threat of disaster. *Nature* 271 : 321–25

Merlivat, L., Memery, L. 1983. Gas exchange across an air-water interface : experimental results and modeling of bubble contribution to transfer. *J. Geophys. Res.* 88 : 707–24

Minas, H. J. 1980. Analyse de diagrammes de facteurs hydrologiques et chimiques (température, salinité, oxygène, sels nutritifs) :

Application à l'étude du système production—régénération dans les résurgences côtières (côtes NW africaines) et les zones à fort mélange vertical (Méditerranee, océan antartique). In *Prod. Primaire et Secondaire, Colloq. Fr.-Sov. Publ. CNEXO No. 10*, pp. 21–36

Muntz, M. M. A., Aubin, E. 1886. *Recherches sur l'Acide Carbonique de l'Air, Mission Scientifique du Cap Horn 1882–1883*, Vol. 3, pp. A3–A82. Paris : Gauthier-Villars

National Research Council Commission on Physical Sciences, Mathematics and Resources. 1983a. *Changing Climate, Report of the Carbon Dioxide Assessment Committee, Board on Atmospheric Sciences and Climate*. Washington, DC : Natl. Acad. Press. 496 pp.

National Research Council Commission on Physical Sciences, Mathematics and Resources. 1983b. *Toward an International Geosphere-Biosphere Program : A Study of Global Changes*. Washington, DC : Natl. Acad. Press. 81 pp.

Nordhaus, W. D., Yohe, G. W. 1983. Future paths of energy and carbon dioxide emissions. See National Research Council Commission on Physical Sciences, Mathematics and Resources 1983a, pp. 87–153

Oeschger, H., Siegenthaler, U., Schotterer, U., Gugelmann, A. 1975. A box diffusion model to study the carbon dioxide exchange in nature. *Tellus* 27 : 168–92

Ostlund, H. G. 1981. Tritium laboratory data release #81-23. *TTO Test Cruise Radiocarbon and Tritium Results*, Univ. Miami, Coral Gables, Fla. 19 pp.

Patel, R. C., Bose, R. J., Atkinson, G. 1973. The CO_2-water system. I. Study of the slower hydration-dehydration step. *J. Solution Chem.* 2 : 357–72

Peng, T. H., Broecker, W. S., Mathieu, G. G., Li., Y. H. 1979. Radon evasion rates in the Atlantic and Pacific Oceans as determined during the GEOSECS program. *J. Geophys. Res.* 84 : 2471–86

Peng, T. H., Broecker, W. S., Freyer, H. D., Trumbore, S. 1983. A deconvolution of the tree ring based $\delta^{13}C$ record. *J. Geophys. Res.* 88 : 3609–20

Peng, T. H., Broecker, W. S., Freyer, H. D. 1984. Revised estimates of atmospheric CO_2 variations based on tree ring ^{13}C record. *Abstr. AGU Chapman Conf. on Nat. Var. in Carbon Dioxide and the Carbon Cycle, Innisbrook, Tarpon Springs, Fla.*

Plass, G. N. 1959. Carbon dioxide and climate. *Sci. Am.* 201 : 41–47

Postma, H. 1964. The exchange of oxygen and carbon dioxide between the ocean and the atmosphere. *Neth. J. Sea Res.* 2 : 258–83

Pytkowicz, R. M. 1983. *Equilibria, Nonequili-*

bria, and Natural Waters, Vols. 1, 2. New York: Wiley-Interscience. 351 pp., 383 pp.

Pytkowicz, R. M., Chen, C. T. A. 1984. Ionic strength, solubility, and gas exchange across the sea surface. *Proc. Conf. Gas-Liq. Chem. Nat. Waters, Brookhaven, N.Y.*, pp. 34.1–34.7

Revelle, R. 1982. Carbon dioxide and world climate. *Sci. Am.* 247(2):35–43

Revelle, R. 1983. Probable future changes in sea level resulting from increased atmospheric carbon dioxide. See National Research Council Commission on Physical Sciences, Mathematics and Resources 1983a, pp. 433–48

Revelle, R., Suess, H. 1957. Carbon dioxide exchange between atmosphere and ocean and the question of an increase of atmospheric CO_2 during the past decades. *Tellus* 9:18–27

Rotty, R. 1977. Global carbon dioxide production from fossil fuels and cement A.D. 1950–A.D. 2000. In *The Fate of Fossil Fuel CO_2 in the Oceans*, ed. R. N. Anderson, A. Malahoff, pp. 167–82. New York: Plenum

Rotty, R. 1982. Distribution of and changes in industrial carbon dioxide production. *Inst. Energy Anal. Rep.*, Oak Ridge, Tenn. 22 pp.

Rotty, R., Marland, G. 1980. Constraints on fossil fuel use. In *Interactions of Energy and Climate*, ed. W. Bach, J. Pankrath, J. Williams, pp. 191–212. Dordrecht, Neth: Reidel

Saltzman, B. 1982. Stochastically-driven climatic fluctuations in the sea-ice, ocean temperature, CO_2 feedback system. *Tellus* 34:97–112

Schlesinger, M. E. 1983. Simulating CO_2-induced climatic change with mathematical climate models: capabilities, limitations and prospects. *Proc. Carbon Dioxide Res. Conf., Berkeley Springs, W. Va., 1982*, pp. III.3–III.139

Seidel, S., Keyes, D. 1983. Can we delay a greenhouse warming? The effectiveness and feasibility of options to slow a build-up of carbon dioxide in the atmosphere. *Off. Policy Resour. Manage. Rep.*, Environ. Prot. Agency, Washington, DC. 180 pp.

Shabecoff, P. 1984. Widespread ills found in forests in eastern U.S. *New York Times*, 24 February 1984, p. 1

Shiller, A. M. 1981. Calculating the oceanic CO_2 increase: a need for caution. *J. Geophys. Res.* 86:11083–88

Shiller, A. M. 1982. Reply to comments of Chen, Millero and Pytkowicz. *J. Geophys. Res.* 87:2086

Shiller, A. M., Gieskes, J. M. 1980. Processes affecting the oceanic distributions of dissolved calcium and alkalinity. *J. Geophys. Res.* 85:2719–27

Siegenthaler, U. 1984. 19th Century measurements of atmospheric CO_2—a comment. *Clim. Change* 6:409–11

SIO reference #84-14. 1984. *CSS Hudson Cruise 82-001, 14 February–6 April 1982*, Vol. 1. *Physical and Chemical Data*. Berkeley: Univ. Calif. 305 pp.

Smith, W. D., Cheek, C. H. 1982. Freon-11 as a tracer of oceanic mixing. *Nav. Res. Lab. 1982 Rev.*, pp. 51–52

Stauffer, B., Siegenthaler, U., Oeschger, H. 1984. Atmospheric CO_2 concentrations during the last 50,000 y from ice core analysis. *Abstr. AGU Chapman Conf. on Nat. Var. in Carbon Dioxide and the Carbon Cycle, Innisbrook, Tarpon Springs, Fla.*

Steinberg, M. 1983. An analysis of concepts for controlling atmospheric carbon dioxide. *Rep. DOE/CH/00016-1*, Dep. Energy Environ. 66 pp.

Stokes, G. M., Barnard, J. C., Pearson, E. W. 1984. Historical carbon dioxide: abundances derived from the Smithsonian spectrobolograms. *Rep. DOE/NBB-0063*, US Dep. Energy. 114 pp.

Stuiver, M. 1978. Atmospheric carbon dioxide and carbon reservoir changes. *Science* 199:253–58

Stuiver, M., Burk, R. L., Quay, P. D. 1984. $^{13}C/^{12}C$ ratios and the transfer of biospheric carbon to the atmosphere. *J. Geophys. Res.* 89:1731–48

Takahashi, T., Broecker, W. S., Bainbridge, A. E. 1981. The alkalinity and total carbon dioxide concentration in the world oceans. In *Carbon Cycle Modelling*, ed. B. Bolin, pp. 271–86. New York: Wiley. 390 pp.

Takahashi, T., Chipman, D., Volk, T. 1983. Geographical, seasonal, and secular variations of the partial pressure of CO_2 in the surface waters of the North Atlantic Ocean: the results of the north Atlantic TTO program. *Proc. Carbon Dioxide Res. Conf., Berkeley Springs, W. Va., 1982*, pp. II.123–II.145

Tyndall, J. 1861. On the absorption and radiation of heat by gases and vapours, and on the physical connexion of radiation, absorption, and conduction—The Bakerian Lecture. *The London, Edinburgh, and Dublin Philos. Mag. and J. of Sci.*, 4th Ser. 22:169–94, 273–85

US Department of Energy (DOE). 1979. Workshop on the global effects of carbon dioxide from fossil fuels, Miami Beach, Florida, March 7–11, 1977. *Rep. CONF-770385*. 122 pp.

US Department of Energy (DOE). 1980. Environmental and societal consequences of a possible CO_2-induced climate change: a research agenda. *Rep. DOE/EV/10019-1*, Vol. 1. 116 pp.

US Department of Energy (DOE). 1982.

Environmental and societal consequences of a possible CO_2-induced climate change: effect of increasing CO_2 on ocean biota. *Rep. DOE/EV/10019-2*, Vol. 2,. 22 pp.

US Department of Energy (DOE). 1983. An analysis of concept for controlling atmospheric carbon dioxide. *Rep. DOE/CH/00016-1*. 66 pp.

Weiss, R. F., Ostlund, H. G., Craig, H. 1979. Geochemical studies of the Weddell Sea. *Deep-Sea Res.* 26: 1093–1120

Weller, G., Baker, D. J. Jr., Gates, W. L., MacCracken, M. C., Manabe, S., Vonder Haar, T. H. 1983. Detection and monitoring of CO_2-induced climate changes. See National Research Council Commission on Physical Sciences, Mathematics and Resources 1983a, pp. 292–382

Wigley, T. M. L. 1983. The pre-industrial carbon dioxide level. *Clim. Change* 5: 315–20

Wong, C. S. 1978. Atmospheric input of carbon dioxide from burning wood. *Science* 200: 197–200

Woodwell, G. M. 1978. The carbon dioxide question. *Sci. Am.* 238(1): 34–43

Woodwell, G. M. 1983. Biotic effects on the concentration of atmospheric carbon dioxide: a review and projection. See National Research Council Commission on Physical Sciences, Mathematics and Resources 1983a, pp. 216–41

Ann. Rev. Earth Planet. Sci. 1986. 14:237–65

COASTAL PROCESSES AND THE DEVELOPMENT OF SHORELINE EROSION

Paul D. Komar and Robert A. Holman

College of Oceanography, Oregon State University, Corvallis, Oregon 97331

INTRODUCTION

In 1971 the US Army Corps of Engineers inventoried the stability of the nation's coastlines, concluding in their final summary, *A Report on the National Shoreline Study*, that of the 135,000 km of coastline, some 33,000 km were "seriously eroding." More recently, Dolan et al (1983) assembled the existing data on shoreline changes for the United States, reaffirming the magnitude of this problem. Although such inventories have not been undertaken for the entire world's shorelines, coastal erosion is undoubtedly a global problem.

Beach erosion is for the most part episodic, with the major retreats in the shoreline occurring during unusually intense storms. On the east coast of the United States, beach erosion and property losses are produced by "nor'easters" and hurricanes; on the west coast, intense storms move down from the Gulf of Alaska, generating waves that attack the coastline. This obvious association of beach erosion with the energy level of storm waves creates an oversimplified impression of the actual physical processes involved in the erosion. Only recently have we begun to understand and appreciate the diversity of processes that are actually involved in coastal erosion. It is now recognized that long-period oscillations may dominate water motions in the nearshore, sometimes containing more energy than the storm-generated waves. These oscillations, as well as a variety of nearshore currents, rearrange the beach sand to cause highly variable rates of shoreline retreat along even short stretches of beach. Although the long-term rise in sea level has been recognized as an important factor in shoreline

237

0084–6597/86/0515–0237$02.00

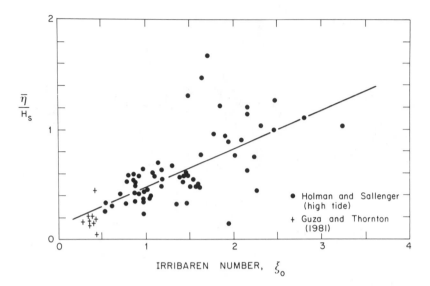

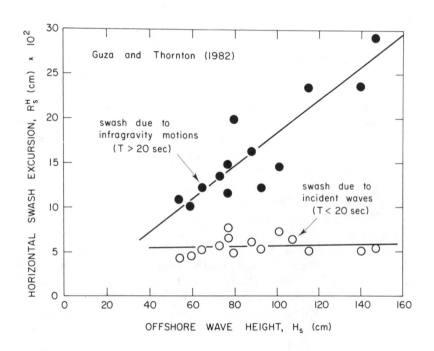

retreat, recent episodes of coastal erosion, particularly in California, have led to the recognition that sea level may rise by tens of centimeters for time spans of only a few months and thus contribute to this erosion. Many factors, therefore, may interact to produce an episode of beach erosion with coastal property losses.

STORM WAVES, RUN-UP, AND BEACH EROSION

The swash of waves on the beachface represents the biting edge of the sea, the zone of shoreline retreat during a storm. Recent investigations have focused on the run-up of the wave swash, obtaining unexpected results that are significant to our understanding of nearshore processes.

It is of course the maximum shoreward extent of wave run-up that has particular significance to erosion of the beach and to property losses. This run-up is found to consist of three primary components [Figure 1 (*right*)]: (*a*) a superelevation of the mean water level above the still-water level of the sea, a rise that is termed *set-up*; (*b*) fluctuations about that mean due to the direct swash of individual waves; and (*c*) a component in the swash oscillations having periods in excess of 20 s, periods beyond the range of normal ocean waves. The maximum run-up height achieved by the water is the summation of these three components.

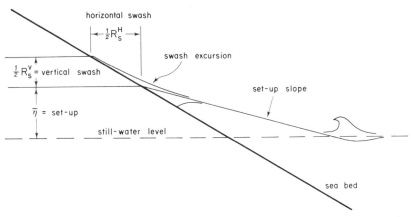

Figure 1 (*Right*) Schematic of the wave set-up $\bar{\eta}$, the rise of the mean water level due to the presence of waves, and the horizontal and vertical components of the swash excursion R_s about that mean due to the direct run-up of the waves and superimposed long-period water motions. (*Upper left*) The set-up divided by the offshore wave height H_s as a function of the Irribaren number of Equation (1). (*Lower left*) The dependence of the horizontal swash excursion distance on the wave height, showing the separate portions of the swash due to incident waves and infragravity water motions (Guza & Thornton 1982).

Field measurements of the wave run-up and set-up have been obtained by unusual techniques. Guza & Thornton (1981, 1982) utilized an 80-m-long dual-resistance wire that was stretched out across the beach profile and held at 3 cm above the sand level by nonconducting supports. In a different approach, Holman & Sallenger (1985) obtained their run-up measurements by taking time-lapse photographs each second over a period of 35 minutes; this "remote-sensing" technique has the advantage that it can document simultaneously the run-up along an extended stretch of shoreline. The swash position is then digitized frame by frame to follow the time-varying position of the waterline. Holman & Guza (1984) have undertaken direct comparisons of the results of their different techniques. In either case, the average of the entire record yields the set-up level, and the variability about this mean represents the time-varying components of the swash action.

Interest in wave set-up was inspired by observations during a hurricane in 1938 that struck the east coast of the United States. The maximum mean water elevation at the shoreline was found to be 1 m higher in an exposed area of the coast, where the wave energy was dissipated as surf, than in the calmer waters of a sheltered region. This difference could not be explained by variations in storm-surge height, so it was instead proposed that the breaking waves caused the elevation in mean water level. This speculation has been verified by theoretical analyses (Longuet-Higgins & Stewart 1964) and detailed laboratory measurements (Bowen et al 1968), but only recently has it been confirmed by field observations (Guza & Thornton 1981, Holman & Sallenger 1985).

The theoretical analysis of Longuet-Higgins & Stewart (1964) determined that the momentum flux (radiation stress) of the breaking waves causes the mean water level or set-up to slope upward toward the land [Figure 1 (right)], a pattern verified by the laboratory measurements of Bowen et al (1968). The set-up above the still-water level therefore reaches a maximum at the shoreline. Guza & Thornton (1981) measured the set-up on an ocean beach of low slope and wide surf zone, finding that the maximum set-up at the shoreline is approximately $0.17H_s$, where H_s is the significant wave height of the incident waves in deep water (the average of the highest one third of the waves). According to this relationship, deep-water waves of 300 cm will produce a set-up of 51 cm at the shoreline. If a storm then occurs, increasing the wave height by 200 cm, the set-up elevation will be increased by an additional 34 cm, reaching a height of 85 cm. If the overall beach slope is 1 : 50 (a common value), the 34-cm rise due to the storm would move the mean shoreline landward by some 17 m. It would appear from these results that changes in set-up from normal to storm conditions would be an important factor in the resulting erosion due to the storm.

The data of Guza & Thornton (1981) were obtained from a beach on the coast of California. Holman & Sallenger (1985) obtained additional measurements from a steeper beach (1 : 10 average slope) in North Carolina under a wide range of wave conditions (heights of 0.4 to 4.0 m) and effective beach slopes. The measured wave set-up at the shoreline amounted to as much as 1.6 m, again demonstrating its significance. They found that a direct correlation between the set-up and wave height was highly scattered, but as shown in Figure 1 (*upper left*), the scatter was greatly reduced if the set-up $\bar{\eta}$, divided by the deep-water wave height H_s, is related to the Irribaren number

$$\xi_0 = \tan \beta/(H_s/L_0)^{1/2} \tag{1}$$

where $\tan \beta$ is the beach slope and L_0 is the deep-water wave length (and thus the ratio H_s/L_0 is the wave steepness).

Guza & Thornton (1982) found in their analyses that the excursion distance of the swash fluctuations about the mean set-up level is also directly dependent on the incident wave height. When expressed as a vertical run-up height R_s^y [Figure 1 (*right*)], it was found that the swash is about $0.7H_s$. Here again, the increase in H_s during storms will produce an increase in the run-up that is in addition to the rise in the set-up. This is expected of course, but a surprising result is that most of the swash excursion was found by Guza & Thornton to occur at periodicities that are much greater than the period range of the incident waves (normally limited to less than 20 s). In their analyses, Guza & Thornton were able to separate these two swash components through spectral analyses of the run-up records, with one component being the direct swash of the incident wave bores and the other being the long-period *infragravity* component. The results are shown in Figure 1 (*lower left*), and they demonstrate the unexpected conclusion that the component due to wave bores has no dependence on the heights of the waves in deep water, whereas the infragravity motions grow with the offshore wave heights. It is apparent from the results of Guza & Thornton that most of the energy of the swash oscillations is in the infragravity range (> 20 s) rather than in the period range of the incident waves, and this is especially so during storm conditions with higher incident waves.

The study of Guza & Thornton was undertaken on a beach of low slope, where the wide surf zone apparently acts as a filter between the offshore waves and the swash at the shoreline. A doubling of the incident wave height, for example, causes the waves to break at approximately twice the water depth, doubling the distance from the shoreline to the breaker zone; the result is that there is little effect on the swash produced by the individual wave bores, which must now travel twice as far. Such a nearshore system is termed a *dissipative* beach, in the sense that its low slope acts to dissipate the energy of the waves. At the other extreme is the *reflective* beach system, typically a steeper beach where the waves break directly at the base of and immediately swash up the beachface. As expected, the swash on a reflective beach is dominated by the incident waves. As is seen throughout this article, the distinction between dissipative and reflective beaches is important to a wide range of physical processes.

The North Carolina beach site of the study of Holman & Sallenger (1985) was intermediate, being steeper than in the study of Guza & Thornton (1982), although a significant surf zone still prevailed. The beach tended to be reflective at high tide and dissipative at low tide, responding to the effective slope of the concave-up beach profile. As with their measurements of wave set-up, Holman & Sallenger found that the swash height R_s^v can be related to the Irribaren number for both the incident-wave component and the portion due to infragravity motions. They determined a critical Irribaren number of approximately $\xi_0 = 1.75$; below this value the swash is dominated by infragravity motions (dissipative beach system), while at higher Irribaren values the swash of incident waves is more important (reflective beach).

The results of the field studies by Guza & Thornton (1981, 1982) and Holman & Sallenger (1985) indicate that during a storm, the run-up of water on the beachface will increase as a result of a rise in both the set-up and the swash oscillations about that mean set-up level. A storm can typically raise the set-up vertically by more than 1 m, moving the shoreline landward by tens of meters. On a low-sloping dissipative beach, there will be a considerable increase in long-period infragravity energy in the nearshore water motions during the storm but little if any actual increase in the incident-wave energy at the shoreline itself. In contrast, on a steeper reflective beach the processes continue to be dominated by the incident waves, with the strong infragravity motions not generally developing.

The responses of the beachface to the changing run-up and set-up are not well known. The beachface is of course eroded back during storms, with the sand moving offshore to form systems of nearshore bars. In the past, this shift has been related empirically to the wave steepness and sediment grain size (Komar 1976, pp. 289–94). However, the actual processes of this cross-shore sand movement remain poorly understood. It is likely that the beachface responds to a rise in set-up much as it changes with a rising tide: The slope steepens during the rise as deposition occurs on the unsaturated upper beachface and erosion occurs on the saturated lower beach (Duncan 1964). The studies of Sallenger & Richmond (1984) and Howd & Holman (1984) document beachface changes that are apparently in response to the long-period infragravity motions of the swash. These studies measured the sediment level at a series of locations that stretched along as well as across the swash zone. Howd & Holman found sediment-level oscillations of up to 6 cm with time scales of 8–10 min, while Sallenger & Richmond measured oscillations on time scales of 6–15 min. The crests of the oscillations were shore-parallel and were found to migrate landward, progressively decreasing in height. Their onshore movement did not necessarily represent a shoreward movement of sand, however, since in some cases they occurred during a net seaward sediment transport. Howd & Holman found that the sand-level oscillations correlated with measured infragravity motions in the run-up, which indicates that the oscillations are forced by the swash action. Employing sediment transport equations, they also modeled the development and migration of the sediment oscillations.

THE CONTRIBUTION OF EDGE WAVES TO THE INFRAGRAVITY MOTIONS

Although the infragravity motions seen in the swash could be long-period waves incident from deep water (which would then reflect from the beach) or could possibly be the result of the grouping of incoming waves, there is a considerable body of evidence that an important portion of this infragravity motion results from edge waves in the nearshore. As depicted in Figure 2, edge waves are trapped in the nearshore by the seaward slope of the bottom and are in a sense held there by wave refraction; when reflected from the shoreline at an angle, they first travel seaward but refract as they go, eventually turning back toward the land to be reflected once more to repeat the process. Their net movement is thereby in the longshore direction, with a longshore wavelength

$$L_e = \frac{g}{2\pi} T_e^2 \sin\left[(2n+1)\beta\right], \tag{2}$$

where T_e is the edge-wave period and $n = 0, 1, 2, \ldots$ is the mode. This mode number is particularly important to the offshore profile of the edge wave [Figure 2 (*bottom*)] and represents the number of zero-crossings in the offshore amplitude variations. In all cases, however, the maximum amplitude occurs at the shoreline. As with normal surface waves, edge waves may either be progressive, moving in one direction along the shoreline, or standing, where in the longshore direction there are alternating antinodes of full amplitude excursion in the swash and nodes where there is no swash contribution [Figure 2 *(top)*].

Edge waves ultimately receive their energy from the wind-generated waves, and this in part accounts for the dependence of the infragravity energy on the incident wave height seen in Figure 1 *(lower left)*. Since the energy is trapped within the nearshore region when it enters edge waves, the infragravity energy can grow to substantial levels. Of particular significance to beach erosion, the amplitude of the edge-wave motions increases toward

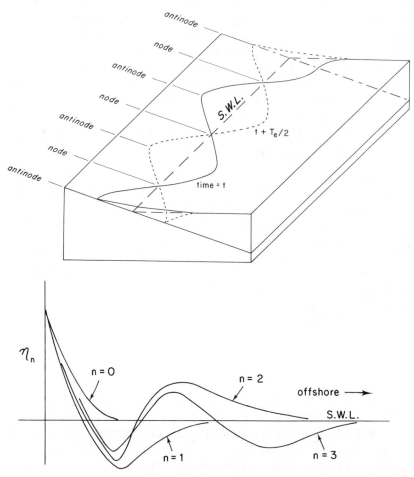

Figure 2 *(Top)* The simplest type of standing edge wave ($n = 0$), consisting of antinodes (where the full swash motion is observed) and nodes (where no motion occurs). *(Bottom)* Offshore profiles of amplitude variations of edge waves $n = 0$ through 3.

the shoreline (where it reaches a maximum) and thereby dominates the incident waves, which decrease in energy by breaking as they cross the surf zone (Holman 1983, Wright & Short 1984).

A number of studies have analyzed the generation of edge waves. Guza & Davis (1974) demonstrated that on steep reflective beaches, the dominant edge waves formed will have a period exactly twice that of the incident waves ($T_e = 2T_i$) and will be mode $n = 0$; these are termed *subharmonic* edge waves. Also important to beach processes are *synchronous* edge waves, which have the same period as the incident waves ($T_e = T_i$). On low-sloping dissipative beaches, one generation mechanism involves the interaction of incident waves of slightly different periods and possibly different approach angles; this mechanism yields long-period edge waves ($T_e > T_i$), just as is observed on dissipative beaches, although the exact periods are not predictable (Gallagher 1971, Bowen & Guza 1978).

In addition to being measured in the run-up at the shoreline, edge waves may be detected by measuring their orbital motions with electromagnetic current meters, and this has been the most commonly used approach (Huntley & Bowen 1973, 1975, Wright et al 1979, Huntley et al 1981, Holman 1981). Spectral analyses of the measured currents can distinguish between the incident-wave orbital velocities and those due to infragravity motions, again on the basis of the periodicities. Figure 3 shows an example of a spectrum from a reflective beach as measured by Huntley & Bowen

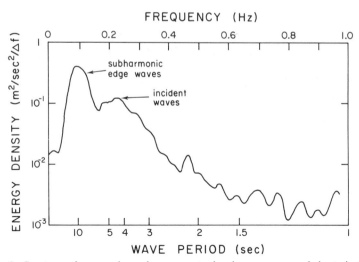

Figure 3   Spectrum of measured nearshore currents, showing an energy peak due to incident waves and a high-energy peak due to the presence of subharmonic edge waves (Huntley & Bowen 1973).

(1973), historically the first field measurements of edge waves; note that the edge-wave energy is at a period twice that of the peak in the incident waves, in agreement with the theoretical development of Guza & Davis (1974) for reflective beaches. Positive identification of this spectral peak as an edge wave involved a demonstration that its energy matched the predicted offshore decay [Figure 2 (*bottom*)] for the mode $n = 0$ edge wave, and that the zero phase relation between the onshore and longshore velocities was consistent with standing edge waves.

Theory predicts that on dissipative beaches, a continuum of edge-wave periods and modes may be generated, so that the resulting superposition can be complex. Current-meter measurements commonly reveal several energy peaks (with periods of up to 300 s) within the infragravity range of the spectra. Interpretation of these spectra is difficult because of the offshore variations in the many possible edge waves: There are cross-shore and alongshore changes in the edge-wave intensities that affect the spectra of their orbital motions. Some of the peaks and valleys in the spectra may thereby result from the position of the current meter and thus may not always accurately display the periods and energies of edge waves in the nearshore. The study of Huntley et al (1981) was the first to overcome such difficulties and to positively identify the source of the infragravity nearshore water motions. The authors were able to do this by using an enormous data set obtained from 19 electromagnetic current meters distributed along a 500-m stretch of beach. Cross-spectra between all instrument pairs demonstrated that the measurements were consistent with the simultaneous presence of several low-mode edge waves.

EDGE WAVES AND BEACH MORPHOLOGY

There is considerable evidence that edge waves have a significant effect on the beach morphology, redistributing the sediments by their orbital and net-drift velocities. Perhaps the clearest example of this is in the formation of beach cusps [Figure 4 (*left*)]. Although there have been many hypotheses put forward to explain the origin of beach cusps (Komar 1976, pp. 265–74), an edge-wave mechanism provides the best general explanation for the occurrence of cusps and is the only hypothesis that offers a prediction of their spacings (and is hence testable). The first direct test was provided by Guza & Inman (1975), who produced beach cusps in a laboratory wave basin under the action of standing subharmonic edge waves. The embayments between the cusps were hollowed out at the positions of maximum edge-wave run-up as seen in Figure 2 (*top*), whereas the cusp horns corresponded to locations of minimum run-up. The resulting cusp spacing was 1.8 m, half the edge-wave length L_e from Equation (2), since there are two cusps per wavelength longshore distance. Kaneko (1985) similarly generated beach cusps in a laboratory wave basin with a wide range of wave conditions, also concluding in favor of the edge-wave mechanism. Guza & Bowen (1981) provided further analyses of the role of edge waves in generating beach cusps; these analyses included predictions of the maximum amplitude to which cusps can grow when formed by edge waves. Sallenger (1979) and Huntley & Bowen (1979) undertook field measurements that demonstrated the role of edge waves in forming beach cusps. Sallenger monitored the formation of beach cusps and showed that their spacings corresponded to the expected values if formed by edge waves. Direct confirmation was provided by the study of Huntley & Bowen; during one of their field studies to measure edge waves, beach cusps were observed to form in the midst of the experiment. These cusps had a mean spacing of 12.7 m; the incident waves had a period of 6.9 s, and the measured edge waves were the $n = 0$ subharmonic ($T_e = 13.8$ s), which from Equation 2 (using the measured beach slope $\tan \beta$) yields a prediction for the cusp spacing of 12.0 m, very close to that observed. A compilation of such measurements is given in Figure 4 (right); the results are in general consistent with formation by subharmonic edge waves, although in some cases generation may have been by synchronous edge waves (where $T_e = T_i$). Although evidence for an edge-wave origin for beach cusps is fairly conclusive, questions remain concerning cusp interactions with the incident waves and the possibility of the cusps continuing to grow and achieving new equilibrium configurations after the edge waves have disappeared (Inman & Guza 1982).

Long-period edge waves on dissipative beaches also appear to be important in affecting the beach morphology. The orbital velocities associated with these infragravity motions are of the same order as the currents associated with the incident waves and with nearshore circulation (Wright & Short 1983). Bowen & Inman (1971) showed theoretically and in wave-basin experiments that standing edge waves cause the beach sand to move to zones where drift currents associated with the edge waves converge and reach a minimum; these zones are in the form of a series of crescentic bars (Figure 5). Holman & Bowen (1982) have expanded this analysis to include interactions of two or more progressive edge waves, finding that two edge waves traveling in the same direction will generate a series of oblique bars extending out from the shoreline; the bar morphologies predicted by this analysis are closely similar to the oblique bars observed on beaches.

More complete reviews of the role of edge waves in affecting beach morphology can be found in Holman (1983) and Bowen & Huntley (1984). Such controls of the beach morphology by edge waves have an indirect effect on erosion experienced at the shoreline by the arrangement of the beach sediments into offshore bars that partly control the patterns of wave breaking and refraction, and by the patterns of wave-induced currents. Orford & Carter (1984) provide an example of equally spaced washovers on a barrier island, which were likely produced by the effects of standing edge waves.

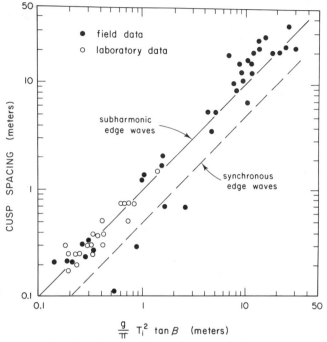

Figure 4 (*Left*) Well-developed cusps on a sandy beach in Mexico (*Sunset Magazine*, June 1971). (*Right*) Cusp spacings versus edge-wave lengths, showing that most agree with an origin due to subharmonic edge waves [data compiled in Inman & Guza (1982) and Kaneko (1985)].

RIP CURRENTS AND BEACH EROSION

Waves reaching a beach generate a variety of nearshore currents, the pattern depending in large part on the angle the waves make with the shoreline. When the wave crests are parallel or nearly parallel to the shoreline, the nearshore currents are dominated by a cell circulation with seaward-flowing rip currents (Figure 6). Bowen (1969a) and Bowen & Inman (1969) have shown that this cell circulation is produced by longshore variations in wave breaker heights, which in turn produce longshore variations in the wave set-up. The set-up will raise the water in the nearshore to higher levels shoreward from positions of large breakers than shoreward of smaller breakers. Water will then flow alongshore toward locations of small breakers and set-up, converging and turning seaward as a rip current. There are definite examples on ocean beaches where the longshore variations of breaker heights are produced by wave refraction over submarine canyons and shoals on the continental shelf, the rip

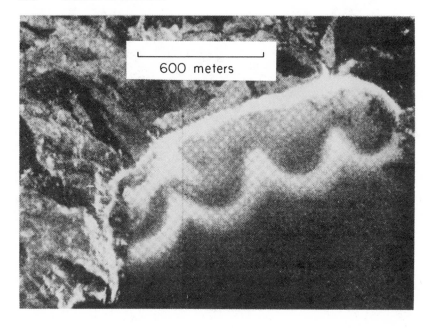

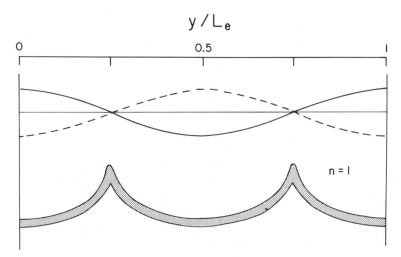

Figure 5   (*Top*) Crescentic bars offshore from a beach near Cape Kalaa, Algeria. (*Bottom*) The expected relationship between crescentic bars and the shoreline swash of a $n = 1$ standing edge wave (Bowen & Inman 1971).

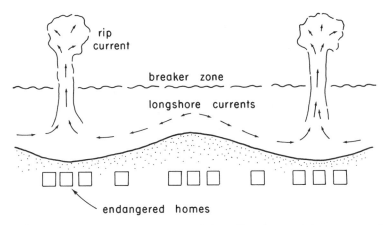

Figure 6   The nearshore current system, consisting of longshore and rip currents, and the typical rhythmic shoreline they produce. At times, erosion within the embayments can endanger homes or other property.

currents being found in the expected locations of smallest breakers. However, rip currents are commonly observed on long, straight beaches lacking the irregular offshore topography necessary to produce wave refraction. In such cases, a number of rip currents generally are observed with a regular spacing (roughly four times the width of the surf zone). Bowen & Inman (1969) have explained such occurrences as resulting from the presence of edge waves in the nearshore. The edge waves interact with the incoming swell waves, summing in certain longshore positions to produce high breakers and interfering in other locations to yield low breakers. Bowen & Inman were able to demonstrate rip-current generation by edge waves in a laboratory wave basin, but the hypothesis remains to be tested on ocean beaches.

Rip currents can achieve high velocities and can combine with the wave action to rearrange the beach sediments. Most commonly, the rip currents transport sand offshore to beyond the breaker zone, hollowing out embayments in the process (Komar 1971, 1983a). A series of rip currents can thereby produce a series of embayments separated by cuspate projections (which typically have spacings of tens to hundreds of meters, much larger than the "beach cusps" discussed earlier). Of significance to beach erosion, rip-current embayments sometimes cut back entirely through the beach and begin to impinge on the adjacent coastal property, becoming sites of erosional losses. Dolan (1971) has shown that such embayments along the barrier islands of the eastern United States can threaten man-made structures and often control locations of barrier-island washovers. Wright & Short (1983) provide an example from the coast of Australia. We have similarly observed that rip-current embayments have an important role in erosion along the Oregon coast (Figure 7; Komar & Rea 1976, Komar 1983b). Exceptionally large embayments become the focal points for property erosion, although most of the erosion actually results from wave attack during intense storms, when large waves can pass through the deeper water of the embayments with little loss of energy.

Figure 7 Embayments cut into the beach and foredunes on Siletz Spit, Oregon, by seaward-flowing rip currents. These embayments led to the loss of one home, while others were saved by the immediate placement of riprap (Komar & Rea 1976).

LONGSHORE CURRENTS AND SAND TRANSPORT

When waves break at an angle to the shoreline, they generate a longshore current that continues to flow along the beach, rather than turning seaward as a rip current. This current is particularly important, in that it combines with the wave action to transport sand for considerable alongshore distances. We become aware of this sand movement mainly when it is blocked by the construction of jetties or a breakwater, which leads to the all too familiar ensuing problems of erosion (Komar 1983c).

Considerable research efforts have gone into attempts to understand and predict this longshore current and sand transport as functions of the causative wave conditions (Komar 1983d). Bowen (1969b) was the first to show that the longshore current is generated by the momentum (or radiation stress) carried by the waves. The onshore component of this momentum produces the set-up already discussed, but when the waves break at an angle to the shoreline, there is also a longshore component of the momentum, which generates a longshore current. This was further developed by Longuet-Higgins (1970a,b), who balanced the longshore momentum against the frictional drag on the resulting current to arrive at an analytical equation predicting the magnitude of this flow. Komar (1979) further simplified the relationship to

$$v_l = 1.19(gH_b)^{1/2} \sin \alpha_b \cos \alpha_b, \tag{3}$$

so that the magnitude of the current v_l can be estimated directly from the measured breaker height H_b and breaker angle α_b. This equation shows excellent agreement with the available field and laboratory measurements of longshore currents.

Equation (3) is for the longshore current measured at the midsurf position, halfway between the shoreline and the breaker zone. It is known that longshore currents generated by oblique waves vary in magnitude across the nearshore zone, increasing with distance from the shoreline, reaching a maximum usually just beyond the midsurf position, and decreasing rapidly outside the breaker zone. Studies such as those of Bowen (1969b), Longuet-Higgins (1970b), and Liu & Dalrymple (1978) provide derivations of this complete velocity profile across the nearshore, and some field testing has been undertaken with reasonable confirmation (Kraus & Sasaki 1979).

The earliest evaluations of the longshore sand transport were based on the measured quantities of sand trapped by jetties (Watts 1953, Caldwell 1956). Such an approach can yield only a long-term average accumulation rate, and most subsequent studies have utilized sand tracers, native beach sand tagged with a fluorescent dye or low-level radioactivity (Komar &

Inman 1970, Knoth & Nummedal 1978, Duane & James 1980). These measurements are shown plotted in Figure 8. The sand-transport rate Q_s is a function of P_l defined by

$$P_l = (ECn)_b \sin \alpha_b \cos \alpha_b, \tag{4}$$

where $(ECn)_b$ is the energy flux or power of the breaking waves; P_l is often termed "the longshore component of wave power" because of the presence of the $\sin \alpha_b$ factor, although there are technical objections to such an interpretation. The straight line fitted to the data in Figure 8 yields

$$Q_s = 6.8 P_l, \tag{5}$$

where Q_s is the total volume transport rate of sand along the shore (in $m^3 \ day^{-1}$), and where P_l is given in units of $W \ m^{-1}$. [Equation (5) is equivalent to the formula given by Komar & Inman (1970), which however is expressed as the immersed-weight sand-transport rate so as to be dimensionally correct.]

Equation (5) permits one to evaluate the longshore sand-transport rate produced by waves breaking at an angle to the shoreline. As originally

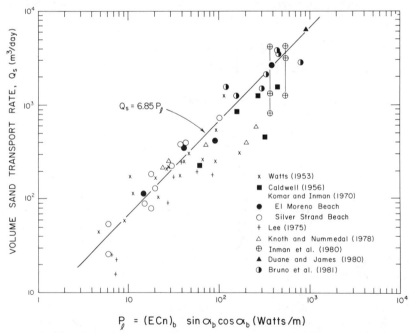

Figure 8 The relationship between the longshore sand-transport rate on beaches and the P_l "longshore component of wave-energy flux" [data compiled by Komar (1983d)].

presented, however, the relationship was purely empirical and so offered little insight into the sand-transport processes. Bagnold (1963) derived a sand-transport relationship based on the concept that the waves provide the principal forces to set sand into motion, but that a superimposed unidirectional current produces the net movement of the sand. Komar & Inman (1970) and Kraus et al (1982) have tested the application of this model to sand transport on beaches, verifying the equation

$$Q_s = 2.5(ECn)_b \frac{v_l}{u_m}, \tag{6}$$

where u_m is the orbital velocity under the breaking waves. This relationship is more general in application than Equation (5), in that the longshore current v_l in Equation (6) can be any nearshore current—tidally induced currents, a part of the cell circulation (Figure 6), wind-driven currents, or the longshore current produced by waves breaking at an angle to the shore. As shown by Komar & Inman (1970), when the longshore current v_l in Equation (6) is generated by an oblique wave approach and hence is given by Equation (3), the general Equation (6) reduces to the empirical Equation (5), which thereby provides an explanation for the success of that empirical relationship.

The above equations now permit evaluations of longshore currents and sand-transport rates on beaches, although the scatter of the data makes such predictions uncertain. Many scientific questions still remain concerning the details of the processes; for example, what are the proportions of suspension and bed load in sand transport along beaches, and what is the profile of the sand transport across the nearshore? Such questions aside, the relationships do permit examinations of longshore sand transport and its blockage by engineering structures, and hence its role in causing coastal erosion.

COMPUTER MODELS OF SHORELINE EROSION

The erosion rate of a segment of shoreline is not related simply to the magnitude of the littoral drift, Q_s, but instead to its change or gradient in the longshore direction, dQ_s/dx, where x is the coordinate axis alongshore. If Q_s is constant alongshore, then $dQ_s/dx = 0$ and the shoreline will be stable. A positive value for dQ_s/dx implies an increasing quantity of sand transport in the longshore direction, and in the absence of other sand sources (such as a river), this increase in Q_s can come only from beach erosion. In the opposite sense, negative values for dQ_s/dx produce deposition and shoreline advance.

Studies such as that of Greenwood & McGillivray (1980) have applied this relationship between dQ_s/dx and the resulting shoreline changes to analyses of observed patterns of coastal erosion. The specific interest of their study was erosion induced by engineering structures along the Lake Ontario waterfront of Toronto. Their approach was to first compute wave-refraction diagrams for the various wave conditions important to the area, determine from these diagrams the changes in wave conditions (wave power and breaker angles) along the shoreline, and then compute Q_s and dQ_s/dx for each shoreline position. It was found that these computed patterns of dQ_s/dx reasonably matched the observed patterns of shoreline erosion and accretion, which thereby provided an explanation as to their cause as well as verified this analysis approach.

A conceptually similar but potentially more powerful analysis procedure is to develop a computer model that simulates the shoreline changes. The derivative dQ_s/dx then enters the analysis as part of a continuity equation for the sand on the beach. In this approach, the smooth shoreline is divided into a number of cells, and the model computes the sand transfer from cell to cell via longshore sand transport and evaluates the changing volumes of sand in the various cells. Erosion results at a particular cell if it loses more sand than it acquires, and the resulting shoreline retreat is evaluated from that net loss. Computer-simulation models can also incorporate other sand contributions and losses (river sources, offshore losses or gains, etc), in addition to evaluations of the longshore sand transport.

Such models have been developed to examine shoreline changes produced by engineering structures (Price et al 1973, Komar et al 1976, Dean 1979, Perlin & Dean 1979, Kraus & Harikai 1983), the related hooked-beach erosion downdrift of a jetty or headland (Rea & Komar 1975), the shoreline erosion induced by offshore dredging (Motyka & Willis 1975), and geomorphology problems such as the growth of a delta (Komar 1973). Figure 9 illustrates the product of such a model; the upper diagram displays the actual shoreline changes produced by jetty construction started in 1900, while the lower diagram shows the shoreline evolution generated by the computer model. It is apparent from this example that such models are useful for predicting shoreline changes that result from man's alteration of the coastal environment [in this example by the construction of a jetty, but also by offshore dredging or by the elimination of a beach-sediment source (such as the damming of a river)].

SEA LEVEL AND COASTAL EROSION

A well-recognized factor in coastal erosion is the water level. In the short term, such changes may be affected by tides and storm surge; in the long term, by the progressive rise in sea level due to melting of glaciers.

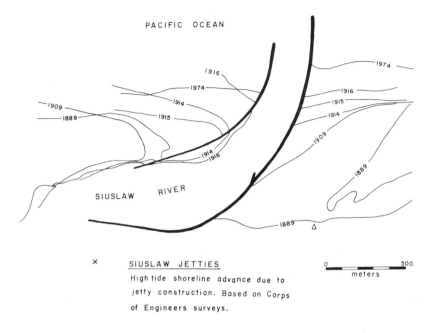

PACIFIC OCEAN

×   SIUSLAW JETTIES
High tide shoreline advance due to
jetty construction. Based on Corps
of Engineers surveys.

0   500
meters

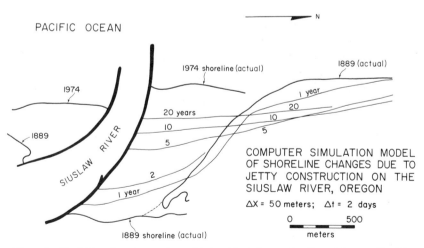

PACIFIC OCEAN

N

1974 shoreline (actual)

1889 (actual)

COMPUTER SIMULATION MODEL
OF SHORELINE CHANGES DUE TO
JETTY CONSTRUCTION ON THE
SIUSLAW RIVER, OREGON

$\Delta X = 50$ meters; $\Delta t = 2$ days

0   500
meters

1889 shoreline (actual)

Figure 9   (*Top*) Shoreline changes brought about by jetty construction at the mouth of the Siuslaw River, Oregon, based on surveys obtained during jetty development. (*Bottom*) A computer model of the shoreline changes. The actual 1889 shoreline is used as the starting configuration in the calculations, whereas the other shorelines were calculated by the computer model (Komar et al 1976).

The impact of a storm on beach erosion can be greatly enhanced if the storm occurs during a time of high spring tides. Wood (1977) has demonstrated that unusually high spring tides have been a significant factor in historic occurrences of coastal flooding and erosion. Termed *perigean spring tides*, these exceptional tides occur when there is both an alignment of the Earth, Moon, and Sun (which produces all spring tides) and the Moon is located at its perigee position (its closest approach to the Earth in its elliptical orbit). Wood found over 100 cases of major flooding were associated with these conditions on the North American coastline in 341 yr (1635–1976).

The advances and retreats of continental glaciers have produced alternating sea-level lowerings and rises. Within the last 20,000 yr sea level has changed by more than 100 m, exerting a considerable influence on our coasts. The timetable of the most recent (and continuing) rise in sea level has been established by dating materials such as submerged peat beds, beach rock, and fossil intertidal animals, material that is known to have formed originally near the shoreline but is now found at depth on the continental shelf (Shepard & Curray 1967, Milliman & Emery 1968). These chronologies indicate that sea level stood some 130 m lower than at present about 20,000 yr ago at the time of the last major glacial advance. As the glaciers melted, there was initially a rapid rise in sea level, averaging about 8 mm yr^{-1}, that has slowed down to only a couple of millimeters per year since about 7000 yr ago.

Important to coastal erosion is the sea-level rise within the last few thousand years. Its continuation is indicated by tide-gauge records, the year-to-year averages of these records yielding measurements of the on-going rise in sea level (Hicks 1972, Emery 1980, Gornitz et al 1982). The "sea-level rise" measured in the tide-gauge record at a particular coastal site is the product of the general worldwide (eustatic) rise in sea level plus any local land-level changes. The curve from New York (Figure 10) is typical of those from much of the east coast of the United States, indicating a rise of approximately 3.0 mm yr^{-1}. It is believed that nearly half of this rate is contributed by land subsidence. At Galveston, Texas, the sea-level rise averages about 6.0 mm yr^{-1}, mostly produced by the substantial sinking of that portion of the Gulf Coast. The coast of Oregon is rising, apparently at about the same rate as the worldwide rise in sea level, which yields a nearly static local sea level (Figure 10). At Juneau, Alaska, the land is rising at a high rate, so that there is a net lowering of the local water level and a retreat of the shoreline.

It is of course this local net sea-level change that is important to erosion at a specific coastal site. Although these values of sea-level rise amount to only a few millimeters per year, the rise is inexorable and cumulative. A rise of 4 mm yr^{-1}, common on the east coast of the United States, amounts to 40 cm over a century, a very significant change, especially in low-lying coastal areas such as the barrier islands (Nummedal 1983). The exact response of the beach and coast to a slowly rising sea level remains poorly understood. Bruun (1962) was the first to develop models of this process, and later investigators have devised more sophisticated analyses (see review in Dean & Maurmeyer 1983). A similar situation exists in the Great Lakes, where decadal variations in lake levels have produced extensive shoreline erosion and property losses. Hands (1983) has demonstrated some agreement between a shoreline-erosion model and measured

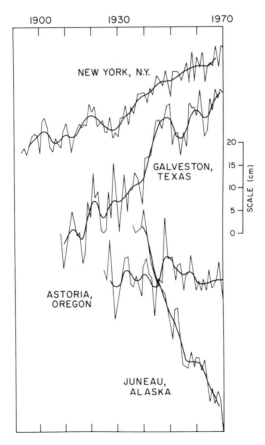

Figure 10   Yearly-average sea levels as determined from tide-gauge records at various coastal sites, illustrating the effects of a slow rise in the water level of the oceans plus local changes of the land masses (Hicks 1972).

erosion rates in the Great Lakes, and Rosen (1978) found basic agreement with erosion along the shore of Chesapeake Bay. Further testing of these models is required, however.

There is some concern that this sea-level rise is a growing problem that will develop into major coastal erosion over the next century. Determining the worldwide sea-level rise from tide-gauge records is difficult as a result of the changing land levels. Hicks (1978) combined the data from 27 tidal stations (eliminating those that were most obviously affected by land-level changes), and the resulting curve for the period 1940–1975 indicated an average sea-level rise of 1.5 ± 0.3 mm yr^{-1}. In similar analyses, Emery (1980) and Gornitz et al (1982) concluded that sea level is rising at an accelerated rate, with the rate during the past decade being some two to three times greater than pre-1970 rates. This increased rate is attributed to the sharp global warming trend underway since the 1960s, which warms the seawater and produces its volume expansion as well as accelerated glacial melting. Gornitz et al predict that sea level will rise by about 40 to 60 cm by the year 2050. In areas such as the east and Gulf coasts of the United States, which will continue to sink, the local apparent rise in sea level could exceed 1 m; such a rise would have a devastating effect on these barrier-island coasts. This prospect is somewhat uncertain, however; analyzing the same data, Barnett (1984) has argued against such conclusions of recently accelerated rates of sea-level rise.

There is a wide variety of shorter-term sea-level variations, some of which may contribute to coastal erosion. These sea-level rises typically last for a few months and are on the order of 10 to 100 cm. Many are responses to seasonal changes in the coastal water temperatures, the atmospheric pressure, and the strength of ocean currents (which affect sea level by their cross-flow geostrophic water-surface slope). Probably the most interesting of these, and the most clearly associated with coastal erosion, are sea-level changes produced by El Niño effects along the eastern margin of the Pacific Ocean. An El Niño is a breakdown of the normal equatorial wind and current patterns (Wyrtki 1975, 1977, Enfield 1981). This breakdown releases water that is normally set up in the western Pacific by the trade winds. The release creates a "wave" of sea-level rise, which first propagates eastward along the equator and then poleward along the eastern ocean margin. Such "waves" have been measured in the tide-gauge records along the west coasts of South America, Mexico, and the United States; these records reveal sea-level changes ("wave" heights) amounting to some 20 to 60 cm that slowly move along the coast, altering water levels at any specific site for several months (Enfield & Allen 1980).

It is probable that the extensive erosion that occurred along the coasts of California and Oregon during 1982–1983 was in part caused by such El

Niño sea-level "waves." On the coast of Oregon, sea level reached a maximum in February 1983, nearly 60 cm higher than the level in May 1982 (Huyer et al 1983). Significant water-level rises were also observed in the California tide-gauge records (D. Enfield, personal communication, 1985). However, there were other El Niño–related meteorological and oceano-graphic changes that contributed to the coastal erosion. Most note-worthy, the entire northern Pacific meteorological patterns shifted farther south than usual; storm tracks entering North America from the Pacific were displaced hundreds of miles to the south of their paths in normal years. Storm systems that normally cross the Oregon-Washington coast to produce the high wave energies characteristic of this region were instead directed at the coast of southern California, with devastating results.

SUMMARY

There is more to coastal-erosion processes than just storm-generated waves attacking the shoreline, although this is generally a major factor. Recent studies have shown that storm waves break farther offshore, and thus much of their energy is dissipated before reaching the shoreline. The energy of the swash is increased during a storm, but on low-sloping (dissipative) beaches most of this energy is in the infragravity range (longer-period motions than those of the incident waves). On steep (reflective) beaches, however, more of the swash energy is directly caused by the incident waves, although even here subharmonic edge waves may be significant. This increase in swash energy during a storm is superimposed on the wave set-up, which has its maximum rise at the shoreline and will commonly shift the swash attack by tens of meters in the landward direction. Although such changes in swash activity during a storm are known to erode back the beachface, the exact processes of this offshore sand transport remain poorly understood.

It is frequently observed that there is a considerable variability in beach erosion in the longshore direction. This may be caused by a number of factors, including the patterns of offshore bars (which affect the wave refraction) and nearshore currents. Rip currents, in particular, can erode back the beach, since they transport sand directly offshore and in the process hollow out embayments into the shoreline. During storms, these embayments have been observed to become the focal points for wave attack and barrier-island washovers.

Man is often a factor in causing or enhancing the coastal erosion. This is most obvious when jetties are constructed (which block the natural longshore transport of sand) but can also occur with the damming of rivers or offshore dredging of sand sources. Such impacts can now be anticipated and quantitatively assessed, thanks to our ability to evaluate longshore currents and sand-transport rates and to develop computer models that actually simulate the processes.

It is usually recognized that the progressive rise in sea level due to melting glaciers (and thermal expansion of sea water) must play an important role in the long-term erosion of a coast. This is demonstrated by the natural "experiment" in the Great Lakes, where high lake levels have produced considerable property losses. There is evidence, although debatable at this stage, that sea level is now rising at a faster rate and could become a major factor in shoreline losses over the next century. Short-term sea-level rises apparently can also play a role in beach erosion; an example is the coastally trapped sea-level "waves" associated with the 1982–1983 El Niño, which likely were a factor in erosion along the coasts of California and Oregon.

A given episode of erosion during a storm at a specific coastal site is likely in response to the combined effects of high wave energies, increased infragravity or incident swash energies, a rise in the water level due to set-up and possibly a wind-driven storm surge, and the intensification of rip currents and other nearshore currents. The susceptibility of the site to erosion by the storm may also have been increased because of long-term or shorter-term rises in sea level, or because of a loss in buffering action of the beach as a result of a jetty interrupting sand movement to the site.

ACKNOWLEDGMENTS

This review was undertaken with support by the NOAA Office of Sea Grant, Department of Commerce, under grant NA81AA-D-00086 (Project R/CP-21). The US Government is authorized to produce and distribute reprints for governmental purposes, notwithstanding any copyright notation that may appear hereon.

Literature Cited

Bagnold, R. A. 1963. Mechanics of marine sedimentation. In *The Sea*, ed. M. N. Hill, 3:507–523. New York: Interscience. 963 pp.

Barnett, T. P. 1984. The estimation of "global" sea level change: a problem of uniqueness. *J. Geophys. Res.* 89(C5):7980–88

Bowen, A. J. 1969a. Rip currents, 1. Theoretical investigations. *J. Geophys. Res.* 74:5467–78

Bowen, A. J. 1969b. The generation of longshore currents on a plane beach. *J. Mar. Res.* 27:206–15

Bowen, A. J., Guza, R. T. 1978. Edge waves and surf beat. *J. Geophys. Res.* 83(C4):1913–20

Bowen, A. J., Huntley, D. A. 1984. Waves, long waves and nearshore morphology. *Mar. Geol.* 60:1–13

Bowen, A. J., Inman, D. L. 1969. Rip currents, 2. Laboratory and field observations. *J. Geophys. Res.* 74:5479–90

Bowen, A. J., Inman, D. L. 1971. Edge waves and crescentic bars. *J. Geophys. Res.* 76:8662–71

Bowen, A. J., Inman, D. L., Simmons, V. P. 1968. Wave "set-down" and "set-up." *J. Geophys. Res.* 73:2569–77

Bruno, R. O., Dean, R. G., Gable, C. G., Walton, T. L. 1981. Longshore sand transport study at Channel Islands Harbor, California. *US Army Corps Eng., Coastal Eng. Res. Cent., Tech. Pap. 81–2.* 48 pp.

Bruun, P. 1962. Sea-level rise as a cause of shore erosion. *J. Waterw., Harbors Coastal Eng. Div. ASCE* 88:117–30

Caldwell, J. M. 1956. Wave action and sand movement near Anaheim Bay, California. *US Army Beach Erosion Board Tech. Memo.* 68. 21 pp.

Dean, R. G. 1979. Diffraction calculation of shoreline planforms. *Proc. Coastal Eng. Conf., 16th,* pp. 1903–17

Dean, R. G., Maurmeyer, E. M. 1983. Models for beach profile response. See Komar 1983e, pp. 151–65

Dolan, R. 1971. Coastal landforms: crescentic and rhythmic. *Geol. Soc. Am. Bull.* 82:177–80

Dolan, R., Hayden, B., May, S. 1983. Erosion of the U.S. shorelines. See Komar 1983e, pp. 285–99

Duane, D. B., James, W. R. 1980. Littoral transport in the surf zone elucidated by an Eulerian sediment tracer experiment. *J. Sediment. Petrol.* 50:929–42

Duncan, J. R. 1964. The effects of water table and tide cycle on swash-backwash sediment distribution and beach profile development. *Mar. Geol.* 2:168–87

Emery, K. O. 1980. Relative sea levels from tide-gauge records. *Proc. Natl. Acad. Sci. USA* 77:6968–72

Enfield, D. B. 1981. El Niño—Pacific eastern boundary response to interannual forcing. In *Resource Management and Environmental Uncertainty,* ed. M. H. Glantz, pp. 213–54. New York: Wiley

Enfield, D. B., Allen, J. S. 1980. On the structure and dynamics of monthly mean sea level anomalies along the Pacific coast of North and South America. *J. Phys. Oceanogr.* 10:557–78

Gallagher, B. 1971. Generation of surf beat by non-linear wave interactions. *J. Fluid Mech.* 49:1–20

Gornitz, V., Lebedeff, S., Hansen, J. 1982. Global sea level trend in the past century. *Science* 215:1611–14

Greenwood, B., McGillivray, D. G. 1980. Theoretical model of the littoral drift system in the Toronto waterfront area, Lake Ontario. *J. Great Lakes Res.* 4:84–102

Guza, R. T., Bowen, A. J. 1981. On the amplitude of beach cusps. *J. Geophys. Res.* 86:4125–32

Guza, R. T., Davis, R. E. 1974. Excitation of edge waves by waves incident on a beach. *J. Geophys. Res.* 79:1285–91

Guza, R. T., Inman, D. L. 1975. Edge waves and beach cusps. *J. Geophys. Res.* 80:2997–3012

Guza, R. T., Thornton, E. B. 1981. Wave set-up on a natural beach. *J. Geophys. Res.* 86(C5):4133–37

Guza, R. T., Thornton, E. B. 1982. Swash oscillations on a natural beach. *J. Geophys. Res.* 87(C1):483–91

Hands, E. B. 1983. The Great Lakes as a test model for profile responses to sea level changes. See Komar 1983e, pp. 167–89

Hicks, S. D. 1972. On the classification and trends of long period sea level series. *Shore and Beach* 40:20–23

Hicks, S. D. 1978. An average geopotential sea level series for the U.S. *J. Geophys. Res.* 83:1377–79

Holman, R. A. 1981. Infragravity energy in the surf zone. *J. Geophys. Res.* 84:6442–50

Holman, R. A. 1983. Edge waves and the configuration of the shoreline. See Komar 1983e, pp. 21–33

Holman, R. A., Bowen, A. J. 1982. Bars, bumps and holes: models for the generation of complex beach topography. *J. Geophys. Res.* 87(C1):457–68

Holman, R. A., Guza, R. T. 1984. Measuring run-up on a natural beach. *Coastal Eng.* 8:129–40

Holman, R. A., Sallenger, A. H. 1985. Set-up and swash on a natural beach. *J. Geophys. Res.* 90(C1):945–53

Howd, P. A., Holman, R. A. 1984. Beach foreshore response to long-period waves. *Proc. Coastal Eng. Conf., 19th,* pp. 1968–82

Huntley, D. A., Bowen, A. J. 1973. Field observations of edge waves. *Nature* 243:160–61

Huntley, D. A., Bowen, A. J. 1975. Field observations of edge waves and their effect on beach material. *J. Geol. Soc. London* 131:68–81

Huntley, D. A., Bowen, A. J. 1979. Beach cusps and edge waves. *Proc. Coastal Eng. Conf., 16th,* pp. 1378–93

Huntley, D. A., Guza, R. T., Thornton, E. B. 1981. Field observations of surf beat, 1. Progressive edge waves. *J. Geophys. Res.* 83:1913–20

Huyer, A., Gilbert, W. E., Pittock, H. L. 1983. Anomalous sea levels at Newport, Oregon, during the 1982–83 El Niño. *Coastal Oceanogr. Climatol. News* 5:37–39

Inman, D. L., Guza, R. T. 1982. The origin of swash cusps on beaches. *Mar. Geol.* 49:133–48

Inman, D. L., Zampol, J. A., White, T. E., Hanes, D. M., Waldorf, B. W., Krastens, K. A. 1980. Field measurements of sand motion in the surf zone. *Proc. Coastal Eng. Conf., 17th*, pp. 1215–34

Kaneko, A. 1985. Formation of beach cusps in a wave tank. *Coastal Eng.* 9 : 81–98

Knoth, J. S., Nummedal, D. 1978. Longshore sediment transport using fluorescent tracer. *Coastal Sediments '77, ASCE*, pp. 383–98

Komar, P. D. 1971. Nearshore cell circulation and the formation of giant cusps. *Geol. Soc. Am. Bull.* 82 : 2643–50

Komar, P. D. 1973. Computer models of delta growth due to sediment input from rivers and longshore transport. *Geol. Soc. Am. Bull.* 84 : 2217–26

Komar, P. D. 1976. *Beach Processes and Sedimentation.* Englewood Cliffs, NJ : Prentice-Hall. 429 pp.

Komar, P. D. 1979. Beach-slope dependence of longshore currents. *J. Waterw. Port Coastal Ocean Div. ASCE* 105 : 460–64

Komar, P. D. 1983a. Rhythmic shoreline features and their origins. In *Mega-Geomorphology*, ed. R. Gardner, H. Scoging, pp. 92–112. Oxford : Clarendon. 240 pp.

Komar, P. D. 1983b. The erosion of Siletz Spit, Oregon. See Komar 1983e, pp. 65–76

Komar, P. D. 1983c. Coastal erosion in response to the construction of jetties and breakwaters. See Komar 1983e, pp. 191–204

Komar, P. D. 1983d. Nearshore currents and sand transport on beaches. In *Physical Oceanography of Coastal and Shelf Seas*, ed. B. Johns, pp. 67–109. Amsterdam : Elsevier. 470 pp.

Komar, P. D., ed. 1983e. *Handbook of Coastal Processes and Erosion.* Boca Raton, Fla : CRC Press. 305 pp.

Komar, P. D., Inman, D. L. 1970. Longshore sand transport on beaches. *J. Geophys. Res.* 75 : 5914–27

Komar, P. D., Rea, C. C. 1976. Erosion of Siletz Spit, Oregon. *Shore and Beach* 44 : 9–15

Komar, P. D., Lizarraga-Arciniega, J. R., Terich, T. A. 1976. Oregon coast shoreline changes due to jetties. *J. Waterw. Harbors Coastal Eng. Div. ASCE* 102 : 13–30

Kraus, N. C., Harikai, S. 1983. Numerical models of the shoreline change at Oarai Beach. *Coastal Eng.* 7 : 1–28

Kraus, N. C., Sasaki, T. O. 1979. Effect of wave angle and lateral mixing on the longshore current. *Coastal Eng. Jpn.* 22 : 59–74 (see also *Mar. Sci. Commun.*, 1979, 5 : 91–126)

Kraus, N. C., Farinato, R. S., Horikawa, K. 1982. Field experiments on longshore sand transport in the surf zone. *Coastal Eng. Jpn.* 24 : 171–94

Lee, K. K. 1975. Longshore currents and sediment transport in west shore of Lake Michigan. *Water Resour. Res.* 11 : 1029–32

Liu, P. L.-F., Dalrymple, R. A. 1978. Bottom frictional stresses and longshore currents due to waves with large angles of incidence. *J. Mar. Res.* 36 : 357–75

Longuet-Higgins, M. S. 1970a. Longshore currents generated by obliquely incident waves, 1. *J. Geophys. Res.* 75 : 6778–89

Longuet-Higgins, M. S. 1970b. Longshore currents generated by obliquely incident sea waves, 2. *J. Geophys. Res.* 75 : 6790–6801

Longuet-Higgins, M. S., Stewart, R. W. 1964. Radiation stress in water waves, a physical discussion with applications. *Deep-Sea Res.* 11 : 529–63

Milliman, J. D., Emery, K. O. 1968. Sea levels during the past 35,000 years. *Science* 162 : 1121–23

Motyka, J. M., Willis, D. H. 1975. The effect of refraction over dredged holes. *Proc. Coastal Eng. Conf., 14th*, pp. 615–25

Nummedal, D. 1983. Barrier islands. See Komar 1983e, pp. 77–121

Orford, J. D., Carter, R. W. G. 1984. Mechanisms to account for the longshore spacing of overwash throats on a coarse clastic barrier in southeast Ireland. *Mar. Geol.* 56 : 207–26

Perlin, M., Dean, R. G. 1979. Prediction of beach planforms with littoral controls. *Proc. Coastal Eng. Conf., 16th*, pp. 1818–38

Price, W. A., Tomlinson, K. W., Willis, D. H. 1973. Predicting the changes in the plan shape of beaches. *Proc. Coastal Eng. Conf., 13th*, pp. 1321–29

Rea, C. C., Komar, P. D. 1975. Computer simulation models of a hooked beach shoreline configuration. *J. Sediment. Petrol.* 45 : 866–72

Rosen, P. S. 1978. A regional test of the Bruun Rule of shoreline erosion. *Mar. Geol.* 26 : M7–M16

Sallenger, A. H. 1979. Beach-cusp formation. *Mar. Geol.* 29 : 23–37

Sallenger, A. H., Richmond, B. M. 1984. High-frequency sediment level oscillations in the swash zone. *Mar. Geol.* 60 : 155–64

Shepard, F. P., Curray, J. R. 1967. Carbon-14 determination of sea level changes in stable areas. *Prog. Oceanogr.* 4 : 283–91

Watts, G. M. 1953. A study of sand movement at South Lake Worth Inlet, Florida. *US Army Beach Erosion Board Tech. Memo.* 42. 24 pp.

Wood, F. J. 1977. *The Strategic Role of Perigean Spring Tides in Nautical History and North American Coastal Flooding, 1635–1976.* Washington, DC : NOAA, US Dep. Commerce. 538 pp.

Wright, L. D., Short, A. D. 1983. Morphodynamics of beaches and surf zones in Australia. See Komar 1983e, pp. 35–64

Wright, L. D., Short, A. D. 1984. Morphodynamic variability of surf zones and beaches: a synthesis. *Mar. Geol.* 56:93–118

Wright, L. D., Chappell, J., Thom, B. G., Bradshaw, M. P., Cowell, P. 1979.

Morphodynamics of reflective and dissipative beach and inshore systems: southeastern Australia. *Mar. Geol.* 32:105–40

Wyrtki, K. 1975. Fluctuations of the dynamic topography in the Pacific Ocean. *J. Phys. Oceanogr.* 5:450–59

Wyrtki, K. 1977. Sea level during the 1972 El Niño. *J. Phys. Oceanogr.* 7:779–87

Ann. Rev. Earth Planet. Sci. 1986. 14 : 267–91

FORECASTING VOLCANIC ERUPTIONS[1]

Robert W. Decker

U.S. Geological Survey, 345 Middlefield Road, Menlo Park, California 94025

> *In the fourth gulf of the eighth circle of hell are those who presumed to foretell the future, their heads fixed face-backward on their necks.*
>
> Dante, *Divine Comedy* (early 14th century)

INTRODUCTION

Forecasting natural events such as landslides, earthquakes, and volcanic eruptions is a difficult problem compounded by conflicting expectations. Society wants accurate warnings of these events, yet the scientific community is not able to provide forecasts as accurate as desired because these natural events are only partly understood. The present situation is an uneasy compromise, with Earth scientists recognizing that public support requires that major efforts be made to forecast potential natural disasters, and the public becoming increasingly aware that probabilistic forecasts—though fraught with uncertainty—are useful in decision making.

Effective forecasting of natural events that could have a major impact on society involves cooperation among three groups who are not always accustomed to working closely with one another: scientists, who are responsible for making the forecasts and for estimating their degree of uncertainty; public officials, who are responsible for the safety and welfare of their constituents; and the news media, who are responsible for accurate communication of information to the public. There is no way to win in a natural disaster; one can only hope to reduce the losses. Close cooperation among these three groups, with each understanding the different problems faced by the others, can lead to significant reductions in public risk. On the other hand, lack of trust, understanding, or cooperation can easily

---

[1] The US Government has the right to retain a nonexclusive royalty-free license in and to any copyright covering this paper.

exacerbate a disaster or potential disaster, or even lead to a false alarm. Some false alarms are bound to occur; that is inherent in probabilistic forecasting. However, with better public understanding of the uncertainties involved, the reduction in losses resulting from correct forecasts can far outweigh the problems of false alarms. Present trends indicate that the accuracy of forecasting is improving, and there is reasonable hope that unless scientists promise—or the public expects—too much too soon, scientific forecasting of natural events such as volcanic eruptions will become a valuable and respected endeavor.

Volcanic eruptions worldwide have killed on average about 640 people per year in the twentieth century through 1982 (Blong 1984) and have caused an estimated total of about $10,000,000,000 (present value) in property damage. Catastrophic eruptions occur at widely scattered intervals. For example, the eruption of Mont Pelée in the West Indies annihilated the 29,000 inhabitants of the port of Saint Pierre in 1902, but other years have been devoid of destructive eruptions. Figure 1 shows the episodic nature of fatalities caused by volcanic eruptions.

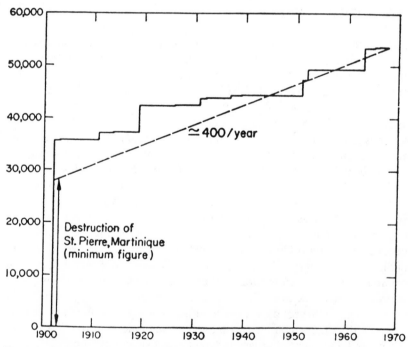

Figure 1   Cumulative plot of deaths caused by volcanic eruptions. More than half the deaths occurred during a single eruption that destroyed the city of St. Pierre. [From Latter (1969), with permission.]

An ideal forecast of volcanic activity would include the location, timing, character, and magnitude of the potential eruption, and a quantitative estimate of the probability of each of these factors. We are far from this goal; nevertheless, the present state of the art is encouraging. As to location, it can be said in general that future eruptions will occur at or near the sites of known volcanoes. The birth of a new volcanic province is a rare event, even in geological time. More specifically, following the subsurface locations of a migrating earthquake swarm beneath a volcano provides a good estimate of where an eruption may occur. Seismologists at the Hawaiian Volcano Observatory have guided the field geologists by radio to within a few hundred meters of the outbreak locations of all the eruptions of Kilauea Volcano since 1979.

Forecasting the approximate timing of a volcanic outbreak has proved successful in several cases over the past decade. Some of these instances are discussed in later sections. Forecasting the timing of the climax and the duration of an eruption has proved more elusive, and from a hazards point of view this information may be much more important than the onset time.

Forecasts regarding the probable character and magnitude of an eruption can be made in a general way based on the tectonic setting of the volcano and on its previous recorded and recent geologic history. Statistically, small-magnitude eruptions occur far more frequently than large ones, but there is no monitoring technique presently available that can anticipate the specific character and magnitude of a forthcoming eruption.

At present most of the probability statements in forecasts are couched in only semiquantitative terms such as high, low, or significant. These words mean different things to different readers. The goal of numerically quantitative probability statements is still many years away.

The hazards from volcanoes are closely related to the character of their eruptions. Effusive eruptions of molten lava, as in most Hawaiian eruptions, pose little danger to people but can be very destructive to property. In this century, only one person has been killed by a volcanic eruption in Hawaii; however, during this same period, 5% of the island of Hawaii has been covered by new lava flows. During explosive eruptions, debris avalanches, lateral blasts, ash flows, and mudflows travel at speeds that cannot be outrun, and thus all of these pose major dangers to both life and property. In the Mont Pelée eruption, hot pyroclastic flows (nuées ardentes) traveling at high speeds killed nearly the entire population of Saint Pierre. (Out of a population of about 29,000 people, there were only two known survivors.)

Forecasting volcanic eruptions has a distinct advantage at this time over forecasting major earthquakes. Magma moves upward from depth to the surface before an eruption begins, and this movement (which may take hours to years) can be detected by present techniques. However, rising magma may stop before reaching the surface, and may instead form a shallow intrusion rather than an eruption.

The use of the word forecasting in this review, rather than prediction, estimation, or anticipation, is deliberate. Weather forecasting is an established science based on probabilities; the forecasts are not certain, but the public understands that even with this limitation, they are much more useful than no forecasts at all. It is this connotation of utility but uncertainty that has become established for the word forecasting that makes it my choice.

Several comprehensive reviews of forecasting volcanic behavior have been published in English in the past 15 years (Gorshkov 1971, Minakami 1971, Civetta et al 1974, Walker 1974, Decker 1978, Fournier d'Albe et al 1979, Martin & Davis 1982, Tazieff & Sabroux 1983, Souther et al 1984). Some of these previous reviews are much more detailed than this one, and they should be consulted by any serious student of the subject.

An anecdote about forecasting volcanic activity says that every active or potentially active volcano should be studied by a geologist to find out what did happen, by a geophysicist to determine what is happening, and by a lobbyist to tell the government what might happen. This review is organized along those lines: an approach based on the historic and prehistoric eruptive record; a monitoring approach based on current geological, geophysical, and geochemical data; and the societal consider-ations of assessing volcanic hazards and reducing volcanic risk.

APPROACH BASED ON ERUPTIVE RECORD

Worldwide, more than 1300 volcanoes have erupted during the past 10,000 yr (Simkin et al 1981). However, the recent geologic histories of fewer than 10% of these potentially destructive volcanoes have been studied in detail. This is unfortunate, for the track record of a potentially active volcano provides the best method of assessing its future volcanic hazards on a long-term basis. These hazards assessments are made by studying the historic records and the geologically recent deposits on and around a volcano to establish the frequency and character of past eruptions. This record is then extrapolated to provide a general forecast of future activity.

A major problem in assessing the hazards from future eruptions is that many of the most destructive eruptions in the past occurred at volcanoes that had been dormant for hundreds to thousands of years (Figure 2). This creates the paradox that often the most potentially dangerous volcanoes have relatively poor records of past eruptive activity.

Another major problem is that even volcanoes with well-documented records of many historical eruptions show both a wide variation in the repose times between eruptions and large variations in the character of these eruptions. For example, Asama Volcano in Japan has erupted thousands of times since its first recorded eruption in A.D. 685 (Kuno 1962). Since 1900, the shortest repose times were less than one day, and the maximum repose time was five years. Most eruptions were moderate explosions of ash, but the disastrous eruption of 1783 involved large ash explosions, pumice falls, pyroclastic flows, and mudflows.

Despite these shortcomings, some long-term hazards assessments based on historical records and geologic mapping have shown remarkable success. For example, Crandell & Mullineaux's (1978) hazards assessment for Mount St. Helens Volcano in the northwestern United States was prophetic. They concluded that Mount St. Helens had been more active and more explosive during the preceding 4500 yr than any other volcano in the contiguous 48 states. In that period the volcano produced viscous lava domes, pumice falls, pyroclastic flows of hot, fluidized rock fragments, lava flows, and mudflows. The average interval between eruptive periods was 225 yr. On the basis of their study of the past behavior of Mount St. Helens

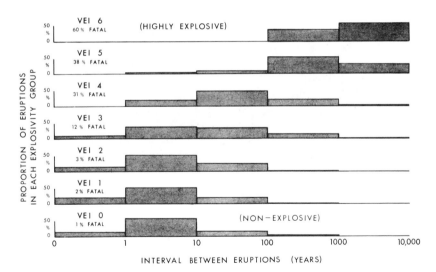

Figure 2 Degree of explosiveness and time intervals between eruptions. For each volcanic explosivity index (VEI) number, eruptions are grouped by increasing time intervals between eruptions. The number of eruptions considered in groups 0 to 6 are, respectively, 354, 338, 2882, 617, 102, 19, and 8. Examples of VEI numbers are Krakatau, 1883 = 6; Mount St. Helens, 18 May 1980 = 5; typical Hawaiian eruption = 0 to 1. [From Simkin et al (1981), with permission.]

(Table 1), Crandell & Mullineaux wrote:

> In the future Mount St. Helens probably will erupt violently and intermittently just as it has in the recent geologic past, and these future eruptions will affect human life and health, property, agriculture and general economic welfare over a broad area. . . . The volcano's behavior pattern suggests that the current quiet interval will not last as long as 1,000 years; instead an eruption is more likely to occur within the next 100 years, and perhaps even before the end of this century.

Rarely does a long-term forecast get so quickly evaluated. A swarm of earthquakes beneath Mount St. Helens began on 20 March 1980, followed by small ash eruptions beginning March 27. The climactic eruption occurred on 18 May 1980. It was apparently triggered by a magnitude 5 earthquake that caused the failure of the north slope of the mountain into a 2.7-km^3 debris avalanche. The north flank of Mount St. Helens had become oversteepened by the intrusion of a large, shallow mass of magma during late March, April, and May that produced a 1.8-km-diameter bulge about 150 m high on the north face of the mountain. Failure of the oversteepened north slope released the pressure on the shallow intrusion and its surrounding hydrothermal system, causing northward-directed blasts of steam and rock fragments that devastated an area of 600 km^2 (Figure 3). This was followed by a 9-hr vertical ash cloud eruption that reached heights in excess of 20 km, causing ash fallout of a few centimeters over much of central and eastern Washington State. Destructive floods and mudflows descended the streams draining westward from Mount St. Helens. These entered the Columbia River and caused severe shoaling of the shipping channel. A few smaller explosive eruptions occurred in 1980, and a lava dome has been growing since late 1980 in the 2-km-wide horseshoe-shaped crater formed by the May 18 avalanche and eruption (Lipman & Mullineaux 1981, Decker & Decker 1981).

Almost every part of Crandell & Mullineaux's (1978) long-term forecast occurred. The eruption did occur in this century, 57 people were killed, and property damage amounted to about $1,000,000,000. However, the magnitude of the avalanche and the lateral blast greatly exceeded their expectations.

At volcanoes with many recorded eruptions, statistical analysis of the time series of eruptions may reveal characteristic patterns that could aid in forecasting the timing of, and in planning for, future eruptions. Wickman (1966) has used this approach to show that some volcanoes show a random pattern in the timing of their historical eruptions, whereas others show patterns of increasing or decreasing probability of eruption as the repose period between eruptions increases. Mauna Loa Volcano in Hawaii has a random pattern, with an average repose time of nearly 4 yr, whereas Hekla Volcano in Iceland has an average repose time of 58 yr and a pattern of

Table 1 Eruptions and dormant intervals at Mount St. Helens since 2500 B.C. (Crandell et al 1975)[a]

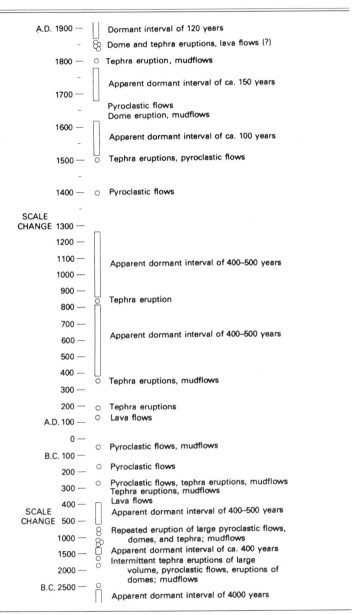

[a] The circles represent specific eruptions that were observed or that have been dated or closely bracketed by radiocarbon age determinations; the vertical boxes represent dormant intervals.

increasing probability of eruption as the repose time increases. Rose & Stoiber (1969), using Wickman's technique, show that the probability of an eruption of Izalco Volcano in El Salvador is 3% a month for several months following an eruption, but that after a repose period of 2 to 3 yr, the probability of an eruption decreases to 2% per month. This result indicates that eruptions of Izalco tend to cluster into groups, with longer periods of repose between the groups of eruptions.

Klein (1982) has recently analyzed the eruption patterns of Kilauea and Mauna Loa volcanoes in Hawaii. He finds that most Hawaiian eruptions are largely random in their timing, with an average repose time at Kilauea of 501 days compared with Mauna Loa's 1412 days. However, large-volume eruptions of both volcanoes tend to be followed by longer repose periods, summit (or flank) eruptions of Kilauea tend to cluster, and the

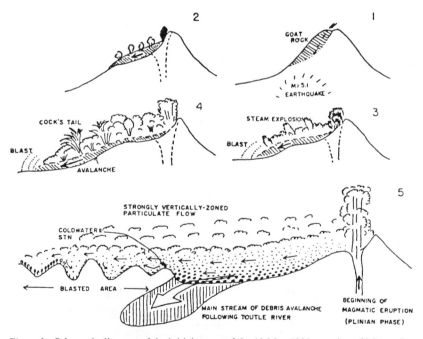

Figure 3   Schematic diagram of the initial stages of the 18 May 1980 eruption of Mount St. Helens. (1) Earthquake triggers avalanche on bulging north face of volcano. (2) Start of debris avalanche. (3) Steam and gas explosions occur as shallow magma body and hydrothermal system are decompressed. (4) Magnitude of explosion increases as hot debris and blast enter Spirit Lake. (5) Blast spreads across topographic ridges; debris avalanche flows down river valley; Plinian explosion of magma begins from the central conduit. [From Ui & Aramaki (1983), with permission.]

longest periods of repose at each volcano appear to be associated with increased eruptive activity at the other volcano.

Although forecasting based on the historical and prehistoric activity of a volcano is effective and inexpensive, Scott (1984) points out several caveats to the method. He notes that the older record becomes more obliterated by subsequent events; this is particularly true in heavily glaciated areas prior to 10,000 yr ago. The wide range of repose intervals makes the average repose interval a very uncertain estimate of the timing of future eruptions. Eruption habits may change and past behavior may then be misleading. Events may occur that are unprecedented at a specific volcano, and changes in the size and shape of a volcano with time may change the locations of hazardous areas.

Before leaving this subject, several classic studies of the recent geologic history of specific volcanoes and volcanic areas should be mentioned: Asama Volcano, Japan (Aramaki 1957), São Miguel Island, Azores (Booth et al 1978), Cascade volcanoes, USA (Crandell & Mullineaux 1975), Oshima Volcano, Japan (Nakamura 1964), Mount Etna, Italy (Romano 1982), and Hekla Volcano, Iceland (Thorarinsson 1967). Different volcanoes and areas present different problems, and these papers show an interesting spectrum of approaches to solving some of these problems.

MONITORING APPROACH

Monitoring an active or potentially active volcano involves ongoing geological, geophysical, and geochemical observations, particularly of changes in data. Newhall (1984b) has recently summarized current volcano monitoring techniques. Important monitoring observations during periods of dormancy may include visual estimations of the quantity, color, height, and drift direction of volcanic fume; temperature of fumaroles; volume and chemistry of fumarolic gases; gases in soils, such as hydrogen, helium, mercury vapor, and radon; seismicity—both earthquake locations and volcanic tremor; geodesy—particularly the elevation and diameter of calderas, craters, and rift zones; electromagnetic data such as self-potential (ground voltages), magnetic and magnetotelluric fields, and conductivity to both direct and alternating currents; and thermal radiation. The rationale of these observations is to establish the "normal" levels of the volcano under study (so-called baseline data), to determine its subsurface structure, and to discover changes that may give some evidence of its dynamics. Any other changing phenomena on the volcano not included in the list above may also merit observation and recording. For example, an area of dying vegetation may be the first evidence of newly heated ground.

Observations during eruptions are even more important. The physical and chemical character, volume, and volume rate of eruption products, as well as their distribution and effects, need to be carefully observed and recorded on a chronological basis. Geophysical and geochemical data may be changing rapidly during eruptions and should be measured continuously or at least at short intervals. The location of seismic swarms can often be a guide to deciding where to focus other observations.

Long-term monitoring of volcanoes has two primary objectives: (a) an understanding of how volcanoes work, and (b) the establishment of patterns, whether empirical or understood, that will aid in forecasting eruptions (Klein 1984). It takes time to establish the necessary observations—perhaps 30 yr on a volcano like Kilauea in Hawaii that erupts frequently; perhaps 300 yr on Mount St. Helens, which has a much longer repose time between periods of eruption.

This need to study many eruptions illustrates the paradox as to why more is known about "safe" volcanoes like Kilauea, which erupts frequently compared with more dangerous volcanoes like Mount Rainier or Mount Fuji. Nevertheless, the monitoring techniques developed in Italy, Japan, Hawaii, and the Soviet Union, and the experience gained by visits to eruptions in Indonesia, Central America, and the Philippines, were extremely valuable to the scientists studying and forecasting the course of the recent eruptions at Mount St. Helens.

How can long-dormant volcanoes be adequately monitored? One solution is to obtain baseline data on these volcanoes and repeat the observations a year or two later. If no changes have occurred, seismometers can be installed to keep tabs remotely on the potentially dangerous volcano. Any substantial change in seismicity, either in number or pattern, would then be a signal to redo other observations.

In this way a single volcano observatory can monitor 10 to 20 volcanoes in a region. Financial support for volcano observatories tends to go up sharply after a major eruption and then to diminish over the next several years as other societal problems appear more urgent. Any long-term volcano monitoring program must therefore be cost-effective.

Progress in volcano monitoring has been spurred in the past few years by the eruptions of Mauna Loa and Mount St. Helens volcanoes, and by the subsurface "unrest" at the calderas of Rabaul, Pozzuoli, and Long Valley. Even though the current rates of seismicity and surface deformation have diminished at these three calderas and there have been no eruptions to date (October 1985), the monitoring observations on these potentially dangerous volcanoes have greatly improved knowledge of their subsurface structure and dynamics. The following sections provide more specific information on each of these monitoring investigations.

Mauna Loa Volcano, Hawaii

Seismicity beneath the summit region of Mauna Loa increased in 1974 and remained at high levels for several months preceding a summit eruption in 1975 (Koyanagi et al 1975). Following this brief eruption, Lockwood et al (1976) forecast on the basis of previous eruption patterns that a northeast rift eruption would occur within a few years. Mauna Loa seismicity remained low, however, and did not increase again until 1980–1981.

During the period from 1976 to 1981, increases of 10–20 cm in the caldera diameter and uplift of 15 cm of the summit area indicated slow reinflation of a shallow magma reservoir at a depth of 3 to 4 km beneath the summit. The volume of inflation during this period was about 20×10^6 m^3. Seismicity increased rapidly in 1983 (Figure 4), and a forecast was published stating that "the probability significantly increases for an eruption of Mauna Loa during the next 2 years" (Decker et al 1983). An eruption from the summit and northeast rift zone began 25 March 1984 and lasted for 3 weeks (Lockwood et al 1985). During that period, basaltic lava flows with a total

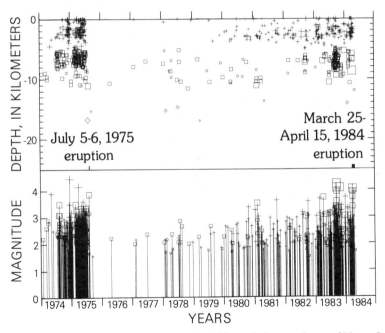

Figure 4   Earthquakes of magnitude greater than 1.5 beneath the summit area of Mauna Loa Volcano preceding the 1975 and 1984 eruptions. [From Lockwood et al (1985), with permission.]

volume of 220×10^6 m^3 covered an area of 48 km^2. Since the eruption volume greatly exceeded the new magma storage volume between 1975 and 1984, it is apparent that most of the erupted magma in 1984 had accumulated beneath Mauna Loa prior to 1975. Flows extended from their rift vent for 25 km toward the city of Hilo but stopped about 5 to 10 km away. Contingency plans were made to evacuate parts of the city, but they did not have to be put into operation.

During the eruption, the summit of Mauna Loa subsided 63 cm and the caldera decreased in diameter by about 30 cm. The pattern of deformation and the composition of the volcanic gases indicate that the magma supplying the eruption had been withdrawn from storage at a depth of 3 to 4 km, the same location where magma had been slowly injected prior to the eruption. The monitoring established that Mauna Loa has a shallow magma reservoir system, an important new concept about Hawaiian volcanoes. It is now evident that the interior magma storage system in Hawaiian volcanoes grows upward in pace with the external height of the lava pile. The caldera, rift zones, and shallow magma reservoirs begin early, on and beneath the seafloor, and all evolve upward with the growing volcano.

Mount St. Helens Volcano, Washington

The first small phreatic eruption at Mount St. Helens occurred on 27 March 1980 and was followed by hundreds of small steam and ash eruptions until the major eruption of 18 May 1980 (Lipman & Mullineaux 1981). The March 27 eruption was preceded by 7 days of intense local seismic activity that clearly signaled the high probability of the first eruption. Although the disastrous climatic eruption of 18 May 1980 occurred without any distinct short-term warnings, the longer-term precursory events were numerous and dramatic. The seismic swarm had continued with high and nearly constant energy release for 60 days. Bursts of volcanic tremor, generally interpreted as indicating magma movement at depth, began on March 31, continued intermittently through April 5, and recurred on April 12 and May 8. Major visible deformation of the north summit area was first seen on March 27 and was monitored with surveying instruments after April 23. The large rates of deformation—about 2 m day^{-1} in an area with dimensions of 1.5 km $\times$ 2 km on the high north flank of the mountain—were of major concern. The close connection in time and space between the earthquake foci and the bulging area led most of the scientists studying the eruption to conclude that a shallow intrusion of magma was taking place beneath an area just north of the summit. Barry Voight (of Pennsylvania State University) suggested in an internal report on May 1 that both a major avalanche and an explosive eruption were possible.

Seven days of volcanic tremor preceded the moderate explosive eruption on May 25. Nine hours of tremor preceded a similar eruption on June 12, and by this time short-term forecasts of possible impending activity were being issued to people working near the volcano. By August 1980, public forecasts were being issued by the US Geological Survey and the University of Washington.

Swanson et al (1983) document that 13 eruptions (5 explosive, and all except 1 involving growth of the lava dome in the 18 May 1980 crater) from June 1980 through December 1982 were predicted tens of minutes to a few hours in advance. The last 7 of these eruptions were predicted between 3 days and 3 weeks in advance. This remarkable record was achieved by monitoring precursory seismicity, deformation of the crater floor and lava dome, and gas emissions.

Figure 5 shows an example of the deformation monitoring, which provided the earliest warning of many eruptions. An active thrust fault on the floor of the crater near the active lava dome was being driven by the growing dome. Daily measurements show the accelerating shortening across the thrust fault prior to the outbreak of new lava on the dome. Deformation of the crater floor by 26 August 1981 indicated that a new pulse of magma was beginning to inflate the dome. A prediction was issued that a dome-building eruption was likely during the two-week period 2–16 September 1981. Increasing seismicity supported the deformation data

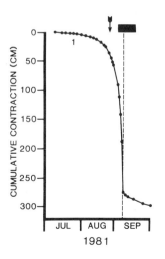

Figure 5 Shortening of the taped distance across an active thrust fault on the floor of the Mount St. Helens crater prior to the September 1981 dome-building eruption. The arrow indicates the date a prediction was issued, the black box is the time period in which the eruption was expected, and the dashed line is the date of the actual eruption. [From Swanson et al (1983), with permission.]

(Figure 6; Malone et al 1983), and on September 6 at 8 A.M. the prediction was updated to state that "a dome-building eruption accompanied by increased fume but little or no ash emission will probably begin within the next 12 to 48 hours." Such an eruption began in midafternoon or early evening on that same day.

Other papers summarizing recent monitoring at Mount St. Helens include Chadwick et al (1983), Dzurisin et al (1983), Casadevall et al (1983), Cashman & Taggart (1983), Melson (1983), Weaver et al (1983), and Waitt et al (1983). Newhall (1982, 1984a) has made quantitative estimates of the intermediate- and long-term risks from volcanic activity at Mount St. Helens.

Rabaul Caldera, Papua New Guinea

Significant increases in seismicity and inflation began at Rabaul in September 1983 (McKee et al 1985a). This followed 12 yr of relatively slow inflation and gradually increasing seismicity. From September 1983 to October 1984, 92,000 shallow earthquakes were recorded beneath the caldera. Figure 7 shows well-located earthquakes representative of this swarm (McKee et al 1985b). The largest earthquake had a magnitude of 5.1, and many were felt by the 69,000 residents who live near the great harbor formed by the caldera.

Deformation measurements show that the caldera was widening and inflating, with two centers of uplift within the ellipse of earthquake epicenters (Figure 8). Maximum cumulative uplift from September 1983 to October 1984 was 63 cm, and the volume change was about 40×10^6 m^3. The pattern of deformation indicates that the northern center of inferred magma intrusion was 1–2 km deep, whereas the southern center was about 3 km deep.

During an eruption at Rabaul in 1937, 506 people were killed and 7500 were evacuated (Lowenstein 1982). The recent crisis represented a period of greatly increased probability of a new eruption at Rabaul, and detailed contingency plans were made to evacuate the area if necessary. By August 1985, however, the rates of seismicity and inflation had diminished. Unless these rates begin to increase rapidly again, the likelihood of an eruption has also diminished, but has not disappeared.

Pozzuoli, Italy

Pozzuoli is a city of 70,000 people located in the Campi Flegrei (fiery fields) west of Naples. The area lies within a complex 12-km-diameter caldera formed 35,000 yr ago by the eruption of about 80 km^3 of pumice and ash (Barberi et al 1984). The latest eruption in the Campi Flegrei occurred in 1538 and formed a large cinder cone called Monte Nuovo. Hot springs and

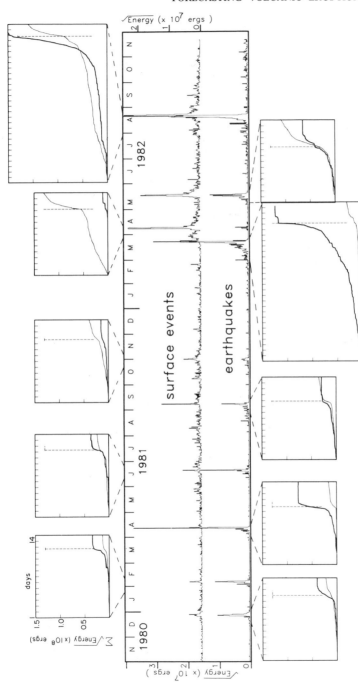

Figure 6 The square root of seismic energy release at Mount St. Helens per half day from November 1980 to November 1982. Surface events are plotted separately from earthquakes. The cumulative square root of seismic energy released during a 2-week period (a 4-week period in the March 1982 and August 1982 eruptions) around each of the eruptions is shown in boxes. Light lines indicate the energy of surface events, and heavy lines represent the energy of shallow earthquakes. Vertical dashed lines indicate the beginning of eruptions. [From Malone et al (1983), with permission.]

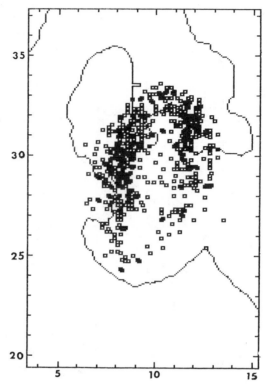

Figure 7   Epicenters of well-located earthquakes beneath Rabaul Caldera from September 1983 to November 1984. About 95% of the hypocenters occurred at depths of 3 km or less. The horizontal and vertical grids are in kilometers. [From McKee et al (1985b), with permission.]

fumaroles occur throughout the area. The caldera has undergone some remarkable uplifts and subsidences through the centuries. Marble columns of a Roman market built at Pozzuoli in the second century B.C. are still standing, and mollusk borings into the columns record both submergences and reemergences since Roman times (Figure 9). Slow subsidence totaling 10 to 12 m predominated from the time the columns were built until the tenth to eleventh centuries A.D., a rate of about 1 cm yr^{-1} (Parascandola 1947, Yokoyama 1971, Caputo 1979, Bianchi et al 1985). This was followed by slow uplift until the sixteenth century. In 1538 rapid uplift of 6 to 10 m was followed by the eruption of Monte Nuovo. Following the eruption, the area sank again to about 2 m below sea level. Beginning in 1969, the area was rapidly uplifted until 1972, gaining about 150 cm in elevation. This inflation was accompanied by moderate microearthquake activity. From

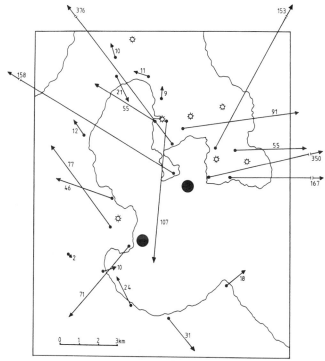

Figure 8   Inflation of Rabaul Caldera as shown by tilt vectors (in microradians) for the period September 1983 to November 1984. The black circles are the centers of inferred maximum uplift. Small, sun-shaped symbols show volcanic centers and peaks. [From McKee et al (1985b), with permission.]

1972 until 1982 the area remained relatively stable, but in 1982 to 1984 another pulse of rapid uplift raised the port of Pozzuoli an additional 160 cm, making most of the shipping docks unusable. The volume of the 1969 to 1985 uplift was about 150×10^6 m^3, with an inferred magma intrusion depth of about 3 to 5 km.

The 1982 to 1984 uplift was accompanied by more severe swarms of shallow earthquakes (up to magnitude $4+$). These small-to-moderate (but nearly continuous) earthquakes slowly shook many of Pozzuoli's weaker buildings into ruins, and thousands of the inhabitants from the city center left the area.

The recent uplift (Figure 10) and earthquake swarm ceased during the fall of 1984. Whether this is the end of the crisis or just another pause in the subsurface activity similar to the quiet period between 1972 and 1982 is not known.

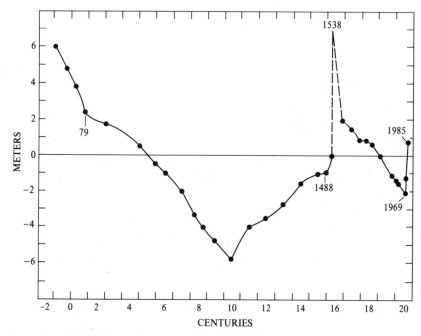

Figure 9 A 2000-yr record of ground-surface subsidence and uplift at Pozzuoli, Italy. These changes are considered to be caused by magma cooling, intruding, or being withdrawn from beneath the Campi Flegrei area. [After Yokoyama (1971), with new data from Roberto Scandone in SEAN (1985).]

Long Valley, California

Long Valley is an elliptical caldera 17 by 32 km wide formed by an enormous eruption of 600 km³ of silicic magma about 700,000 yr ago (Bailey et al 1976). The area is located on the eastern fault scarp of the Sierra Nevada range, a region of ongoing tectonic activity. The most recent volcanic activity in the caldera is a line of rhyolite domes that erupted about 500 to 600 yr ago (Miller et al 1982).

Seismic activity beneath the caldera began to increase in 1979 but was not considered unusual until a swarm of four earthquakes of magnitude 6 to 6+ occurred during 25–27 May 1980 (Ryall & Ryall 1981, 1983). These were followed by intermittent swarms of thousands of smaller earthquakes (Figure 11), particularly during 1980, 1982, and early 1983. Increases in temperature and volume of hot springs in the area were also reported (Miller et al 1982).

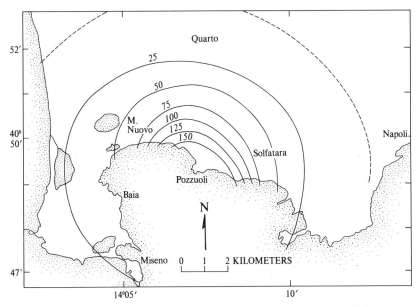

Figure 10 Contours of uplift (in centimeters) at Campi Flegrei and Pozzuoli, Italy, from January 1982 to January 1985. (Data from F. Barberi and G. Luongo, with permission.)

Uplift of the interior of the caldera occurred over a 30-km-wide area, with a maximum measured increase in elevation of 40 cm. This inflation and changes in horizontal distance across the caldera have been modeled to estimate an intrusion of $100–200 \times 10^6$ m^3 of magma at a depth of 5 to 10 km beneath the surface (Savage & Clark 1982, Hill et al 1984, Rundle & Whitcomb 1984; Figure 12).

Since 1983, seismic activity in Long Valley has decreased. Nevertheless, the area has remained one of the most seismically active localities in California into 1985.

Although most geologists studying the unrest at Rabaul, Pozzuoli, and Long Valley conclude that magma has recently been intruded at shallow depths beneath these calderas and that the probability of surface eruptions was, and to a lesser extent still is, greater than during periods of quiet, they do not have any quantitative estimates of the probability of a new eruption occurring at these locations. Newhall et al (1984) conclude on the basis of a major search of the world literature that both eruptions and periods of unrest without eruptions at large silicic volcanic calderas during the past 100 yr are not uncommon. They found that during an average year, 3 to 4 silicic calderas worldwide exhibit some form of unrest (i.e. some noticeable

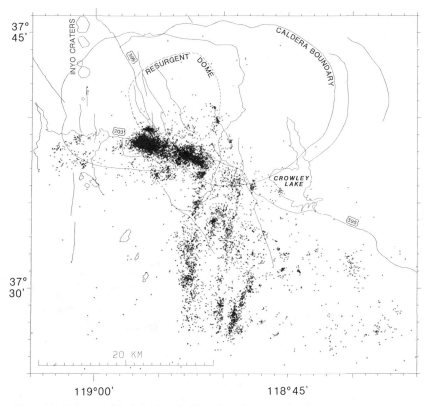

Figure 11 Epicenters (black dots) of 13,000 earthquakes located by the US Geological Survey beneath Long Valley Caldera (the large oval) and the Eastern Sierra Escarpment (south of Long Valley) during the period June 1982 through July 1984. Most of the hypocenters occur at depths of less than 10 km (Cockerham & Pitt 1984, Hill 1984).

change in seismicity, ground deformation, or fumarolic activity), and that 45% of these periods of unrest (70 cases out of 149) eventually led to eruptions. They caution, however, that precursory activity is better recorded for volcanoes that have erupted than for those that have not. The actual figure for unrest at silicic calderas that leads to eruptions is therefore probably much less than 45%.

BASIC RESEARCH

Forecasting based on empirical data may work on a limited basis, but in the long run it will not approach the potential accuracy that may be achieved by understanding how volcanoes work. Basic research into the origin,

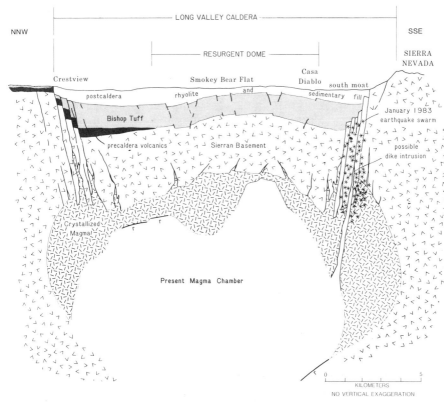

Figure 12 Schematic cross section through Long Valley Caldera, showing the structure of the caldera floor and the underlying magma chamber as inferred from geological and geophysical data. The area of maximum inflation occurs on the south portion of the resurgent dome. The star symbols represent earthquake hypocenters. [From Hill et al (1985), with permission.]

evolution, structure, and dynamics of volcanic systems should provide this understanding.

Volcanoes provide a natural laboratory whose roots reach deep into the Earth's interior and whose ashes are often hurled into the stratosphere. Understanding volcanoes requires an eclectic approach by scientists from many disciplines: geologists mapping both active volcanoes and the insides of extinct volcanoes exposed by erosion, geophysicists monitoring active volcanoes and probing the structure and dynamics of the Earth's interior, geochemists analyzing volcanic gases and studying the composition and evolution of magmas and the mantle, and meteorologists studying the effects of volcanic gases and aerosols on the atmosphere. Some volcano observatories should have basic research as their primary function. Many of the monitoring measurements made on active volcanoes can be used not only for pattern recognition but also for testing of theoretical models of subsurface structure and dynamics.

In my opinion, there are ultimate limits to the accuracy of forecasting volcanic eruptions. I think that forecasts will always be probabilistic rather than deterministic. For example, we may be able someday to accurately measure the state of stress within Kilauea Volcano. However, each time magma fractures the carapace of the volcano and forms a new shallow intrusion or eruption, it changes the strength of the volcano. These changes may be small and subtle, but they could affect the timing, location, and possibly the character of the next eruption. I cannot envision how these small changes in the volcano's overall strength can be determined, except in hindsight.

SOCIETAL CONSIDERATIONS

The hazards and the areas at risk from volcanic eruptions can be estimated by geologists (Blong 1984). However, the problems of reducing the risks to life and property from volcanic eruptions are much more complex and involve the interaction of geologists, sociologists, government officials, journalists, businessmen, and the public (United Nations 1976, Sheets & Grayson 1979, Shimozuru 1981, Fiske 1984). Approaches to these problems of reducing risk include informing the public of hazards without causing undue panic or complacency, land-use planning, hazards zoning (Crandell et al 1984), contingency planning, providing shelters and emergency supplies, and eruption insurance.

Special problems are raised by lava diversion plans (Lockwood & Torgerson 1980, Lockwood & Romano 1985) or other possible modifications of volcanic behavior. Natural disasters are considered "acts of God" by most legal systems, and God cannot be sued for damages. However, if humans divert a lava flow, or perhaps even attempt to divert a flow, they may be held responsible for any subsequent damage done by that flow. Although lava diversion in some cases appears to be technically feasible, I think that with our meager present knowledge, "acts of God" should be left to God. This does not imply that research on these problems should be abandoned. Only by increasing our knowledge may we someday be able to interfere with beneficial results.

Most well-known volcanic eruptions owe their notoriety to the deaths and destruction they have caused. Stories of successful forecasts that lead to saving thousands of lives are less well known. One such example occurred in 1983. Colo Volcano, on a small island in Indonesia, was shaken by an earthquake swarm that began on July 14, and small explosive eruptions began on July 18 (Katili & Sudradjat 1984). On the basis of the past behavior of Colo and similar volcanoes in Indonesia, geologists from the Indonesian Volcanological Survey recommended to the local governmental officials that the island be evacuated. The officials concurred, and all 7000 inhabitants were evacuated by boat. The climax eruption occurred on July 23, sweeping the island with hot pyroclastic flows. All livestock, housing, and coconut plantations were destroyed. It will take years to rebuild the island's economy, but the people survived.

ACKNOWLEDGMENTS

An earlier draft of this paper was significantly improved by the thoughtful reviews of my colleagues Christopher Newhall and Donal Mullineaux. Thanks Chris and Don.

Literature Cited

Aramaki, S. 1957. The 1783 activity of Asama Volcano. *Jpn. J. Geol. Geogr.* 28:11–33

Bailey, R. A., Dalrymple, G. B., Lanphere, M. A. 1976. Volcanism, structure, and geochronology of Long Valley Caldera, Mono County, California. *J. Geophys. Res.* 81:725–44

Barberi, F., Corrado, G., Innocenti, F., Luongo, G. 1984. Phlegraean Fields 1982–1984: brief chronicle of a volcano emergency in a densely populated area. *Bull. Volcanol.* 45-2:175–85

Bianchi, R., Coradini, A., Federico, C., Giberti, G., Sartoris, G., Scandone, R. 1985. Modeling of surface ground deformation in the Phlegraean Fields volcanic area, Italy. *Bull. Volcanol.* In press

Blong, R. J. 1984. *Volcanic Hazards.* New York: Academic. 424 pp.

Booth, B., Croasdale, R., Walker, G. P. L. 1978. A quantitative study of five thousand years of volcanism on São Miguel, Azores. *Philos. Trans. R. Soc. London Ser. A* 288:271–319

Caputo, M. 1979. Two thousand years of geodetic and geophysical observation in the Phlegraean Fields near Naples. *Geophys. J. R. Astron. Soc.* 56:319–28

Casadevall, T., Rose, W., Gerlach, T., Greenland, L. P., Ewert, J., et al. 1983. Gas emissions and the eruptions of Mount St. Helens through 1982. *Science* 221:1383–85

Cashman, K. V., Taggart, J. E. 1983. Petrologic monitoring of 1981 and 1982 eruptive products from Mount St. Helens. *Science* 221:1385–87

Chadwick, W. W. Jr., Swanson, D. A., Iwatsubo, E. Y., Heliker, C. C., Leighley, T. A. 1983. Deformation monitoring at Mount St. Helens in 1981 and 1982. *Science* 221:1378–80

Civetta, L., Gasparini, P., Luongo, G., Rapolla, A., eds. 1974. *Physical Volcanology.* Amsterdam: Elsevier. 333 pp.

Cockerham, R. S., Pitt, A. M. 1984. Seismic activity in Long Valley Caldera area, California: June 1982 through July 1984. See Hill et al 1984, 2:493–526

Crandell, D. R., Mullineaux, D. R. 1975. Technique and rationale of volcano-hazards appraisals in the Cascade Range, northwestern United States. *Environ. Geol.* 1:23–32

Crandell, D. R., Mullineaux, D. R. 1978. Potential hazards from future eruptions of Mount St. Helens Volcano, Washington. *US Geol. Surv. Bull. 1383-C.* 26 pp.

Crandell, D. R., Mullineaux, D. R., Rubin, M. 1975. Mount St. Helens Volcano: recent and future behavior. *Science* 187:438–41

Crandell, D. R., Booth, B., Kusumadinata, K., Shimozuru, D., Walker, G. P. L., Westercamp, D. 1984. Source-book for volcanic hazards zonation. *UNESCO Nat. Hazards Ser.* Paris: UNESCO. 97 pp.

Decker, R. W. 1978. State of the art in volcano forecasting. In *Geophysical Predictions, Stud. Geophys.,* pp. 47–57. Washington, DC: Natl. Acad. Press. See also 1974, *Bull. Volcanol.* 37:372–93

Decker, R. W., Decker, B. B. 1981. The eruptions of Mount St. Helens. *Sci. Am.* 244(3):68–80

Decker, R. W., Koyanagi, R. Y., Dvorak, J. J., Lockwood, J. P., Okamura, A. T., et al. 1983. Seismicity and surface deformation of Mauna Loa Volcano, Hawaii. *Eos, Trans. Am. Geophys. Union* 64:545–47

Dzurisin, D., Westphal, J. A., Johnson, D. J. 1983. Eruption prediction aided by electronic tiltmeter data at Mount St. Helens. *Science* 221:1381–83

Fiske, R. S. 1984. Volcanologists, journalists, and the concerned local public: a tale of two crises in the eastern Caribbean. In *Explosive Volcanism: Inception, Evolution, and Hazards, Natl. Res. Counc. Stud. Geophys.*, pp. 170–76. Washington, DC: Natl. Acad. Press

Fournier d'Albe, E. M., Tazieff, H., Booth, B., Baker, P. E., Guest, J. E., et al. 1979. Volcano prediction. *J. Geol. Soc. London* 136:321–59

Gorshkov, G. S. 1971. Prediction of volcanic eruptions and seismic methods of location of magma chambers—a review. *Bull. Volcanol.* 35-1:198–211

Hill, D. P. 1984. Monitoring unrest in a large silicic caldera, the Long Valley–Inyo Craters volcanic complex in east-central California. *Bull. Volcanol.* 47-2:371–95

Hill, D. P., Bailey, R. A., Ryall, A. S., eds. 1984. *Proceedings of Workshop on Active Tectonic and Magmatic Processes Beneath Long Valley Caldera, Eastern California, US Geol. Surv. Open-File Rep. 84-939*, 2 vols. 942 pp.

Hill, D. P., Bailey, R. A., Ryall, A. S. 1985. Active tectonic and magmatic processes beneath Long Valley Caldera, eastern California. *J. Geophys. Res.* 90:11111–20

Katili, J. A., Sudradjat, A. 1984. The devastating 1983 eruption of Colo Volcano, Una-Una Island, central Sulawesi, Indonesia. *Geol. Jahrb., Reihe A* 75:27–47

Klein, F. W. 1982. Patterns of historical eruptions at Hawaiian volcanoes. *J. Volcanol. Geotherm. Res.* 12:1–35

Klein, F. W. 1984. Eruption forecasting at Kilauea Volcano, Hawaii. *J. Geophys. Res.* 89:3059–74

Koyanagi, R. Y., Endo, E. T., Ebisu, J. S. 1975. Reawakening of Mauna Loa Volcano, Hawaii. *Geophys. Res. Lett.* 2:405–8

Kuno, H. 1962. *Japan, Taiwan and Marianas, Pt. 11. Catalogue of the Active Volcanoes of the World.* Rome: Int. Volcanol. Assoc. 332 pp.

Latter, J. H. 1969. Natural disasters. *Adv. Sci.* 25:362–80

Lipman, P. W., Mullineaux, D. R. 1981. The 1980 eruptions of Mount St. Helens, Washington. *US Geol. Surv. Prof. Pap. 1250.* 844 pp.

Lockwood, J. P., Romano, R. 1985. Lava diversion proved in 1983 tests at Etna. *Geotimes* 30:10–12

Lockwood, J. P., Torgerson, F. A. 1980. Diversion of lava flows by aerial bombing—lessons from Mauna Loa Volcano, Hawaii. *Bull. Volcanol.* 43-4:727–41

Lockwood, J. P., Koyanagi, R. Y., Tilling, R. I., Holcomb, R. T., Peterson, D. W. 1976. Mauna Loa threatening. *Geotimes* 21(6): 12–15

Lockwood, J. P., Banks, N. G., English, T. T., Greenland, L. P., Jackson, D. B., et al. 1985. The 1984 eruption of Mauna Loa Volcano, Hawaii. *Eos, Trans. Am. Geophys. Union* 66:169–71

Lowenstein, P. L. 1982. Problems of volcanic hazards in Papua New Guinea. *Geol. Surv. Papua New Guinea Rep. 82/7.* 61 pp.

Malone, S. D., Boyko, C., Weaver, C. S. 1983. Seismic precursors to the Mount St. Helens eruptions in 1981 and 1982. *Science* 221:1376–78

Martin, R. C., Davis, J. F., eds. 1982. *Status of Volcanic Prediction and Emergency Response Capabilities in Volcanic Hazard Zones of California, Calif. Div. Mines Geol. Spec. Publ. 63.* 275 pp.

McKee, C. O., Lowenstein, P. L., de Saint Ours, P., Talai, B., Mori, J. J., Itikarai, I. 1985a. Rabaul Caldera, Papua New Guinea: volcanic hazards, surveillance, and eruption contingency planning. *J. Volcanol. Geotherm. Res.* 23:195–237

McKee, C. O., Lowenstein, P. L., de Saint Ours, P., Talai, B., Mori, J. J., Itikarai, I. 1985b. Seismic and ground-deformation crisis at Rabaul Caldera, September 1983 to October 1984. *Volcano News* 1985(19–20): 3

Melson, W. G. 1983. Monitoring the 1980–1982 eruptions of Mount St. Helens: composition and abundances of glass. *Science* 221:1387–91

Miller, C. D., Mullineaux, D. R., Crandell, D. R., Bailey, R. A. 1982. Potential hazards from future volcanic eruptions in the Long Valley–Mono Lake area, east-central California and southwest Nevada—a preliminary assessment. *US Geol. Surv. Circ. 877.* 10 pp.

Minakami, T., ed. 1971. The surveillance and prediction of volcanic activity. *UNESCO Earth Sci. Ser. No. 8.* Paris: UNESCO. 116 pp.

Nakamura, K. 1964. Volcano-stratigraphic study of Oshima Volcano, Izu. *Bull. Earthquake Res. Inst. Univ. Tokyo* 42:649–728

Newhall, C. G. 1982. A method for estimating intermediate- and long-term risks from volcanic activity, with an example from Mount St. Helens, Washington. *US Geol. Surv. Open-File Rep. 82-396.* 59 pp.

Newhall, C. G. 1984a. Semiquantitative assessment of changing volcanic risk at Mount St. Helens, Washington. *US Geol. Surv. Open-File Rep. 84-272.* 29 pp.

Newhall, C. G. 1984b. Short-term forecasting

of volcanic hazards. *Proc. Geol. Hydrol. Hazards Train. Program, US Geol. Surv. Open-File Rep.* 84-760, pp. 507–92

Newhall, C. G., Dzurisin, D., Mullineaux, L. S., 1984. Historical unrest at large Quaternary calderas of the world. See Hill et al 1984, 2:714–42

Parascandola, A. 1947. *I Fenomeni Bradisismici del Serapeo di Pozzuoli.* Naples: Genovese

Romano, R., ed. 1982. *Mount Etna Volcano, Mem. Soc. Geol. Ital.* 23. 205 pp.

Rose, W. I. Jr., Stoiber, R. E. 1969. The 1966 eruption of Izalco Volcano, El Salvador. *J. Geophys. Res.* 74:3119–30

Rundle, J. B., Whitcomb, J. H. 1984. A model for deformation in Long Valley, California. *J. Geophys. Res.* 89:8287–8302

Ryall, A., Ryall, F. 1981. Attenuation of *P* and *S* waves in a magma chamber in Long Valley Caldera, California. *Geophys. Res. Lett.* 8:557–60

Ryall, A., Ryall, F. 1983. Spasmodic tremor and possible magma injection in Long Valley Caldera, eastern California. *Science* 219:1432–33

Savage, J. C., Clark, M. M. 1982. Magmatic resurgence in Long Valley Caldera, California: possible cause of the 1980 Mammoth Lakes earthquakes. *Science* 217:531–33

Scott, W. E. 1984. Assessments of long-term volcanic hazards. *Proc. Geol. Hydrol. Hazards Train. Program, US Geol. Surv. Open-File Rep.* 84-760, pp. 447–98

SEAN (Scientific Event Alert Network). 1985. *SEAN Bull.* 10(2):6. Washington, DC: Smithson. Inst.

Sheets, P. D., Grayson, D. K. 1979. *Volcanic Activity and Human Ecology.* New York: Academic. 644 pp.

Shimozuru, D. 1981. Volcano surveillance in Japan: an intricate cooperative program. *Episodes* 1981(1):23–27

Simkin, T., Siebert, L., McClelland, L., Bridge, D., Newhall, C., Latter, J. H.

1981. *Volcanoes of the World.* Stroudsburg, Penn: Smithsonian Inst./Hutchinson Ross. 232 pp.

Souther, J. G., Tilling, R. I., Punongbayan, R. S. 1984. Forecasting eruptions in the Circum-Pacific. *Episodes* 7(4):10–18

Swanson, D. A., Casadevall, T. J., Dzurisin, D., Malone, S. D., Newhall, C. G., Weaver, C. S. 1983. Predicting eruptions at Mount St. Helens, June 1980 through December 1982. *Science* 221:1369–75

Tazieff, H., Sabroux, J. C., eds. 1983. *Forecasting Volcanic Events.* Amsterdam: Elsevier. 635 pp.

Thorarinsson, S. 1967. The eruptions of Hekla in historical times, a tephrochronological study. In *The Eruption of Hekla,* ed. T. Einarsson, G. Kjartansson, S. Thorarinsson, pp. 1–170. Reykjavik: Soc. Sci. Islandica

Ui, T., Aramaki, S. 1983. Volcanic dry avalanche deposit in 1980 eruption of Mt. St. Helens, U.S.A. *Bull. Volcanol. Soc. Jpn.* 28:289–99

United Nations. 1976. *Disaster Prevention and Mitigation,* Vol. 1. *Volcanological Aspects.* Geneva: Off. United Nations Disaster Relief Coord. 38 pp.

Waitt, R. B. Jr., Pierson, T. C., MacLeod, N. S., Janda, R. J., Voight, B., Holcomb, R. T. 1983. Eruption-triggered avalanche, flood, and lahar at Mount St. Helens—effects of winter snowpack. *Science* 221:1394–97

Walker, G. P. L. 1974. Volcanic hazards and the prediction of volcanic eruptions. *Geol. Soc. London Misc. Publ.* 3:23–41

Weaver, C. S., Zollweg, J. E., Malone, S. D. 1983. Deep earthquakes beneath Mount St. Helens: evidence for magmatic gas transport? *Science* 221:1391–94

Wickman, F. E. 1966. Response period patterns of volcanoes. *Ark. Mineral. Geol.* 4(7–11):291–367 (in 5 parts)

Yokoyama, I. 1971. Pozzuoli event in 1970. *Nature* 229:532–33

EPILOGUE   While this article was in press, the eruption of Nevado del Ruiz in Colombia on 13 November 1985 caused the worst volcanic disaster since the eruption of Mont Pelée in 1902.

A relatively small eruption of magma triggered the release of a large volume of melt water from the glacier-covered summit of the 5400-m-high volcano. The flood of water and debris swept down the steep canyons on the volcano's flanks, particularly the Rio Lagunillas. The resulting mudflows killed about 25,000 people, mainly in the town of Armero, 50 km east of and nearly 5000 m below the summit of Nevado del Ruiz.

This disaster emphasizes the great need to put the words in this review article into action. Volcanic eruptions will continue to occur, but more of their dangers can be anticipated, and the risk can be reduced.

Ann. Rev. Earth Planet. Sci. 1986. 14:293–322

RUPTURE PROCESS OF SUBDUCTION-ZONE EARTHQUAKES

Hiroo Kanamori

Seismological Laboratory, California Institute of Technology, Pasadena, California 91125

INTRODUCTION

This review is primarily concerned with the rupture process of large subduction-zone earthquakes determined by various seismological methods, and with its interpretation in terms of an asperity model. It is not possible to make a thorough and extensive review on the subject because of the limited length. Consequently, this review is inevitably biased toward the works in which I was directly involved through collaborations with various investigators.

The distribution of large earthquakes along subduction zones has a distinct pattern. Great earthquakes occur in South America, Alaska, the Aleutians, and Kamchatka. In contrast, earthquakes along the Marianas are smaller. The seismicity in other subduction zones is intermediate between these two groups (see Figure 1). Although this regional variation is now generally accepted, it was not until an appropriate method for quantification of large earthquakes was developed that the regional variation was clearly recognized. In view of its fundamental importance in seismology, we first review the quantification method.

QUANTIFICATION OF EARTHQUAKES

Since the physical process underlying an earthquake is very complex, we cannot express every detail of an earthquake by a single parameter. Nevertheless, it would be useful if we could find a single number that represents the overall physical size of an earthquake. This was the very

293

0084–6597/86/0515–0293$02.00

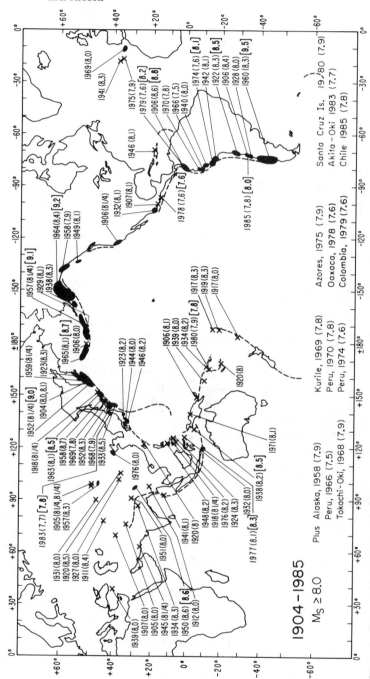

Figure 1 Great and large earthquakes for the period from 1904 to 1985. The surface-wave magnitude M_s is given in parentheses, and the moment magnitude M_w is given in brackets. Shaded areas indicate the rupture zones (modified from Kanamori 1977c).

philosophy of Richter's (1935) earthquake magnitude scale. He defined a local magnitude scale M_L for southern California using the amplitude of seismic waves recorded by the Wood-Anderson seismograph. Although M_L is a purely empirical scale without any direct relation to the physical parameters of the earthquake source, it has proved to be extremely useful in studying seismicity.

Gutenberg and Richter (e.g. Gutenberg 1945, Gutenberg & Richter 1956) extended the magnitude scale to earthquakes worldwide and developed various empirical scales. One of the most widely used scales is the surface-wave magnitude M_s, which is defined by the amplitude of surface waves with a period of about 20 s. (For more details on magnitude scales, see Geller & Kanamori 1977, Bath 1981, Abe 1981, Chung & Bernreuter 1981, Kanamori 1983.)

An important empirical relation is the magnitude-energy relation obtained by Gutenberg & Richter (1956):

$$\log E_s = 1.5 M_s + 11.8, \tag{1}$$

where E_s is the energy (in ergs) radiated from an earthquake source in the form of elastic waves. This relation was determined empirically too, and it is subject to large uncertainty. Nevertheless, it is widely used in seismology.

In Figure 1, the surface-wave magnitude M_s is given in parentheses. The 1906 Colombia earthquake has the largest M_s (8.6). However, the size of its rupture zone (indicated by the shaded area in Figure 1) is not the largest. The 1960 Chilean earthquake and the 1964 Alaskan earthquake, for example, have much larger rupture zones, yet the M_s values for these events are smaller. If M_s is related to the energy E_s through (1), and if the energy release per unit rupture area is approximately constant, it is rather strange to have smaller M_s values for events with much larger rupture zones. This problem had been recognized by several investigators, but no systematic investigation was made until long-period seismic waves began to be used for seismic source studies.

The difficulty in using M_s for quantification of large earthquakes arises from the fact that the period of surface waves (20 s) used for the determination of M_s is much shorter than the time scale of faulting associated with great earthquakes, for which the fault length can be as long as 1000 km. The duration of faulting is approximately equal to the fault length L divided by the rupture velocity V (which is usually about 2–3 km s^{-1}). The time scales of faulting are therefore about 40, 100, and 400 s for events with rupture lengths of 100, 250, and 1000 km, respectively. Hence, the 20-s surface waves cannot represent the entire rupture process of great earthquakes, and the result is that the M_s scale is saturated (as seen in Figure 1).

We can circumvent this difficulty by using longer-period surface waves for quantification purposes. Aki (1966a,b) used long-period Love waves to determine the seismic moment M_0 of the 1964 Niigata earthquake. The seismic moment M_0 is equal to μDS, where D is the average slip on the fault, S is the fault area, and μ is the rigidity of the material surrounding the fault (see Steketee 1958, Maruyama 1963, Burridge & Knopoff 1964). The amplitude of long-period (longer than the duration of faulting) seismic waves generated by an earthquake is proportional to the seismic moment.

Although the seismic moment represents the size of an earthquake only at very long periods, it has proved to be a useful parameter for overall quantification of earthquakes, particularly for great earthquakes. It can be also related to E_s with some assumptions. Kanamori (1977a) shows that

$$E_s = M_0/(2\mu/\Delta\sigma), \qquad (2)$$

where $\Delta\sigma$ is the average stress drop in earthquakes. Various observations indicate that $2\mu/\Delta\sigma$ is roughly 2×10^4. With these assumptions, we can estimate E_s from M_0. Since M_0 can be reliably determined from observational data, and since the energy is a fundamental physical quantity, this method is useful for quantification purposes. Once E_s is determined, it is possible to convert it to a magnitude scale by using (1) inversely, i.e.

$$M_w = (\log E_s - 11.8)/1.5 \qquad (3)$$

or

$$M_w = (\log M_0 - 16.1)/1.5. \qquad (4)$$

M_w is a magnitude scale defined in terms of energy or moment through (3) or (4). It is important to note that unlike other magnitude scales, this scale is not empirical as long as the estimate of the energy through (2) is correct. Since this magnitude is derived from either energy or seismic moment, it is often called either the energy magnitude or the moment magnitude (see also Purcaru & Berckhemer 1978, Hanks & Kanamori 1979). For earthquakes smaller than 8, M_w generally agrees with M_s, which suggests that both of the assumptions used in estimating E_s from M_0 through (2) and relation (1) are reasonable. Since M_0 does not saturate as the size of the rupture zone increases, M_w does not saturate. In this regard, M_w is a better parameter for quantification of earthquakes than M_s. In Figure 1, M_w values for very large earthquakes are given in brackets. The 1960 Chilean earthquake has the largest M_w (9.5), and the 1964 Alaskan earthquake has the second largest M_w (9.2).

Since M_w is determined from very-long-period waves, it does not necessarily represent the size of an earthquake at short periods (e.g. 1–10 s). However, many studies suggest that earthquake source spectra have certain common features, so that the source parameters determined at long periods, such as the seismic moment, can generally represent the entire spectrum reasonably well (Aki 1967, Kanamori & Anderson 1975, Geller 1976).

The problem of quantification of earthquakes is not very simple. Many parameters are required for a complete description of an earthquake source. However, if we are to use a single parameter to represent the gross size of earthquakes, M_w seems to be the most useful one.

CHARACTERISTICS OF SUBDUCTION ZONES AND SEISMICITY

Figure 1 depicts the variation of seismicity between different subduction zones more clearly with the unsaturated magnitude scale M_w than with M_s. Subduction zones in southern Chile, Colombia, Alaska, the Aleutians, and Kamchatka have great earthquakes with $M_w > 8.7$. Subduction zones in the Marianas, Mexico, Tonga Kermadec, and the New Hebrides do not have great earthquakes; most of the major earthquakes in these zones are smaller than 8.2. Since Figure 1 shows the data for the period from 1904 to 1985, this pattern may not be completely representative of subduction-zone seismicity. However, historical data (which are not shown in Figure 1) indicate that the pattern is generally representative of long-term seismicity of subduction zones. A notable exception is Sumatra. A recent study by Newcomb & McCann (1984) shows that two large earthquakes ($M_w = 8\frac{3}{4}$, $8\frac{1}{4}$ to $8\frac{1}{2}$) occurred in Sumatra in 1833 and 1861.

Since most large subduction-zone earthquakes represent the slip between the subducting oceanic plate and the overriding upper plate, the difference in the level of seismicity can be interpreted as the difference in the strength of mechanical coupling between the subducting and overriding plates (Kanamori 1971, 1977b, Kelleher et al 1974, Uyeda & Kanamori 1979). Subduction zones with great earthquakes are strongly coupled, and conversely those without are weakly coupled. Uyeda & Kanamori (1979) called the strongly coupled subduction zones the "Chilean-type" subduction zones, and the weakly coupled zones the "Mariana-type" subduction zones, since these two subduction zones are most representative of strongly and weakly coupled zones, respectively.

Since interplate interaction is a key element that determines tectonic features at plate boundaries, one would expect significant tectonic differences between the Chilean-type and the Mariana-type subduction zones. Figure 2 schematically shows characteristic features of the Chilean- and Mariana-type subduction zones.

The differences in various features are summarized in the following (for more details, see Uyeda 1982, 1984).

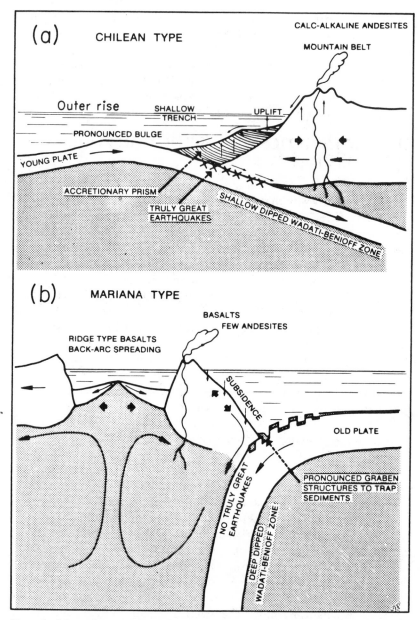

Figure 2 Schematic comparison between the Chilean- and the Mariana-type subduction zones (slightly modified from Uyeda 1984).

1. BENIOFF-ZONE DIP ANGLES   The dip angle of the Benioff (or Wadati-Benioff) zone is very shallow in the Chilean-type subduction zone and steep in the Mariana-type subduction zone. Since the width of the interplate contact zone is likely to increase as the dip angle decreases (Kelleher et al 1974), this difference is consistent with the difference in the strength of interplate coupling. The trench is deeper at Mariana-type zones than at Chilean-type zones.

2. EXISTENCE OF THE BACK-ARC BASIN   The back-arc basin is commonly found for the Mariana-type subduction zones, but it is rare in the Chilean-type zones. A possible reason for this is that the increased horizontal compressional stress associated with the strong plate coupling in the Chilean-type zones inhibits back-arc opening.

3. OUTER RISE   Watts & Talwani (1975) show that the shape of the outer rise (see Figure 2) varies from place to place, reflecting the difference in the magnitude of the compressive stress in the oceanic lithosphere. However, more recent studies indicate that the outer rise is a universal feature of the trench-arc system and that no obvious regional variation exists (Caldwell et al 1976). If the magnitude of the compressive stress increases with the strength of interplate coupling, one would expect a more pronounced outer rise off the Chilean-type subduction zones. The data available now, however, are inconclusive.

In the outer-rise zone, many normal-fault and thrust events occur (Stauder 1968, Chapple & Forsyth 1979). Normal-fault events are generally shallower than thrust events, and they are considered to result from the bending of the oceanic plate before it subducts beneath the upper plate. Ward (1983) examined the depth of outer-rise events and suggests that the depth of transition from normal to thrust events varies from place to place. In general, it is shallower at the Chilean-type zones than at the Mariana-type zones, which suggests that the horizontal compressive stress is larger for the Chilean-type zones.

4. VOLCANIC ROCK TYPES   In general, andesites are more abundant in the Chilean-type subduction zones, whereas basalts are more common in the Mariana-type zones (e.g. Miyashiro 1974, Gill 1981). Although the distribution of volcanic rocks is rather complex and there are many exceptions, this general trend can be explained in terms of the different degree of plate coupling. At the Chilean-type zones, the higher compressive stress in the upper plate would increase interaction between the ascending magma and the crust, producing more andesitic volcanic rocks (e.g. Coulon & Thorpe 1981). However, the detailed mechanism is presently unknown.

5. UPLIFT OF SHORELINES   There is a striking difference in the uplift rate of shorelines between Chilean-type and Mariana-type subduction zones. Figure 3 shows the maximum height of the Holocene shoreline (5000 ± 1000 yr old) around the Pacific compiled by Yonekura (1983). The uplift of the shorelines in the Chilean-type zones such as Chile and Alaska is very large, but almost no uplift is observed for the Marianas and Tonga subduction zones (both Mariana-type zones). A notable exception is the New Hebrides. Because of its moderate seismicity, the plate coupling there is not considered to be very strong. However, the uplift is almost as large as in Chile. This can be explained by the collision of buoyant topographic features against the New Hebrides arc (Taylor et al 1980). Such collisions seem to play an important role in controlling the mechanical interaction between the plates.

PLATE MOTION AND SEISMICITY

Since great subduction-zone earthquakes are ultimately caused by strain accumulation due to plate motion, seismicity is expected to correlate with plate parameters such as the absolute velocity, convergence rate, plate age,

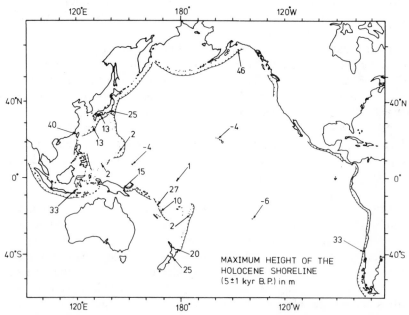

Figure 3   Maximum height of the Holocene shoreline (5 ± 1 × 10³ yr old) in meters in and around the Pacific (Yonekura 1983).

Benioff-zone dip angle, and length of the downgoing slab. In order to investigate this problem, it is necessary to quantify the seismicity of each subduction zone. One difficulty is that the instrumental data are available only for roughly the past 80 yr.

Ruff & Kanamori (1980) assumed that the level of seismic activity of an indivudual subduction zone can be represented by the magnitude M_w of the largest earthquake recorded for that subduction zone. Implicit in this assumption is that at least one large earthquake characteristic of each subduction zone occurred during the past 80 yr. Ideally, it is best to integrate the seismic energy released during the time period considered and determine the energy release rate per unit time and unit length of the subduction zone. Unfortunately, it is difficult to do this accurately because of the lack of reliable data. In an attempt to do this approximately, Ruff & Kanamori (1980) considered the overall level of seismicity during the preinstrumental period and modified M_w. The modified magnitude M'_w differs only slightly from M_w, but it represents the overall seismicity of subduction zones better than M_w.

Peterson & Seno (1984) carefully evaluated the magnitude and the seismic moment values listed in various seismicity catalogs and estimated the moment release rate (MRR) per unit time and unit length along the arc. Although considerable error is involved in converting the magnitude to the seismic moment, their estimates of the moment release rate are the best presently available.

Figure 4 compares MRR (determined by Peterson & Seno 1984) with M'_w (determined by Ruff & Kanamori 1980) for various subduction zones. Since the division of subduction zones is slightly different between the two studies, some adjustments are made in this comparison, as explained in the figure caption. In general, $\log(MRR)$ correlates very well with M'_w; this correlation suggests that M'_w is a good parameter to represent the level of seismicity in each subduction zone. The regression line shown in Figure 4 gives $MRR = 10^{(1.2M'_w + 18.2)}$ dyne-cm/(100 km–100 yr). In this sense, it is more appropriate to interpret M'_w as a parameter that represents the rate of seismic moment release rather than as the magnitude of the characteristic earthquake in the region.

Ruff & Kanamori (1980) correlated M_w (or M'_w) with various plate parameters. For example, Figure 5a shows the relation between M'_w and the plate convergence rate V. In general, as the convergence rate increases, one would expect stronger interplate interaction and, therefore, higher seismicity. Although Figure 5a shows a generally positive correlation, the scatter is very large, which suggests that other factors are also important in controlling seismicity.

Another important plate parameter is the age of the subducting plate.

Since the older plates are more dense, they have a stronger tendency to sink spontaneously, thereby decreasing the strength of mechanical coupling (Molnar & Atwater 1978, Vlaar & Wortel 1976, Wortel & Vlaar 1978). Figure 5b shows the relation between M'_w and the age T of the subducting plate. The correlation is negative, as expected, but the scatter is very large; this result suggests that the plate age is not the sole controlling factor of seismicity.

Figure 5 indicates that the convergence rate V and the plate age T together might be controlling seismicity. Ruff & Kanamori (1980) performed a three-parameter regression analysis in the form $M'_w = aT + bV + c$, where a, b, and c are constants. Using the data listed in Table 1 of Ruff & Kanamori (1980), they obtained a relation,

$$M'_w = -0.00953T + 0.143V + 8.01 \tag{5}$$

where T is in million years and V is in cm yr^{-1}.

Figure 6 compares the observed M'_w values with those calculated from T and V through (5), showing a good correlation between the observed and

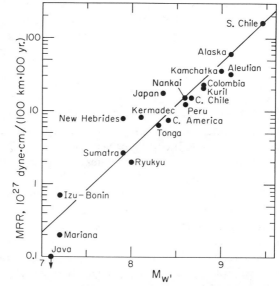

Figure 4   The relation between the seismic moment release rate MRR [in 10^{27} dyne-cm/(100 km–100 yr)] and M'_w. The data are taken from Peterson & Seno (1984) and Ruff & Kanamori (1980). Peterson & Seno's regionalization of the subduction zones is modified as follows (the first and the second names in the parentheses refer to Peterson & Seno's and Ruff & Kanamori's data, respectively): (Peru-south, Peru), (average of Central America and Mexico, Central America), (Aleutian-east and Aleutian-west, Aleutian), (Nankai, SW Japan), (Japan, NE Japan).

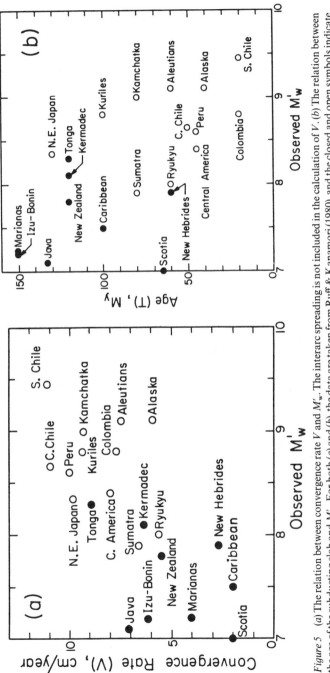

Figure 5   (*a*) The relation between convergence rate *V* and M'_w. The interarc spreading is not included in the calculation of *V*. (*b*) The relation between the age of the subducting slab and M'_w. For both (*a*) and (*b*), the data are taken from Ruff & Kanamori (1980), and the closed and open symbols indicate subduction zones with and without active back-arc opening, respectively.

calculated M'_w values. Also note that the subduction zones with active back-arc opening plot in the lower-left corner of Figure 6, which indicates good correlation between low seismicity and back-arc opening. Ruff & Kanamori (1980) tried similar correlations between M'_w and other plate parameters. They found that among all the three-parameter combinations considered, the M'_w-T-V combination yields the best correlation.

It should be noted that (5) is obtained empirically without any particular physical model. It is possible that other parameters are equally important. For example, Uyeda & Kanamori (1979) and Peterson & Seno (1984) suggest that the absolute velocity of the upper plate is important in determining the strength of plate coupling. In any case, despite the lack of clear physical models, (5) provides a useful scheme for interpreting global seismicity in terms of a simple plate interaction model.

An alternative approach would be to build a specific physical model and test it by using M_w and other plate parameters. Hager et al (1983) used a global convective flow model of the Earth's mantle (Hager & O'Connell 1981) to determine the flow patterns and pressure distributions in the mantle. They found that the model that allows whole mantle flow yields

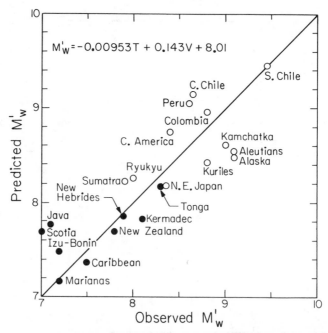

Figure 6 The relation between M'_w calculated from T and V using the relation $M'_w = -0.00953T + 0.143V + 8.01$ and the observed M'_w. Closed and open symbols indicate sub-duction zones with and without active back-arc opening, respectively.

flow patterns that are consistent with the observed Benioff-zone geometries. Higher pressure gradients between the mantles beneath the trench and the back-arc area computed for this model correlate well with higher M_w. Since a larger pressure gradient tends to press the downgoing slab more tightly against the upper plate, this correlation is reasonable. The correlation is especially good when it is taken for the following three parameters: pressure gradient, age of the subducting plates, and M_w. This example demonstrates that seismicity data can provide important constraints on the dynamics of the Earth's mantle.

ASPERITY MODEL

As shown in the preceding sections, the global variation of seismicity can be interpreted in terms of the variation of strength of mechanical coupling. However, so far it is not clear what is causing this regional variation of strength.

The width of the lithospheric interface has been correlated with the maximum length of rupture zones by Isacks et al (1968) and Kelleher et al (1974). In this case, the width of the contact zone is the primary factor determining the coupling strength. Kanamori (1971, 1977b) interpreted the variation of interplate coupling in terms of a weakening of the lithospheric interface. Topographic features on the subducting seafloor, such as seamounts and fracture zones, may be controlling the coupling strength (Kelleher & McCann 1976, 1977). However, until recently little was known about the stress distribution on the fault plane of great earthquakes.

Lay & Kanamori (1980) studied body waves and surface waves of large earthquakes in the Solomon Islands region in an attempt to determine the stress distribution on the thrust plane. They found that relatively short-period (about 10 s) seismic body waves are radiated from only small parts of the entire rupture plane, which generates longer-period (about 200 s) surface waves and over which the aftershocks occur (Figure 7a). They interpreted the results in terms of an asperity model, as shown by Figure 7. This asperity model is an outgrowth of laboratory experiments on rock friction. Byerlee (1970) and Scholz & Engelder (1976) suggested that the two sides of a fault are held together by areas of high strength, which they termed asperities. Extending this model to earthquake faults, we call the areas on the fault plane from where relatively short-period seismic body waves are radiated the (fault) asperities; it is assumed that the stronger spots are responsible for high-frequency seismic radiation. However, the fault asperities can be areas of geometrical irregularity. From seismic observations alone, it is not possible to determine their physical properties. The parts of the fault plane where only long-period seismic radiation occurs or where processes with longer time scales than that of seismic radiation (e.g. creep, aftershock expansion, etc) occur are called the weak zones.

Lay & Kanamori (1980) examined seismograms of events from other subduction zones and found that the complexity of body waveforms is very different. For example, the events that occur in the Kurile Islands generally radiate body waves with very complex waveforms, and these waveforms suggest a rather heterogeneous and complex asperity distribution on the fault plane (see Figure 7b).

Lay & Kanamori (1981) and Lay et al (1982) extended these results to a more general picture (see Figure 7c). In the typical Chilean-type subduction zones, the fault plane is uniformly strong (all asperity), and the size of the rupture zone is determined by major transverse tectonic structures such as fracture zones and ridges (Mogi 1969a, Kelleher & Savino 1975). In subduction zones such as the Aleutian Islands and the Solomon Islands, asperities are relatively large, but they are surrounded by weak zones. The sizes of earthquakes are not as large as those that occur in Chile and Alaska, but failure of one asperity often triggers failure of adjacent asperities, which results in large multiple events or earthquake multiplets (two or more distinct events closely spaced in time). In subduction zones such as the Kurile Islands, asperities become smaller and more heterogeneous in size. In the Mariana-type subduction zones, the fault plane is uniformly weak (no asperities), and no large earthquakes occur.

Ruff & Kanamori (1983a) used seismic body waves to directly determine the asperity distributions on the fault plane of several large earthquakes. Figure 8 illustrates this method.

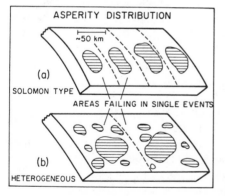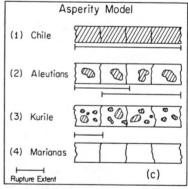

Figure 7 Representation of the interplate interface in subduction zones by an asperity model. Hatched areas indicate asperities. (*a*) A model for the Solomon Islands region. The asperities are of comparable size. (*b*) A model for a more heterogeneous interface in regions such as Japan and the Kurile Islands. (*c*) A schematic representation of asperity distributions for different regions (from Lay & Kanamori 1980, 1981, Lay et al 1982).

Figure 8*a* shows a simple fault with length *L* and area *S*. At point A in the near field, the displacement would look like a ramp function (Figure 8*b*). The total displacement is equal to half the offset *D* of the fault. The build-up time τ is approximately (within a factor of two) equal to L/V, where *V* is the rupture velocity. At point B on the other side of the fault, the displacement time function is reversed in polarity (Figure 8*c*). At a point in the far field, the displacements from both sides of the fault interfere. As a result, the time function in the far field is given by the time derivative of the near-field displacement (Figure 8*d*). The area under the bell-shaped function is proportional to the seismic moment, and the width is approximately equal to τ. This function is often called the far-field source time function, and it is used to characterize the source complexity. We can determine the fault dimension and the seismic moment from the duration and the area of the far-field time function.

Figure 8*f* shows a more complex case that involves two asperities. The near-field and far-field displacements for this case are shown in Figures 8*g* and 8*h*. The far-field time function has two pulses, each representing one of the asperities. From the far-field time function like the one shown in Figure 8*h*, we can determine the asperity distribution. However, because of (*a*) the distortion of the waveform during propagation between the source and the

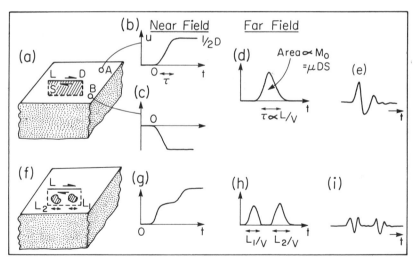

Figure 8   Fault model and far-field body-wave time function. (*a*) A simple fault model with uniform strength (i.e. single asperity). (*b*) The near-field displacement at point A. (*c*) The near-field displacement at point B. (*d*) The far-field displacement, often called the far-field source time function. (*e*) The far-field displacement viewed through a seismograph (seismogram). (*f*) A complex fault model with two asperities. (*g*) The near-field displacement. (*h*) The far-field displacement. (*i*) Seismogram.

station and (b) the instrument response, the observed waveforms are very complex (Figures 8e,i), and the determination of the far-field time function is extremely difficult. Furthermore, for very large earthquakes such as the 1964 Alaskan earthquake, relatively few usable seismograms exist because most of the seismograms were off-scale. Ruff & Kanamori (1983a) used diffracted P waves to circumvent this difficulty.

Figure 9 compares the waveforms of four large earthquakes: the 1964 Niigata earthquake ($M_w = 7.6$), the 1963 Kurile Islands earthquake ($M_w = 8.5$), the 1965 Rat Islands earthquake ($M_w = 8.7$), and the 1964 Alaskan earthquake ($M_w = 9.2$). The Niigata earthquake is the smallest in this group. Since the fault length of this earthquake is about 60 km, the duration of faulting is about 25 s, and the body waveform viewed through the passband of the long-period WWSSN seismograph (i.e. 5–50 s) is a simple impulse (Figure 9a), which suggests a simple source (Figure 10a), at least as viewed through this passband. In contrast, the waveform of the Kurile Islands earthquake is far more complex and suggests a complex asperity distribution (Figure 10b). The waveforms of the Alaskan earthquake are unique. Not only is the amplitude very large, but also the period is very long. This long-period wave indicates that the asperity responsible for the body-wave radiation is very large. By removing the effects of the wave propagation and the instrument from the body waves, Ruff & Kanamori (1983a) determined that the length scale of the asperity is about 200 km (Figure 10d). The pattern of

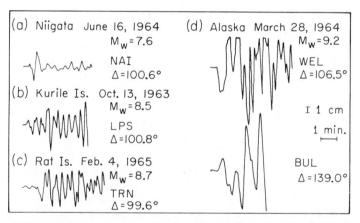

Figure 9 Representative seismograms for four large earthquakes, plotted with the same amplitude and time scales. For the Alaskan earthquake, the seismogram at $\Delta = 106.5°$ is off-scale, and another seismogram from a station in the shadow zone is shown. Note the large amplitude and the long period of the Alaskan earthquake seismograms (modified from Ruff & Kanamori 1983a).

asperity distribution for the Rat Islands earthquake is intermediate between the Kurile Islands and the Alaskan earthquakes (Figure 10c). Hartzell & Heaton (1985) determined far-field time functions from *P* waves of 63 large, shallow subduction-zone earthquakes that occurred in the circum-Pacific belt. They used the seismograms recorded at Pasadena by a Benioff long-period seismograph. Since the passband of this instrument is about 2.5–50 s, the complexity of the source can be resolved only on this time scale. Viewed through this passband, the seismograms showed no obvious global trends in the character of far-field time functions. However, most of the subduction zones do behave characteristically, and this finding suggests regional variations of asperity distribution. For example, the time functions for earthquakes from central Chile, Peru, the Solomon Islands, and the New Hebrides are simple and smooth. The Solomon Islands earthquakes, in particular, are characterized by unusually broad, simple time functions (Figure 11). These results are consistent with the expansion patterns of aftershocks, as is shown later.

More recent studies by Beck & Ruff (1984), Schwartz & Ruff (1985), and Kikuchi & Fukao (1985) have succeeded in resolving important details of the source process. Figure 11 compares the results obtained by different investigators (Hartzell & Heaton 1985). Although the details differ between different studies, the general features as shown in Figure 10 seem to be well established.

If the stress drop $\Delta\sigma_a$ at the asperity does not vary from event to event, the magnitude of the tectonic stress drop $\Delta\sigma$ is approximately equal to $\Delta\sigma_a\,(S_a/S)$, where S is the total area of the fault plane and S_a is the total area of the asperities. We conclude that in the context of the asperity model, the regional variation of the strength of mechanical coupling (i.e. $\Delta\sigma$) is a manifestation of the regional variation of asperity size and its distribution.

This type of asperity model is a gross generalization of the actual stress distribution on the fault plane, but it also provides simple physical interpretations of some seismicity patterns (e.g. swarms, quiescence, fore-shock) before large earthquakes (e.g. Kanamori 1981).

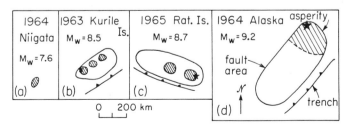

Figure 10   Interpretation of Figure 9 in terms of the asperity model (modified from Ruff & Kanamori 1983a). Compare Figures 9 and 10 with Figures 8*i* and 8*f*.

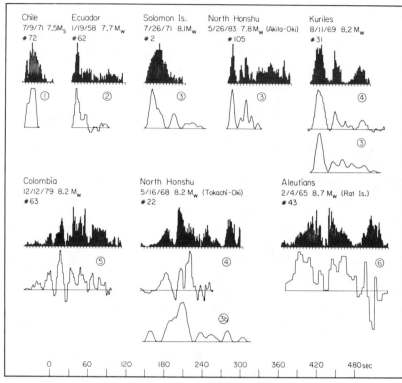

Figure 11   Comparisons of far-field time functions obtained by different investigators. The shaded functions are from Hartzell & Heaton (1985). Sources are as follows: (1) Malgrange et al (1981); (2) Beck & Ruff (1984); (3) M. Kikuchi (written communication); (3b) Kikuchi & Fukao (1985); (4) Schwartz & Ruff (1985); (5) S. Beck & L. Ruff (unpublished work); (6) Ruff & Kanamori (1983a) (after Hartzell & Heaton 1985).

Another class of heterogeneous fault models is the barrier model proposed by Das & Aki (1977). In this model, strong spots (barriers) on the fault plane do not fail during the mainshock. The differences between the asperity model and the barrier model are discussed by Madariaga (1979) and Rudnicki & Kanamori (1981). It is difficult to distinguish between these two models from seismic data alone.

EXPANSION PATTERN OF AFTERSHOCK AREA

Further evidence for regional variations in mechanical heterogeneities of fault zones, here characterized by asperity distributions, came from expansion patterns of aftershock areas. Although the mechanism of aftershock occurrence is not fully understood, the spatial distribution of aftershocks and its expansion pattern seem to provide clues to the mechanical property of the source region. Mogi (1968, 1969b) noted significant variations of expansion patterns of aftershock areas and interpreted them in terms of regional variations of tectonic structures.

Tajima & Kanamori (1985a) developed a method to objectively define the aftershock area by using spatio-temporal patterns of seismic energy release; they then used this method to examine the aftershock area expansion patterns of about 50 earthquakes. Four examples are shown in Figure 12. For the 1964 Alaskan earthquake, no significant expansion occurred, while the aftershock area of the 1978 Miyagi-Oki, Japan, earthquake expanded nearly threefold in area during the period from 1 day to 100 days after the mainshock. Tajima & Kanamori (1985b) defined the 100-day linear expansion ratio η_l (100) by the ratio of the maximum linear dimension of the aftershock area observed at 100 days after the mainshock to that of the 1-day aftershock area. Figure 13, which shows η_l (100) plotted at the respective mainshock epicenter, demonstrates a distinct regional

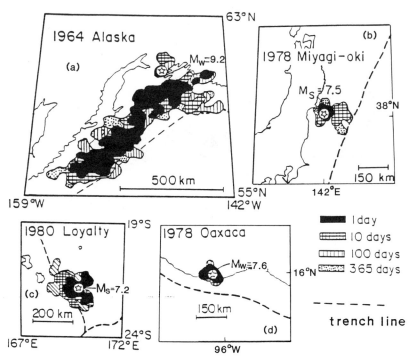

Figure 12 The aftershock areas determined at four different times: 1 day, 10 days, 100 days, and 1 yr (slightly modified from Tajima & Kanamori 1985b).

variation. In the western Pacific subduction zones (northeastern Japan, Ryukyu, the Philippines, New Hebrides, and Tonga), the expansion ratios are generally large. On the other hand, in the northern Pacific subduction zones and along the Central American trench, the ratios are generally small. The events in the Solomon Islands (Nos. 18 and 27) are doublets and are indicated by an open circle (large expansion); however, this large linear expansion could be due to the occurrence of the second event of the doublet. If each event of the doublet is considered separately, they would have been classified as events with a small expansion ratio. In the South American subduction zones, the expansion ratios vary from event to event.

These results can be explained in terms of the asperity model, as shown by Figure 14. The rupture during the mainshock mostly involves asperities. After the main rupture is completed, the stress change caused by the mainshock gradually propagates outward into the surrounding weak zones. This stress propagation manifests itself as an expansion of aftershock activity. If the fault zone consists of large asperities abutting each other (e.g. Chilean-type), little expansion occurs (Figure 14b). If the average size of asperities is relatively small, then the strength of interplate coupling is moderate. If these asperities are densely distributed and separated by small weak zones, no extensive aftershock expansion is expected (Figure 14c). The Mexican subduction zone is a typical example of this type. In contrast, if smaller asperities are sparsely distributed (Mariana-type), large expansion

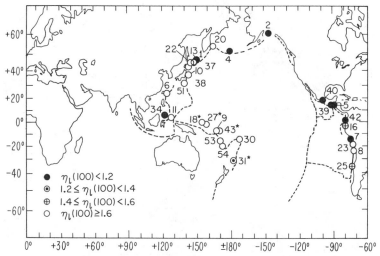

Figure 13 Plot of the 100-day linear expansion ratio η_l (100). The number beside each symbol corresponds to the event number listed in Tajima & Kanamori (1985b). The asterisk indicates the first event of a doublet (after Tajima & Kanamori 1985b).

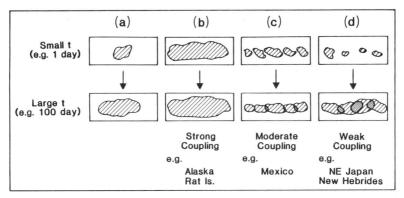

Figure 14 Schematic figure showing typical expansion patterns of aftershock area. The hatched areas indicate aftershock areas. The aftershock areas at small *t* (time) are considered asperities on a fault zone. (*a*) A typical expansion pattern. The aftershock area expands from the asperity to the surrounding weak zone. (*b*) A fault zone with a uniform, large asperity. The aftershock area does not expand, and interplate coupling is strong. (*c*) Densely distributed small asperities. The aftershock zone expands little, and interplate coupling is moderate. (*d*) Sparsely distributed small asperities. An aftershock area expands significantly, often overlapping with that of the adjacent event. Interplate coupling is weak (after Tajima & Kanamori 1985b).

ratios are to be expected, and the aftershock areas of adjacent events may overlap each other (e.g. northeastern Japan, the Philippines, New Hebrides, and Tonga-Kermadec; Figure 14*d*).

The pattern shown in Figure 13 is generally consistent with that of asperity distribution inferred from the locations of great earthquakes and from the body-wave studies, and it renders support to the asperity model.

THE ASPERITY

In the preceding discussion, we introduced asperities as spots of increased strength on the fault plane without specifying what they physically are. As shown earlier, the age of the subducting plate and the plate convergence rate seem to have a strong influence on plate coupling. On the other hand, the coupling strength is also controlled by the heterogeneity on the fault plane. However, the relation between plate parameters such as the age, the convergence rate, and the fault-plane heterogeneities is not obvious. Ruff & Kanamori (1983b) discussed this subject in detail and suggested various possibilities, but at present there is no definite answer to this question.

One important factor, however, seems to be the trench sediment. The amount of trench sediments is very different between different subduction zones because it is controlled by factors such as the supply rate from the

314   KANAMORI

adjacent land and the disposal rate. In some subduction zones, sediments riding on the oceanic plate are scraped off and deposited to form accretionary prisms, while other subduction zones seem to be subducting most of the sediments without forming accretionary prisms (see, e.g., Scholl et al 1977, Hilde 1983, Uyeda 1984). Hilde & Sharman (1978) suggest that horst-and-graben structures, which are often seen on the seafloor near the trench, may be acting as a carrier of the sediments beneath the upper plate, as illustrated in Figure 15. The horst-and-graben structure is thought to be a result of plate bending before subduction. Ruff (1985) proposed that thick excess sediments at subduction zones form a uniform interplate contact plane enhancing the coupling, whereas seafloors with horst-and-graben structures develop a heterogeneous contact plane that decreases the strength of mechanical coupling. Figure 16 shows the global distribution of subduction zones with excess trench sediments (ETS) and with horst-and-graben structures (HGS). Also shown in this figure are the epicenters of large earthquakes. In general, the subduction zones that have experienced very large earthquakes are associated with ETS, and those without large earthquakes are characterized by HGS. We note several exceptions. The Sumatran subduction zone is characterized by ETS, but no great earthquakes have occurred there in this century. Newcomb & McCann (1984), however, documented two great Sumatran earthquakes that occurred in 1833 and 1861. The subduction zone off southwestern Japan has not experienced events with $M_w \geq 8.2$ during this century, but it is the site of two $M_w = 8.1$ earthquakes in this century and many historical great earthquakes. The subduction zone off the Washington-Oregon coast of North America (Juan de Fuca subduction zone) has not experienced great earthquakes for at least the past 150 yr; however, its long-term seismic potential is unknown and is presently a matter of debate (see Heaton & Kanamori 1984).

Another interesting aspect of this model is that it may provide a link between the M_w'-T-V relation and the asperity distribution. There is a general correlation between the dip angle and the age of the oceanic plate

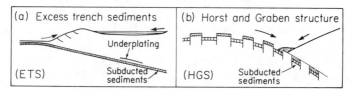

Figure 15   Two trench morphologies. (*a*) Excess trench sediments. Sediments are scraped off on subduction and form an accretionary prism. (*b*) Horst-and-graben structure. A well-developed horst-and-graben structure provides a mechanism to carry down the sediments (after Ruff 1985).

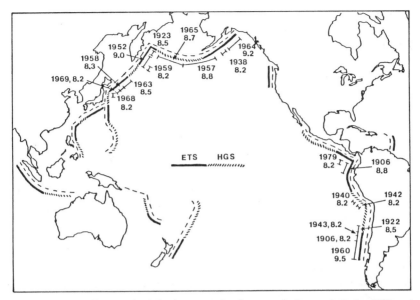

Figure 16 Classification of subduction zones by the two trench morphologies (ETS and HGS). Also shown are M_w values and the approximate rupture zones of major earthquakes (after Ruff 1985).

(Molnar & Atwater 1978). Therefore, older plates may develop more pronounced horst-and-graben structures, since they are bent more sharply before subduction, which results in smaller asperities and weaker coupling.

RELATED PROBLEMS

Evaluation of Seismic Potential

The M'_w-T-V relation given by (5) could be used to evaluate the seismic potential of subduction zones that have not experienced great earthquakes in historical times. Since the age T of the subducting plate and the convergence rate V are known for most subduction zones, M'_w can be estimated from (5).

Heaton & Kanamori (1984) applied (5) to the subduction zone off the Oregon-Washington coast (Juan de Fuca subduction zone), where no great earthquake is known to have occurred for at least the past 150 yr. Furthermore, instrumentally determined seismicity during the last 80 yr is extremely low. Hence, if we extrapolate the past seismicity to the future, we would expect a very low seismic potential. However, the estimated T and V are about 8 Myr and 3.5 cm yr^{-1}, respectively, which would yield a M'_w of 8.4 through (5); this M'_w value suggests a high seismic potential. Since (5) is obtained entirely empirically, this value of M'_w should not be taken at face value. However, it does mean that the Juan de Fuca subduction zone is similar, with respect to the plate-boundary characteristics, to other subduction zones that have experienced great earthquakes, and further investigations into this problem are warranted.

One of the unique features of the Juan de Fuca subduction zone is the very young age of the subducting plate. Since the subduction zones used to determine (5) have a subducting plate older than 10 Myr, there is some question as to whether (5) applies to the very young Juan de Fuca plate. A recent study by Singh et al (1985), however, demonstrates that the 1932 Jalisco, Mexico, earthquake ($M_s = 8.2$) occurred on the boundary between the very young (9 Myr) Rivera plate and the North American plate. This boundary is geometrically similar to that between the Juan de Fuca plate and the North American plate.

Aseismic and Seismic Slip

At the Mariana-type subduction zones, seismicity is so low that the slip associated with earthquakes cannot accommodate the total plate motion; this suggests that a substantial amount of slip is occurring aseismically. Kanamori (1977b) estimated the amount of seismic slip at various subduction zones and concluded that the ratio η of seismic slip to the total plate motion is approximately 1, 0.25, and 0 for southern Chile, the Kurile Islands, and the Marianas, respectively. However, these estimates are subject to large uncertainties because of (a) errors in the estimates of seismic slip and (b) incomplete data on the repeat times. Subsequently, Seno & Eguchi (1983), Sykes & Quittmeyer (1981), and Peterson & Seno (1984) have made independent analyses of the data. The results of these four studies are in general agreement, though they differ in details, particularly for subduction zones for which repeat-time data are incomplete. On the basis of these studies, Kanamori & Astiz (1985) attempted another interpretation of the data. Figure 17 shows the values of η as a function of the age of the subducting plate. These values probably depend on various factors such as the convergence rate, the upper plate velocity, the roughness of the oceanic and upper plate, and the structure of the sediment. Figure 17 shows only the effect of T on η.

Subduction zones with a relatively young subducting plate (ranging from 20 to 60 Myr old) have η values very close to 1, and thus interplate slip for these subduction zones is predominantly seismic. For subduction zones with a very old plate, such as the Marianas and northern Japan, η is very small, and therefore the plates are almost decoupled. It is interesting to note that the two largest lithospheric normal-fault earthquakes, the 1933 Sanriku and 1977 Sumbawa earthquakes, occurred in lithospheres that are

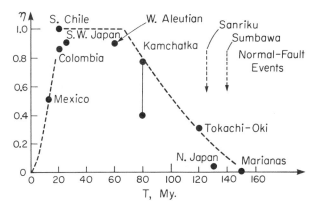

Figure 17 The ratio of seismic slip to total plate motion as a function of the age of the subducting plate (after Kanamori & Astiz 1985). The two largest lithospheric normal-fault earthquakes (1933 Sanriku and 1977 Sumbawa earthquakes) occurred in lithospheres that are 120 and 140 Myr old, respectively.

120 to 140 Myr old; this suggests that these events result from increased tensional stress caused by plate decoupling.

As $T \to 0$, the interplate deformation is most likely to be aseismic as a result of high temperatures, and one would expect η to approach 0 as $T \to 0$. Because of the large uncertainty in the data, however, this result should be considered only tentative.

Radiation of High-Frequency Seismic Waves

The asperity model discussed above was based upon seismic data whose time scale is longer than 10 s. Since the strength of plate coupling and fault heterogeneity (asperities) are most likely to affect the radiation of high-frequency (e.g. > 1 Hz) seismic waves, one would expect regional variations in the radiation efficiency of high-frequency waves from earthquakes. A better understanding of high-frequency characteristics of earthquakes is important for predicting earthquake strong ground motions.

In view of the recent increase in the number of large man-made structures, such as high-rise buildings and offshore drilling platforms, estimation of strong ground motions from large earthquakes in subduction zones is becoming increasingly important. However, observed high-frequency waveforms from large earthquakes are very complex, so that a detailed analysis such as the one illustrated by Figure 8 is not feasible. An alternative approach is to determine the frequency spectrum of far-field time functions. Houston & Kanamori (1986) used digital seismograms, which recently became available, to determine the spectra of a few large

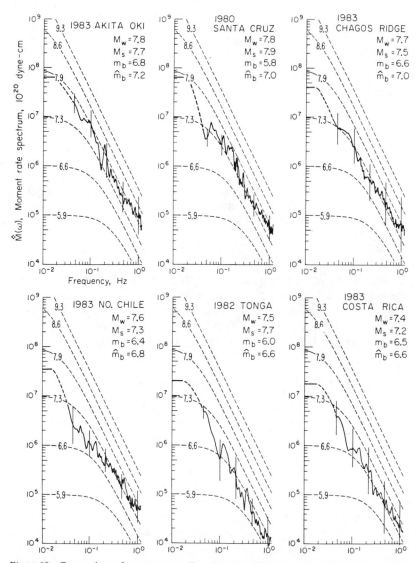

Figure 18 Comparison of source spectra (i.e. spectrum of the source time function, such as the one shown by Figure 8*h*) of six large earthquakes. The dashed curves, which are computed for a theoretical source model, are shown for reference. Note that the spectral amplitude of the 1982 Tonga earthquake at 1 s (10^0 Hz) is significantly lower than that for other events.

subduction-zone earthquakes with comparable magnitudes. The results are shown in Figure 18.

Among the events shown in Figure 18, the event in the Tonga Islands occurred in a Mariana-type subduction zone. Comparing the long-period and short-period spectral amplitudes, we see that the Tonga earthquake has proportionally less high-frequency energy than the other events, possibly because of the weak mechanical coupling at the Tonga subduction zone. Since the presently available data are very limited in quantity, more definitive conclusions must await further studies.

CONCLUSIONS

The distribution of great earthquakes and fault-zone mechanical hetero-geneities inferred from seismic waveforms and expansion patterns of aftershock areas suggest that the nature and strength of interplate coupling at subduction zones vary significantly from place to place. This regional variation can be characterized best by the use of an asperity model, which assumes that a fault zone consists of two parts: mechanically strong spots herein called the asperities, and weaker zones surrounding the asperities. Strongly coupled subduction zones have a fault zone that consists of large and uniform asperities. As the asperity size decreases with increasing area of weak zones, the fault zone becomes more heterogeneous and the strength of plate coupling decreases.

Tectonic features associated with strongly coupled subduction zones such as the Chilean and Alaskan subduction zones include shallow oceanic trenches, shallow dipping Benioff zones, generally andesitic volcanism, and the absence of active back-arc basins. Weakly coupled subduction zones such as the Marianas have tectonic features more or less opposite in nature. This difference is intuitively consistent with the difference in the magnitude of horizontal compressive stress in the trench-arc system associated with the difference in the degree of plate coupling. However, the details of the physical mechanism remain to be investigated.

The magnitude M'_w of the largest "characteristic" earthquake of a subduction zone correlates well with the age T of the subducting plate and the convergence rate V in the form $M'_w = aT + bV + c$ ($a = -0.00953$, $b = 0.143$, $c = 8.01$, T in Myr, V in cm yr^{-1}). This relation, though purely empirical, is useful in predicting M'_w for subduction zones that have not experienced large earthquakes in historical time.

ACKNOWLEDGMENTS

I thank Luciana Astiz and Holly Eissler for reviewing the manuscript. Larry Ruff provided me with some figures before publication. This work was partially supported by the National Science Foundation under grant EAR-8116023 and the US Geological Survey under contract 14-08-0001-G-979. This article is contribution No. 4252 of the Division of Geological and Planetary Sciences.

Literature Cited

Abe, K. 1981. Magnitude of large shallow earthquakes from 1904 to 1980. *Phys. Earth Planet. Inter.* 27:72–92

Aki, K. 1966a. Generation and propagation of G waves from the Niigata earthquake of June 16, 1964. Part 1. A statistical analysis. *Bull. Earthquake Res. Inst. Univ. Tokyo* 44:23–72

Aki, K. 1966b. Generation and propagation of G waves from the Niigata earthquake of June 16, 1964. Part 2. Estimation of earthquake moment, from the G wave spectrum. *Bull. Earthquake Res. Inst. Univ. Tokyo* 44:73–88

Aki, K. 1967. Scaling law of seismic spectrum. *J. Geophys. Res.* 72:1217–31

Bath, M. 1981. Earthquake magnitude—recent research and current trends. *Earth Sci. Rev.* 17:315–98

Beck, S. L., Ruff, L. J. 1984. The rupture process of the great 1979 Colombia earthquake: evidence for the asperity model. *J. Geophys. Res.* 89:9281–91

Burridge, R., Knopoff, L. 1964. Body force equivalents for seismic dislocations. *Bull. Seismol. Soc. Am.* 54:1901–14

Byerlee, J. D. 1970. Static and kinetic friction of granite under high stress. *Int. J. Rock Mech. Min. Sci.* 7:577–82

Caldwell, J. G., Haxby, W. F., Karig, D. E., Turcotte, D. L. 1976. On the applicability of a universal elastic trench profile. *Earth Planet. Sci. Lett.* 31:239–46

Chapple, W. M., Forsyth, D. W. 1979. Earthquakes and bending of plates at trenches. *J. Geophys. Res.* 84:6729–49

Chung, D. H., Bernreuter, D. L. 1981. Regional relationship among earthquake magnitude scales. *Rev. Geophys. Space Phys.* 19:649–63

Coulon, C., Thorpe, R. S. 1981. Role of continental crust in petrogenesis of orogenic volcanic associations. *Tectonophysics* 77:79–93

Das, S., Aki, K. 1977. Fault planes with barriers: a versatile earthquake model. *J. Geophys. Res.* 82:5658–70

Geller, R. J. 1976. Scaling relations for earthquake source parameters and magnitudes. *Bull. Seismol. Soc. Am.* 66:1501–23

Geller, R. J., Kanamori, H. 1977. Magnitude of great shallow earthquakes from 1904 to

1952. *Bull. Seismol. Soc. Am.* 67:587–98

Gill, J. 1981. *Orogenic Andesites and Plate Tectonics.* New York: Springer-Verlag. 390 pp.

Gutenberg, B. 1945. Amplitudes of surface waves and magnitudes of shallow earthquakes. *Bull. Seismol. Soc. Am.* 35:3–12

Gutenberg, B., Richter, C. F. 1956. Magnitude and energy of earthquakes. *Ann. Geofis.* 9:1–15

Hager, B. H., O'Connell, R. J. 1981. A simple global model of plate motions and mantle convection. *J. Geophys. Res.* 86:4843–67

Hager, B. H., O'Connell, R. J., Raefsky, A. 1983. Subduction, back-arc spreading and global mantle flow. *Tectonophysics* 99:165–89

Hanks, T. C., Kanamori, H. 1979. A moment magnitude scale. *J. Geophys. Res.* 84:2348–50

Hartzell, S. H., Heaton, T. H. 1985. Teleseismic time functions for large, shallow subduction zone earthquakes. *Bull. Seismol. Soc. Am.* 75:965–1004

Heaton, T., Kanamori, H. 1984. Seismic potential associated with subduction in the northwestern United States. *Bull. Seismol. Soc. Am.* 74:933–41

Hilde, T. W. C. 1983. Sediment subduction versus accretion around the Pacific. *Tectonophysics* 99:381–97

Hilde, T. W. C., Sharman, G. F. 1978. Fault structure of the descending plate and its influence on the subduction process. *Eos, Trans. Am. Geophys. Union* 59:1182 (Abstr.)

Houston, H., Kanamori, H. 1986. Source spectra of great earthquakes: teleseismic constraints on rupture process and strong motion. *Bull. Seismol. Soc. Am.* In press

Isacks, B., Oliver, J., Sykes, L. 1968. Seismology and the new global tectonics. *J. Geophys. Res.* 73:5855–99

Kanamori, H. 1971. Great earthquakes at island arcs and the lithosphere. *Tectonophysics* 12:187–98

Kanamori, H. 1977a. The energy release in great earthquakes. *J. Geophys. Res.* 82:2981–87

Kanamori, H. 1977b. Seismic and aseismic slip along subduction zones and their tectonic implications. In *Island Arcs,*

Deep Sea Trenches and Back-Arc Basins,
Maurice Ewing Ser., ed. M. Talwani, W. C.
Pitman III, 1:163–74. Washington, DC:
Am. Geophys. Union
Kanamori, H. 1977c. Quantification of earth-
quakes. Nature 271:411–14
Kanamori, H. 1981. The nature of seismicity
patterns before large earthquakes. In
Earthquake Prediction—An International
Review, Maurice Ewing Ser., ed. D. W.
Simpson, P. G. Richards, 4:1–19. Wash-
ington, DC: Am. Geophys. Union
Kanamori, H. 1983. Magnitude scale and
quantification of earthquakes. Tectono-
physics 93:185–99
Kanamori, H., Anderson, D. L. 1975.
Theoretical basis of some empirical relations in seismology. Bull. Seismol. Soc. Am.
65:1073–95
Kanamori, H., Astiz, L. 1985. The 1983
Akita-Oki earthquake (M_w = 7.8) and its
implications for systematics of subduction
earthquakes. Earthquake Predict. Res. 3:
305–17
Kelleher, J., McCann, W. 1976. Buoyant
zones, great earthquakes, and unstable
boundaries of subduction. J. Geophys. Res.
81:4885–96
Kelleher, J., McCann, W. 1977. Bathymetric
highs and the development of convergent
plate boundaries. In Island Arcs, Deep Sea
Trenches and Back-Arc Basins, Maurice
Ewing Ser., ed. M. Talwani, W. C. Pitman
III, 1:115–22. Washington, DC: Am.
Geophys. Union
Kelleher, J., Savino, J. 1975. Distribution of
seismicity before large strike-slip and
thrust-type earthquakes. J. Geophys. Res.
80:260–71
Kelleher, J., Savino, J., Rowlett, H., McCann,
W. 1974. Why and where great thrust
earthquakes occur along island arcs. J.
Geophys. Res. 79:4889–99
Kikuchi, M., Fukao, Y. 1985. Iterative
deconvolution of complex body waves
from great earthquakes—the Tokachi-Oki
earthquake of 1968. Phys. Earth Planet.
Inter. 37:235–48
Lay, T., Kanamori, H. 1980. Earthquake
doublets in the Solomon Islands. Phys.
Earth Planet. Inter. 21:283–304
Lay, T., Kanamori, H. 1981. An asperity
model of great earthquake sequences. In
Earthquake Prediction—An International
Review, Maurice Ewing Ser., ed. D. W.
Simpson, P. G. Richards, 4:579–92. Wash-
ington DC: Am. Geophys. Union
Lay, T., Kanamori, H., Ruff, L. 1982. The
asperity model and the nature of large
subduction zone earthquakes. Earthquake
Predict. Res. 1:3–71
Madariaga, R. 1979. On the relation between

seismic moment and stress drop in the
presence of stress and strength hetero-
geneity. J. Geophys. Res. 84:2243–50
Malgrange, M., Deschamps, A., Madariaga,
R. 1981. Thrust and extentional faulting
under the Chilean coast: 1965, 1971
Aconcagua earthquakes. Geophys. J. R.
Astron. Soc. 66:313–31
Maruyama, T. 1963. On the force equivalents
of dynamical elastic dislocations with
reference to the earthquake mechanism.
Bull. Earthquake Res. Inst. Univ. Tokyo
41:467–86
Miyashiro, A. 1974. Volcanic rock series in
island arcs and active continental margins.
Am. J. Sci. 274:321–55
Mogi, K. 1968. Sequential occurrences of
recent great earthquakes. J. Phys. Earth
16:30–36
Mogi, K. 1969a. Relationship between the
occurrence of great earthquakes and tectonic structures. Bull. Earthquake Res.
Inst. Univ. Tokyo 47:429–41
Mogi, K. 1969b. Some features of recent
seismic activity in and near Japan, (2)
Activity before and after great earthquakes. Bull. Earthquake Res. Inst. Univ.
Tokyo 47:395–417
Molnar, P., Atwater, T. 1978. Inter-arc
spreading and cordilleran tectonics as
alternates related to the age of subducted
oceanic lithosphere. Earth Planet. Sci. Lett.
41:330–40
Newcomb, K. R., McCann, W. R. 1984.
Seismic history and seismotectonics of the
Sunda arc. Preprint (Lamont-Doherty
Geol. Obs., Palisades, N.Y.)
Peterson, E. T., Seno, T. 1984. Factors affecting seismic moment release rates in subduction zones. J. Geophys. Res. 89:10233–
48
Purcaru, G., Berckhemer, H. 1978. A magni-
tude scale for very large earthquakes.
Tectonophysics 49:189–98
Richter, C. F. 1935. An instrumental earth-
quake magnitude scale. Bull. Seismol. Soc.
Am. 25:1–32
Rudnicki, J. W., Kanamori, H. 1981. Effects
of fault interaction on moment, stress drop
and strain energy release. J. Geophys. Res.
86:1785–93
Ruff, L. J. 1985. Do trench sediments affect
subduction zone seismicity? In Earthquake
Source Mechanics, Maurice Ewing Ser., ed.
S. Das, Vol. 5. Washington, DC: Am.
Geophys. Union. In press
Ruff, L., Kanamori, H. 1980. Seismicity
and the subduction process. Phys. Earth
Planet. Inter. 23:240–52
Ruff, L., Kanamori, H. 1983a. The rupture
process and asperity distribution of three
great earthquakes from long-period dif-

fracted P-waves. Phys. Earth Planet. Inter. 31:202–30

Ruff, L., Kanamori, H. 1983b. Seismic coupling and uncoupling at subduction zones. Tectonophysics 99:99–117

Scholl, D. W., Marlow, M. S., Cooper, A. K. 1977. Sediment subduction and offscraping at Pacific margins. In Island Arcs, Deep Sea Trenches and Back-arc Basins, Maurice Ewing Ser., ed. M. Talwani, W. C. Pitman III, 1:199–210. Washington DC: Am. Geophys. Union

Scholz, C. H., Engelder, J. T. 1976. The role of asperity indentation and ploughing in rock friction—1. Asperity creep and stick-slip. Int. J. Rock. Mech. Min. Sci. Geomech. 13:149–54 (Abstr.)

Schwartz, S., Ruff, L. 1985. The 1968 Tokachi-Oki and the 1969 Kurile Islands earthquakes: variability in the rupture process. J. Geophys. Res. 90:8613–26

Seno, T., Eguchi, T. 1983. Seismotectonics of the western Pacific region. In Geodynamics of the Western Pacific–Indonesian Region, Geodyn. Ser., ed. T. W. C. Hilde, S. Uyeda, 11:5–40. Washington, DC: Am. Geophys. Union/Geol. Soc. Am.

Singh, S. K., Ponce, L., Nishenko, S. P. 1985. The great Jalisco, Mexico, earthquake of 1932 and the Rivera subduction zone. Bull. Seismol. Soc. Am. 75:1301–13

Stauder, W. 1968. Tensional character of earthquake foci beneath the Aleutian trench with relation to sea floor spreading. J. Geophys. Res. 73:7693–7701

Steketee, J. A. 1958. On Voltera's dislocations in a semi-infinite elastic medium. Can. J. Phys. 36:192–205

Sykes, L. R., Quittmeyer, R. C. 1981. Repeat times of great earthquakes along simple plate boundaries. In Earthquake Prediction—An International Review, Maurice Ewing Ser., ed. D. W. Simpson, P. G. Richards, 4:217–47. Washington, DC: Am. Geophys. Union

Tajima, F., Kanamori, H. 1985a. Global survey of aftershock area expansion patterns. Phys. Earth Planet. Inter. 40:77–134

Tajima, F., Kanamori, H. 1985b. Aftershock area expansion and mechanical heterogeneity of fault zone within subduction zones. Geophys. Res. Lett. 12:345–48

Taylor, F. W., Isacks, B. L., Jouannic, C., Bloom, A. L., Dubois, J. 1980. Coseismic and Quaternary vertical tectonic movements, Santo and Malekula Islands, New Hebrides island arc. J. Geophys. Res. 85:5367–81

Uyeda, S. 1982. Subduction zones: an introduction to comparative subductology. Tectonophysics 81:133–59

Uyeda, S. 1984. Subduction zones: their diversity, mechanism and human impacts. GeoJournal 8:381–406

Uyeda, S., Kanamori, H. 1979. Back-arc opening and the mode of subduction. J. Geophys. Res. 84:1049–61

Vlaar, N. J., Wortel, M. J. R. 1976. Lithospheric aging, instability, and subduction. Tectonophysics 32:331–51

Ward, S. N. 1983. Body wave inversion: moment tensors and depths of oceanic intraplate bending earthquakes. J. Geophys. Res. 88:9315–30

Watts, A. B., Talwani, M. 1975. Gravity anomalies seaward of deep-sea trenches and their tectonic implications. Geophys. J. R. Astron. Soc. 36:57–90

Wortel, M. J. R., Vlaar, N. J. 1978. Age-dependent subduction of oceanic lithosphere beneath western South America. Phys. Earth Planet. Inter. 17:201–8

Yonekura, N. 1983. Late Quaternary vertical crustal movements in and around the Pacific as deduced from former shoreline data. In Geodynamics of the Western Pacific–Indonesian Region, Geodynamics Ser., ed. T. W. C. Hilde, S. Uyeda, 11:41–50. Washington, DC: Am. Geophys. Union/Geol. Soc. Am.

Ann. Rev. Earth Planet. Sci. 1986. 14 : 323–50

GEOCHEMISTRY OF TEKTITES AND IMPACT GLASSES

Christian Koeberl

Institute of Geochemistry, University of Vienna, A-1094 Vienna, Austria

INTRODUCTION

Glass produced by melting during impact events is called impact glass or impactite. Although impact cratering produces a great variety of shock-metamorphosed rocks, all of which are called impactites by various authors (to a differing extent), we restrict the use of this term here to material completely converted into glass. The degree of metamorphism seen in different classes of impact materials (starting with those showing simple shock effects, shock lithification, shatter cones, diaplectic glasses, coesite, stishovite, lechatelierite, impact breccias, glass) depends on the peak pressure and temperature the rock has experienced. As a result of mixing, brecciation, and similar processes, impact glasses also tend to include fragments of less shocked or less homogeneous metamorphosed rocks. We limit our discussion here, however, to the more homogeneous members of the terrestrial impactite family. [A discussion of lunar impact glasses was given by Delano et al (1981).]

Another group of natural glasses are tektites, which at first glance are similar to obsidian. Most tektites are very homogeneous, often aerodynamically shaped (spherically symmetric) objects that are several centimeters in size and are found in various areas (referred to as tektite strewn fields). In reflected light they are usually black, but in transmitted light they show variations of grey, brown, and green. The surface features (gouges, furrows, grooves, pits, striations) clearly distinguish them from terrestrial volcanic glasses and have probably originated from solution etching and weathering effects on Earth (Baker 1963, LaMarche et al 1984) and from surface stresses occurring during their cooling. Tektites found in geographically adjacent

323

0084–6597/86/0515–0323$02.00

areas are related to each other with respect to their chemistry, age, and petrological and physical characteristics.

The origin of tektites has been a long-standing dispute for almost two centuries. Several reviews on tektites have been published (O'Keefe 1963, 1976, Ottemann 1966, Taylor 1973) featuring different theories of their origin. Although most workers in the field already favor a terrestrial origin, we present here several additional aspects invalidating the lunar hypothesis and giving further clues to our view of tektites as terrestrial impact glasses.

TEKTITES

Strewn Fields

To date, only four tektite strewn fields are known. These are the Australasian strewn field (the largest), the North American strewn field, the Ivory Coast strewn field, and the moldavite strewn field (in fact two subfields, located in Bohemia and Moravia, Czechoslovakia). Tektites of the Australasian strewn field include australites, thailandites, indochinites, philippinites or rizalites, javanites, and billitonites, while the North American tektites consist of two main groups—the bediasites (Texas) and the Georgia tektites. It is a matter of taste as to whether the glasses found at the Zhamanshin (USSR) crater constitute a fifth strewn field. The geographical and geological characteristics of the first four strewn fields have been discussed in the literature to a great extent (e.g. O'Keefe 1963, 1976, and references therein, Barnes 1963, 1964, Chapman & Larson 1963, Cohen 1963, Weiskirchner 1967, Bouška et al 1968, Chapman 1971, Bouška 1972, Rost 1972a,b, Konta 1972, Chalmers et al 1976, Glass et al 1979, Glass 1984).

In addition to numerous samples from the well-known strewn fields, sometimes single finds from outside these areas have been reported, although none of these finds is conclusive evidence for extending the list of strewn fields. The findings reported include a tektite from Cuba (Garlick et al 1971), which is in some respects (major element chemistry and age) similar to bediasites; two tektites from Liberia, which are similar to the Australasian tektites (Preuss 1969); a tektite from South West Africa (Saul & Cassidy 1970); tektites from Louisiana (King 1970), which seem to be related to tektites from the Southeast Asian (northern part of Australasian field) strewn field (but are not similar in age or chemistry to North American tektites); and tektites from northern Thailand (Yabuki et al 1981). A suggestion concerning the extension of the moldavite strewn field (O'Keefe 1983) has been made with the incorporation of moldavite findings from outside the Moravian and Bohemian subfields, such as a moldavite from Stainz (Styria, Austria) (described by Sigmund 1911). Recently,

Koeberl (1985a) has shown that this must be a displaced Bohemian moldavite. A genuine addition to the North American strewn field was the discovery of tektite fragments in deep-sea deposits at Barbados (Glass et al 1984). Microtektites also have been found at the same site.

ZHAMANSHIN CRATER   The most recent and important addition to the field of tektites and impact glasses was the discovery of glasses at Zhamanshin crater. This crater is situated 200 km north of Aralsk, near the Irghiz River, 60 km east of the southern end of the Ural Mountains (USSR). The circular depression is 5.5–6.3 km in diameter, centrally filled with loess and lake sediments. The margin consists of Paleozoic rock outcrops, which are overlain by a rim of brecciated Paleogene and Mesozoic rocks (with impact glasses). The crater is situated in Cretaceous and Paleogene sediments overlying crystalline bedrocks. The Cretaceous sediments consist of Albian and Cenomanian sandstones, dark grey Turonian claystone, and light grey marls and limestone, while the Paleogene sediments incorporate Eocene clays, upper Eocene grey quartz sand and clays of several formations, Oligocene clays, sands, conglomerates, and quartzites. The bedrocks consist of metamorphosed lower Paleozoic rocks (phyllite and quartz-sericite schist), with ultrabasic vein intrusions in the western area of the crater. From this short description we can see that the geological setting of the crater is rather complicated, and that we can expect no simple chemistry for its impact glasses and tektites.

Several classes of impact glasses have been found at Zhamanshin crater, and these are termed irghizites (for which a similarity to tektites has been inferred) and zhamanshinites (the impact glasses, with the following subdivisions according to their silica content: silica-rich zhamanshinites, zhamanshinites, and basic zhamanshinites). The distribution of these glasses is restricted to the crater and its near vicinity, and therefore it does not seem pertinent to refer to the irghizite-zhamanshinite complex as the fifth strewn field. We note, however, that irghizites (despite a larger inhomogeneity) resemble tektites, and that here the occurrence of rocks and glasses at various degrees of shock metamorphism supplies important genetic clues. Recently, a blue glass variety showing an interesting chemical composition has been described (Koeberl et al 1986b).

Further details on the Zhamanshin crater and the geological setting of the impact materials can be found in papers by Florenskij (1975a,b, 1977), Florenskij et al (1979a,b), Florenskij & Dabizha (1980), Bouška et al (1981), and Masaitis et al (1984).

Mineralogical and Petrographical Characteristics

The mineralogy and petrology of tektites has been extensively reviewed by Chao (1963) and O'Keefe (1976), so we restrict ourselves here to some of

Table 1   Electron microprobe analyses of two-colored moldavites, in wt% (from King & Bouška 1968, Bouška & Ulrych 1984)

Sample	SiO_2	TiO_2	Al_2O_3	FeO	MgO	CaO	Na_2O	K_2O
Lipí								
(olive-brown)	77	—	12.6	2.6	—	1.6	0.5	3.6
(light green)	81	—	9.7	0.9	—	2.9	0.5	3.5
Ratiborova Lhotka								
(yellowish								
olive-green)	78.4	0.14	9.74	1.90	2.32	3.65	0.21	3.07
(intensive green)	79.0	0.13	9.45	1.74	2.66	3.72	0.21	3.14
Třebanice								
(olive-brown)	75.1	0.14	12.03	2.95	2.10	3.54	0.27	3.53
(light green)	79.8	0.17	9.41	1.56	2.07	3.46	0.27	3.42

the more important investigations. Most tektites show flow structures, schlieren, lechatelierite (Barnes 1963, 1964), and bubbles; in particular, Konta & Mráz (1969) and Konta (1972) have shown a connection between the abundance of bubbles and lechatelierite in moldavites. Deformation of bubbles often occurs and indicates viscous flow during or after cooling. Examination of large numbers of samples shows that inclusions, bubbles, relict grains, and minerals are common not only in Muong Nong–type tektites (see below) but also in normal splash-form tektites. For example, studies of moldavite petrology have revealed that small-scale inhomogeneities are fairly abundant (Weiskirchner 1969, Barnes 1969, Konta 1972, Jung & Weiskirchner 1979). Recently, the interesting variety of two-colored moldavites (King & Bouška 1968) has been described by Bouška et al (1982) and Bouška & Ulrych (1984). These moldavites contain lechatelierite grains (of which some size fractions are more abundant than others) and rather sharp boundaries between zones of different color. It was shown by these authors that the color of the moldavites is a function of the iron content (and probably also of the redox state), with higher iron abundances in the darker green or olive areas (Table 1). This is in contrast to the color/chemistry dependence in the colored layers of Muong Nong–type tektites (see below). A possible explanation of the color differences may involve nonadiabatic terms during compression and thus may reflect the inhomogeneity of the source material.

Major Element Chemistry

For most classes of tektites, the major element chemistry is well known. The largest amount of data is available for tektites from the Australasian strewn field, mainly from Taylor and co-workers (see Taylor 1960, 1962, 1965, 1966, 1968, Taylor et al 1961, Taylor & Sachs 1961, 1964), Chapman & Scheiber (1969), Chapman (1971), and Mason (1979). Some additional references are given in earlier reviews (Chao 1963, Schnetzler & Pinson 1963, Taylor 1968, O'Keefe 1976). Important findings have been the discovery of chemical variations and subgroups such as HMg-tektites (where "H" stands for "high"), HCa-australites, HCa-philippinites, and normal Australasian tektites (Chapman & Scheiber 1969), although the criteria leading to the exact subdivisions have been disputed (Mason 1979). The correlation between geographical distribution and chemistry as stated by Chapman (1971) was not confirmed completely by Mason (1979).

Data for the different groups of tektites are given in Table 2. In most groups the inverse relation between SiO_2 and the other constituents (sometimes except CaO) is valid, but in a few cases (such as the HCa-australites or the HMg-australites) Na_2O and K_2O show even a positive correlation.

A large amount of data has been assembled for North American tektites by Chao (1963), King (1964, 1966), Cuttitta et al (1967) (see Table 3), and O'Reilly et al (1983). The chemistry of these tektites points to a common origin for bediasites (Texas) and Georgia tektites, with the latter being the high-silica end members of the North American tektites. Major-element analyses of tektite fragments found in deep-sea deposits at Barbados have been reported by Glass et al (1984).

Early data on moldavites have been reviewed by Chao (1963) and Schnetzler & Pinson (1963), but since then many more data have become available. The most important analyses are given in Bouška & Povondra (1964), Schnetzler & Pinson (1964), Philpotts & Pinson (1966), Konta & Mráz (1969), Rost (1972a,b), Bouška et al (1973), and Luft (1983) (see Table 3). Delano & Lindsley (1982) used data from some of these sources and Langmuir et al's (1978) method of mixing calculations to study the compositional variations among the moldavites; Langmuir et al's method was previously applied by Delano et al (1982) to the study of Australasian tektites and microtektites. They came to the conclusion that the chemical variations may be explained by the mixing of at least two components, one of which is argillaceous (e.g. shale) and the other more calcareous (see Table 4). From chemical, isotopic, and experimental approaches, Graup et al (1981) and Luft (1983) suggested that a special class of tertiary argillaceous sands (OSM upper freshwater molasse) is very close to the parent material of the moldavites. Further discussion of Sm-Nd and Rb-Sr isotopic systematics (Shaw & Wasserburg 1982) and rare earth element (REE) patterns (Koeberl et al 1985c) led to the conclusion that the sample OSM 050 (a silty surface-deposited sand from near Unterfiningen, Liezheimer

Table 2 Compositional ranges for major elements in different classes of Australasian tektites (in wt%) from Chapman & Scheiber (1969). Numbers in parentheses under the class designation denote the number of samples analyzed

	HNa/K-australites (8)	Normal australites (3)	Normal philippinites (2)	Normal indochinites (2)	HAl-australites (3)	LCaHAl-philippinites (5)	HCa-australites (5)	HCa-philippinites (5)	HMg-tektites (10)
SiO_2	62.2–63.9	70.4–72.4	71.2–71.6	72.9–73.3	66.9–68.5	67.2–69.2	68.9–79.7	68.6–74.9	64.8–77.0
Al_2O_3	15.8–17.1	12.9–14.3	13.0–13.4	13.1–13.5	15.4–16.1	15.2–17.7	9.3–14.3	8.9–11.0	10.7–13.3
FeO	5.54–6.88	4.67–4.97	4.55–5.12	4.47–4.49	5.27–5.43	4.87–5.87	3.57–4.75	3.87–4.55	3.85–8.63
MgO	3.57–4.68	2.16–2.23	2.33–2.42	2.00–2.04	2.44–2.60	2.45–3.15	1.31–2.53	1.87–2.43	1.83–7.95
CaO	4.49–6.00	2.94–3.48	3.31–3.34	2.17–2.41	3.18–3.83	1.37–2.37	1.83–5.48	4.34–9.77	1.79–3.73
K_2O	0.90–1.12	2.41–2.62	2.33–2.47	2.36–2.40	2.37–2.57	2.33–2.81	2.14–2.21	1.77–2.33	1.34–2.56
Na_2O	2.75–3.91	1.32–1.56	1.31–1.50	1.17–1.27	1.04–1.27	0.97–1.18	1.00–1.24	0.91–1.28	0.62–1.38
TiO_2	0.53–0.65	0.80–0.83	0.78–0.82	0.72–0.89	0.87–0.93	0.85–1.00	0.49–0.81	0.50–0.63	0.66–0.77

Table 3 Major element data for tektite groups other than Australasian tektites. All ranges given in wt% and taken from footnoted references. The numbers in parentheses below the name denote the number of samples analyzed

	Georgia tektites[a] (9)	Bediasites[b] (21)	Moldavites Bohemia[c] (61)	Moravia[c] (21)	Ivory Coast tektites[d] (15)	Irghizites[e] (31)	Si-rich zhaman-shinites[f] (12)	Zhaman-shinites[g] (16)	Basic zhaman-shinites[g] (4)
SiO$_2$	79.8–83.6	71.9–80.2	75.5–85.1	74.9–81.4	67.0–69.3	70.0–79.4	62.9–88.1	52.4–56.7	39.0–39.9
Al$_2$O$_3$	9.50–11.7	11.2–17.6	7.32–11.4	9.44–13.8	15.8–17.1	9.45–13.6	4.80–21.2	19.6–22.2	14.1–14.5
FeO	1.83–3.14	2.29–5.75	1.08–2.93	1.72–3.50	6.03–6.80	4.24–6.81	1.98–8.05	4.68–8.15	10.5–11.0
MgO	0.37–0.69	0.37–0.95	1.34–2.74	1.13–2.06	2.64–3.93	2.16–3.76	0.34–1.16	1.82–3.23	12.1–12.5
CaO	0.40–0.69	0.49–0.96	1.21–3.96	0.95–3.17	0.71–1.61	1.75–2.85	0.55–2.16	6.68–9.07	10.0–10.3
Na$_2$O	1.00–1.53	1.20–1.84	0.20–0.89	0.40–1.08	1.54–2.44	0.85–1.22	0.57–1.84	3.24–4.55	2.86–2.94
K$_2$O	2.22–2.51	1.60–2.43	2.23–3.81	2.83–3.81	1.70–2.07	1.58–2.14	0.10–3.07	1.22–1.85	2.01–2.12
TiO$_2$	0.42–0.60	0.59–1.05	0.24–0.74	0.31–1.40	0.52–0.60	0.69–0.99	0.23–1.10	0.17–0.25	4.82–5.07

[a] Cuttitta et al 1967.
[b] Chao 1963.
[c] Bouška et al 1973, Bouška & Povondra 1964, Schnetzler & Pinson 1964, Luft 1983, Cohen 1963, Konta & Mráz 1969, Philpotts & Pinson 1966.
[d] Chapman & Scheiber 1969, Cuttitta et al 1972, Shaw & Wasserburg 1982.
[e] Bouška et al 1981, Taylor & McLennan 1979.
[f] Bouška et al 1981, Florenskij & Dabizha 1980, Taylor & McLennan 1979, Koeberl et al 1985a, C. Koeberl & K. Fredriksson, in preparation.
[g] Bouška et al 1981.

Forst, West Germany) is the top candidate for a moldavite parent material (although the actual parent material may have been consumed totally during the tektite production process). A comparison of the analytical data for the sample OSM 050 with the composition margins for the argillaceous component of Delano & Lindsley's (1982) calculations (see Table 4) shows good agreement. However, a problem that must still be resolved is the discrepancy in the $\delta^{18}O$ data (Luft 1983).

For Ivory Coast tektites, the available data are relatively sparse. Early data were reviewed by Chao (1963) and Schnetzler & Pinson (1963). Five samples have been analyzed by Chapman & Scheiber (1969), and a very detailed study of seven samples (including many trace elements) was made by Cuttitta et al (1972). In addition, three samples have been analyzed by Shaw & Wasserburg (1982). An interesting difference between Ivory Coast tektites and other tektites is that the Na_2O/K_2O ratio approaches 1 for the Ivory Coast tektites, whereas other tektite groups have $Na_2O/K_2O < 1$. Comparisons with glasses from the Bosumtwi crater (Ghana) strongly support a genetic connection (Cuttitta et al 1972, Shaw & Wasserburg 1982).

Analyses of glasses from the Zhamanshin crater seem to be of great importance. Data given by Ehmann et al (1977), Fredriksson et al (1977), Taylor & McLennan (1979), Florenskij & Dabizha (1980), Bouška et al (1981), Shaw & Wasserburg (1982), Koeberl et al (1985a), and C. Koeberl & K. Fredriksson (in preparation) show that the glasses have a wide range of compositions; thus, several classes have been created, although not all authors agree about the designation of the subdivisions. Microprobe studies (Koeberl et al 1985a) have shown that the irghizites and zhamanshinites are more inhomogenous than other tektites, with the only exception

Table 4 Compositional limits for the two components of the moldavite parent material (as calculated by Delano & Lindsley 1982) compared with the values for OSM 050 tertiary sand (Luft 1983). The latter are very close to the values for component 1

Ratio	Component 1	Component 2	OSM 050
CaO/TiO_2	0–3.5	14–55	0.96
Al_2O_3/MgO	11–25	2.4–4.5	14.1
FeO/MgO	2.0–5.1	0.3–0.7	4.73
Sr/MgO	0.012–0.027	0.0045–0.0065	0.0096
K_2O/MgO	3.5–7.9	0.9–1.5	2.82
Na_2O/MgO	0.5–1.6	0–0.15	0.86
Rb/Sr	1.2–1.3	0.60–0.85	1.4
SiO_2/MgO	70–175	20–30	125.2

being the Muong Nong–type indochinites (see below). Data for some characteristic groups are given in Table 3. Attempts by Bouška et al (1981) and Koeberl et al (1985a) to evaluate the parent rocks of the glasses indicate that at least two precursory components have been involved, one a relatively chemically unaltered acidic bedrock (eventually in the form of young clastic sediments) and the other a clay or sandstone, with the possible addition of a diluting compound such as quartz. It was not possible to determine the exact source rock as a result of the lack of thorough analyses of the bedrocks present at the site and the complex geological setting of the crater.

Trace Elements

Trace elements are the most important sources of information on the genesis and the history of tektites. Despite this, there are almost no trace element data available for many groups. Only the australites have been studied in depth, largely as a result of the efforts of Taylor & co-workers (Taylor 1960, 1962, 1965, 1966, 1968, Taylor & Sachs 1961, 1964, Taylor & McLennan 1979; also Chapman & Scheiber 1969). Recently, this situation has improved for moldavites (Rost 1972a, Bouška et al 1973, Bouška & Řanda 1976, Schock 1976, Luft 1983) and North American tektites (early work by Chao 1963; also Haskin et al 1982, Weinke & Koeberl 1985, Dod et al 1985), but the number of samples analyzed from the Ivory Coast strewn field remains rather small (Schnetzler et al 1967, Pinson & Griswold 1969, Chapman & Scheiber 1969, Cuttitta et al 1972).

There is also a lack of trace element data for the Zhamanshin crater material—this problem is readily apparent because of the large number of different tektite and impactite classes present. Studies by Ehmann et al (1977), Fredriksson et al (1977), Bouška et al (1981), and Florenskij & Dabizha (1980) have included only a limited number of trace elements, while the most complete studies available to date (Taylor & McLennan 1979, C. Koeberl & K. Fredriksson, in preparation) together incorporate only eight samples—a low number for four groups of glasses. From their trace element data, Taylor & McLennan (1979) [and also Van Patter et al (1981) and Glass (1979)] proposed a similarity, and possibly a genetic relationship, between Australasian tektites and zhamanshinites and irghizites. This conclusion was challenged both with some not quite correct arguments by O'Keefe (1980) (but see also Taylor & McLennan 1980) and on the basis of isotopic (Shaw & Wasserburg 1982) and chemical data (Koeberl et al 1985a,c, C. Koeberl & K. Fredriksson, in preparation) as well as age (Storzer & Wagner 1979). Thus, it seems unlikely that there is any genetic connection between these two events. It was clearly demonstrated, however, that the same process took place.

Important arguments can be derived from the chondrite-normalized abundances of the rare earth elements. These have already been noted in earlier reviews on trace element significance in tektites (Haskin & Gehl 1963, Taylor 1966, Taylor & Kaye 1969), but recently this problem has attracted more interest. Extensive investigations by Bouška & Řanda (1976), Taylor & McLennan (1979), Haskin et al (1980, 1982), Bouška et al (1981), and Koeberl et al (1985c) have confirmed the earlier suggestions that the REE abundance patterns are very similar to terrestrial sediments and have solidified further conclusions on the similarity between certain parent rocks and tektites. Schnetzler et al (1967) noted a close relationship between the REE patterns of Ivory Coast tektites and Bosumtwi crater glasses (and similarities between these materials and the local country rocks). The REE patterns of Australasian tektites, North American tektites, and terrestrial post-Archean sediments (shale, loess, etc) are almost indistinguishable, as has been noted by several authors (e.g. Bouška & Řanda 1976, Taylor & McLennan 1979, Haskin et al 1980, 1982, Koeberl et al 1985c), but since there is a uniform REE pattern present in post-Archean sediments (Taylor & McLennan 1981), we must be cautious in extending any identification too far unless there is ample evidence. A striking correlation between REE abundances in moldavites and the OSM 050 tertiary sand sample (see above) has been noted by Koeberl et al (1985c); this correspondence further supports a connection between these two materials.

The study of REE patterns in Zhamanshin material yielded (despite the low number of samples studied) interesting results (Taylor & McLennan 1979, Bouška et al 1981, Koeberl et al 1985c) that support the conclusions drawn from major element studies concerning the mixture of parent rocks.

However, for some other elements, which are rare and/or difficult to determine (but otherwise have genetic significance), the picture is depressing. Table 5 is an attempt to present data for all tektite groups and the whole periodic system of elements. Where possible, more recent data are used, and (if available) averages from a number of samples are given. For elements such as Sc, V, Cr, Mn, Co, Ni, Cu, Rb, Sr, Zr, Cs, Ba, the REE, Hf, Th, and U, data are abundant and readily available for most groups. However, for many other elements the numbers given in Table 5 are the only ones available. For Rh, In, Te, and Pt, I was unable to locate any reliable data.

Li data were sparse until the publication of the papers by Shukla & Goel (1979), Koeberl & Berner (1983), and Koeberl et al (1984e). These authors noted significant similarities between tektites and terrestrial sediments. Some Be data were reported in connection with new ^{10}Be isotopic studies (Pal et al 1982). Only recently has the situation improved for other elements of the second row of the periodic system. Boron in tektites was reported by

Table 5 Trace element data for all tektite groups. All data in ppm. If differing data have been found, they are reported in the same column[a]

	Australites	Indochinites	Philippinites	Moldavites	Ivory Coast tektites	Bediasites	Georgianites	Irghizites	Si-rich zhamanshinites
Li	40.0[6]	47.1[6]	—	42.0[6]	43[4]	—	—	—	26[6]
Be	2.2[18]	2.2[18]	2.6[18]	1.9[18]	1.2[4]	—	—	—	—
B	19[9]	24[9]	39[9]	19[9]	11[4], 15[9]	18[9]	—	—	—
C	100[20]	60[20]	55[20]	160[20]	—	150[20]	—	—	—
F	36[7], 18[20]	46[7], 34[20]	27[20]	39[20]	—	14[21], 12[20]	—	61[8]	72[8]
P	300[22]	—	—	—	202[4]	—	—	—	—
S	2.9[20], 7.2[20]	3[20]	4.3[20]	—	—	5.7[20]	—	—	—
Cl	4.3[3], 7[20]	1.4[3], 8[20]	3.6[3], 15[20]	2.1[3], 10[20]	—	2.8[3], 11[20]	—	128[8]	170[8]
Sc	13[1]	10.5[1]	11.5[1]	4.93[2]	17[4]	13[11]	6.9[11]	8.8[1]	15[1]
V	83[1]	63[1]	84[1]	32[13]	132[4]	—	46[30]	46[1]	117[1]
Cr	145[1], 72[14]	63[1], 90[14]	100[1]	22[2]	327[4]	49[11]	22[11]	170[1], 260[8]	92[1], 49[8]
Mn	670[13]	690[13]	—	620[13]	500[4]	300[21]	—	600[8]	700[8]
Co	25[1]	11[1]	13.5[1]	4.45[2]	21[4]	13.5[11]	6.7[11]	73[1], 91[8]	16[1], 8.6[8]
Ni	105[1], 43[14]	19[1], 44[14]	50[1]	17[2], 15[14]	129[4]	—	—	1200[1]	40[1]
Cu	6.5[1]	4[1]	10[1]	5.4[17]	17[4]	12[17]	3.2[17]	24[1]	36[1]
Zn	2.0[13]	5.7[13]	—	26[13]	—	—	—	—	16[27]
Ga	8.4[22], 0.9[24]	8.2[24]	7.1[24]	5.5[25]	9.8[24]	11[17]	9[17]	—	—
Ge	0.06[24]	0.21[24]	0.4[24]	0.09[25]	0.35[24]	0.13[26]	0.23[26]	—	—
As	—	—	—	—	—	0.9[21]	—	—	4.7[8], 13[10]
Se	0.0019[7]	0.00074[7]	—	—	—	0.21[3], 0.12[21]	—	—	0.0018[7]
Br	0.18[3]	0.23[3]	0.13[3]	0.09[3]	—	—	—	—	0.4[8]
Rb	80[6]	130[6]	—	130[2], 117[6]	69[4]	66[11]	66[11]	—	118[8]
Sr	200[22]	90[26]	100[26]	134[2]	303[4]	125[11]	160[11]	38[8]	630[10], 127[27]
Y	31[1]	29[1]	28[1]	2[17]	17[4]	25[17]	20[17]	17.5[1]	33.8[1]
Zr	264[1]	252[1]	280[1]	225[2]	116[4]	230[11]	160[11]	351[1]	272[1]

Table 5 (*continued*)

	Australites	Indochinites	Philippinites	Moldavites	Ivory Coast tektites	Bediasites	Georgianites	Irghizites	Si-rich zhamanshinites
Nb	18.7[1]	19.6[1]	18.8[1]	—	—	20[17]	—	12.5[1]	16.8[1]
Mo	0.3[1]	0.3[1]	0.3[1]	—	—	—	—	0.40[1]	0.73[1], 3.6[10]
Ru	0.30[13]	0.28[13]	—	0.29[13]	—	—	—	—	0.21[13]
Pd	0.001[15]	—	—	—	—	—	—	—	—
Ag	—	—	—	—	1.0[4]	—	—	—	—
Cd	0.01[15]	—	—	—	—	2[17]	—	—	—
Sn	1.3[1]	0.9[1]	1.7[1]	0.61[2]	—	—	—	0.75[1]	3.1[1]
Sb	0.06[7]	0.5[7]	0.33[29]	0.19[3]	—	0.05[21]	—	0.25[8]	0.50[8]
I	0.22[3]	0.56[3]	0.26[3]	—	—	0.17[3]	—	—	—
Cs	5.7[1]	6.5[1]	6.7[1]	13.1[2]	2.3[4]	2.0[11]	1.0[11]	2.6[1]	7.8[1], 3.7[8]
Ba	356[1]	360[1]	390[1]	705[2]	650[5]	470[11]	620[11]	527[1]	347[1]
La	36.9[1]	36.5[1]	36.2[1]	30.1[2]	—	35[11]	19.6[11]	19.7[1]	34[1], 18.2[8]
Ce	78.6[1]	73.1[1]	74.8[1]	53.8[2]	45.2[5]	76[11]	43[11]	44.2[1]	83.6[1], 50.1[8]
Pr	9.0[1]	8.80[1]	8.76[1]	—	—	—	—	4.53[1]	8.98[1]
Nd	35.0[1]	33.2[1]	34.5[1]	24.1[2]	24.9[5]	20[21]	—	18.7[1]	34.1[1], 16.8[8]
Sm	6.10[1]	6.60[1]	6.74[1]	4.96[2]	4.59[5]	7.2[11]	4.03[11]	3.78[1]	6.95[1], 4.1[8]
Eu	1.17[1]	1.22[1]	1.28[1]	0.88[2]	1.13[5]	1.58[11]	0.97[11]	0.80[1]	1.56[1], 0.99[8]
Gd	5.34[1]	5.24[1]	5.92[1]	3.20[2]	3.57[5]	—	—	3.46[1]	6.47[1]
Tb	0.84[1]	0.85[1]	0.88[1]	0.57[2]	—	0.97[11]	0.59[11]	0.60[1]	0.97[1], 0.69[8]
Dy	5.17[1]	5.58[1]	5.44[1]	2.74[12]	2.76[5]	3.6[21]	—	3.45[1]	5.63[1], 3.8[8]
Ho	0.97[1]	1.03[1]	1.12[1]	0.71[2]	—	—	—	0.67[1]	1.12[1]
Er	2.90[1]	2.91[1]	3.32[1]	—	1.36[5]	—	—	1.89[1]	3.29[1]
Tm	—	—	—	0.24[2]	—	—	—	—	—
Yb	2.80[1]	2.90[1]	3.01[1]	1.83[2]	1.68[5]	3.0[11]	1.73[11]	1.90[1]	3.4[1], 1.98[8]
Lu	0.44[12]	0.54[12]	0.39[28]	0.24[2]	0.12[28]	0.47[11]	0.29[11]	0.26[1]	0.35[8]

Hf	7.1[1]	6.95[1]	7.95[1]	5.71[2]	6.7[11]	4.7[11]	8.66[1]	6.3[1]
Ta	1.2[13]	1.6[13]	—	0.54[2]	0.60[21]	—	0.82[8]	0.53[8]
W	0.39[1]	0.29[1]	0.54[1]	1[2]	—	—	0.17[1]	0.73[1]
Re	0.000005[15], 0.00021[15]	—	0.00038[16]	—	—	—	—	—
Os	0.002[15], 0.0003[13], 0.00005[15]	0.0005[13]	—	0.002[13]	—	—	—	0.001[13]
Ir	0.002[15], 0.00018[13], 0.00002[15]	0.0008[13]	—	0.00007[16], 0.00008[13]	—	—	—	0.00012[13]
Au	0.0022[13]	0.002[13]	—	0.006[13]	—	—	—	0.0067[10], 0.0029[13]
Hg	0.0016[17]	—	—	—	—	—	—	—
Tl	—	—	0.08[1]	6[23]	—	—	—	0.43[1]
Pb	5.0[1]	1.6[1]	10.5[1]	14[4]	—	—	2.9[1]	26[1]
Bi	—	—	0.03[1]	—	—	—	0.04[1]	0.12[1]
Th	13.7[1]	14.0[1]	15.6[1]	10.4[2]	7.6[11]	4.9[12]	5.98[1]	10.3[1], 7.8[8]
U	2.1[1]	2.07[1]	3.49[1]	2.35[2]	2.0[11]	1.25[11]	1.02[1]	2.97[1], 1.50[8]

ᵃ References:
1. Taylor & McLennan 1979.
2. Luft 1983.
3. Becker & Manuel 1972.
4. Cuttitta et al 1972.
5. Schnetzler et al 1967.
6. Koeberl et al 1984e.
7. Koeberl et al 1984d.
8. C. Koeberl & K. Fredriksson, in preparation.
9. Mills 1968.
10. Bouška et al 1981.
11. Haskin et al 1982.
12. Koeberl et al 1985c.
13. C. Koeberl, unpublished data.
14. Koeberl et al 1985d.
15. Morgan 1978.
16. Morgan et al 1979.
17. O'Keefe 1976.
18. Pal et al 1982.
19. Yellin et al 1983.
20. Moore et al 1984.
21. Weinke & Koeberl 1985.
22. Taylor 1966.
23. Tilton 1958.
24. Cohen 1960.
25. Cohen 1963.
26. Schnetzler & Pinson 1963.
27. Van Patter et al 1981.
28. Bouška & Randa 1976.
29. Tanner & Ehmann 1967.
30. Dod et al 1985.

Mills (1968). The first determinations of fluorine in tektites and impact glasses were reported by Koeberl et al (1983, 1984d) and Moore et al (1984); these abundances are completely dissimilar to those of any volcanic glasses. Data on other halogens (Becker & Manuel 1972, Moore et al 1984, Weinke & Koeberl 1985, C. Koeberl & K. Fredriksson, in preparation) show that splash-form tektites have experienced a loss of these extremely volatile elements, since the abundances are even lower than in terrestrial sediments. Other volatile element abundances (such as As, Se, Ag, Cd, In, Sb, Te, Hg, Tl) have not been investigated in detail [although recently a few single analyses have been reported; see, for example, Cuttitta et al 1972 (Ag), Weinke & Koeberl 1985 and C. Koeberl & K. Fredriksson, in preparation (As, Sb), Koeberl et al 1984d (Se, Sb), Morgan 1978 (Cd), and Taylor & McLennan 1979 (Tl)], so further conclusions are not possible at present.

Siderophile trace element data (such as for the Pt-group metals) are similarly sparse but have been used sometimes to show meteoritic contamination (see below). In any case, more trace element data, especially for some "difficult" elements, are needed.

Muong Nong–Type Tektites

This interesting subgroup has recently attracted much attention. Occurring at the northern end of the Australasian strewn field, they are of unusual blocky, chunky appearance (in contrast to splash forms) and contain many large bubbles [with gas inclusions of about half the atmospheric pressure (Müller & Gentner 1968, Jessberger & Gentner 1972)] and abundant crystalline inclusions such as corundum, rutile, quartz, zircon, chromite, and cristobalite (e.g. Glass 1970, 1972b, Barlow & Glass 1976, Glass & Barlow 1979). Some specimens even show brecciation (Futrell & Fredriksson 1983). The most important deviation from "normal tektite behavior" is in the chemical composition. From early studies by Chapman & Scheiber (1969) (seven samples) and Müller & Gentner (1973), it became evident that Muong Nong–type indochinites are considerably enriched in volatile elements. Recent investigations on a larger number of samples by the author and co-workers have confirmed these findings and have added some interesting results (Koeberl et al 1984a–c,f,g, 1985b, Weinke & Koeberl 1984, Koeberl 1985b). Apart from the halogens, other volatile trace elements (such as Zn, As, Rb, Sb) are enriched. B and Cu enrichments had already been noted by Chapman & Scheiber (1969), who accordingly named this subgroup in their analyses the "HCu,B-indochinites."

Table 6 gives an average from the Muong Nong–type indochinite analyses available to date. Moore et al (1984) analyzed a single sample for F, Cl, S, and C; their results yielded a very high fluorine content, which is probably an exception, compared with the data for 19 samples of Koeberl et al (1984a). Sulfur seems to be enriched by a factor of about three compared to splash-form tektites.

Since Muong Nong–type indochinites show layers of different colors (light and dark), the reasons for these differences have been the subject of several investigations. Barnes & Pitakpaivan (1962) reported analyses of whole samples of different colors that showed differences in their chemistry

Table 6   Major and trace element data (average from 19 samples) for Muong Nong–type indochinites and their colored layers. Data from the author (see references), except as indicated. Major elements in wt%; trace elements in ppm. Total iron content reported as FeO

	Muong Nong tektite average	MN8319 dark layer	MN8319 light layer		Muong Nong tektite average	MN8319 dark layer	MN8319 light layer
SiO_2	78.30	80.44	76.38	Cs	5.09	4.4	5.2
Al_2O_3	10.18	9.33	12.40	Ba	340	340	—
FeO	3.75	3.25	4.38	La	24.4	26.6	30.9
MgO	1.43	1.35	1.71	Ce	60	68	78
CaO	1.21	1.26	1.33	Nd	21.7	34	75
K_2O	2.41	2.49	2.62	Sm	4.82	4.70	5.90
Na_2O	0.92	1.25	0.91	Eu	0.84	0.76	0.65
TiO_2	0.63	0.57	0.66	Tb	0.69	0.9	1.9
				Dy	4.40	3.1	7.4
Li	42.1	—	—	Yb	2.31	2.4	3.4
Be	3.70	—	—	Lu	0.44	0.42	0.51
B	47.7, 55.3[a]	—	—	Hf	8.13	7.8	10.5
C	70[b]	—	—	Ta	1.18	1.9	—
F	97	—	—	W	0.9	—	—
S	10[b]	—	—	Th	11.0	12.5	17
Cl	170[b]	—	—	U	2.48	2.4	4.7
Sc	7.70	7.72	10.0				
V	72[a]	—	—				
Cr	61, 77[a]	61	70				
Mn	675, 770[a]	640	670				
Co	12.6, 12.6[a]	10.9	14.5				
Ni	50, 33[a]	—	—				
Cu	13.5, 16.4[a]	—	—				
Zn	67	—	—				
As	5.0	1.4	7.2				
Br	4.1	—	—				
Rb	110	160	140				
Sr	100	—	—				
Y	27[a]	—	—				
Zr	290, 270[a]	700	—				
Sb	0.82	—	—				

[a] Chapman & Scheiber 1969 (seven samples).
[b] Moore et al 1984 (one sample).

(darker colors corresponded to more silica). A microprobe study by Yagi et al (1982) failed to detect differences between the layers, but Weinke & Koeberl (1984) and Koeberl (1985b) demonstrated a remarkable difference—the abundances of most elements except Si (including Fe, Ti, Al, Mg, Sc, Cr, Co, As, Cs, the REE, Hf, Th, and U) are larger in the light layers (see Table 6). This behavior is different from the two-colored moldavites (Table 1). The petrology and the chemistry [also with respect to the changed ferric/ferrous ratios (Koeberl et al 1984g)], the enrichment in volatiles, the gas-rich bubbles, the layered structures, and the unmelted high temperature inclusions all point to a lower temperature of origin than for splash-form tektites. We see clearly from the Zhamanshin crater glasses that a medium-sized impact can produce glasses of different homogeneity and chemistry, and a similar process (except on a larger scale) acted in the case of the Muong Nong–type tektites. At present we do not know exactly if the Muong Nong–type variety occurs only at the Australasian strewn field, since there also seem to be Muong Nong–type tektites among the moldavites (Rost 1966).

Age, Isotopes, and Other Properties

Age determinations and isotope studies have been reviewed by Zähringer (1963) and O'Keefe (1976). An important result is the determination that the moldavites and the Ries crater have exactly the same age. The same conclusion applies as well for the Ivory Coast tektites and the Bosumtwi crater. Interesting results concerning the Australasian strewn field have also been inferred from age data. Fleischer et al (1969) state that the HNa/K-australites (see Table 2) of Chapman & Scheiber (1969) have an age of 4 Myr, which is much older than the other australites. This has recently been confirmed by Storzer (1985). Recently, Storzer & Wagner (1980) reported different ages for the australites (0.83 Myr) and the rest of the Australasian strewn field (indochinites, philippinites: 0.69 Myr). No chemical evidence has yet been found to support the resulting suggestion of two impact events.

Oxygen isotope data (e.g. Taylor & Epstein 1969) and lead isotope data (Tilton 1958) are consistent with a terrestrial origin, and recently Shaw & Wasserburg (1982) have demonstrated from Sm-Nd and Rb-Sr data that tektites have been melted from sediments derived from old terrestrial crust. More evidence related to a terrestrial sedimentary origin (possibly from the continental margins) comes from ^{10}Be, ^{26}Al, and ^{53}Mn data (Pal et al 1982, Tera et al 1983, Englert et al 1984, Yiou et al 1984). Nd and Sr isotope data have been used recently to strengthen the relationship between microtektites and tektites from Barbados and other North American tektites (Ngo et al 1985).

Some interesting studies have involved new techniques, such as the proton probe study (PIXE) by Van Patter et al (1981), the rapid instrumental neutron activation analysis (RINAA) by Koeberl & Grass (1983) and Koeberl et al (1984d), and the electron magnetic resonance (EMR) studies by Weeks et al (1980). EMR investigations were aimed at determining the fusing conditions of a particular glass, which turned out to be of rather low p_{O_2} for tektites. Different fusing conditions were demonstrated between tektites and Libyan Desert Glass. The conclusion obtained by Evans & Leung (1979) from Mössbauer spectroscopy—that tektites could not have formed from terrestrial melts—is not conclusive, since the authors do not consider the effect of shock homogenization and melting. Other authors incorporate these effects and conclude that melting of terrestrial materials may very well account for tektite production (see, for example, Franke & Heide 1977, Frenzel & Ottemann 1978, Florenskij et al 1978, Weeks et al 1980, Luft 1983). Grass et al (1983) concluded that their Mössbauer results are consistent with a terrestrial origin. Florenskij et al (1978) have argued convincingly from their extensive X-ray and Mössbauer studies (in contrast to Evans & Leung 1979) that the structural properties of the cations in tektites can only be explained by formation in high-temperature melts, followed by quenching.

A point of genetic significance is the water content of tektites. Gilchrist et al (1969) have described a method to determine water content using infrared spectrometry; this method yielded an average value of 0.012 wt% H_2O in tektites (and also indicated an inhomogeneous distribution of the water in the glass), which is far below the average obsidian value (and many orders of magnitude above the H_2O content of lunar rocks—which is essentially zero!). The water content of irghizites is higher (0.05 wt%; King & Arndt 1977) and is consistent with the larger inhomogeneity of the glass and the lower peak pressures and temperatures of origin. Luft (1983) measured water in moldavites (0.007–0.013 wt%) and in several experimental melts, concluding that the H_2O-rich tertiary OSM sands could very well be transformed into H_2O-poor glasses during melting. He also summarized evidence (as shown by Frenzel & Ottemann 1978) that the extremely rapid melting (and, depending on the pressure, homogenization) of glass results in an increased SiO_2 content, which points to a tektite origin from certain sediments.

Microtektites

Microscopic spheres (diameter less than 1 mm) found in deep-sea sediment cores from throughout the Indian Ocean, the Philippine Sea, and the western Pacific Ocean (Australasian microtektites), in the eastern equatorial Atlantic Ocean (Ivory Coast microtektites), and in the Gulf of Mexico, Caribbean Sea, and equatorial Pacific Ocean (North American microtektites) have been identified as tektite material (e.g. Glass 1967, 1968, 1969, 1972a, Senftle et al 1969, Cassidy et al 1969, Frey et al 1970, Glass et al 1973, 1979, 1984, Frey 1977). The very thorough work by Frey (1977) provided, on the basis of extensive trace element studies, convincing evidence that the microtektites and tektites of the respective strewn fields are in fact genetically related. Magnetic studies by Senftle et al (1969) indicated that most of the iron in microtektites occurs as Fe^{2+}, as is also the case for tektites. The probability of an association between microtektites and geomagnetic reversals (such as a connection between the Australasian microtektite layer and the Brunhes/Matuyama geomagnetic reversal boundary, or between the Ivory Coast microtektites and the beginning of the Jaramillo event) has been discussed by Glass et al (1979) and Glass (1982), among others. Chalmers et al (1976) challenged a connection between microtektites and tektites, arguing that the stratigraphic ages of the australites (7000–20,000 yr) and the microtektites (700,000 yr) are not consistent. However, this would also mean that all other (well-documented) age determinations of the australites (fission track, K/Ar, etc) are in error, and this seems very unlikely. Other arguments against Chalmers et al's (1976) claim are discussed by Glass (1978).

Since the discovery of the Ir anomaly at the Cretaceous/Tertiary boundary, which is most probably related to the impact of an Earth-crossing asteroid, other Ir anomalies have been sought. Ganapathy (1982) and Asaro et al (1982) report finding an Ir anomaly in Caribbean deep-sea core sediments at the end of the Eocene, the peak of which almost coincides with the maximum distribution of North American microtektites in the same core. Ganapathy (1982) provided an explanation for the small displacement between the Ir peak and the microtektite peak, and thus there may be an association. However, Keller et al (1983) challenged the interpretation of the Eocene event as an extinction. Glass (1985), however, has shown that most of Keller et al's assumptions are not conclusive.

An important recent discovery is that of microirghizites (Fredriksson & Glass 1983, Glass et al 1983). The pale yellow to opaque black glass particles (size smaller than 1 mm) have been recovered from a stream deposit near the Zhamanshin crater. They overlap the irghizites and zhamanshinites in composition but show a lower average SiO_2 content and greater variations in composition, just as is the case with the other microtektites. Compared with the irghizites there seems to be a larger meteoritic contamination, a conclusion drawn from the higher Ni content (up to 0.41%) and differences in the Al_2O_3 and FeO vs SiO_2 trends. Chemical variations and mineral inclusions contradict the proposal (O'Keefe 1984) that these particles originated from the impact of a homogeneous glass projectile. The discovery of these glasses is important because they are the first microtektites found on land.

IMPACT GLASSES

In many different terrestrial impact craters, glass has been found that originated during the impact of the meteoritic projectile by melting of the country rocks. These glasses depict various degrees of homogeneity (see the Introduction). The important connection between tektites, impact glasses, and impact cratering has been demonstrated at the Zhamanshin crater. The impact glasses from this crater have already been discussed in the previous sections. Detailed investigations of impact glasses are, in general, rarer than for tektites. A review of many systematic, petrological, and physical properties of impact glasses has already been given by Stöffler (1984), which permits us to concentrate here on geochemical questions. At many impact craters we do not find a "perfect" impact glass, but rather several different melted rocks or breccias. Of the known impact glasses, Henbury glass and Darwin glass are among the best analyzed. Multielement studies have been reported by Taylor & Solomon (1964), Taylor & Kolbe (1965), Taylor (1966, 1967), Chapman et al (1967), Taylor & McLennan (1979), and Koeberl et al (1985a,c, 1986a). A crater related to Darwin glass was recently discovered (Ford 1972, Fudali & Ford 1979) embedded in a heavily faulted synclinal series of lightly metamorphosed Siluro-Devonian slates, argillites, and quartzites. Before the discovery of this crater, Taylor & Solomon (1964) had inferred from trace element studies of the glass that the parent material might have been an argillaceous sandstone. REE patterns (first reported by Koeberl et al 1985c) are also consistent with this interpretation. In contrast, the origin of and connection between the Henbury impact glass and the craters has been well known for a long time (Taylor & Kolbe 1965).

The proposed connection between Ivory Coast tektites and the Bosumtwi crater in Ghana (Jones et al 1981) is strengthened by the results of the analyses of the impact glass. Many trace element abundances (such as those for Ga, Cs, Rb, Li, Co, Ba, V, Sr, Sc, Y, Zr) are similar for the Bosumtwi impact glasses and the Ivory Coast tektites (e.g. Schnetzler et al 1967, Cuttitta et al 1972), although there are also a few disagreements [e.g. the K/Rb ratio and the absolute Cr and Ni abundances (although the Cr/Ni ratio remains almost identical)].

The Aouelloul impact crater (diameter about 250 m) is situated in the Adrar Desert in Mauritania. It was discovered by A. Pourquié during a flight over that area and was later visited on ground by T. Monod (Monod & Pourquié 1951). Analyses of the impact glass associated with the crater have been reported by Campbell-Smith & Hey (1952), and major element and some trace element abundances have been determined by Chao et al (1966a), Cressy et al (1972), Annell (1976), and Koeberl et al (1985a,c, 1986a). Koeberl et al (1985a,c) concluded from major element correlations and REE data that some kind of sandstone, possibly the local Zli sandstone, may be a parent rock.

Libyan Desert Glass (LDG) is a high-silica natural glass of mostly yellow color that is found scattered along the southwestern margin of the Great Sand Sea in western Egypt (Spencer 1939, Barnes & Underwood 1976, Fudali 1981, Weeks et al 1984). The glass contains inclusions such as zircon, baddeleyite, rutile, quartz, and titanomagnetite (Kleinmann 1969). It (as well as the local Nubian sandstone) is almost entirely silica ($\sim 98\%$ SiO_2). Early analyses were in error, especially regarding the CaO and Na_2O content (reported by Spencer 1939 to be ~ 0.3 and 0.34%, respectively). Modern data reveal a much lower abundance: a CaO content of 0.01% (Barnes & Underwood 1976) and a Na abundance of 25–60 ppm (Weeks et al 1984, Koeberl 1985c). Extensive trace element studies have been reported by Weeks et al (1984) for four samples and by Koeberl (1985c) for six samples. The exact origin for LDG is not known—some authors doubt an impact origin—and although some nearby craters, such as the B.P. structure and the Oasis structure (French et al 1974, Barnes & Underwood 1976), have been associated with LDG formation, no connection between any crater and LDG has yet been established. Presently, comparisons with the exact ground rocks are not possible. A cometary impact model has been proposed by Seebaugh & Strauss (1984), but this leaves open many questions. However, Fudali (1981) considers an impact origin as the most likely explanation.

Another interesting material is Lonar glass. The Lonar crater is situated in the Deccan Trap basalts of India and is of circular shape, 1830 m in diameter and (with a shallow lake at the crater floor) 150 m deep. The crater and the impact glass are of special interest because this is the only known terrestrial impact crater in basalt, and thus it provides a unique opportunity for comparison with lunar craters. Chemical connections between the glasses and the basalt demonstrate that lunar glasses may also very often be related to their parent rocks (Fredriksson et al 1979). Morgan (1978) showed that some trace element depletions (such as Se or Re) and the degree of shock seem to be correlated in Lonar glasses.

Besides the multielement studies cited above, some authors have devoted their analyses to specific elements, with data for a few impact glasses also reported. These investigations include studies for Li (Shukla & Goel 1979, Koeberl & Berner 1983, Koeberl et al 1984e), B (Mills 1968), F (Koeberl et al 1984d), and U and Th (Yellin et al 1983).

PROJECTILE IDENTIFICATION

In many studies of lunar impact materials, attempts have been made to determine the nature and composition of the projectile from the abundance pattern of certain diagnostic trace elements (e.g. the platinum-group metals). Later, comparable studies have been made of terrestrial impact materials. Morgan et al (1975) demonstrated the feasibility of their method from analysis of samples from the Wabar and Meteor (Arizona) craters (where the composition of the projectiles is known). Consequently, terrestrial impact glasses and other impact materials have been subjected to projectile identification analysis. Morgan et al (1975) and Koeberl & Kiesl (1983) have shown that in Aouelloul glass, there is a considerable enrichment of siderophile elements above the indigenous level of the local sandstone [aside from Ni-Fe spherules (Chao et al 1966b)]; this pattern can be matched by a group III iron meteorite.

Most analyses, however, have dealt with impact melt rocks or breccias. Craters investigated and the inferred projectiles include the following: Rochechouart (IIA iron; Janssens et al 1977), Gow Lake (iron?; Wolf et al 1980), Mistastin (iron?; Morgan et al 1975, Wolf et al 1980), Clearwater East (C1 or C2 chondrite; Palme et al 1978a,b, 1979), Brent (L or LL chondrite; Palme et al 1978b, 1981), Lake Wanapitei (C or LL chondrite; Wolf et al 1980), Lappajärvi (volatile-rich chondrite; Göbel et al 1980, Reimold 1982), Nicholson Lake (nakhlite, ureilite; Wolf et al 1980), and Manicouagan (achondrite?; Palme et al 1978a,b).

The projectile identification for the Zhamanshin crater has not yet yielded conclusive results. Bouška et al (1981) and Glass et al (1983) proposed a chondritic projectile, but their study is not based on Pt-group siderophiles, so this conclusion may not be significant. Palme et al (1981) and C. Koeberl (unpublished data) investigated siderophile trace elements and found hints for an iron projectile; however, no firm conclusion is yet possible, since the (usually necessary) data for country rocks are not available.

Ivory Coast tektites show a meteoritic component (Palme et al 1978a, 1981), which points to an iron projectile. No contamination is present in the country rocks from the Bosumtwi crater. The situation is also vague for Australasian tektites. Ni-Fe spherules in philippinites (Chao et al 1964) were analyzed by Ganapathy & Larimer (1983) for some siderophile elements; from Ir, W, Au, As, and Co distributions they concluded that a terrestrial in situ reduction origin for the spherules was most likely. However, elements such as As and W may possibly be contaminated from country-rocks, and the other siderophiles (usually not present in terrestrial rocks) do not deviate very much from the meteoritic range. Analyses of an australite (Morgan 1978) and of three other Australasian tektites (Koeberl & Kiesl 1983) that are slightly enriched in siderophiles show hints of a chondritic projectile, but no known meteorite class gives a good fit. For the Ries crater and the moldavites, more data are available. From Fe-Cr-Ni particles and veinlets in a compressed zone below the Ries crater bottom, El Goresy & Chao (1976) inferred a stony meteorite projectile. Analyses of Ries material by Morgan et al (1979) and Horn et al (1983) and of moldavite data by Koeberl & Kiesl (1983) showed indications for an achondritic projectile (perhaps an aubrite).

CONCLUSIONS

Mindful of lunar data and with Occam's razor in hand, it should be clear by now that the complete chemistry, petrology, and other arguments leave only a terrestrial impact origin for tektites (and of course for impact glasses!) as a sound theory.

At present, however, we do not know the exact tektite production process. This ignorance is mainly due to our poor knowledge of the behavior of rocks subjected to very high pressures (shock waves) and very high temperatures over very brief time scales. In fact, a more promising approach to this problem is to further study the physics and the chemical effects of large impacts. Models formulated by Urlin (1966), David (1972), Jones & Sandford (1977), O'Keefe & Ahrens (1982), and Melosh (1982), for example, point the direction. Measurements by Florenskij et al (1978) showed that melting was followed by a quenching process. If the shock-melted material is lofted from the Earth (through the tunnel created in the atmosphere by the projectile), it will be quenched by short exposure to vacuum [which also explains the low p_{O_2} of formation (Weeks et al 1980)]. The material will then reenter the atmosphere (ablation melting!) and be distributed over a wide area (O'Keefe & Ahrens 1982, Melosh 1982). However, this scenario is quite sketchy, and it may take a while before we fully understand the processes taking place in large impacts.

Literature Cited

Ahrens, L., ed. 1968. *Origin and Distribution of the Elements*. Oxford: Pergamon. 1178 pp.

Annell, C. 1976. Unpublished analyses cited in O'Keefe 1976, pp. 125, 138

Asaro, F., Alvarez, L. W., Alvarez, W., Michel, H. V. 1982. Geochemical anomalies near the Eocene/Oligocene and Permian/Triassic boundaries. See Silver & Schultz 1982, pp. 517–28

Baker, G. 1963. Form and sculpture of tektites. See O'Keefe 1963, pp. 1–24

Barlow, R. A., Glass, B. P. 1976. Crystalline inclusions in Muong-Nong tektites. *Meteoritics* 11 : 248–49

Barnes, V. E. 1963. Tektite strewn fields. See O'Keefe 1963, pp. 25–50

Barnes, V. E. 1964. Variation of petrographical and chemical characteristics of indochinite tektites within their strewn field. *Geochim. Cosmochim. Acta* 28 : 893–913

Barnes, V. E. 1969. Petrology of moldavites. *Geochim. Cosmochim. Acta* 33 : 1121–34

Barnes, V. E., Pitakpaivan, K. 1962. Origin of

indochinite tektites. *Proc. Natl. Acad. Sci. USA* 48:947–55

Barnes, V. E., Underwood, J. R. 1976. New investigations of the strewn field of Libyan Desert Glass and its petrography. *Earth Planet. Sci. Lett.* 30:117–22

Becker, V. J., Manuel, O. K. 1972. Chlorine, bromine, iodine, and uranium in tektites, obsidians, and impact glasses. *J. Geophys. Res.* 77:6353–59

Bouška, V. 1972. Geology of the moldavite-bearing sediments and the distribution of moldavites. *Acta Univ. Carol.* 1972(1):1–29

Bouška, V., Povondra, P. 1964. Correlation of some physical and chemical properties of moldavites. *Geochim. Cosmochim. Acta* 28:783–94

Bouška, V., Řanda, Z. 1976. Rare earth elements in tektites. *Geochim. Cosmochim. Acta* 40:486–88

Bouška, V., Ulrych, J. 1984. Electron microprobe analyses of two-coloured moldavites. *J. Non-Cryst. Solids* 67:375–81

Bouška, V., Faul, H., Naeser, C. W. 1968. Size, shape, and colour distribution of moldavites. *Acta Univ. Carol. Geol.* 1968(4):277–86

Bouška, V., Benada, J., Řanda, Z., Kunciř, J. 1973. Geochemical evidence for the origin of moldavites. *Geochim. Cosmochim. Acta* 37:121–31

Bouška, V., Povondra, P., Florenskij, P. V., Řanda, Z. 1981. Irghizites and zhamanshinites: Zhamanshin-crater, USSR. *Meteoritics* 16:171–84

Bouška, V., Ulrych, J., Kaigl, J. 1982. Electron microprobe analyses of two-coloured moldavites. *Acta Univ. Carol. Geol.* Konta Vol. 1–2:13–28

Campbell-Smith, W., Hey, M. H. 1952. The silica glass from the crater of Aouelloul (Adrar, Western Sahara). *Bull. Inst. Fr. Afr. Noire (IFAN, Dakar)* 14:762–76

Cassidy, W. A., Glass, B. P., Heezen, B. C. 1969. Physical and chemical properties of Australasian microtektites. *J. Geophys. Res.* 74:1008–25

Chalmers, R. O., Henderson, E. P., Mason, B. 1976. Occurrence, distribution, and age of Australian tektites. *Smithson. Contrib. Earth. Sci.* 17:1–46

Chao, E. C. T. 1963. The petrographic and chemical characteristics of tektites. See O'Keefe 1963, pp. 51–94

Chao, E. C. T., Dwornik, E. J., Littler, J. 1964. New data on the nickel-iron spherules from South-East Asian tektites and their implications. *Geochim. Cosmochim. Acta* 28:971–80

Chao, E. C. T., Merrill, C. W., Cuttitta, F., Annell, C. 1966a. The Aouelloul crater and Aouelloul glass of Mauritania, Africa. *Eos, Trans. Am. Geophys. Union* 47:144 (Abstr.)

Chao, E. C. T., Dwornik, E. J., Merrill, C. W. 1966b. Nickel-iron spherules from Aouelloul glass. *Science* 154:759–65

Chapman, D. R. 1971. Australasian tektite geographic pattern, crater and ray of origin, and theory of tektite events. *J. Geophys. Res.* 76:6309–38

Chapman, D. R., Larson, H. K. 1963. On the lunar origin of tektites. *J. Geophys. Res.* 68:4305–58

Chapman, D. R., Scheiber, L. C. 1969. Chemical investigation of Australasian tektites. *J. Geophys. Res.* 74:6737–76

Chapman, D. R., Keil, K., Annell, C. 1967. Comparison of Macedon and Darwin glass. *Geochim. Cosmochim. Acta* 31:1595–1603

Cohen, A. 1960. Trace element relationships and terrestrial origin of tektites. *Nature* 188:653–54

Cohen, A. J. 1963. Asteroid- or comet-impact hypothesis of tektite origin: the moldavite strewn fields. See O'Keefe 1963, pp. 189–211

Cressy, P. J., Schnetzler, C. C., French, B. M. 1972. Aouelloul glass: aluminum-26 limit and some geochemical comparisons with Zli sandstone. *J. Geophys. Res.* 77:3043–51

Cuttitta, F., Clarke, R. S., Carron, M. K., Annell, C. S. 1967. Martha's Vineyard and Georgia tektites: new chemical data. *J. Geophys. Res.* 72:1343–49

Cuttitta, F., Carron, M. K., Annell, C. S. 1972. New data on selected Ivory Coast tektites. *Geochim. Cosmochim. Acta* 36:1297–1309

David, E. 1972. The tektite production process. *Fortschr. Mineral.* 49:154–82

Delano, J. W., Lindsley, D. H. 1982. Chemical systematics among the moldavite tektites. *Geochim. Cosmochim. Acta* 46:2447–52

Delano, J. W., Lindsley, D. H., Rudowski, R. 1981. Glasses of impact origin from Apollo 11, 12, 15, and 16: evidence for fractional vaporization and more highland mixing. *Proc. Lunar Planet. Sci. Conf., 12th*, pp. 339–70

Delano, J. W., Lindsley, D. H., Glass, B. P. 1982. Nickel, chromium, and phosphorous abundances in HMg and bottle-green microtektites from the Australasian and Ivory Coast strewn fields. *Lunar Planet. Sci.* 13:164–65

Dod, B. D., Sipiera, P. P., Povenmire, H. 1985. Electron microprobe and INAA analyses of major and trace element content for six georgiaites. *Lunar Planet. Sci.* 16:187–88

Ehmann, W. D., Stroube, W. B., Ali, M. Z., Hossain, T. I. M. 1977. Zhamanshin crater glasses: chemical composition and comparison with tektites. *Meteoritics* 12:212–14

El Goresy, A., Chao, E. C. T. 1976. Evidence of the impacting body of the Ries crater—the discovery of Fe-Cr-Ni veinlets below the crater bottom. *Earth Planet. Sci. Lett.* 31:330–40

Englert, P., Pal, D. K., Tuniz, C., Moniot, R. K., Savin, W., et al. 1984. Manganese-53 and beryllium-10 contents of tektites. *Lunar Planet. Sci.* 15:250–51

Evans, B. J., Leung, L. K. 1979. Mössbauer spectroscopy of tektites and other natural glasses. *J. Phys. (Paris)* 40(C2):489–91

Fleischer, R. L., Price, P. B., Woods, R. T. 1969. A second tektite fall in Australia. *Earth Planet. Sci. Lett.* 7:51–52

Florenskij, P. V. 1975a. Meteoritnyi krater Zhamanshin (Severnoje Priaral'je); jego tektity i impaktity. *Izv. Akad. Nauk SSSR Ser. Geol.* 10:73–86

Florenskij, P. V. 1975b. Irgizity - tektity iz meteoritnogo kratera Zhamanshin (Severnoje Priaral'je). *Astron. Vestn.* 9:237–44

Florenskij, P. V. 1977. Der Meteoritenkrater Zhamanshin (nördliches Aralgebiet, UdSSR) und seine Tektite und Impaktite. *Chem. Erde* 36:83–95

Florenskij, P. V., Dabizha, A. I. 1980. *Meteoritnyi Krater Zhamanshin.* Moscow: Nauka. 128 pp.

Florenskij, P. V., Dikow, J. P., Gendler, T. S. 1978. Die strukturchemischen Besonderheiten der Tektite - das Ergebnis von Schmelz- und Abschreckvorgängen. *Chem. Erde* 37:109–18

Florenskij, P. V., Dabizha, A. I., Aaloe, A. O., Gorshkov, E. S., Mikljajev, V. I. 1979a. Geologo-geofizicheskaja kharakteristika meteoritnogo kratera Zhamanshin. *Meteoritika* 38:86–98

Florenskij, P. V., Perelygin, V. P., Bazhenov, M. L., Lkhagvasuren, D. D., Stecenko, S. G. 1979b. Kompleksnoje opredelenije vozrasta meteoritnogo kratera Zhamanshin. *Astron. Vestn.* 13:178–86

Ford, R. J. 1972. A possible impact crater associated with Darwin glass. *Earth Planet. Sci. Lett.* 16:228–30

Franke, H., Heide, K. 1977. Einige historische und glaschemische Aspekte des Tektitproblems. *Chem. Erde* 36:299–311

Fredriksson, K., Glass, B. P. 1983. Microirghizites from a sediment sample from the Zhamanshin impact structure. *Lunar Planet. Sci.* 14:209–10

Fredriksson, K., deGasparis, A., Ehmann, W. D. 1977. The Zhamanshin structure: chemical and physical properties of selected samples. *Meteoritics* 12:229–31

Fredriksson, K., Brenner, P., Dube, A., Milton, D., Mooring, C., Nelen, J. A. 1979. Petrology, mineralogy and distribution of Lonar (India) and lunar impact breccias and glasses. *Smithson. Contrib. Earth Sci.* 22:1–13

French, B. M., Underwood, J. R., Fisk, E. P. 1974. Shock metamorphic features in two meteorite impact structures, Southeast Libya. *Geol. Soc. Am. Bull.* 85:1425–28

Frenzel, G., Ottemann, J. 1978. Über Blitzgläser vom Katzenbuckel Odenwald, und ihre Ähnlichkeit mit Tektiten. *Neues Jahrb. Mineral. Monatsh.* 1978:439–46

Frey, F. A. 1977. Microtektites: a chemical comparison of bottle-green microtektites, normal microtektites, and tektites. *Earth Planet. Sci. Lett.* 35:43–48

Frey, F. A., Spooner, C. M., Baedecker, P. A. 1970. Microtektites and tektites: a chemical comparison. *Science* 170:845–47

Fudali, R. F. 1981. The major element chemistry of Libyan Desert Glass and the mineralogy of its precursor. *Meteoritics* 16:247–59

Fudali, R. F., Ford, R. J. 1979. Darwin glass and Darwin crater: a progress report. *Meteoritics* 14:283–96

Futrell, D. S., Fredriksson, K. 1983. Brecciated Muong-Nong type tektites. *Meteoritics* 18:15–17

Ganapathy, R. 1982. Evidence for a major meteorite impact on the Earth 34 million years ago: implication on the origin of the North American tektites and Eocene extinction. See Silver & Schultz 1982, pp. 513–16

Ganapathy, R., Larimer, J. W. 1983. Nickel-iron spherules in tektites: non-meteoritic in origin. *Earth Planet. Sci. Lett.* 65:225–28

Garlick, G. D., Naeser, C. W., O'Neill, J. R. 1971. A Cuban tektite. *Geochim. Cosmochim. Acta* 35:731–34

Gilchrist, J., Thorpe, A. N., Senftle, F. E. 1969. Infrared analysis of water in tektites and other glasses. *J. Geophys. Res.* 74:1475–83

Glass, B. P. 1967. Microtektites in deep-sea sediments. *Nature* 214:372–74

Glass, B. P. 1968. Glassy objects (microtektites?) from deep sea sediments near the Ivory Coast. *Science* 161:891–93

Glass, B. P. 1969. Chemical composition of Ivory Coast microtektites. *Geochim. Cosmochim. Acta* 33:1135–47

Glass, B. P. 1970. Zircon and chromite crystals in a Muong-Nong type tektite. *Science* 169:766–69

Glass, B. P. 1972a. Bottle green microtektites. *J. Geophys. Res.* 77:7057–64

Glass, B. P. 1972b. Crystalline inclusions in a Muong-Nong type indochinite. *Earth Planet. Sci. Lett.* 16:23–28

Glass, B. P. 1978. Australasian microtektites and the stratigraphic age of the australites. *Geol. Soc. Am. Bull.* 89:1455–58

Glass, B. P. 1979. Zhamanshin crater, a

possible source of Australasian tektites? *Geology* 7:351–53

Glass, B. P. 1982. Possible correlations between tektite events and climatic changes. See Silver & Schultz 1982, pp. 251–56

Glass, B. P. 1984. Tektites. *J. Non-Cryst. Solids* 67:333–44

Glass, B. P. 1985. No evidence for multiple late Eocene tektite events. *Meteoritics* 20: In press

Glass, B. P., Barlow, R. A. 1979. Mineral inclusions in Muong-Nong type indochinites: implications concerning parent material and process of formation. *Meteoritics* 14: 55–67

Glass, B. P., Baker, R. N., Storzer, D., Wagner, G. A. 1973. North American microtektites from the Caribbean Sea and their fission track age. *Earth Planet. Sci. Lett.* 19:184–92

Glass, B. P., Swincki, M. B., Zwart, P. A. 1979. Australasian, Ivory Coast, and North American tektite strewn fields: size, mass, and correlation with geomagnetic reversals and other earth-events. *Proc. Lunar Planet. Sci. Conf., 10th*, pp. 2535–45

Glass, B. P., Fredriksson, K., Florenskij, P. V. 1983. Microirghizites recovered from a sediment sample from the Zhamanshin impact structure. *Proc. Lunar Planet. Sci. Conf., 13th/J. Geophys. Res. Suppl.* 88: B319–30

Glass, B. P., Burns, C. A., Lerner, D. H., Sanfilippo, A. 1984. North American tektites and microtektites from Barbados, West Indies. *Meteoritics* 19:228

Göbel, E., Reimold, U., Baddenhausen, H., Palme, H. 1980. The projectile of the Lappajärvi impact crater. *Z. Naturforsch.* 35A:197–203

Grass, F., Koeberl, C., Wiesinger, G. 1983. Mössbauer spectroscopy as a tool for the determination of Fe^{3+}/Fe^{2+} ratios in impact glasses. *Meteoritics* 18:305–6

Graup, G., Horn, P., Köhler, H., Müller-Sohnius, D. 1981. Source material for moldavites and bentonites. *Naturwissenschaften* 68:616–17

Haskin, L. A., Gehl, M. 1963. Rare earth distribution in tektites. *Science* 139:1056–58

Haskin, L. A., Gonzales-Garcia, A., Kleinmann, B., Haskin, M. A., Braverman, M., Roca, H. 1980. A trace element study of tektites and materials from their possible parent craters. *Lunar Planet. Sci.* 11:410–12

Haskin, L. A., Braverman, M., King, E. A. 1982. Trace element analyses of some North American tektites. *Lunar Planet. Sci.* 13:302–3

Horn, P., Pohl, J., Pernicka, E. 1983. Siderophile elements in the graded fall-back unit

from Ries crater, Germany. *Meteoritics* 18:317

Janssens, M.-J., Hertogen, J., Takahashi, H., Anders, E., Lambert, P. 1977. Rochechouart meteorite crater: identification of projectile. *J. Geophys. Res.* 82:750–58

Jessberger, E., Gentner, W. 1972. Mass spectrometric analysis of gas inclusions in Muong-Nong glass and Libyan Desert Glass. *Earth Planet. Sci. Lett.* 14:221–25

Jones, E. M., Sandford, M. T. 1977. Numerical simulation of a very large explosion at the Earth's surface with possible application to tektites. In *Impact and Explosion Cratering*, ed. D. J. Roddy, R. O. Pepin, R. B. Merrill, pp. 1009–24. New York: Pergamon

Jones, W. B., Bacon, M., Hastings, D. A. 1981. The Lake Bosumtwi impact crater, Ghana. *Geol. Soc. Am. Bull.* 92:342–49

Jung, D., Weiskirchner, W. 1979. Researches concerning the solubility of moldavites. *Meteoritics* 14:439

Keller, G., D'Hondt, S., Vallier, T. L. 1983. Multiple microtektite horizons in upper-Eocene marine sediments: no evidence for mass extinctions. *Science* 221:150–52

King, E. A. 1964. New data on Georgia tektites. *Geochim. Cosmochim. Acta* 28:915–19

King, E. A. 1966. Major element composition of Georgia tektites. *Nature* 210:828–29

King, E. A. 1970. Tektites from near Glenmora, Rapides Parrish, Louisiana? *Meteoritics* 5:205–6

King, E. A., Arndt, J. 1977. Water content of Russian tektites. *Nature* 269:48–49

King, E. A., Bouška, V. 1968. Electron microprobe analysis and petrology of two-coloured moldavite from Lipí-Slávče (Bohemia), Czechoslovakia. *Int. Geol. Congr., 23rd*, 13:37–41

Kleinmann, B. 1969. The breakdown of zircon observed in the Libyan Desert Glass as evidence of its impact origin. *Earth Planet. Sci. Lett.* 5:497–501

Koeberl, C. 1985a. A moldavite from Stainz (Styria, Austria): the moldavite strewn field revisited. *Lunar Planet. Sci.* 16:447–48

Koeberl, C. 1985b. Geochemistry of Muong Nong type tektites VII: chemistry of dark and light layers—First results. *Lunar Planet. Sci.* 16:449–50

Koeberl, C. 1985c. Trace element chemistry of Libyan Desert Glass. *Meteoritics* 20: In press

Koeberl, C., Berner, R. 1983. Lithium in tektites and impactites. *Lunar Planet. Sci.* 14:385–86

Koeberl, C., Grass, F. 1983. Rapid instrumental neutron activation analysis

348   KOEBERL

(RINAA): application for tektite and impactite analysis. *Meteoritics* 18:325–26

Koeberl, C., Kiesl, W. 1983. The usage of siderophile trace elements for determining the class of tektite-producing cosmic primary bodies. *Meteoritics* 18:326–27

Koeberl, C., Kiesl, W., Kluger, F., Weinke, H. H. 1983. The determination of fluorine in tektites and impactites. *Lunar Planet. Sci.* 14:381–82

Koeberl, C., Kluger, F., Kiesl, W., Weinke, H. H. 1984a. Geochemistry of Muong-Nong type tektites I: fluorine and bromine. *Lunar Planet. Sci.* 15:445–46

Koeberl, C., Berner, R., Kluger, F. 1984b. Geochemistry of Muong Nong type tektites II: lithium, beryllium, and boron. *Lunar Planet. Sci.* 15:441–42

Koeberl, C., Kluger, F., Berner, R., Kiesl, W. 1984c. Geochemistry of Muong Nong type tektites III: selected trace element abundances. *Lunar Planet. Sci.* 15:443–44

Koeberl, C., Kiesl, W., Kluger, F., Weinke, H. H. 1984d. A comparison between terrestrial impact glasses and lunar volcanic glasses: the case of fluorine. *J. Non-Cryst. Solids* 67:637–48

Koeberl, C., Berner, R., Grass, F. 1984e. Lithium in tektites and impact glasses: a discussion. *Chem. Erde* 43:321–30

Koeberl, C., Kluger, F., Kiesl, W. 1984f. Geochemistry of Muong Nong type tektites IV: selected trace element correlations. *Proc. Lunar Planet. Sci. Conf., 15th/J. Geophys. Res. Suppl.* 89:C351–57

Koeberl, C., Kluger, F., Kiesl, W. 1984g. Geochemistry of Muong Nong type tektites V: unusual ferric/ferrous ratios. *Meteoritics* 19:253–54

Koeberl, C., Kluger, F., Kiesl, W. 1985a. Zhamanshin and Aouelloul impact glasses: major element chemistry, correlation analyses, and parent material. *Chem. Erde* 44:47–65

Koeberl, C., Kluger, F., Kiesl, W. 1985b. Geochemistry of Muong-Nong type tektites VIII: short discussion of some correlation diagrams. *Lunar Planet. Sci.* 16:451–52

Koeberl, C., Kluger, F., Kiesl, W. 1985c. Rare earth element abundances in some impact glasses and tektites and potential parent materials. *Chem. Erde* 44:107–21

Koeberl, C., Kluger, F., Kiesl, W. 1986a. Trace element correlations as clues to the origin of tektites and impactites. *Chem. Erde* 45:1–21

Koeberl, C., Dikov, Y. P., Nazarov, M. A. 1986b. Blue glass from Zhamanshin impact crater. *Lunar Planet. Sci.* 17: In press

Konta, J. 1972. Quantitative petrographical and chemical data on moldavites and their mutual relations. *Acta Univ. Carol.* 1972(1):31–45

Konta, J., Mráz, L. 1969. Chemical composition and bulk density of moldavites. *Geochim. Cosmochim. Acta* 33:1103–11

LaMarche, P. H., Rauch, F., Lanford, W. A. 1984. Reaction between water and tektite glass. *J. Non-Cryst. Solids* 67:361–69

Langmuir, C. H., Vocke, R. D. Jr., Hanson, G. H., Hart, S. R. 1978. A general mixing equation with applications to Icelandic basalts. *Earth Planet. Sci. Lett.* 37:380–92

Luft, E. 1983. *Zur Bildung der Moldavite beim Ries-Impakt aus Tertiären Sedimenten.* Stuttgart: F. Enke Verlag. 202 pp.

Masaitis, V. L., Boiko, Y. I., Izokh, E. P. 1984. Zhamanshin impact crater (western Kazakhstan): additional geological data. *Lunar Planet. Sci.* 15:515–16

Mason, B. 1979. Chemical variation among Australasian tektites. *Smithson. Contrib. Earth Sci.* 22:14–26

Melosh, H. J. 1982. The mechanics of large meteoroid impacts in the Earth's oceans. See Silver & Schultz 1982, pp. 121–27

Mills, A. A. 1968. Boron in tektites. See Ahrens 1968, pp. 521–31

Monod, T., Pourquié, A. 1951. Le cratere d'Aouelloul (Adrar, Sahara occidental). *Bull. Inst. Fr. Afr. Noire (IFAN, Dakar)* 13:293–304

Moore, C. B., Canepa, J. A., Lewis, C. F. 1984. Volatile nonmetallic elements in tektites. *J. Non-Cryst. Solids* 67:345–48

Morgan, J. W. 1978. Lonar crater glasses and high-magnesium australites: trace element volatilization and meteoritic contamination. *Proc. Lunar Planet. Sci. Conf., 9th*, pp. 2713–30

Morgan, J. W., Higuchi, H., Ganapathy, R., Anders, E. 1975. Meteoritic material in four terrestrial meteorite craters. *Proc. Lunar Sci. Conf., 6th*, pp. 1609–23

Morgan, J. W., Janssens, M.-J., Hertogen, J., Gros, J., Takahashi, H. 1979. Ries impact crater, southern Germany: search for meteoritic material. *Geochim. Cosmochim. Acta* 43:803–15

Müller, O., Gentner, W. 1968. Gas content in bubbles of tektites and other natural glasses. *Earth Planet. Sci. Lett.* 4:406–10

Müller, O., Gentner, W. 1973. Enrichment of volatile elements in Muong Nong type tektites: clues for their formation history. *Meteoritics* 8:414–15

Ngo, H. H., Wasserburg, G. J., Glass, B. P. 1985. Nd and Sr isotopic compositions of tektite material from Barbados and their relationship to North American tektites. *Geochim. Cosmochim. Acta* 43:1865–67

O'Keefe, J. A., ed. 1963. *Tektites.* Chicago: Univ. Chicago Press. 228 pp.

O'Keefe, J. A. 1976. *Tektites and Their Origin.* New York: Elsevier. 254 pp.

O'Keefe, J. A. 1980. Comments on "Chemical relationships among irghizites, zhaman-shinites, Australasian tektites, and Henbury impact glass." *Geochim. Cosmochim. Acta* 44:2151–52

O'Keefe, J. A. 1983. The moldavite strewn field. *Eos, Trans. Am. Geophys. Union* 64:257 (Abstr.)

O'Keefe, J. A. 1984. Comments on a paper by R. Ganapathy and J. Larimer "Ni-Fe spherules in tektites: non-meteoritic in origin." *J. Non-Cryst. Solids* 67:371–74

O'Keefe, J. D., Ahrens, T. J. 1982. The interaction of the Cretaceous/Tertiary extinction bolide with the atmosphere, ocean, and solid Earth. See Silver & Schultz 1982, pp. 103–20

O'Reilly, T. C., Haskin, L. A., King, E. A. 1983. Element correlations among North American tektites. *Lunar Planet. Sci.* 14:580–81

Ottemann, J. 1966. Zusammensetzung und Herkunft der Tektite und Impaktite. *Fortschr. Chem. Forsch.* 7:409–44

Pal, D. K., Tuniz, C., Moniot, R. K., Kruse, T. H., Herzog, G. F. 1982. Beryllium-10 in Australasian tektites: evidence for a sedimentary precursor. *Science* 218:787–89

Palme, H., Janssens, M.-J., Takahashi, H., Anders, E., Hertogen, J. 1978a. Meteoritic material at five large impact craters. *Geochim. Cosmochim. Acta* 42:313–23

Palme, H., Wolf, R., Grieve, R. A. F. 1978b. New data on meteoritic material at terrestrial impact craters. *Lunar Planet. Sci.* 9:856–58

Palme, H., Göbel, E., Grieve, R. A. F. 1979. The distribution of volatile and siderophile elements in the impact melt of East Clearwater (Quebec). *Proc. Lunar Planet. Sci. Conf., 10th*, pp. 2465–92

Palme, H., Grieve, R. A. F., Wolf, R. 1981. Identification of the projectile at the Brent crater, and further consideration of projectile types at terrestrial craters. *Geochim. Cosmochim. Acta* 45:2417–24

Philpotts, J. A., Pinson, W. H. 1966. New data on the chemical composition and origin of moldavites. *Geochim. Cosmochim. Acta* 30:253–66

Pinson, W. H., Griswold, T. B. 1969. The relationship of nickel and chromium in tektites: new data on Ivory Coast tektites. *J. Geophys. Res.* 74:6811–15

Preuss, E. 1969. Verschleppte Tektite in Liberia. *Naturwissenschaften* 56:512

Reimold, U. 1982. The Lappajärvi meteorite crater, Finland: petrography, Rb-Sr, major and trace element geochemistry of the impact melts and basement rocks. *Geochim. Cosmochim. Acta* 46:1203–25

Rost, R. 1966. A Muong-Nong type moldavite from Lhenice in Bohemia. *Acta Univ.*

Carol. Geol. 1966(4):235–42

Rost, R. 1972a. *Vltaviny a Tektity*. Prague: Academia. 241 pp.

Rost, R. 1972b. Basic characteristics of moldavites. *Acta Univ. Carol. Geol.* 1972(1):47–58

Saul, J. M., Cassidy, W. A. 1970. A possible new tektite occurrence in South West Africa. *Meteoritics* 5:220

Schnetzler, C. C., Pinson, W. H. 1963. The chemical composition of tektites. See O'Keefe 1963, pp. 95–129

Schnetzler, C. C., Pinson, W. H. 1964. A report on some recent major element analyses of tektites. *Geochim. Cosmochim. Acta* 28:793–806

Schnetzler, C. C., Philpotts, J. A., Thomas, H. H. 1967. Rare earth and barium abundances in Ivory Coast tektites and rocks from the Bosumtwi crater area, Ghana. *Geochim. Cosmochim. Acta* 31:1987–93

Schock, H. H. 1976. Möglichkeiten und Grenzen der Anwendbarkeit der Aktivierungsanalyse für geowissenschaftliche Problemstellungen. *Fortschr. Mineral.* 53:187–270

Seebaugh, W. R., Strauss, A. M. 1984. A cometary impact model for the source of Libyan Desert Glass. *J. Non-Cryst. Solids* 67:511–19

Senftle, F. E., Thorpe, A. N., Sullivan, S. 1969. Magnetic properties of microtektites. *J. Geophys. Res.* 74:6825–33

Shaw, H. F., Wasserburg, G. J. 1982. Age and provenance of the target materials for tektites and possible impactites as inferred from Sm-Nd and Rb-Sr systematics. *Earth Planet. Sci. Lett.* 60:155–77

Shukla, P. N., Goel, P. S. 1979. Lithium in tektites and natural glasses. *Geochim. Cosmochim. Acta* 43:1865–67

Sigmund, A. 1911. Neue Mineralvorkommen in Steiermark und Niederösterreich 15. Moldavit von Stainz. *Mitt. Naturwiss. Ver. Steiermark* 48:241–43

Silver, L. T., Schultz, P. H., eds. 1982. *Geological Implications of Impacts of Large Asteroids and Comets on the Earth*. Boulder, Colo: Geol. Soc. Am. Spec. Pap. 190. 528 pp.

Spencer, L. J. 1939. Tektites and silica-glass. *Mineral. Mag.* 25:425–41

Stöffler, D. 1984. Glasses formed by hypervelocity impact. *J. Non-Cryst. Solids* 67:465–502

Storzer, D. 1985. The fission track age of HNa/K australites revisited. *Meteoritics* 20: In press

Storzer, D., Wagner, G. A. 1979. Fission track dating of Elgygytgyn, Popigai and Zhamanshin impact craters: no sources for Australasian or North American tektites. *Meteoritics* 14:541–42

350   KOEBERL

Storzer, D., Wagner, G. A. 1980. Australites older than indochinites—evidence from fission track plateau dating. *Naturwissenschaften* 67:90–91

Tanner, J. T., Ehmann, W. D. 1967. The abundance of antimony in meteorites, tektites and rocks by neutron activation. *Geochim. Cosmochim. Acta* 31:2007–26

Taylor, H. P., Epstein, S. 1969. Correlations between $^{18}O/^{16}O$ ratios and chemical compositions of tektites. *J. Geophys. Res.* 74:6834–44

Taylor, S. R. 1960. Abundance and distribution of alkali elements in australites. *Geochim. Cosmochim. Acta* 20:85–100

Taylor, S. R. 1962. The chemical composition of australites. *Geochim. Cosmochim. Acta* 26:685–722

Taylor, S. R. 1965. Similarity in composition between Henbury impact glass and australites. *Geochim. Cosmochim. Acta* 29:599–601

Taylor, S. R. 1966. Australites, Henbury impact glass and subgreywacke: a comparison of the abundances of 51 elements. *Geochim. Cosmochim. Acta* 30:1121–36

Taylor, S. R. 1967. Composition of meteorite impact glass across the Henbury strewn field. *Geochim. Cosmochim. Acta* 31:961–68

Taylor, S. R. 1968. Geochemistry of Australian impact glasses and tektites (australites). See Ahrens 1968, pp. 533–41

Taylor, S. R. 1973. Tektites: a post-Apollo view. *Earth Sci. Rev.* 9:101–23

Taylor, S. R., Kaye, M. 1969. Genetic significance of the chemical composition of tektites: a review. *Geochim. Cosmochim. Acta* 33:1083–1100

Taylor, S. R., Kolbe, P. 1965. Geochemistry of Henbury impact glass. *Geochim. Cosmochim. Acta* 29:741–54

Taylor, S. R., McLennan, S. M. 1979. Chemical relationships among irghizites, zhamanshinites, Australasian tektites, and Henbury impact glasses. *Geochim. Cosmochim. Acta* 43:1551–65

Taylor, S. R., McLennan, S. M. 1980. Authors' reply. *Geochim. Cosmochim. Acta* 44:2153–57

Taylor, S. R., McLennan, S. M. 1981. The composition and evolution of the continental crust: REE evidence from sedimentary rocks. *Philos. Trans. R. Soc. London Ser. A* 301:381–99

Taylor, S. R., Sachs, M. 1961. Abundance and distribution of alkali elements in Victorian australites. *Geochim. Cosmochim. Acta* 25:223–28

Taylor, S. R., Sachs, M. 1964. Geochemical evidence for the origin of australites. *Geochim. Cosmochim. Acta* 28:235–64

Taylor, S. R., Solomon, M. 1964. The geochemistry of Darwin glass. *Geochim. Cosmochim. Acta* 28:471–94

Taylor, S. R., Sachs, M., Cherry, R. D. 1961. Studies of tektite composition I: inverse relation between SiO_2 and other major constituents. *Geochim. Cosmochim. Acta* 22:155–63

Tera, F., Brown, L., Klein, J., Middleton, R., Mason, B. 1983. Beryllium-10 and aluminum-26 in tektites. *Meteoritics* 18:405–6

Tilton, G. R. 1958. Isotopic composition of lead from tektites. *Geochim. Cosmochim. Acta* 14:323–30

Urlin, V. D. 1966. Melting at ultra-high pressures in a shock wave. *Sov. Phys. JETP* 22:341–46

Van Patter, D. M., Swann, C. P., Glass, B. P. 1981. Proton probe analysis of an irghizite and a high-magnesium Java tektite. *Geochim. Cosmochim. Acta* 45:229–34

Weeks, R. A., Nasrallah, M., Arafa, S., Bishay, A. 1980. Studies of fusion processes of natural glasses by electron magnetic resonance spectroscopy. *J. Non-Cryst. Solids* 38/39:129–34

Weeks, R. A., Underwood, J. R., Giegengack, R. 1984. Libyan Desert Glass: a review. *J. Non-Cryst. Solids* 67:593–619

Weinke, H. H., Koeberl, C. 1984. Geochemistry of Muong Nong type tektites VI: major element determinations and inhomogeneities. *Meteoritics* 19:333–35

Weinke, H. H., Koeberl, C. 1985. Trace elements in two bediasite tektites. *Meteoritics* 20: In press

Weiskirchner, W. 1967. Zur Petrographie moldavitführender Sedimente Südböhmens und Westmährens. *Fortschr. Mineral.* 44:148

Weiskirchner, W. 1969. Inhomogeneities in moldavites. *Meteoritics* 4:297

Wolf, R., Woodrow, A. B., Grieve, R. A. F. 1980. Meteoritic material at four Canadian impact craters. *Geochim. Cosmochim. Acta* 44:1015–22

Yabuki, H., Shima, M., Yabuki, S. 1981. Tektite(?) from northern Thailand. *Sci. Pap. Inst. Phys. Chem. Res. (Jpn.)* 75:41–47

Yagi, K., Kuroda, Y., Koshimizu, K. 1982. Chemical composition and fission track age of some Muong Nong type tektites. *Proc. Symp. Antarct. Meteorites (NIPR, Tokyo), 7th,* pp. 162–70

Yellin, J., Perlman, I., Gentner, W., Müller, O. 1983. High precision elemental abundances for tektites and crater glasses—thorium, uranium, and potassium. *J. Radioanal. Chem.* 76:35–47

Yiou, F., Raisbeck, G. M., Klein, J., Middleton, R. 1984. $^{26}Al/^{10}Be$ in terrestrial impact glasses. *J. Non-Cryst. Solids* 67:503–9

Zähringer, J. 1963. Isotopes in tektites. See O'Keefe 1963, pp. 137–49

Ann. Rev. Earth Planet. Sci. 1986. 14:351–76

CLIMATIC RHYTHMS RECORDED IN STRATA

Alfred G. Fischer

Department of Geological Sciences, University of Southern California, University Park, Los Angeles, California 90089-0714

INTRODUCTION

We live by rhythms, in a world of predictably recurring tides, suns, moons, and seasons. This *calendar frequency band* reflects the rotational and orbital cycles of Earth and Moon. No such rhythms appear to exist at shorter time-spans, to govern the lives of microbes or the behavior of molecules. Are there significant rhythms of lower frequency, beyond human experience? The 22-yr *Hale cycle* in solar magnetic polarity and its 11-yr sunspot hemicycle seem curiously paralleled by rhythmic tree growth (Zeuner 1952, Pearson 1978, Schove 1983) and by isotope variations in such growth bands (Libby 1983). Cyclic orbital variations affect the distribution of solar energy in the *Milankovitch frequency band*, between about 10 and 500 kyr (Milankovitch 1941). The solar system oscillates through the *midgalactic plane*, with a period of about 30 Myr, and it *orbits through the galaxy* in about 300 Myr. Have these cycles driven climatic and oceanic changes throughout Earth history? Do the rocks contain a record of the Earth's response? If so, what can we learn from that response about the nature of former atmospheres and oceans? Can the response patterns furnish a geochronology? Have the drivers changed with time? Do the rocks retain records of other cycles?

In this review we examine the evidence for rhythmicities that can be read from sequences of sedimentary strata, at scales from the hand specimen to the sequences exposed in mountainsides or boreholes. This limits our inquiry to frequencies between 1 and 1,000,000 yr—that is, from the calendar band through the Milankovitch band.

Stratigraphic sequences in which distinctive strata appear in repetitive patterns are common at these frequencies. Many such sequences are *autocyclic* (Beerbower 1964), resulting from local processes of mechanical

351

sedimentation, such as the alternations of channel and overbank deposits in alluvial systems or the alternations of turbidites and pelagic background sediments in the flysch facies. Most of the sedimentological literature (e.g. Duff et al 1967) deals with such phenomena. Our concern here is not with this type of cyclicity, but rather with *allocyclic* stratigraphic oscillations that reflect rhythmical changes in climate and hydrography (i.e. rhythmic changes in the behavior of the global atmosphere and ocean). Because of the great sensitivity of organisms to environmental factors, the sediments that

one would look to for evidence of such oscillations are those largely composed of skeletal matter (the marls, limestone, and siliceous oozes) and those that record alternating presence and absence of macrobenthos in alternately burrow-mixed and unburrowed (laminated) strata. Repetitive bedding patterns, either simple or compound, are striking in sediments of these types (Figure 1). Can these patterns be tied to certain known driving cycles? Gilbert's (1895) view that they reflect the precessional cycle and can be used for geochronometry was simplistic, but enough progress has been made to warrant a review.

Orbital Variations

Because much of this review revolves around climatic "Milankovitch cycles" forced by the Earth's orbital variations, we turn briefly to these variations, which are discussed in more detail in many works (e.g. Vernekar 1972, Imbrie & Imbrie 1979, Berger 1978). Several hundred years of astronomical observations have left no doubt that the Earth's orbit is subject to cyclic variations. While the annual energy receipt of solar radiation over the Earth as a whole is not changed, the distribution of this energy, by latitude and by season, is affected by three variables: the Earth's rotational axis changes its inclination by up to 3° in an *obliquity cycle* of 41,000 yr; the Earth's orbit changes from almost circular to more elliptical in two *eccentricity cycles*, of which one has a quasi-period of 100,000 yr and

---

Figure 1 Representative pelagic and hemipelagic bedding rhythms. Scale bar is 1 m (from Fischer et al 1985). (*a*) Lower Cretaceous, Verdon Gorge, southern France. Shown as evenly spaced shale-limestone couplets (obliquity cycle?). (*b*) Bridge Creek Limestone Member, Greenhorn Formation (Cenomanian-Turonian), west of Pueblo, Colorado: Marl-limestone couplets of mixed productivity-dilution origin are observed. The spacing suggests an obliquity cycle and a change of sedimentation rate. (*c*) Fort Hays Limestone Member, Niobrara Formation (Coniacian), west of Pueblo, Colorado. Chalky limestone with thin shale breaks is observed, and the couplets are bundled into sets by variations in limestone thickness, which suggests a productivity cycle driven by climatic precession. (*d*) Maastrichtian at Zumaya, Spain, showing a hemipelagic, calcareous interval in a flysch sequence. Couplets of shale and limestone or marlstone are grouped into bundles of five; such bundling implies a productivity cycle, possibly enhanced by a dissolution cycle, driven by the climatic precession. (*e,f*) Paleocene "red Danian" at Zumaya, Spain. Coccolith-globigerinid limestones alternate with red marl and clay. This calcareous interlude in a flysch setting probably reflects climatic precession expressed in a combination of productivity and dissolution cycles. (*g,h*) Albian Fucoid Marl at Moria and Piobbico, Italy. Pelagic marls can be seen in which precessional couplets are bundled by the short cycle of eccentricity; these couplets are in turn grouped into superbundles by a longer carbonate oscillation–redox cycle that reflects the long cycle of orbital eccentricity (see Figure 2C). The Jacob's staff measures 1.5 m. (*i*) Smoky Hill Chalk Member, Niobrara Formation (Campanian), western Kansas. Marly chalks and chalky marls of epicratonic hemipelagic type are observed, showing the climatic precession signature of couplets grouped into sets of five. Reproduced with permission of SEPM.

the other a period of 413,000 yr (Figure 2*A*); and finally, the Earth's axis *precesses* with a period of 26,000 yr (Figure 2*A*).

When the *obliquity* is low (i.e. the axis is more nearly normal to the ecliptic), more energy is delivered to the equator and less to the poles, which might be expected to steepen the mean annual latitudinal temperature gradient. At the same time, seasonality is decreased.

Eccentricity and *precession* affect climate by interaction. As a result of drag on the elliptical orbit, the period of the precession relative to the elliptical orbit (for example, to perihelion) may be shortened to as little as 14,000 yr and has two modes, one at 19,000 and the other at 23,000 yr. At such intervals, one of the hemispheres will face the Sun at midsummer and will experience exceptionally short, hot summers and long, cold winters while the other hemisphere enjoys long, cool summers and short, mild winters. In the opposite phase of the precessional cycle, the roles of the hemispheres are reversed. The intensity of this precessional effect is proportional to the eccentricity of the orbit (i.e. it waxes and wanes with the cycles in orbital eccentricity).

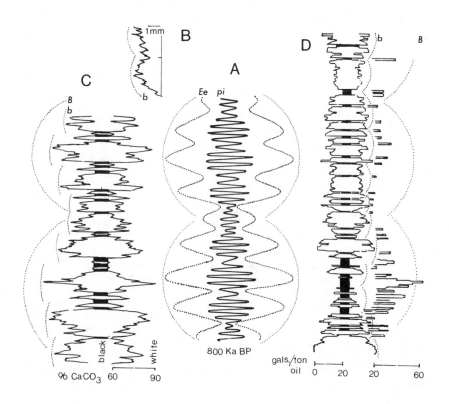

Milankovitch (1941) was the first to calculate the effects on insolation distribution in detail. This work has since been refined by Berger (1978), whose calculations of the eccentricity and the "precession index" or "climatic precession Δe sin ω, an expression of the combined effects of eccentricity and precession on insolation, are shown in Figure 2A. While insolation patterns are only one ingredient of climate, their effects should be apparent in climatic periodicities.

Pleistocene Ice Rhythms

Milankovitch, as others before him, sought in these variations the cause of the repeated advances and retreats of the Pleistocene ice sheets (Imbrie & Imbrie 1979), but the "Milankovitch" patterns did not match those of Pleistocene history as then reconstructed from the continental record. However, Emiliani (1955) discovered Milankovitch periodicities in the

---

Figure 2 Comparison of cycles plotted to approximately similar time scales. (*A*) Milankovitch parameters plotted for the last 800,000 yr (after Berger 1978). Symbols are as follows: pi, precession index or "climatic precession" Δe sin ω (*e*: eccentricity; ω: longitude of perihelion); Ee, the long and short cycles of orbital eccentricity; Ka, thousand years. (*B*) Variation in thickness of sulfate member in a series of 200,000 varves for the Permian Castile Anhydrite, New Mexico (after Anderson 1984). The precessional 20,000-yr cycle rides on a longer one (tentatively identified with the 100,000-yr short cycle of eccentricity), and the overall slope suggests the presence of a yet longer cycle. (*C*) Rhythmic structure of Fucoid Marl, Cretaceous (Albian), Piobbico core, Italy. A 4.5-m sequence is estimated from mean depositional rates to represent about 900,000 yr. The decimeter-scale couplets in carbonate content (marls, limestones) are attributed to a productivity cycle driven by precession. They are grouped into 50-cm bundles (b) by a redox cycle, in which the marls periodically are black, laminated, and less calcareous sapropels. This is attributed to the stagnation of the seafloor during monsoonal climates driven by the short eccentricity cycle. Such bundles are grouped into superbundles (B) by a modulation in the thickness and incidence of sapropels and by a reciprocal variation in the limestone purity. These superbundles are taken to reflect the long cycle in orbital eccentricity. The curve is based on densitometer scans of darkness (essentially proportional to carbonate content). See also Figure 1*g,h*. Data replotted from Fischer et al (1985). (*D*) Rhythmic structure in Parachute Creek Member, Eocene Green River Formation, in US Bureau of Land Management triangulation station shale core hole #2, Garfield County, Colorado, 425–730 ft. Inverted mirror plot of oil yield is displayed (Trudell et al 1983), with 20 gal ton^{-1} in center; curve at right shows distribution of high yields. Couplets consist of dark, varved, kerogen-rich calcitic carbonate ("oil shale") of lacustrine origin, alternating with light-colored, lean, more coarsely banded dolomitic beds interpreted as a record of intervals dominated by the playa state. Such couplets seem to be generally grouped into bundles (b) of about five, and these seem grouped by richness of oil shales into longer sets (superbundles, B) of four to five bundles. The lower of the two superbundles is the rich "Mahogany zone." The period of couplets was calculated by Bradley (1929) from outcrops of the overlying superbundle by extrapolating varve counts of "oil shale" members; the result was a period of 21,600 yr (precession). Accordingly, bundles represent the short cycle and superbundles the long cycle of eccentricity. For a longer record of bundles and superbundles, see Figure 4.

marine Pleistocene isotope record. Marine foraminiferal oxygen isotope ratios turned out to be monitors of global ice volume, and the enormous sets of data now collected and analyzed by time-series methods (Hays et al 1976, Imbrie & Imbrie 1979, Briskin & Harrell 1980, Imbrie et al 1984) leave no doubt that the glacial regimes were forced by the orbital periodicities. Just how these variations in insolation patterns influence climate remains an active topic of research and debate, but the general consensus is that in glacial times the climatic effects of insolation patterns are strongly amplified and modified by the albedo of the expanding ice sheets and by other feedback effects. Did orbital variations drive climatic oscillations in nonglacial times, without such amplification?

Milankovitch Rhythmicity in the Pre-Pleistocene Record

Gilbert (1895) suggested that oscillations in the carbonate content of hemipelagic beds in the Cretaceous of Colorado reflected an allocyclic, rhythmic process, namely climatic rhythms driven by precession. On the basis of this, he ventured a geochronology, suggesting that the marine Late Cretaceous deposits of the Rocky Mountain region represent about 1000 precessional cycles, or 20 Myr—surely not less than 10 and not more than 40 Myr. The subsequent radiometric timing is about 30 Myr, well within Gilbert's error limits. Similar patterns were subsequently found to be widely distributed in pelagic and hemipelagic systems, as well as in platform carbonates.

Gilbert recognized only a simple rhythm and chose precession for lack of another plausible cause. We must consider the possibilities of other Milankovitch cycles, of additional cycles induced by the response of the Earth to Milankovitch forcing, and of cycles wholly unrelated to the orbital parameters. As we shall see, the sedimentary record contains a hierarchy of rhythms. When overlapped and interspersed with random events, these can form rather complex patterns. We discuss here only the relatively clean ones. Gilbert assumed constancy of the orbital cycles, which remained to be tested. Crucial throughout are the periods of the rhythms observed.

Three direct approaches to measurement of rhythm periods are (a) varve counts, (b) direct radiometric determinations from cyclic sedimentary sequences, and (c) extrapolations of cycles to segments of geological time (such as stages or polarity chrons) for which radiometric durations have been estimated. Only varve counts have provided figures of sufficient accuracy to answer the question of whether the periodicities of ancient cycles are those of the present orbital variations. Radiometric dates obtained from within pre-Pleistocene cyclic sequences carry uncertainties that render them useless (Fischer 1980). Extrapolated periodicities are also limited by uncertainties, but they are often sufficient to allow a choice between alternative Milankovitch periods—say, precession versus obliquity. Other factors come to one's help (Fischer et al 1985). The character of the precession index is such that precessional cycles must wax and wane in strength with the eccentricity. Furthermore, if the full range of precessions is recorded, the number of precessional events in an eccentricity group must lie between three and eight, averaging a fraction under five. Note, however, that some settings may record only the high-amplitude signals (Fischer et al 1985). The obliquity cycle has a more steady beat and can be expected to yield simple, regular sedimentary oscillations. Thus, distinctive patterns come to be useful criteria for the identification of driving cycles.

We now turn to the stratigraphic record, discussing in turn varves, rhythmicity in pelagic and hemipelagic sediments, rhythmic emergence in carbonate platforms, and the rhythmic transgression patterns of Carboniferous cyclothems.

VARVES AND VARVE COUNTS OF LONGER CYCLES

Varves are by definition annual layers and are normally couplets composed of two members of different character that reflect seasonal change (Anderson 1964). The name was initially applied by G. de Geer to the banded muds laid down in lakes ponded by glaciers and left widely distributed over northern Europe and North America in the wake of the retreating Pleistocene ice sheets (Zeuner 1952). Such varves are normally some millimeters thick and consist of a thicker, silty summer layer and a thin, clayey winter layer. The Holocene varve chronology assembled by G. de Geer and his students and followers (de Geer 1940), a stratigraphic classic, is too noisy to record cycles between the annual and Milankovitch bands, and is too short to test for Milankovitch cyclicity.

Somewhat similar laminites of silt and clay have been reported from other parts of the geological record and were commonly interpreted as varves (Zeuner 1952); however, some of these are from turbidite sequences and represent the distal deposits of turbidity currents. Thus most ancient detrital varves—even those associated with glacial deposits—are suspect. However, varving is not limited to detrital alternations: In many settings, a fairly steady background supply of clay and fine organic matter is enriched in summer by the biogenic products of organic blooms (i.e. diatom frustules or carbonate precipitated by photosynthetic activity). In settings of seasonal rainfall, the supply of detritus may vary as well. Generally such seasonal layers are only fractions of a millimeter thick, and on most bottoms they are destroyed by burrowing organisms. They are preserved, however, where oxygen deficiency or salinity inhibits bottom life, as happens in many lakes and certain marine basins. Such varve records provide detailed geochronologies and the best data on the periods of slower rhythms. Periodicities in a number of varve sequences, determined by spectral analyses, are shown in Figure 3.

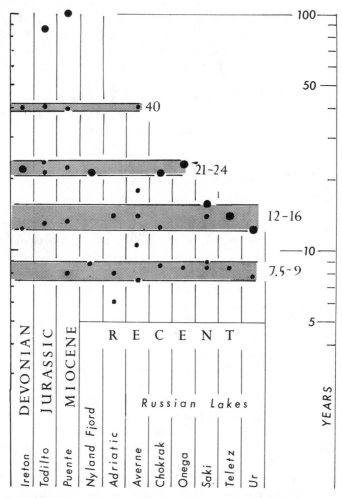

Figure 3 Periodicities in varve thickness in various varve sequences as revealed in power spectra calculated by Koopmans and by Ware and reported in Anderson (1961). The Todilto varves are from an evaporite setting, the Ireton and Puente varves are marine, the Nyland Fjord varves are estuarine, and the remainder are lacustrine. Periodicities cluster in four bands: 7.5–9 yr, 12–16 yr, 21–24 yr (Hale solar magnetic cycle), and 40 yr. Beyond this, the number of series is too small to show correlations. Large dots represent the most prominent cyclicity in their series.

Lake Varves

Many lakes trap sediment at rates high enough to record annual variations. Deep lakes and salinity-stratified lakes are liable to seasonal or permanent oxygen deficiency in the hypolimnion, which inhibits bottom life and preserves varving. While most lakes fill up too quickly to provide long time series, the large lakes developed in rift valleys and other tectonic basins last for millions of years and record long varve sequences. Periodicities in a number of Holocene Russian lake varve sequences, measured by V. B. Shostakovitch and analyzed in Anderson (1961), are shown in Figure 3. Above the 5-yr level, such periodicities are gathered into four frequency bands (7.5–9, 12–16, 21–24, and 40 yr). The sunspot cycle is poorly recorded, while the Hale cycle is prominent in two of the sequences studied.

EOCENE LACUSTRINE RECORD   The Eocene Green River Formation accumulated in a lake-and-playa complex that filled synorogenic intramontane basins in the middle Rocky Mountains of Colorado, Utah, and Wyoming. Its "oil shales" are hypolimnial, kerogen-rich carbonates (Bradley 1929, 1931, Surdam & Stanley 1978), which locally yield spectacular fish faunas. At a microscopic level, carbonate-rich layers thought to record summer blooms alternate with laminae of kerogen. In sequential varve thickness, Bradley recognized the 11-yr sunspot cycle, which appears to be the chief cause of the lamination apparent to the unaided eye (personal observation). Bradley also claimed to have found a 50-yr cyclicity, but no time-series studies have been made. At intervals of several meters, beds that are predominantly varved and kerogen rich alternate with lighter, coarsely color-banded dolomitic marls. These record times when the lake frequently dried up to a playa (Surdam & Stanley 1978), and the 2–3 m couplets thus represent oscillations in lake level. By extrapolating varve thicknesses, Bradley calculated the mean period of this rhythm at 21,600 yr and identified it with the precessional cycle.

Investigations of the "oil shale" potential of the Green River Formation led to the drilling of many cores, which were analyzed for oil yield at 1-ft intervals by the Bureau of Mines and succeeding agencies (Figures 2D, 4A). The cycles described by Bradley from the upper Parachute Creek Member can be traced from the vicinity of Rifle, Colorado, westward to near the Green River in Utah—a distance on the order of 150 km. While most regularly developed some 50 m above the kerogen-rich "Mahogany zone," they occur in less regular form throughout the deposits of Lake Uinta—the southern of the two major lakes in this region during the Eocene. Furthermore, they seem grouped into bundles of 3–6 couplets, just as precessional events are grouped by the short (100,000 yr) cycle of eccentricity. These bundles in turn seem grouped into larger sets (or "superbundles") of 3.5–5 bundles, reflected in the oil-yield potential of the constituent "oil shale" beds. These patterns are in need of Fourier analysis, but if the couplets are of precessional origin and the bundles represent the short cycle of eccentricity, the superbundles must reflect the long cycle of eccentricity. The Parachute Creek Member contains about 9 superbundles, and if their mean period was 413,000 yr (as it is now), it represents about 3.7 Myr.

The phase relations of these cycles to Milankovitch cycles remain unclear. Did the drop in lake levels result from hot perihelial summers? Or did those perihelial summers bring heavy monsoonal precipitation, leading to raised water levels and "oil shale" deposition?

Crude inspection does not reveal an obliquity cycle, but the complexities of the patterns shown in Figure 4, with multiple points of symmetry and reversing skewness of the oil-yield peaks, suggest the interference of a number of different waves, which can only be properly revealed by time-series analysis.

TRIASSIC LACUSTRINE RECORD   The Triassic Lockatong lake in the Newark rift valley, extending from New York through New Jersey into Pennsylvania, also rose and fell and varied chemically with climate (Van Houten 1964, Olsen 1984). Varving characterizes the black shale facies, deposited in deep water. Alternation of such shales with drab-to-red playa deposits yields a period of 21,000 yr by extrapolation of varve thickness. By the same token, alternations of chemical and detritus-rich sequences occur on a 100,000-yr schedule, and red (highly oxidized) sequences occur at 400,000-yr intervals. No time-series analyses have been made.

Marine Varves

Marine varves are uncommon. In pelagic deposits the mean size of sedimentary particles generally exceeds the annual accumulation, so that no annual record can develop. Varves are therefore largely limited to settings of more rapid sedimentation (marginal to continents). Here, only

---

Figure 4   (*A*) Oil yields of the Parachute Creek Member, Eocene Green River Formation, Equity Oil Co. core hole BX-1, Garfield County, Colorado (redrawn from Trudell et al 1970). Shaded portion (with "oil shale" members of couplets shown in black) is the part in which Bradley (1929) provided varve calibration that identified the couplets with precession. Couplets seem to be organized into bundles (b), here identified with the short cycle of eccentricity, and bundles seem grouped into nine superbundles (B) that are identified with the long cycle of orbital eccentricity (compare with the more detailed and inverted plot of Figure 2*D*). (*B*) Variations in varve thickness: a 1300-yr record from the Jurassic Todilto evaporite (after Anderson & Kirkland 1960). Dominant features are the 10–13 yr sunspot cycle and an amplitude cycle of about 170 yr.

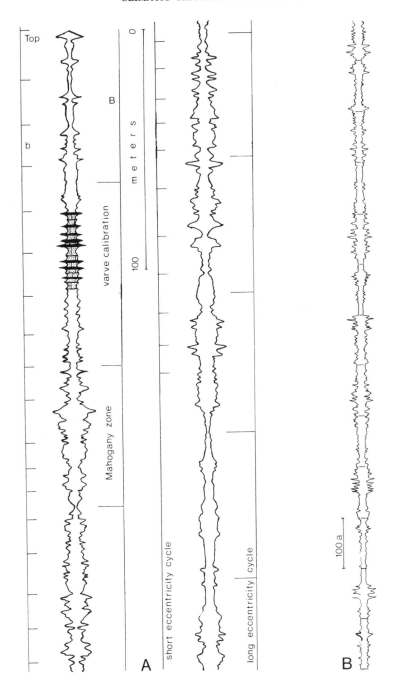

bottoms below wave base, protected from currents (including turbidity currents) and free of bioturbation, are likely to preserve a varve record. Silled basins having anoxic bottom waters, such as the Santa Barbara Basin off California, accumulate varves of millimeter scale; these varves consist of a summer layer rich in diatoms and a winter layer of clay and organic matter.

A 200-yr time series of such varves from the Santa Barbara Basin (Soutar & Crill 1977) shows remarkable parallelism of varve thickness with tree-ring curves of spruces in the mountains of southern California, and lags rainfall curves of the region by 1–2 yr. While these curves were not analyzed, they appear cyclic, with periods on the order of the Hale and sunspot cycles. An 1800-yr time series, studied by decade rather than by year, provided biologically important data on the comings and goings of pelagic fish species (anchovy, sardine, and hake) but revealed no striking patterns in varve rhythms (Soutar & Isaacs 1969). Analyses of this sequence for organic carbon and uranium by E. K. Kalil and I. R. Kaplan (reported in Libby 1983) have yielded Fourier periodicities at 53–55, 70–71, 81–82, 95, 118–21, and 156–61 yr. Whether the varve sequences of the California border basins are long and continuous enough to shed light on the Milankovitch frequency band remains to be seen.

In Holocene varves of the Adriatic, Seibold & Wiegert (1960) found periodicities of 6, 8, and 14 yr (Figure 3). Varving in the Miocene Monterey Formation of California was discussed by Bramlette (1946) and Anderson (1964), who suggest that a longer cycle exists in the 100–200 yr range. The Miocene Puente Formation (Riveroll & Jones 1954, Anderson 1961) shows a strong periodicity at 100 yr and weaker signals in each of the four frequency bands containing the main lake varve periods (Figure 3). Laminae in Cretaceous deep-sea cores from the Gulf of Mexico and the Atlantic were interpreted as varves by Cotillon & Rio (1984) and suggest periodicities of 3.4 and 23.6 yr. The Devonian Ireton Shale of Alberta (Anderson 1961; Figure 3) shows a strong cyclicity at 22 yr (Hale cycle) and lesser ones at 12 and 40 yr.

Evaporite Varves

While no varving has been recognized in modern evaporite settings, various ancient deep evaporite basins accumulated regularly laminated carbonate and gypsum deposits in response to seasonal aridity. Anderson & Kirkland (1960) studied such a record from the Jurassic Todilto Formation of New Mexico (Figure 4B). Spectral analysis (Anderson 1961; Figures 3, 4) shows periodicities at 13, 21, 23, and 41 yr. Very strong periodicities also exist at 86 and 180 yr. Richter-Bernburg (1963) found somewhat similar patterns (not spectrally analyzed) in the Permian Zechstein Anhydrite and Salt of

Germany. In parts of this, anomalous varves reflect the 11-yr sunspot cycle, while the 22-yr Hale cycle dominates periodicites in the Miocene Sicilian anhydrites. Udden's (1924) report of a sunspot cycle in the Permian Castile Anhydrite of New Mexico was not confirmed in a spectral study by Anderson (Schove 1983), but Anderson (1982) did find some periodicities between 100 and 400 yr and a strong periodicity at about 2700 yr. In addition, he found some magnificent data in the Milankovitch band (Anderson 1982, 1984): Varves wax and wane with a period that varies between 14,000 and 26,000 yr and averages roughly 20,000 yr—the very pattern of the Earth's axial precession (Figure 2*B*). Furthermore, this cycle rides on another that is at least 100,000 yr long and is most likely the short cycle of eccentricity; it has an overall slope that suggests the presence of a yet longer cycle—presumably the long eccentricity cycle.

Summary of Varve Data

Varve studies have yielded detailed local chronologies for very short segments of the geological record and provide calibration for longer rhythms. The 11-yr sunspot cycle is rarely found, but many sequences show periodicities in the intervals of 7.5–9, 12–16, and 21–24 (Hale cycle) yr, and at 40 yr. Strong periodicities between 80 and 200 yr have also been found, but there are too few such long series to define a pattern. Extrapolation of intermittent varving to the duration of long lacustrine sequences has provided the strongest evidence for the identification of rhythms in the Milankovitch frequency band. Every suitably long varved lake sequence studied to date has yielded a strong cycle with precessional (roughly 20,000 yr) timing, grouped into 100,000-yr bundles (the short eccentricity cycle). In the Green River Formation, such bundles are grouped in turn into the 400,000-yr superbundles of the long eccentricity cycle. Limited as they are, these varve studies lead to three important conclusions: (*a*) the solar magnetic cycle has existed at least since Devonian time; (*b*) the precessional and eccentricity cycles have existed at least since Permian time, more than 200 Myr; and (*c*) over that time span, these cycles have maintained approximately the same relative and absolute timing. The obliquity cycle was not found in any of these series, though evidence of its presence in the Jurassic and Cretaceous is given below.

CYCLIC BEDDING IN PELAGIC AND HEMIPELAGIC SEDIMENTS

Pelagic sediments, characteristic of oceanic domains, consist chiefly of the skeletons of planktonic organisms (calcium carbonate or silica) and of an admixture of detrital clay and volcanic dust (chiefly supplied by wind).

Opaline silica, supplied since Cambrian time by radiolarians and since Cretaceous time by diatoms, has accumulated in siliceous oozes. Aragonitic calcium carbonate has been contributed by pelagic mollusks (such as ammonites in the Paleozoic and Mesozoic, and pteropods from Cretaceous time on). Calcitic calcium carbonate has been supplied by coccolithophorids since the Jurassic, supplemented by calpionellid infusorians in the late Jurassic and early Cretaceous, and by globigerinacean foraminifera since Cretaceous time.

Hemipelagic sedimentation occurs on oceanic margins, where the same components are more heavily diluted or interspersed with water-borne muds supplied from the continents. Fluctuations in the proportion of biogenic components are widespread. Turbidity current deposits characterize certain hemipelagic settings and are surely in part autocyclic, but sensitivity to sea-level fluctuations (Shanmugam & Moiola 1982), such as those of the Pleistocene, suggests that these deposits too may carry a rhythmic climatic message.

Productivity, Dilution, Dissolution, Redox Cycles

Sequences of such pelagic and hemipelagic oozes typically show a stratification resulting from oscillations in the proportions of the biogenic fraction to the clay. Such oscillations come about as a result of four variables (Einsele 1982). The rate of skeletal supply varies with the productivity of the responsible organisms, leading to *productivity cycles.* Variations in the clay supply result in dilution of the biogenic components and are therefore termed *dilution cycles.* In addition, skeletal components at greater depths may undergo dissolution on the seafloor, depending on the intensity of the lysocline and the depth of the aragonite and calcite compensation depths. Changes in these lead to *dissolution cycles.* Furthermore, the degree to which the sediments on the seafloor are oxidized changes with the supply of organic matter, the rate of burial, and the oxidizing capacity of bottom waters and biota, leading to *redox cycles.*

Diagenetic Overprints

In poorly consolidated sediments, such as oozes retrieved in cores from the ocean floor or chalks exposed on land, stratification is generally faint and gradational. With increasing consolidation, the bedding becomes more sharply defined, and the contrasts between the beds become greater: The primary differentiation has become enhanced by diagenesis. Carbonate is moved from the more clayey layers, where skeletal matter is dissolved, to the purer carbonate layers, where calcite cement is precipitated (Arthur et al 1984, Ricken 1985). As a result, the layers initially richest in carbonate are further enriched and undergo little compaction, whereas the more clayey members of the sequence may collapse into shale seams. The process may (but need not) lead to the development of stylolites. Siliceous sediments are presumably modified in similar ways by diagenesis.

Sujkowski (1958) and Hallam (1964) have suggested that in some initially homogeneous sediments, rhythmic bedding may be developed entirely by diagenesis as a result of a "rhythmic unmixing" of phases, presumably controlled by diffusion processes such as those that form concretions. However, the development of extensive layered fabrics that imitate bedding so perfectly is not readily conceived in the absence of some primary bedding controls. Such controls need not be restricted to clay content, and they may include differential solubilities in the carbonate fraction, such as the calcite-aragonite cycles discovered by Droxler (1984).

We therefore proceed on the assumption that the bedding (stratofabric) of sediments, while modified by diagenesis, nevertheless harks back to changes in the depositional environment, and that oscillations in bedding imply oscillations in primary sedimentation.

Precession-Caused Couplets in Eccentricity-Driven Bundles

The "Umbrian pelagic facies," exposed in the Apennines of central Italy, contains a remarkable sequence of pelagic sediments extending from the mid-Jurassic to the mid-Tertiary. Rhythmicity has been investigated in the Early to Middle Cretaceous portion of this sequence, which consists of the Tithonian-Barremian Maiolica Limestone, a lithified coccolith-calpionellid ooze; the Aptian-Albian Fucoid Marl, originally pelagic clays, marls, and coccolith-globigerinid oozes; and the Cenomanian Scaglia Bianca limestone, a lithified coccolith-globigerinid ooze (Arthur & Premoli Silva 1982). All of these also contain minor radiolarites. In the upper Maiolica and the upper Scaglia Bianca, limestone beds separated by thin shale breaks are grouped into bundles defined by a thickening of shales and by the occurrence of siliceous sapropelic shales or thin radiolarites. In the intervening Fucoid Marl, beds of limestone alternate with beds of marl, and these couplets are commonly bundled (Figure 1g,h) by thinning of the limestones and change of marls to laminated sapropels (Arthur & Premoli Silva 1982, Fischer et al 1985).

Schwarzacher & Fischer (1982) and Fischer & Schwarzacher (1984) made a time-series analysis of variations in bedding thickness. Assuming a 4-Myr duration for the Cenomanian (Kennedy & Odin 1982), they arrived at a timing of 24,000 yr for the Scaglia Bianca couplets (precession?) and of 118,000 yr (short cycle of eccentricity?) for the bundles. When these data are combined with the Maiolica data and the bundle is normalized to 100,000 yr, a couplet period of 21,244 yr is obtained. Studying the upper Fucoid

Marl with different techniques and assumptions, deBoer (1982a,b) and deBoer & Wonders (1984) obtained a couplet/bundle ratio of 4.3, which corresponds to a couplet period of 23,300 yr.

The tie of these oscillations to the orbital cycles is confirmed by T. Herbert's work (Figure 2C) on a 4.5-m sequence from the Piobbico (Italy) core (Fischer et al 1985). Rock types vary from white limestone through drab marl to black laminated sapropel. Darkness is inversely proportional to calcite content, varying from 50% in the sapropels to 90% in white limestone. The marl-limestone couplets of a carbonate cycle are bundled by a longer redox cycle in which the marls are sapropels (shown black in center of mirror plot). The brightness of marls and limestones, determined by microdensitometer scans of color-controlled diapositives, yields the curve shown. The mean sedimentation rate for the Fucoid Marl is about 5 m Myr^{-1}, which suggests that this segment represents about 900,000 yr. This curve has a plane of symmetry, along which the sapropels are least well developed. The astronomical eccentricity curve and the precession index (Figure 2A) of the last 800,000 yr also have a plane of symmetry, which corresponds to the eccentricity low of 400,000 yr ago. Aligned by this symmetry plane, the bundles are seen to match the short eccentricity cycles, the individual couplets can be roughly matched to the precession cycle, and the whiteness (purity) of limestones mirrors the long (413,000 yr) cycle of eccentricity.

Along with deBoer (1982a,b), Fischer et al (1985) interpret the couplets as a result of carbonate productivity. The paradox of equating the sapropels with times of lowest marine productivity is explained by the character of their organic component, which is extremely refractory material of terrestrial derivation (deBoer 1982a,b, L. Pratt, personal communication): The sapropels represent times when strong water stratification inhibited local fertility and reduced the rate of planktogenic skeletal matter while the wind-supplied terrestrial fraction continued to accumulate, and up to 2% terrestrial organic matter (enough to color the sediment black) persisted because of anaerobism in the stagnant bottom waters. Like the Quaternary sapropels of the eastern Mediterranean (Rossignol-Strick 1985), these Albian ones appear to mark times of perihelial summers that caused a monsoon-related increase in freshwater supply, resulting in intense water stratification and redox cycle in this part of the Tethyan seaway.

While this model appears to fit most of the Fucoid Marl rhythmicity, it does not apply to the underlying Maiolica and overlying Scaglia Bianca: In these, the black shales at the bundle boundaries are rich in marine organic matter (Arthur & Premoli Silva 1982, Van Graas et al 1983), marking times of high organic and radiolarian (but not carbonate) productivity. This bundling seems to have the same timing, but the precession index cycle expressed itself in upwelling rather than in stratification of the water column.

Bundling is common in pelagic and hemipelagic sediments, occurring for example in the Coniacian Fort Hays Limestone (Figure 1c) and the Campanian Smoky Hill Chalk (Figure 1i) of Colorado and Kansas, in the Maastrichtian (Figure 1d) and Danian (Figure 1e,f) of northern Spain, and in the Annona and Arkola chalks of Arkansas (Fischer et al 1985, Bottjer, in preparation).

Redox cycles occur abundantly in the Cretaceous sediments of the Atlantic deep-sea floor (e.g. McCave 1979, Arthur & Premoli Silva 1982). Many of these are bundled, but the interrupted core record leaves the nature of many of these sequences in doubt. Here, too, the black intervals fall into two groups: those characterized by terrestrial organic matter, and those of marine composition and indicative of high marine productivity (Tissot et al 1979). In addition, there are resedimented black turbidites (de Graciansky et al 1979). By extrapolating to stage lengths, using the Kennedy & Odin (1982) scale, Cotillon & Rio (1984) calculated the periods of Valanginian couplets in the Gulf of Mexico, Atlantic Ocean, and Vocontian Basin (France) at 16,000, 15,200 and 16,000 yr, respectively. Their averages for the Hauterivian stage are 26,000, 16,500 and 20,250 yr, respectively. These couplets are bundled in sets of 4–6. By using "cyclograms" (which plot the thickness of couplets, thickness trends, discontinuities, and skewness), Cotillon (1984) and Cotillon & Rio (1984) feel able to correlate bundle sequences from the Gulf of Mexico to France, but these plots may not convince skeptical readers.

In summary, it appears that the precession index has been widely recorded in marine sediments and is "fingerprinted" by the grouping of individual precessional events into sets that correspond to the short eccentricity cycle. The eccentricity cycle must have global validity, whereas the precessional effects of the Northern and Southern Hemisphere may be expected to be 180° out of phase.

Simple Obliquity-Driven Oscillations

In contrast to the bundled Cenomanian of Umbria, the Cenomanian of the Maritime Alps as exposed at Peille near Nice (Thomel 1976) shows a simple and fairly regular oscillation in which hemipelagic mudstones contain thin limestone beds spaced at about 1-m intervals. A few double limestones suggest that perhaps not all of the 92 limestones identified by Thomel should be counted in the same rhythm, but the assumption of 4 Myr for the Cenomanian (Kennedy & Odin 1982) implies a mean timing of this rhythm in the 40,000–50,000 yr bracket. This suggests that the obliquity cycle was the driver for what appears to have been a dilution rhythm (Fischer et al

1985). A more extensive development of this rhythm is found in the basin of Lower Saxony, northern Germany (Schneider 1964). Here, dark olive-drab shales alternate with somewhat lighter marls in couplets that average about 50 cm in thickness and do not appear bundled. This pattern extends from somewhere in the Valanginian through the Albian stage. A similar cycle of different lithologies characterizes the Cenomanian (A. G. Fischer, personal observation). By extrapolating mean couplet thicknesses measured in outcrop to the full thicknesses of stages as determined in nearby wells, Schneider (1964) compiled an estimate of the number of cycles for the Hauterivian through Albian stages. The radiometry of the day suggested a mean duration of 26,000 yr. Modern time scales give longer figures: The Kennedy & Odin (1982) scale yields 32,200 yr, and the Harland et al (1982) scale gives 44,500 yr, thus bracketing the obliquity period.

The classical Jurassic of the Dorset coast (England) contains two long shaly sequences with fairly regular interbeds of limestone—one the Lower Lias, the other in the Kimmeridge Clay. House (1985) has compiled the number of couplets in each of six Liassic and five Kimmeridgian ammonite zones. By partitioning Jurassic time evenly to 60 ammonite zones, he calculates a mean zonal duration of 1–1.25 Myr and arrives at a mean cycle duration of 37,000–48,000 yr, a range that brackets the obliquity cycle.

Another example of simple oscillations (Figure 1b) is furnished by the Bridge Creek Member of the Greenhorn Formation (Cenomanian-Turonian boundary) in Colorado (Pratt 1984, Fischer et al 1985). The period of this rhythm lies somewhere between Kauffman's (1977) estimate of 80,000 yr and Fischer's (1980) equally extreme estimate of 26,000 yr: It seems a likely candidate for the obliquity cycle.

Summary and Remarks on Pelagic-Hemipelagic Sequences

Some pelagic sequences show the same precession-eccentricity patterns—couplets, bundles, and superbundles—that occur in the lacustrine record. Others differ in apparently following the more simple beat of the obliquity cycle. Most but not all of these are from higher latitudes, where that cycle may well dominate the climate.

Many long sequences, such as those of Umbria, of the Cretaceous in the Western Interior of North America (Fischer et al 1985), and of the Atlantic Tertiary (Dean et al 1978, Gardner et al 1984), are compound. Some segments appear to reflect the precession-eccentricity cycles; other parts may show the simple structure of the obliquity cycle; and in yet other parts, the couplets follow in chaotic patterns that may reflect either random events, disruption by discontinuities, or a complex overprinting of various cycles, operating via different pathways (for example, overprinting productivity cycles with dissolution cycles). The averages of such sequences yield meaningless figures. Hope lies in dissecting the complex segments by Fourier analysis, but this is complicated by uncertainties in the time dimension.

RHYTHMIC BEDDING IN PLATFORM LIMESTONES

The occurrence of rhythmic bedding in platform limestones was reviewed by Schwarzacher (1975). The periods of platform rhythms are even more elusive than are those of pelagic carbonates because (a) the biostratigraphy of platforms yields lower resolution than that of pelagic deposits, and (b) platforms are particularly hiatus-prone.

Platform cycles of the Alpine Triassic were recognized and petrographically studied by Sander (1936). Schwarzacher (1948, 1954) first noted the now well-established ratio of five couplets to a bundle. Fischer (1964) established the origin of the cycle as one of periodic emergence, with the cycle leading from a weathered zone through intertidal sabkha-type sediments into subtidal (lagoonal) limestones, succeeded most commonly by another disconformity, but passing in some cases through a regressive sabkha phase into another transgression. Intertidal-subtidal couplets, averaging 5 m in thickness, are bundled into sets of about five. The number of such cycles in the Norian-Rhaetian has not been properly established, but it would seem to lie around 300, and some cycles may not have been preserved. The period is surely in the Milankovitch band, and the pattern suggests the precession index. A eustatic oscillation seems highly likely as the cause, but no evidence for this has so far been seen outside the European Tethys. The most significant aspect of this cycle is its evidence for rapid, Milankovitch-related oscillations in sea level at a time for which no direct evidence of glaciation has been found. It raises the question of whether ice caps were restricted to only very special parts of Earth history, as generally believed, or whether limited polar glaciations—and sea-level oscillations driven by them—were a relatively normal part of Earth history.

Other examples of rhythmic platform limestones are discussed by Schwarzacher (1975) and in the volume edited by Einsele & Seilacher (1982). The rhythmic bedding described by Seibold (1952) from the Upper Jurassic of southern Germany has been restudied by Ricken (1985). He interprets the setting as a carbonate ramp built by carbonate mud transport from adjacent banks and reefs. Very slight primary differences in clay content were greatly enhanced by diagenesis. Ricken intimates that the regularity of bedding is a diagenetic artifact, but the question of rhythmicity remains unsettled.

Paleontological-stratigraphic studies in the Lower Devonian Helderberg

Limestone platform of the New York Appalachians have convinced Anderson et al (1984) that the platform consists of a large number of upward-shoaling cycles, which they term PACs (punctuated aggradational cycles) and which they claim to trace across facies boundaries. While this view appears to conflict with the currently accepted facies models, it merits serious consideration. The question of periodicity was not raised.

CARBONIFEROUS CYCLOTHEMS

The Carboniferous System, especially the Middle and Upper Pennsylvanian, is distinguished in many places by cyclic sequences (Ross & Ross 1985) in which various types of sediment alternate in regular and repeated patterns. Nowhere are these cycles more striking than in the interior of North America. In Illinois, a typical cycle begins with alluvial channel sandstones, followed by an underclay on which rests a coal, in turn succeeded by marine shales containing a limestone and possibly ending in brackish or limnic shale. To such sequences, which record transgressive cycles, Wanless & Weller (1932) applied the name *cyclothem*. Whereas Weller favored a tectonic driver, Wanless & Shepard (1936) appealed to eustatic fluctuations engendered by the rhythmic growth and decay of the Carboniferous ice sheets, a view that has recently been championed by Crowell (1978).

The term cyclothem has since been used for many kinds of sedimentary cycles, such as the autocyclic alluvial coal measure sequences of the Allegheny Basin. The eustatic cyclothems are best developed in the Late Pennsylvanian (Missourian, Virgilian) of Kansas, where they are younger than the typical Illinois cyclothems. Here the marine portion contains from three to five limestones. The transgressive peak is represented by the second marine shale, usually a sapropel with pelagic fauna (Heckel 1977), and the succeeding (third) marine limestone. Moore (1936, 1949) took this cycle to be a composite of several of the Illinois cycles and termed it a megacyclothem, but Heckel (1977) holds it to be the product of a single transgressive-regressive pulse in which the black shale represents the deepest deposits, in anaerobic waters below a pycnocline. Accordingly, the Moore megacyclothem is merely more highly articulated than are the type cyclothems of Illinois. The Missourian and Virgilian contain 19 or 20 such cyclothems. Harland et al (1982) suggest 10 Myr for this time interval, and the period of the cyclothem would thus appear to be of the order of 500,000 yr: If Milankovitch-related, these cyclothems represent the long cycle of eccentricity, and the up to five limestones within them reflect the minor fluctuations of the 100,000-yr short eccentricity cycle.

Ramsbottom (1979) has recognized a cyclic hierarchy at three levels in the British Carboniferous. The lowest of these levels (his cyclothem) is of the eustatic type, and its period lies in the 200,000–500,000 yr range. His second level (or mesothem) shows a progression of cyclothems from muddy to more sandy. With a timing of 1–3.6 Myr, it falls outside the scale of phenomena considered here.

SUMMARY AND IMPLICATIONS

Oscillatory patterns in stratigraphy show that orbital variations drove pre-Pleistocene climates in various ways. The record is best expressed in lacustrine sediments, which reflect variations in lake level, and in evaporite sediments, which reflect variations in salinity. These are essentially closed systems in which varving provides a time base. Low-latitude evaporites and midlatitude lakes have recorded precessional couplets bundled into sets corresponding to the short cycle of eccentricity as far back as Triassic and Permian time. They have not changed their absolute or relative length appreciably during this interval. The very long lake record of the Eocene Green River Formation also shows the presence of the long eccentricity cycle.

The pelagic-hemipelagic record, a more open system, is also more complex. Oscillations include variations in the proportion of skeletal matter and in degree of oxidation. As in lake sediments, a distinctive bundling of couplets reflects the precession index (precessional signals modulated by the cycles of eccentricity). In addition, there are sequences dominated by the simple obliquity cycle. Deposits of the same stage marched to the precessional cycle in some areas and to the obliquity cycle in others. High-latitude sequences may be more prone to the obliquity cycle and lower latitudes to the precessional cycle, but in some sequences both occur intermixed. In addition, pelagic-hemipelagic sequences may contain segments of highly irregular stratal succession.

The Triassic Alpine platform sequences suggest a Milankovitch-timed eustatic cycle at a time that has generally been considered as free of glaciers, throwing doubts on that dogma. The Late Carboniferous, a time of large-scale glaciation, shows large, rhythmic marine transgressions and regressions, which resulted in the well-known cyclothems. In these, the shorter rhythms (precession, short eccentricity) seem to have been suppressed in favor of the long eccentricity rhythm. This "red shift" seems more pronounced than in the Pleistocene and suggests that Carboniferous climates reacted with greater inertia.

While Milankovitch rhythmicity has been established in principle, the relevant studies have been limited to specific sequences in which rhythmicity is particularly striking. More quantitative approaches are needed.

The basic data need to be collected in the form of quantitative analyses and instrumental scans, particularly of continuous cores. Graphic plots of such information can be compared directly if plotted to the same time scale, but for more complex sequences, the only hope rests in time-series analysis. The main problem with this procedure is the uncertainty of detailed geochronologies in the absence of varves. Bootstrap methods of tuning the time axis to the more obvious cycles and of compositing records to correct for gaps in any one stratigraphic sequence have proved helpful in the Pleistocene. In spectral analysis, Walsh spectra, freed of the requirement of sampling at equal time intervals, may prove advantageous in some cases over Fourier spectra.

For geology, the time of geochronological utility (cyclochronology) may be close at hand. Establishing the length of stages in terms of obliquity cycles or eccentricity bundles would offer a basis for correcting the radiometric time scale, which remains crude at this level. Of particular interest will be the establishment of cyclochronology and magnetic chronology for sequences tied to the seafloor anomalies. P. Cotillon and his coworkers believe that they can correlate general cycle patterns across the Atlantic, but very detailed correlations of cyclicities beyond a single basin have yet to be attained. For sedimentology and historical geology, the identification of orbital rhythms spotlights the ultimate causes of cyclic facies changes and promises to illuminate the long-range behavior of the atmosphere and hydrosphere, in particular the pathways by which signals in the receipt of solar energy are transmitted to depositional systems. It also poses alternatives for the interpretation of local facies. For Earth magnetism, cyclochronology, from varves up to higher levels, provides a basis for studying the rates and patterns of secular variation.

For astronomy and climatology, varve sequences provide longer time series than do tree rings for studying climatic and solar behavior, but such studies have only begun. The 22-yr Hale cycle of solar magnetism is more commonly recorded than is the sunspot cycle. The expression of these cycles through climate remains puzzling, as does the presence of other rhythms in the 20–200 yr domain. Rhythms may also be present between this domain and the Milankovitch frequency band—a terra incognita above the reach of microscopic varve studies and below the resolving power of field stratigraphy. While the Hale, precession, and eccentricity cycles appear to have maintained the same general timing over several hundred million years, the data remain crude: More information may reveal patterns of change. Outside the scale considered here, hints of a 2-Myr rhythm (Ramsbottom's mesothem) raise the possibility of orbital variations that have escaped notice.

Allocyclic sedimentary rhythms—driven by orbital reaction of the Earth with its moon, its sibling planets, and the Sun, and transmitted through climate—are real. They are more complex than imagined by Gilbert, but they pervade the sediments, constituting a frontier for exploring the functioning and history of the Earth and of the solar system.

ACKNOWLEDGMENTS

My work on sedimentary rhythms has been carried on intermittently over 22 years with the help of various grants of the National Science Foundation (Division of Earth Sciences and Office of International Studies). It was also aided by an NSF Postdoctoral Fellowship to the University of Innsbruck, a John Simon Guggenheim Memorial Fellowship, a lectureship to Italian universities by the Consiglio Nazionale delle Ricerche, and a CNR grant to I. Premoli Silva and G. Napoleone. From the many persons who have contributed so much to this effort, I shall only list those most directly involved: Bruno d'Argenio, Isabella Premoli Silva, Giovanni Napoleone, Walther Schwarzacher, Michael Arthur, Lisa Pratt, and Timothy Herbert. Glenn Mason of the University of Wyoming's Western Research Center was particularly helpful in providing data on the Green River Formation. It is a pleasure to thank these organizations and companions-in-research.

Literature Cited

Anderson, E. J., Goodwin, P. W., Sobieski, T. H. 1984. Episodic accumulation and the origin of formation boundaries in the Helderberg Group of New York State. *Geology* 12:120–23

Anderson, R. Y. 1961. Solar-terrestrial climatic patterns in varved sediments. *Ann. NY Acad. Sci.* 95:424–35

Anderson, R. Y. 1964. Varve calibration of stratification. *Kans. State Geol. Surv. Bull.* 169(1):1–20

Anderson, R. Y. 1982. A long geoclimatic record from the Permian. *J. Geophys. Res.* 87:7285–94

Anderson, R. Y. 1984. Orbital forcing of evaporite sedimentation. See Berger et al 1984, 1:147–62

Anderson, R. Y., Kirkland, D. W. 1960. Origin of varves, and cycles of Jurassic Todilto Formation, New Mexico. *Am. Assoc. Pet. Geol. Bull.* 44:27–52

Arthur, M. A., Dean, W. E., Bottjer, D., Scholle, P. A. 1984. Rhythmic bedding in Mesozoic-Cenozoic pelagic carbonate sequences: the primary and diagenetic origin of Milankovitch-like cycles. See Berger et al 1984, 1:191–222

Arthur, M. A., Premoli Silva, I. 1982. Development of widespread organic carbon-rich strata in the Mediterranean Tethys. In *Nature and Origin of Cretaceous Carbon-Rich Facies*, ed. S. O. Schlanger, M. B. Cita, pp. 7–54. New York: Academic. 229 pp.

Beerbower, J. R. 1964. Cyclothems and cyclic depositional mechanisms in alluvial plain sedimentation. *Kans. State Geol. Surv. Bull.* 169(1):31–42

Berger, A. 1978. Long-term variations of caloric insolation resulting from the Earth's orbital elements. *Quat. Res.* 9:139–67

Berger, A., Imbrie, J., Hays, J., Kukla, G., Saltzman, B., eds. 1984. *Milankovitch and Climate*. Dordrecht/Boston/Lancaster: Reidel. 510 pp.

Bradley, W. H. 1929. The varves and climate of the Green River epoch. *US Geol. Surv. Prof. Pap. 158*, pp. 87–110

Bradley, W. H. 1931. The origin of the oil shale and its microfossils of the Green River Formation in Colorado and Utah. *US Geol. Surv. Prof. Pap. 168*. 58 pp.

Bramlette, M. N. 1946. The Monterey Formation of California and its siliceous rocks. *US Geol. Surv. Prof. Pap. 212*, pp. 1–57

Briskin, M., Harrell, J. 1980. Time-series

analysis of the Pleistocene deep-sea record. *Mar. Geol.* 36: 1–22

Cotillon, P., Rio, M. 1984. Cyclic sedimentation in the Cretaceous of Deep Sea Drilling Sites 535 and 540 (Gulf of Mexico), 534 (central Atlantic) and in the Vocontian Basin (France). In *Initial Reports of the Deep Sea Drilling Project*, 77: 339–76. Washington, DC: Govt. Print. Off.

Cotillon, P. 1984. Tentative world-wide correlation of early Cretaceous strata by limestone-marl cyclicities in pelagic deposits. *Bull. Geol. Soc. Den.* 33: 91–102

Crowell, J. C. 1978. Gondwana glaciation, cyclothems, continental positioning and climate change. *Am. J. Sci.* 278: 1345–72

Dean, W. E., Gardner, J. V., Cepek, P., Seibold, E. 1978. Cyclic sedimentation along the continental margin of northwest Africa. In *Initial Reports of the Deep Sea Drilling Project*, 41: 965–89. Washington, DC: Govt. Print. Off.

deBoer, P. L. 1982a. Cyclicity and storage of organic matter in Middle Cretaceous pelagic sediments. See Einsele & Seilacher 1982, pp. 456–75

deBoer, P. L. 1982b. Some remarks about the stable isotope composition of cyclic pelagic sediments from the Cretaceous of the Apennines (Italy). In *Nature and Origin of Cretaceous Carbon-Rich Facies*, ed. S. O. Schlanger, M. B. Cita, pp. 129–43. New York: Academic. 229 pp.

deBoer, P. L., Wonders, A. A. H. 1984. Astronomically induced rhythmic bedding in Cretaceous pelagic sediments near Moria (Italy). See Berger et al 1984, 1: 177–90

de Geer, G. 1940. Geochronologia Suecica, Principles. *K. Sven. Vetensk. Akad. Handl. Stockholm* (3), Vol. 18, No. 6. 360 pp., 60 pls.

de Gracianski, P. C., Auffret, G. A., Dupeuble, P., Mortadert, L., Miller, C. 1979. Interpretation of depositional environments of the Aptian-Albian black shales on the north margin of the Bay of Biscay. In *Initial Reports of the Deep Sea Drilling Project*, 43: 877–907

Droxler, A. 1984. PhD thesis. Univ. Miami, Coral Gables, Fla.

Duff, P. M., Hallam, A., Walton, E. K. 1967. *Cyclic Sedimentation*. Elsevier. 280 pp.

Einsele, G., Seilacher, A., eds. 1982. *Cyclic and Event Stratification*. Berlin/Heidelberg/New York: Springer-Verlag. 536 pp.

Einsele, G. 1982. Limestone-marl cycles (periodites): diagnosis, significance, causes—a review. See Einsele & Seilacher 1982, pp. 8–53

Emiliani, C. 1955. Pleistocene temperatures. *J. Geol.* 63: 538–78

Fischer, A. G. 1964. The Lofer cyclothems of the Alpine Triassic. *Kans. State Geol. Surv. Bull.* 169(1): 107–49

Fischer, A. G. 1980. Gilbert—bedding rhythms and geochronology. In *The Scientific Ideas of G. K. Gilbert, Geol. Soc. Am. Spec. Pap.*, ed. E. Yochelson, 183: 93–104

Fischer, A. G., Herbert, T., Premoli Silva, I. 1985. Carbonate bedding cycles in Cretaceous pelagic and hemipelagic sediments. In *Fine-Grained Deposits and Biofacies of the Cretaceous Western Interior Seaway: Evidence of Cyclic Sedimentary Processes, Soc. Econ. Paleontol. Mineral. Guideb. Ser.*, ed. L. M. Pratt, E. Kauffman, F. Zelt, pp. 1–10

Fischer, A. G., Schwarzacher, W. 1984. Cretaceous bedding rhythms under orbital control? See Berger et al 1984, 1: 163–75

Gardner, J. V., Dean, W. E., Wilcox, C. R. 1984. Carbonate and organic-carbon cycles and the history of upwelling at Deep Sea Drilling Project Site 532, Walvis Ridge, South Atlantic Ocean. In *Initial Reports of the Deep Sea Drilling Project*, 75: 905–21. Washington, DC: Govt. Print. Off.

Gilbert, G. K. 1895. Sedimentary measurement of geologic time. *J. Geol.* 3: 121–27

Hallam, A. 1964. Origin of the limestone-shale rhythm in the Blue Lias of England: a composite theory. *J. Geol.*: 157–68

Harland, W. B., Cox, A. V., Llewellyn, P. G., Pickton, C. A. G., Smith, A. G., Walters, R. 1982. *A Geologic Time Scale*. Cambridge: Cambridge Univ. Press. 131 pp.

Hays, J. D., Imbrie, J., Shackleton, N. J. 1976. Variations in the Earth's orbit: pacemaker of the ice ages. *Science* 194: 1121–32

Heckel, P. H. 1977. Black shale in Pennsylvanian cyclothems. *Am. Assoc. Pet. Geol. Bull.* 61: 1045–68

House, M. R. 1985. A new approach to an absolute timescale from measurements of orbital cycles and sedimentary microrhythms. *Nature* 315: 721–25

Imbrie, J., Imbrie, K. P. 1979. *Ice Ages: Solving the Mystery*. Short Hills, NJ: Enslow. 224 pp.

Imbrie, J., Hays, J. D., Martinson, D. G., McIntyre, A., Mix, A. C., et al. 1984. The orbital theory of Pleistocene climate: support from a revised chronology of the marine delta ^{18}O record. See Berger et al 1984, 1: 269–306

Kauffman, E. G. 1977. Cretaceous facies, faunas and paleoenvironments across the Western Interior Basin. *Mt. Geol.* 14: 75–274

Kennedy, W. J., Odin, G. S. 1982. The Jurassic and Cretaceous time scale in 1981. In *Numerical Dating in Stratigraphy*, ed. G. S. Odin, pp. 556–92. London/New York: Wiley

Libby, L. M. 1983. *Past Climates*. Austin: Univ. Tex. Press. 143 pp.

McCave, I. N. 1979. Depositional features of

organic-carbon-rich black and green mud-
stones at DSDP Sites 386 and 387, western
North Atlantic. In *Initial Reports of the
Deep Sea Drilling Project*, 43:411–16.
Washington, DC: Govt. Print. Off.

Milankovitch, M. 1941. *Kanon der Erdbe-
strahlung und seine Anwendung auf das
Eiszeitenproblem, Akad. Royale Serbe 133.*
633 pp.

Moore, R. C. 1936. Stratigraphic classification of the Pennsylvanian rocks of Kansas.
Kans. State Geol. Surv. Bull. 22:26–35

Moore, R. C. 1949. Divisions of the Pennsylvanian System in Kansas. *Kans. State
Geol. Surv. Bull. 83.* 203 pp.

Olsen, P. E. 1984. Periodicity of lake-level
cycles in the Late Triassic Lockatong
Formation of the Newark Basin (Newark
Supergroup, New Jersey and Pennsylvania). See Berger et al 1984, 1:129–46

Pearson, R. 1978. *Climate and Evolution.*
London/New York/San Francisco: Academic. 244 pp.

Pratt, L. M. 1984. Influence of paleoenviron-
mental factors on preservation of organic
matter in the mid-Cretaceous Greenhorn
Formation, Pueblo, Colorado. *Am. Assoc.
Pet. Geol. Bull.* 68:1146–59

Ramsbottom, W. H. C. 1979. Rates of transgression and regression in the Carboniferous of NW Europe. *J. Geol. Soc. London*
136:147–53

Richter-Bernburg, G. 1963. Solar cycle and
other climatic periods in varvitic evaporites. In *Problems in Paleoclimatology*, ed. A.
E. M. Nairn, pp. 510–21. London/New
York/Sydney: Interscience. 705 pp.

Ricken, W. 1985. Epicontinental marl-limestone alternations: event deposition and
diagenetic bedding (Upper Jurassic, southwest Germany). In *Sedimentary and Evolutionary Cycles*, ed. U. Bayer, A. Seilacher,
pp. 127–62. Berlin/Heidelberg/New York/
Tokyo: Springer-Verlag. 465 pp.

Riveroll, D. D., Jones, B. C. 1954. Varves and
foraminifera of a portion of the upper
Puente Formation (Upper Miocene),
Puente, California. *J. Paleontol.* 28:121–
31

Ross, C. A., Ross, J. R. P. 1985. Late Paleozoic depositional sequences are synchronous and worldwide. *Geology* 13:194–97

Rossignol-Strick, M. 1985. Mediterranean
Quaternary sapropels, an immediate response of the African monsoon to variations of insolation. *Palaeogeogr. Palaeo-
climatol. Palaeoecol.* 49:237–63

Sander, B. 1936. Beiträge zur Kenntnis der
Anlagerungsgefüge (rhythmische Kalke
und Dolomite aus Tirol). *Tschermaks
Mineral. Petrogr. Mitt.* 46:27–209

Schneider, F. K. 1964. Erscheinungsbild und
Entstehung der rhythmischen Bankung
der altkretazischen Tongesteine Nord-

westfalens und der Braunschweiger Bucht.
Fortschr. Geol. Rheinl. Westfalen 7:353–82

Schove, D. J. 1983. *Sunspot Cycles.* Stroudsburg, Pa: Hutchinson-Ross. 397 pp.

Schwarzacher, W. 1948. Über die sedi-
mentäre Rhythmik des Dachsteinkalkes
von Lofer. *Geol. Bundesanst. Wien Verh.*
1947(10–12):175–88

Schwarzacher, W. 1954. Die Grossrhythmik
des Dachsteinkalkes von Lofer. *Tschermaks Mineral. Petrogr. Mitt.* 4(5):44–54

Schwarzacher, W. 1975. *Sedimentation Models and Quantitative Stratigraphy.* Elsevier. 377 pp.

Schwarzacher, W., Fischer, A. G. 1982.
Limestone-shale bedding and perturba-
tions of the Earth's orbit. See Einsele &
Seilacher 1982, pp. 72–95

Seibold, E. 1952. Chemische Untersuch-
ungen zur Bankung im unteren Malm
Schwabens. *Neues Jahrb. Geol. Palaeontol.
Abh.* 95:337–70

Seibold, E., Wiegert, R. 1960. Untersuchung
des zeitlichen Ablaufs der Sedimentation
im Malo Jezero (mittl. Adria) auf Period-
izitäten. *Z. Geophys.* 26:87–103

Shanmugam, G., Moiola, R. J. 1982. Eustatic
control of turbidites and winnowed turbidites. *Geology* 10:231–35

Soutar, A., Crill, P. A. 1977. Sedimentation
and climatic patterns in the Santa Barbara
Basin during the 19th and 20th centuries.
Geol. Soc. Am. Bull. 88:1161–72

Soutar, A., Isaacs, J. D. 1969. History of fish
populations inferred from fish scales in
anaerobic sediments off California. *Calif.
Mar. Res. Comm. CalCOFI* 13:63–70

Sujkowski, Z. L. 1958. Diagenesis. *Am. Assoc.
Pet. Geol. Bull.* 42:469–72

Surdam, R. C., Stanley, K. O. 1978. Lacustrine sedimentation during the culmi-
nating phase of Eocene Lake Gosiute,
Wyoming (Green River Formation). *Geol.
Soc. Am. Bull.* 90:93–110

Thomel, G. 1976. *Compte-rendu des excur-
sions dans le Sud Est de la France et
annexes; Journées nicoises de biostrati-
graphie.* Centre d'études Mediterr.
Muséum d'Histoire Naturelle, Nice, p. 11,
27–41

Tissot, B., Deroo, G., Herbin, J. P. 1979.
Organic matter in the Cretaceous sedi-
ments of the North Atlantic. In *Deep
Drilling Results in the Atlantic Ocean:
Continental Margins and Paleoenviron-
ments, Maurice Ewing Ser.*, ed. M. Talwani,
W. W. Hay, W. B. F. Ryan, 3:362–74.
Washington, DC: Am. Geophys. Union

Trudell, L. G., Beard, T. N., Smith, J. W. 1970.
Green River Formation lithology and oil
shale correlations in the Piceance Creek
Basin, Colorado. *US Bur. Mines Rep.
Invest. 7357.* 212 pp.

Trudell, L. G., Smith, J. W., Beard, T. N.,

Mason, G. M. 1983. Primary oil-shale resources of the Green River Formation in the eastern Uinta Basin, Utah. *Dep. Energy Rep. DOE/LC/RI-82-4 (DE83009081)*. 58 pp.

Udden, J. A. 1924. Laminated anhydrite in Texas. *Geol. Soc. Am. Bull.* 35: 347–54

Van Graas, G., Viets, T. C., de Leeuw, J. W., Schenck, P. A. 1983. A study of the soluble and insoluble organic matter from the Livello Bonarelli, a Cretaceous black shale deposit in the central Apennines (Italy). *Geochim. Cosmochim. Acta* 47: 1051–59

Van Houten, F. B. 1964. Cyclic lacustrine sedimentation, Upper Triassic Lockatong Formation, central New Jersey and adjacent Pennsylvania. *Kans. State Geol. Surv. Bull.* 169(2): 497–31

Vernekar, A. 1972. Long-period global variations of incoming solar radiation. *Meteorol. Monogr.*, Vol. 12, No. 34

Wanless, H. R., Shepard, F. P. 1936. Sea level and climatic changes related to late Paleozoic cycles. *Geol. Soc. Am. Bull.* 47: 1177–1206

Wanless, H. R., Weller, J. M. 1932. Correlation and extent of Pennsylvanian cyclothems. *Geol. Soc. Am. Bull.* 43: 1003–16

Zeuner, F. E. 1952. *Dating the Past*. London: Methuen. 495 pp., 24 pls. 3rd ed.

Ann. Rev. Earth Planet. Sci. 1986. 14 : 377–415

TEMPERATURE DISTRIBUTION IN THE CRUST AND MANTLE

Raymond Jeanloz

Department of Geology and Geophysics, University of California, Berkeley, California 94720

S. Morris

Department of Mechanical Engineering, University of California, Berkeley, California 94720

1. INTRODUCTION

Understanding the temperature distribution within the Earth has been a central goal of geological research for over a century. Petrological studies, for example, are carried out in order to determine the temperatures and pressures at which igneous and metamorphic rocks formed (or last equilibrated). Tectonics and structural geology, in turn, are concerned with the consequences of the interior temperature distribution, such as the large-scale deformations suffered by the crust over geological time scales. The sources of energy that heat the planetary interior must be established from geochemical investigations of trace-element concentrations in rocks and from broadly based studies of the way in which the Earth was initially formed. Finally, the mechanisms by which heat is transferred out of the interior are subjects of geophysical observations (e.g. heat-flow measurements) and modeling.

A major theme of this article is that the temperature distribution in the Earth can only be understood to the extent that the dynamics of the planetary interior (the sources, sinks, and transport mechanisms of thermal energy) are well characterized. Hence, observational constraints on the interior temperature distribution can yield powerful insights into the

377

0084–6597/86/0515–0377$02.00

thermal and dynamic evolution of the Earth. For example, the high temperatures of the mantle make advection (convective transport) the dominant mechanism of global heat transfer. High temperatures lead to fluidlike behavior of the mantle over geological times because the effective viscosity of rock is extremely sensitive to temperature. Consequently, our planet is characterized by vigorous geological activity—convection in the mantle and plate tectonics at the surface. In contrast, terrestrial planets in which conduction of heat is likely to play a more dominant role, such as the Moon or Mercury, exhibit little recent geological activity at the surface.

After reviewing the major experimental constraints involved, we conclude that the average temperature as a function of depth is well known throughout the crust and mantle. In order to interpret the basic form of the geotherm, we consider the two dominant mechanisms by which heat is transported in the Earth: conduction through the rock and advection by convective flow. Although advection is the primary mode of outward heat transfer on a global scale, vertical conduction becomes important wherever the convective flow is mainly horizontal rather than vertical. At the surface, for example, vertical displacements are negligible in comparison with the large-scale horizontal motions of plate tectonics. Most of the temperature increase with depth (~ 2–20 K km^{-1}) occurs in these regions (the thermal boundary layer) in which vertical heat transfer is dominantly by conduction. Where the more efficient process of advection takes precedence in moving heat upward (that is, throughout most of the interior of the convecting mantle), the average temperature changes little with depth (less than 0.5 K km^{-1}).

Modeling of heat transfer in the top thermal boundary layer leads to detailed estimates of geotherms through the continental and oceanic lithosphere. Thus, we confirm that the temperature distributions through continental lithosphere and through oceanic lithosphere more than 60 Myr old are virtually indistinguishable. Geological observations on rates of erosion at the surface demonstrate, however, that crustal uplift can play an important role in modifying continental geotherms.

We touch on only a few conclusions regarding the thermal evolution of the Earth's interior, a topic of current research. Among the most important findings is that the average temperature through the Archean crust and mantle was similar to that at present. Four billion years ago, the average temperature of the upper mantle was probably slightly higher than now, but by no more than 200–300 K. For comparison, the lateral variation of temperature at any given depth in the upper mantle has decreased from about 1600 K 3.5 Gyr ago to about 1200 K (the estimated temperature difference beneath ridges and trenches) at present. We emphasize that current limitations in understanding the constitution of the lower mantle result in significant uncertainties in the thermal response time of the planetary interior. Hence, further elucidation of the Earth's thermal evolution depends in large part on unraveling the composition and physical properties of the deep mantle.

2. EXPERIMENTAL CONSTRAINTS ON THE GEOTHERM IN THE CRUST AND MANTLE

The current, horizontally averaged temperature as a function of depth (or *geotherm*) in the crust and mantle is largely inferred from geophysical and petrological observations. These observations are interpreted on the basis of laboratory studies of both solid-state and melting equilibria, with high-pressure experiments playing an especially significant role in analyzing the major features of the Earth's deep structure as derived from seismology. Hence, the solidity of the inner core and of the lower mantle, as well as the pressure at the 400-km seismological discontinuity, provides the strongest constraint on the temperature throughout the deep mantle. In contrast, we do not attempt to use the 670-km discontinuity or the seismologically anomalous D'' region at the base of the mantle to infer temperatures, because these features are not well understood at present (Jeanloz & Thompson 1983, Lay & Helmberger 1983). For the outer regions of the Earth, the seismological observation that the upper mantle is dominantly solid is complemented by petrological data and heat-flow measurements in deriving the temperature distribution.

The major experimental constraints that are currently available for the geotherm are summarized in Figure 1 and are detailed in the following paragraphs. In this section, we emphasize observational data, putting off theoretical considerations of the geotherm to Section 3. Nevertheless, for the sake of comparison, some plausible estimates of the geotherm are included in Figure 1, with uncertainties probably being less than 1000–1500 K throughout most of the mantle. Actually, this is not a large uncertainty, because the lateral variation of temperature is likely to be of comparable magnitude, especially in the upper mantle and crust.

Temperature in the Outer Core

The size, structure, density, and elastic properties of the Earth's core are determined from seismological observations (Jacobs 1975, Dziewonski & Anderson 1981). These observations demonstrate that the core, which extends 3490 km (or about half the radial distance from the Earth's center), is significantly denser than the surrounding mantle. Iron is considered to be the primary constituent of the core because it is by far the most abundant element in the cosmos with elastic properties and density approaching

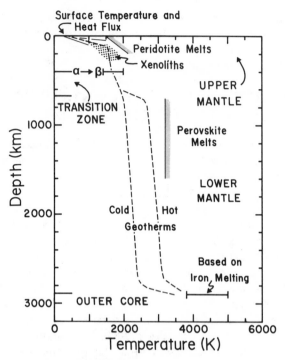

Figure 1 Summary of experimental constraints on the present, horizontally averaged temperature as a function of depth (geotherm) through the crust and mantle. Temperatures are relatively well determined in the crust and upper mantle, with a characteristic geotherm being illustrated by the dashed line in this region. Differences between geotherms beneath continents and geotherms beneath ocean basins older than 70 Myr are not evident at this scale (see Figure 8). Two possible geotherms are shown for the lower mantle ("hot" and "cold" dashed lines), reflecting greater uncertainties in the average temperature at these depths (as discussed in Section 4). The basic form of the geotherm, consisting of adiabatic regions in which the temperature only increases slightly with depth and of thermal boundary layers in which large increases in temperature occur over a few hundred kilometers depth interval, is discussed in Section 3.

those of the core at the corresponding pressures and temperatures (Birch 1952, 1964, Brett 1976, Stevenson 1981). In addition, about 10 wt% of a lighter component is required in the iron to match the observed properties of the core, with sulfur and oxygen currently favored as the dominant alloying components. Note that an iron-rich alloy satisfies the independent constraint that the core must consist of a metal in order that the Earth's magnetic field can be produced by a dynamo mechanism (Merrill & McElhinny 1983). In contrast, the mantle surrounding the core is known to be a relatively poor electrical conductor, and its seismic properties match those of silicates and oxides (Birch 1964, Banks 1972, Anderssen et al 1979, Jeanloz & Thompson 1983).

The core consists of a molten region (outer core having zero rigidity), which is sandwiched between the solid mantle and inner core (1220 km radial extent). Therefore, the temperature at the top of the outer core must exceed the melting point of the iron alloy comprising the core while being less than the temperature required to melt the silicates and oxides of the lower mantle. An estimate of 2800 K for the minimum temperature that would satisfy these conditions has been previously derived from an extrapolation of relatively low-pressure data on the melting of iron alloys (Jeanloz & Richter 1979; cf. Birch 1972).

More recently, Brown & McQueen (1980, 1982) and Brown et al (1984) have increased this estimated lower bound by demonstrating that both iron sulfide and iron melt above 4000 K at pressures of 150–250 GPa (1.5–2.5 Mbar) in shock-wave experiments. These experiments involve impacting a sample with a projectile traveling at a velocity of several kilometers per second, with the result that a temperature rise of several thousand kelvins accompanies the pressure pulse (shock wave) that traverses the sample. From their data, Brown and coworkers deduce that FeS and Fe melt at about 5000 and 5800($\pm$500) K, respectively, at the pressure of the inner core–outer core boundary (330 GPa). These melting temperatures are uncertain, in part because the shock temperatures are not directly measured but must be calculated. Nevertheless, the values provide a good estimate for the range of acceptable temperatures at the solid-liquid interface between the inner and outer core, and the effects of plausible alloying components other than sulfur being present are not likely to change this conclusion (Jeanloz 1986).

The temperature distribution throughout the core can be derived by noting that the generation of the magnetic field requires that most or all of the outer core is convecting vigorously (e.g. Merrill & McElhinny 1983). As discussed below, the geotherm within a convecting region follows an adiabat, so that the temperature at the top of the core can be estimated by an adiabatic extrapolation of the values inferred for the inner core–outer core boundary. Such an extrapolation is well constrained by existing high-pressure data, and a minimum temperature of 4400($\pm$600) K has been determined in this way for the top of the outer core (Brown & McQueen 1982, Spiliopoulos & Stacey 1984, Jeanloz 1986).

Temperature in the Lower Mantle

The dominant minerals of the upper mantle—olivine [$(Mg,Fe)_2SiO_4$], pyroxene [$(Mg,Fe,Ca)SiO_3$], and garnet [$(Mg,Fe,Ca)_3Al_2Si_3O_{12}$]— all transform to the silicate-perovskite phase of composition

$(Mg,Fe,Ca)(Si,Al)O_3$ (or to a perovskite and oxide assemblage) when taken experimentally to the pressures and temperatures of the lower mantle (see Jeanloz & Thompson 1983). In the upper mantle these minerals occur as garnet-bearing peridotite, which consists of about 50–65% olivine by volume, along with 20–40% pyroxene (usually both clinopyroxene and orthopyroxene, as described below), and less than 20% garnet and other minor phases. (For our purposes, more detailed distinctions between rock types, such as lherzolite or harzburgite, are unnecessary.) At the pressures of the lower mantle, however, this peridotite transforms to a rock that consists of about 70–90% silicate perovskite. Therefore, the high-pressure perovskite phase is considered to be the most abundant mineral of the Earth's lower mantle, and a measurement of its melting point at elevated pressures should provide an upper limit on the geotherm at depth.

Recent experiments employing continuous laser heating into a diamond-anvil cell demonstrate that silicate perovskite can be formed and then observed to melt at pressures exceeding 20 GPa (Jeanloz & Heinz 1984). The occurrence of melting is confirmed by observing the presence of glass at high pressures, after the sample is quenched to room temperature by turning off the heating laser. Heinz & Jeanloz have bracketed the melting point of $(Mg_{0.9}Fe_{0.1})SiO_3$ perovskite between 25 and 60 GPa in this way, and they find a value of $2900(\pm 300)$ K independent of pressure (D. L. Heinz, unpublished work; Jeanloz & Heinz 1984). Possible sources of systematic error would place the true melting temperature in the upper range of the quoted uncertainty, so a value of 3200 K represents a good estimate for the maximum temperature in the upper half of the lower mantle.

400-km Discontinuity

The seismologically observed discontinuity of elastic properties at 400 km depth can be readily explained in terms of the transformation of olivine to its high-pressure spinelloid polymorph, β-phase (Akimoto et al 1976, Jeanloz & Thompson 1983, Weidner et al 1984, Weidner 1985). The assumption, based on petrological studies of the upper mantle (e.g. the mineralogical content of xenoliths), is that olivine is a major phase of the upper mantle (Ringwood 1975, Yoder 1976, Aoki 1984). As shown by Akaogi & Akimoto (1979), for example, the olivine transformation is clearly evident when candidate upper-mantle rocks such as garnet peridotite are experimentally taken to high pressures. From the detailed work on olivine summarized by Akimoto et al (1976), a sudden reaction from the assemblage of olivine (α-phase) plus minor amounts of the high-pressure γ-spinel phase to β-phase plus γ-spinel is expected to occur at pressures corresponding to the 400-km discontinuity and at a temperature of $1700(\pm 300)$ K (Jeanloz &

Thompson 1983). The geotherm temperature at the top of the transition zone is therefore estimated to lie between 1400 and 2000 K (Figure 1).

Xenoliths and Peridotite Melting

Xenoliths (peridotite and other rock fragments brought up volcanically from the mantle) contain direct information on the minerals and temperatures that occur at depths as great as 200–250 km. In particular, explosive eruptions such as kimberlites provide a geologically rare but important sampling of the upper mantle (Dawson 1980). Pressures and temperatures at which the xenoliths last equilibrated can be inferred from the compositions of coexisting minerals. It must be assumed, of course, that the minerals did achieve chemical equilibrium at some point. There is good evidence, however, that eruption is sufficiently rapid that the mineral compositions achieved at depth are essentially frozen-in during ascent. For example, the presence of diamond in xenoliths implies rapid extraction from depths exceeding 150 km because diamond is unstable relative to graphite at pressures less than 5 GPa. [The transformation of diamond to graphite is rapid at elevated temperatures (Kennedy & Kennedy 1976, Wakatsuki & Ichinose 1982).]

Among the primary indicators of temperature is the two-pyroxene geothermometer, which is based on the fact that pyroxenes of composition between $Mg_2Si_2O_6$ and $CaMgSi_2O_6$ unmix in the solid state below $1500°C$. Two separate pyroxene minerals are formed: enstatite, of orthorhombic structure and composition near $MgSiO_3 (= Mg_2Si_2O_6)$, and diopside, which is monoclinic and has a composition near $CaMgSi_2O_6$. Small amounts of calcium and magnesium can enter the enstatite (orthopyroxene) and diopside (clinopyroxene) structures, respectively. Diopside coexisting with enstatite becomes increasingly magnesium rich with increasing temperature, and because this effect has been experimentally calibrated, a measurement of the composition of diopside in a xenolith can be directly translated into a temperature (Boyd 1973).

Similarly, experiments demonstrate that increasing (but minor) amounts of aluminum can be included into the enstatite structure with increasing pressure if the pyroxene is in equilibrium with $Mg_3Al_2Si_3O_{12}$ garnet (MacGregor 1974). The experimental calibration of this geobarometer allows the Al content of enstatites to be used to determine the pressure, and hence the corresponding depth of equilibration, for many xenoliths.

In detail, there are some difficulties in applying these thermobarometers. For example, both pressure and temperature affect the pyroxene compositions in interrelated ways, and the effects of minor components present in the natural samples are not well understood (e.g. Mercier & Carter 1975). Nevertheless, recent comparisons of several geothermometer/geobarometer calibrations indicate that pressure and temperature estimates sufficiently reliable for our purposes can be made for a large number of xenoliths (Finnerty & Boyd 1984, Nickel & Green 1985). As shown in Figure 1, temperatures of 1000–1800 K are found at pressures corresponding to depths of 100–250 km.

Consistent with this range of temperatures is the independent petrological constraint provided by the fusion curve of peridotite. There is no evidence for there being more than a few percent partial melt in the upper mantle, and there is reason to believe that no larger amount can be retained in the mantle over geological time periods (Shankland et al 1981, Dziewonski & Anderson 1981, Stolper et al 1981). As peridotitic rocks are thought to make up the bulk of the upper mantle, the solidus (or at most, the temperature for 10–20% partial melting) of dry peridotite provides a maximum bound on the geotherm in the upper mantle, as shown in Figure 1 (Ringwood 1975, Yoder 1976, Basaltic Volcanism Study Project 1981, Nickel & Green 1985). Similarly, lava temperatures at the Earth's surface are typically found to be less than 1250°C. These values undoubtedly reflect the adiabatic upwelling of hot regions in the mantle and are therefore compatible with the upper bound to the geotherm that is provided by the melting of peridotite.

Surface Temperature and Heat Flux

Temperatures of 1000–2000 K at 100 km depth are also consistent with measured values of the increase of temperature with depth [~ 10–20 K km^{-1} is a typical range near the surface (Oxburgh 1980, Sclater et al 1980)]. As discussed below, these values correspond to conductive surface heat fluxes of about 25–70 mW m^{-2} for observed thermal conductivities of rock ($k \sim 2.5$–3.5 W K^{-1} m^{-1}). The high value at the surface for the temperature gradient with depth, combined with the constraints on mantle temperatures below 100 km depth, requires that there be a sharp bend in the geotherm near the top of the mantle, as shown in Figure 1 (e.g. Turcotte & Schubert 1982). This bend is identified with a transition from relatively inefficient, conductive heat transfer at the surface to relatively efficient, advective heat transfer at depth.

3.   HEAT TRANSFER MECHANISMS

Three mechanisms of heat transfer need be considered: conduction, radiation, and advection (see, for example, Carslaw & Jaeger 1959, Turcotte & Schubert 1982). Of these, only the processes of conduction and radiation can transmit heat from one piece of matter to another. In contrast, advection results from the movement of a block of matter from one place to another: Heat is thus transported by a convective flow. As we shall see, heat is transferred within the mantle essentially by the convective flow alone.

Although radiative transport was once thought to be of dominant importance at high temperatures in the mantle, spectroscopic investigations demonstrate that mantle minerals are too opaque for radiative transfer to be effective (e.g. Shankland et al 1974, 1979, Basaltic Volcanism Study Project 1981). Instead, in this optically thick ("opaque") limit, both radiative and lattice conduction can be combined into a single, effective thermal conductivity that is only slightly dependent on temperature (see Clark 1957, Shankland et al 1974, 1979, Basaltic Volcanism Study Project 1981). One location at which heat can only be transmitted upward by conduction and radiation is the Earth's surface: Since no matter crosses the surface, vertical advection does not occur.

We turn now to heat transfer by advection. The essential significance of the high temperatures in the mantle (Figure 1) is that the effective viscosity falls rapidly with increasing temperature. Thus, rather than the elastic and brittle behavior characteristic of the crust, the hot interior of the planet deforms by viscous creep (Weertman & Weertman 1975, Schubert 1979). The high internal temperatures ultimately lead to the occurrence of convection within the planet, and hence to geological activity near the surface (cf. Tozer 1965, 1972).

The effect of temperature on rheological properties can be illustrated in terms of the Maxwell relaxation time (Figure 2):

$$\tau_M = \mu/\eta. \tag{1}$$

This ratio of the elastic shear modulus (μ) and the effective viscosity (η) defines a characteristic time for the deformation of materials: The response to a given shear stress is elastic over time periods much shorter than τ_M and viscous over time periods much longer than τ_M. The relaxation time is sensitive to temperature by way of the temperature dependence of the viscosity,

$$\eta = \eta_0 \exp(Q/kT). \tag{2}$$

Equation 2 describes the experimentally observed creep-deformation rates of rocks and minerals, with the activation energy Q being about 400 kJ mol^{-1} (e.g. Kirby 1983, Poirier 1985). For the present purposes, η_0 and μ can be taken to be constants (k is the Boltzmann constant). Since the bulk of the mantle is at temperatures of about 2000 K, it clearly can behave like a viscous fluid, even over geologically short time periods (Figure 2). Vertical rebound of the Earth's surface over the last 10,000 yr, caused by the melting of the large Pleistocene ice sheets, is taken as direct observational evidence for the viscous response of the mantle to unloading of the crust (e.g. Peltier 1981).

That the high internal temperature and associated low viscosity lead to vigorous convection is demonstrated by considering the Rayleigh number, which is a quantitative measure of the intensity of convection:

$$\mathrm{Ra} = \frac{\alpha g \rho \Delta T D^3}{\kappa \eta}. \tag{3}$$

The Rayleigh number is simply a dimensionless ratio of buoyancy forces driving convection to the dissipative effects opposing convection: α, g, ρ, κ, and D are the volume coefficient of thermal expansion, gravitational

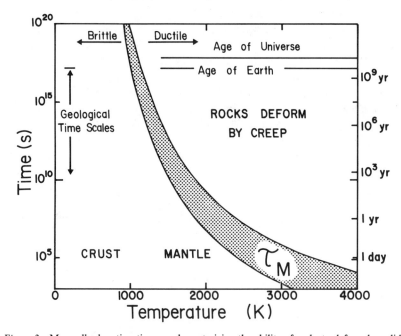

Figure 2 Maxwell relaxation time τ_M, characterizing the ability of rocks to deform by solid-state creep, is shown as a function of temperature on a semi-logarithmic scale. For a given shear stress, τ_M is the ratio of nonrecoverable (creep) to recoverable (elastic) deformation that is achieved over a time period shown along the vertical axis (cf. Equation 1). The curve is based on experimentally measured deformation properties, and its width indicates the variability and uncertainty in the rheological properties of rock. The temperature dependence of τ is given by the activation energy Q (Equations 1 and 2). The preexponential factor τ_0 can be determined from experimental data or from the statistical mechanical relation $\tau_0 = h/kT$, the ratio of the Planck constant to the product of the Boltzmann constant and the temperature (Christian 1975).

acceleration, density, thermal diffusivity, and depth extent of the fluid, respectively. The temperature difference between the top and bottom surfaces of the fluid is given by ΔT. Thus, the internal temperature enters into the Rayleigh number both through ΔT (with the surface of the Earth, the top of the convecting fluid, being at $0°C$) and through the viscosity. As is apparent from Figure 2, the latter effect of temperature is especially important for showing that the Rayleigh number, and hence the strength of convection, increases rapidly with increasing temperature.

Rayleigh numbers that might be achieved in the mantles of terrestrial planets are illustrated in Figure 3, which summarizes the dependence of Ra on internal temperature and depth D. For temperatures sufficiently high that Ra exceeds the critical value of approximately 1000, the buoyancy forces overcome the resistance to convection (e.g. Turcotte & Schubert 1982). According to the results shown in the figure, temperature is the

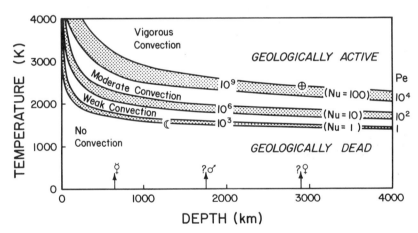

Figure 3   Contours of Rayleigh number as a function of internal temperature and depth extent (Equations 2, 3). Properties appropriate for terrestrial mantles are assumed, and shading indicates the uncertainties in the contour locations. Below the critical Rayleigh number (about 10^3), convection is absent, and increasingly vigorous convection is associated with increasing Rayleigh number above this value. Corresponding values of Nusselt number (Nu) and Peclet number (Pe) are shown toward the right. As described in the text, Pe and Nu describe the effectiveness of heat transfer associated with convection (Equations 4, 9, and 11). Because of the strong dependence of viscosity on temperature (Equation 2), planets with high internal temperatures, like the Earth, are characterized by vigorous convection of the interior and hence by near-surface geological activity. Lower temperatures for the lunar interior (Hood et al 1982, Sonett 1982) suggest that convection is less significant in the Moon, which is consistent with the lack of geological activity over the last 2–3 Gyr (Taylor 1982). Internal temperatures for Venus, Mars, and Mercury are unknown, although surface features suggest that these range from relatively active to inactive, respectively (Basaltic Volcanism Study Project 1981, Phillips & Malin 1983).

dominant factor determining the intensity of convection, with depth playing a secondary role. Hence, a strong correlation is expected between geological activity (the observable, near-surface manifestation of underlying convection) and the internal temperatures of terrestrial planets. For comparison, the Rayleigh number of the Earth's mantle is estimated to lie between 10^5 and 10^9, which implies that convection is vigorous.

That there is convection, at least in the upper portions of the Earth's mantle, is confirmed by the direct observation of plate tectonic motions. Large-scale relative displacements of lithospheric plates are associated with the formation of new crust at midocean ridges and with the sinking of lithosphere into the mantle at subduction zones. That is, conservation of mass implies that the horizontal movement of plates at velocities $u \sim 10^{-2}$ m yr^{-1} is associated with upwelling and downwelling motions within the mantle underlying ridges and trenches (cf. Stevenson & Turner 1979).

What is of particular importance for this discussion is that the rates of internal convection and the associated plate tectonic velocities at the surface are significant when it comes to the transfer of heat within the Earth. This conclusion is demonstrated by the fact that the Peclet number

$$\text{Pe} = \frac{lu}{\kappa} \tag{4}$$

has a value much greater than 1 for distances (l) exceeding 10^3 m [the thermal diffusivity $\kappa = 10^{-6}$ m^2 s^{-1} (e.g. Jeanloz & Thompson 1983)]. The Peclet number is the ratio of conductive to advective time scales required to transport heat a distance l:

$$\tau_{\text{conduction}} = \frac{l^2}{\kappa}, \tag{5}$$

$$\tau_{\text{advection}} = \frac{l}{u}. \tag{6}$$

Thus, convection (large-scale mass transfer) is strong enough in the Earth that the resulting advection of heat is far more effective on a global scale than is heat transfer by conduction (Figure 4): for a distance $l = 1000$ km and a plate tectonic velocity $u \cong 10^{-2}$ m yr^{-1}, the Peclet number is about 300. In short, the bulk of the planet could not be cooled by conduction even over the age of the Universe. Only for distances less than about 10 km is conduction competitive with advection at plate tectonic velocities (Pe $\lesssim$ 1). Because of the importance of advection to large-scale heat transfer in the Earth, we consider next the temperature distributions associated with convective flow.

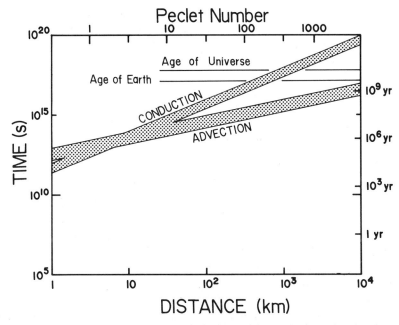

Figure 4   Relative efficiencies of heat transfer in the Earth by conduction or by advection are summarized in terms of the time required to transmit a thermal perturbation over a given distance (logarithmic scale; see Equations 5 and 6). The corresponding Peclet number (Equation 4) is given on the upper scale. For this figure, convective velocities are assumed to be between 1 mm yr^{-1} and 1 cm yr^{-1}, and the conduction length scale is taken to be between $\sqrt{\kappa t}$ and $\sqrt{4\kappa t}$, where t and κ ($\kappa = 10^{-6}$ m^2 s^{-1}) are time and thermal diffusivity, respectively (see Carslaw & Jaeger 1959).

Advection: Mantle Geotherms

The geotherm through the interior of a convecting region is constructed by assuming that outside the thermal boundary layers, the temperature increases isentropically with depth. The assumption is valid if (*a*) there are no entropy sources in the fluid and (*b*) any initial entropy variations have had time to decay. Condition (*a*) is met because the Peclet number is large and viscous dissipation is relatively small. A large Peclet number for the Earth's interior means that a parcel of rock in the flow field of the mantle moves approximately adiabatically (without loss of heat by conduction or radiation). Radioactive decay violates (*a*), but numerical calculations (e.g. McKenzie et al 1974) show that the effect is to make the temperature gradient slightly less than adiabatic, as described below.

   In contrast, it is much harder to say if condition (*b*) is met in the Earth. Numerical calculations of single-layer incompressible convection (Daly

1980) suggest that the flow approaches steady state on a time scale given in Equation 16 below. At this stage, we can only assume that the adjustment time for the *compressible* flow is of the same order. We shall see in Section 4 that this adjustment time is shorter than the age of the Earth if there is a single layer of convection in the mantle: Then (*b*) is valid, and the mantle should be adiabatic. If there are two layers of convection, however, the lower mantle may adjust on a time scale longer than the age of the Earth. Under this circumstance, initial deviations from adiabaticity may still appear in the mantle geotherm.

Whether the lower mantle is adiabatic or not is in principle testable by seismology; current interpretations of the data are in accord with an adiabatic geotherm (Dziewonski & Anderson 1981). We therefore assume that entropy is constant with depth through the interior (cf. Jarvis & McKenzie 1980). (Ignoring the effects of local viscous dissipation, we need not distinguish between adiabatic and constant-entropy conditions.) It is only the superadiabatic (greater than adiabatic) temperature increase with depth that drives the convective flow, however: The adiabatic increase in temperature with depth is not included in the potential temperature ΔT of Equation 3.

One way of concentrating on just the temperatures that are dynamically important is to exclude the effects of compression. Thus, fluid dynamical calculations pertaining to the mantle are typically carried out assuming that the fluid is incompressible (e.g. McKenzie et al 1974). Because entropy is thermodynamically a function of only volume and temperature (for a given composition and phase), entropy is uniquely a function of temperature when no compression is allowed for [i.e. at constant volume (the effects of phase transitions are considered below)]. The consequence is that temperature is constant along an adiabat, and only the nonadiabatic temperatures that drive the flow (the potential temperatures) are considered. To translate the incompressible-flow temperatures to absolute values of temperature, one need only add the adiabatic increase of temperature with depth.

The temperature distribution for an incompressible convecting fluid is summarized in Figure 5. Because the Earth's mantle is thought to be partially heated from below and partially from within, it lies between the two end-member situations that are illustrated. Although estimates of the relative internal heating (mainly by radioactive decay) and heating from the underlying core are uncertain, current thinking is that about one fourth of the total rate of heat loss observed at the surface originates in the core (e.g. Verhoogen 1980). For example, Jeanloz (1983) quotes high-pressure data on iron alloys from which the thermal conductivity at the conditions of the outermost core can be estimated. A minimum value of the heat flux out of the core can then be deduced if one ignores advective heat transfer in the core, and the result corresponds to about one fourth the total heat loss out of the Earth. Alternatively, models of the energy requirements and efficiency of the magnetic dynamo suggest that the heat loss from the core is about 8–25% of the total surface loss (Gubbins & Masters 1979, Verhoogen 1980). Note that if there is no thermal boundary layer at the base of the mantle, the lowest temperature estimated for the top of the core (Figure 1) extrapolates adiabatically to a value exceeding the solidus of perovskite from 1000 to 1600 km depth. (The extrapolation of the temperature is illustrated below in Figure 6.) This implies that there must be a thermal boundary layer at the base of the mantle, and therefore some heat must be entering the mantle from the core.

As pointed out above, the fluid convects when it is subjected to a sufficiently large potential temperature difference ΔT in Figure 5 (thin,

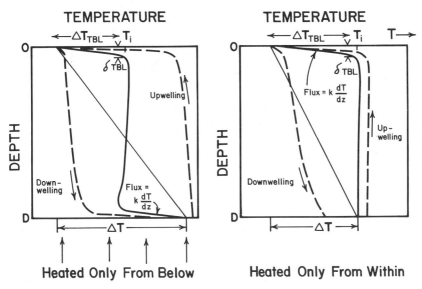

Figure 5   Schematic illustration of the temperature distribution in a vigorously convecting fluid that is either heated fully from below (left panel; temperature at base is higher than that at top by ΔT) or fully heated from within (right panel; rate of heat production given by $H = k\Delta T/D^2$, where k is the thermal conductivity and D is the depth extent of the fluid). The horizontally averaged temperature is shown by the solid curve, whereas temperatures in the upwelling and downwelling regions are illustrated by the dashed curves. The relationship of the properties of the thermal boundary layers [such as the thickness (δ_{TBL}) and temperature offset (ΔT_{TBL})] to the heat flux across top and bottom surfaces and to the average interior temperature of the fluid (T_i) are discussed in the text. This figure summarizes results from the work of McKenzie et al (1974) on incompressible fluids of constant viscosity with Pe $\gg$ 1 and Re $= uD\rho/\eta \ll 1$.

straight line with depth) to ensure that the Rayleigh number exceeds the critical value of about 10^3. For the two cases shown, ΔT is defined either by the temperature difference imposed from below or, in the case of uniform internal heating at the rate H, by the equivalent potential temperature

$$\Delta T = \frac{HD^2}{k}, \tag{7}$$

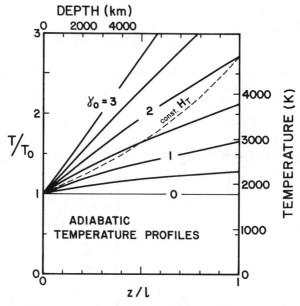

Figure 6 Isentropic temperature distributions (equilibrium adiabats) as functions of depth z (solid curves). The temperature and depth are nondimensionalized by the uncompressed (zero-pressure) temperature T_0 and the length scale over which density increases significantly $[l \equiv (\partial \ln \rho / \partial z)_s^{-1} = K_s / \rho g]$. The adiabatic bulk modulus (K_s), density (ρ), and gravitational acceleration (g) are taken from seismological models of the Earth's internal properties (Dziewonski & Anderson 1981). Corresponding dimensionalized values appropriate for the mantle are shown on the upper and right-hand scales. The relative change in adiabatic temperature with compression is given by the nondimensional, thermodynamic Grüneisen parameter $\gamma = (\partial \ln T / \partial \ln \rho)_s = \alpha K_s / \rho C_P$, which is assumed to decrease upon compression according to $\gamma = \gamma_0 (\rho_0 / \rho)$. (See Birch 1952, 1968, Jeanloz & Richter 1979; α and C_P are the volume coefficient of thermal expansion and specific heat at constant pressures, respectively, and subscript zero indicates zero-pressure conditions.) Thus, the isentropic temperature increases are proportional to γ_0, as shown; values between 1 and 2 are appropriate for the zero-pressure Grüneisen parameter of the mantle (Jeanloz & Thompson 1983, Knittle et al 1985). For contrast, the dashed curve shows the isentrope under the assumption that it is given by a constant scale height $H_T = (\partial \ln T / \partial z)_s^{-1}$, as is often done in fluid mechanics (e.g. Jarvis & McKenzie 1980). Although the approximation of a constant scale height does not exactly yield an isentrope, the deviation is small in the present application.

as shown in the figure. [Here $k = \kappa \rho C_P$ is the thermal conductivity, with C_P being the specific heat (Turcotte & Schubert 1982).]

Once active convection is underway, the original temperature difference ΔT is redistributed to occur in the conductive thermal boundary layers (Figure 5). In these regions, upward transfer of heat by advection is no longer effective because the vertical flow velocity decreases to zero as the fluid turns to move along the horizontal boundary at either the top or bottom. Therefore, vertical heat transfer is dominantly by conduction. (We note that horizontal advection is comparable to vertical conduction, however.) For a given outward heat flux J, the temperature gradient with depth is given by Fourier's law:

$$\frac{J}{k} = \frac{dT}{dz} \cong \frac{\Delta T_{\text{TBL}}}{\delta_{\text{TBL}}}, \tag{8}$$

where ΔT_{TBL} and δ_{TBL} are the temperature offset and thickness characterizing the thermal boundary layer, respectively, as shown in Figure 5 (cf. Turcotte & Schubert 1982). Note that the effect of convection is to enhance the heat flux out of the surface relative to what it would have been were there no convection ($J = k\Delta T/D$). This effect is described by the Nusselt number, a dimensionless measure of the heat flow:

$$\text{Nu} = \frac{k(\Delta T_{\text{TBL}}/\delta_{\text{TBL}})}{k(\Delta T/D)} \cong \frac{D}{\delta_{\text{TBL}}}. \tag{9}$$

Dimensional analysis shows that Nu depends only on the dimensionless buoyancy ratio Ra (Turcotte & Schubert 1982); this is strictly true only for isoviscous, incompressible flows (Figure 5), but deviations are considered of secondary importance for analyzing heat transfer on a global scale. Without convection, the heat flow is purely conductive and Nu = 1; in contrast, Nu $\gg$ 1 when there is vigorous convection (see Figure 3). As there is no bottom boundary layer for the internally heated fluid (because there is no heat flux to be conducted through the bottom surface), a factor of 1 to 2 should be included in Equation 9, but it is ignored for simplicity.

The significance of the thermal boundary layer thickness δ_{TBL} is that it describes the depth interval over which conduction is a significant mode of vertical heat transfer. If we specifically consider the cold boundary layer at the top of the mantle, δ_{TBL} corresponds to the depth to which the mantle has been cooled from the surface by conduction. In time t this is a distance $\sqrt{\kappa t}$ (according to Equation 5); thus, putting time in terms of the distance and velocity that a plate travels at the surface ($t = l/u$), we obtain

$$\delta_{\text{TBL}} \cong \left(\frac{\kappa l}{u}\right)^{1/2}. \tag{10}$$

For the Pacific plate, for example, $u \cong 3 \times 10^{-9}$ m s^{-1}, and the thermal boundary layer grows to a thickness δ_{TBL} of the order of 10^2 km over the length of the plate now observed ($l \cong 10^7$ m, which implies that $t \cong 3 \times 10^{15}$ s).

The important conclusion is that in a vigorously convecting system (Pe $\gg$ 1), the temperature field has a length δ_{TBL} that is determined by the flow and that is significantly less than the depth extent D. This can be shown by combining Equations 4 and 10 (cf. Equation 9):

$$\frac{\delta_{TBL}}{D} \approx \left(\frac{l}{D}\right)^{1/2} \text{Pe}^{-1/2}, \qquad (11)$$

where $(l/D)^{1/2}$ is of order one for plates at the Earth's surface. For the mantle, δ_{TBL} is sometimes interpreted as a thermal measure of lithospheric thickness (e.g. in the ocean-plate cooling model described below), but this is distinct from (although related to) the usual rheological definition of the lithosphere (e.g. Turcotte & Schubert 1982). The distance δ_{TBL} is evident as the depth at which the geothermal gradient rapidly changes beneath the surface (Figure 1).

At a Rayleigh number much larger than the critical value, as is appropriate for the Earth's mantle, the horizontally averaged temperature profile is very nearly adiabatic through the interior of the convecting fluid (i.e. away from the conducting boundary layers). This is strictly true only for a fluid heated from below (McKenzie et al 1974, Jarvis & McKenzie 1980, Jarvis & Peltier 1982). In contrast, if the fluid is heated solely from within, the interior temperature is slightly subadiabatic with depth; that is, in the incompressible fluid calculations illustrated in Figure 5, the temperature decreases slightly with depth. The main point here is that in neither case does the horizontally averaged temperature increase superadiabatically (i.e. more rapidly than an adiabat) with depth through the interior. For the Earth, deviations of the average temperature from an adiabat are expected to be small in the region(s) away from thermal boundary layers, amounting to a decrease of no more than 200–300 K throughout the depth of the whole mantle.

Similarly, the effects of phase transformations and high-pressure reactions occurring deep in the mantle are small. These involve an adiabatic contribution from the heats of reaction and hence can be evaluated thermodynamically from calorimetric measurements and experimentally observed phase equilibria: They are estimated to increase the geotherm temperature by about 300 K in the transition zone (Figure 1; Jeanloz & Thompson 1983, Navrotsky & Akaogi 1984). Throughout the lower mantle, however, phase transformations are not likely to add more than

150 K to the geotherm, and they may instead decrease it by a comparable amount (Jeanloz & Richter 1979).

Therefore, if we ignore these smaller effects, the geothermal gradient through the interior of the mantle is given by the adiabatic value

$$\left.\frac{\partial T}{\partial z}\right|_s = \frac{\alpha T g}{C_P} \tag{12}$$

(see, for example, Birch 1952). Evaluating Equation 12 for the Earth's interior demonstrates that the adiabatic change in temperature across the whole mantle is only about 1000 K, as illustrated in Figure 6. The reason for this small value is that the mantle is relatively incompressible: The length scale over which density increases significantly with depth is large ($l \sim 10^4$ km), and the total compression across the mantle is less than 40% (Dziewonski & Anderson 1981). Thus, most of the temperature change occurs in the thermal boundary layers rather than in the advective central region of the mantle, as is illustrated by the model geotherms in Figure 1.

Horizontal temperature variations, however, are greatest in the central depth interval and tend to vanish at the top and bottom surfaces. These horizontal temperature differences are of approximately the magnitude ΔT, but the actual magnitude of the interior temperature T_i depends on the nature of the heating. As shown in Figure 5, $T_i \sim \Delta T/2$ for heating from below and $T_i \sim \Delta T$ for heating from within. (The fact that the top surface is at 0°C and not 0 K is of negligible effect in this context.)

The temperature distributions under ridges and along subduction zones in the mantle are schematically given by the upwelling and downwelling temperature profiles in Figure 5. Since melts emerge from the midocean ridges at temperatures of about 1500 K, whereas the top of the lithosphere is subducted at a temperature of 300 K, the difference between upwelling and downwelling temperatures near the Earth's surface is approximately 1200 K (cf. Figure 1). Although detailed modeling is required to extrapolate this value to depth, horizontal temperature differences of 1000 K or more can be expected within the upper mantle (e.g. Turcotte & Schubert 1982). In the deeper mantle, these temperature variations might be expected to be reduced, in accord with the temperature distributions calculated for convecting fluids that are partially heated from within (cf. Figure 5). Quantitative estimates are highly uncertain, however. The temperature dependence of the rock-deformation processes (see Figure 2), which is not known directly for the lower mantle or from laboratory data at the corresponding pressures, the possibility of mantle layering (discussed below), and the uncertain ratio of internal heat generation of the mantle to heating from the core all play a role in such estimates. One line of evidence that does suggest the occurrence of smaller horizontal temperature variations in the lower mantle than in the upper mantle comes from recent seismological studies. These indicate that lateral heterogeneity in elastic-wave velocities may be less important below 700 km depth than above this level (Dziewonski 1984, Woodhouse & Dziewonski 1984; see also Hager et al 1985). Nevertheless, there is little point in trying to refine present estimates of the average mantle geotherm to better than a few hundred kelvins because horizontal variations in temperature at a given depth are at least of this magnitude.

Conduction: Lithosphere Geotherms

A conduction model is appropriate for the lithosphere as it resides within the thermal boundary layer at the surface. At a depth of δ_{TBL} the mantle is cooling upward, but for much shallower depths the temperature is steady in time because it reflects a balance between heat loss to the surface and the combined input from local radioactive generation and heat flux from below. Thus, for depths significantly less than δ_{TBL}, we can assume steady-state conditions in deriving a geotherm and investigating its sensitivity to heat-source distributions and spatial variations of thermal conductivity. Such a model is appropriate for old lithosphere for which δ_{TBL} is much greater than the crustal thickness. In contrast, the time-dependent case, in which δ_{TBL} is small relative to the depths considered, is described in the following section on the cooling of the oceanic lithosphere.

One-dimensional, steady-state conduction geotherms for the lithosphere are illustrated in Figure 7. These account for heat production (H) by radioactive decay in the crust and mantle, which is taken to decrease exponentially with depth (z) through the crust [i.e. $H = H_0 \exp(-z/\lambda)$; Lachenbruch 1970]. In the figure, the temperature at depth (T) is normalized to the temperature and heat flux at the surface (T_0 and J_0, respectively), as well as to the near-surface thermal conductivity (k_0). Different geotherms are indicated as a function of the ratio $H_0\lambda/J_0$, which describes the fraction of the heat that is produced in the crust relative to the "reduced" heat flux entering the crust from the mantle, under the assumption that the crustal thickness is much greater than λ. Physically, $H_0\lambda/J_0$ is determined by the flow in the underlying mantle. Also, an important factor that is included in the figure is that the thermal conductivity increases substantially from the crust to the mantle (Jeanloz & Thompson 1983, England & Thompson 1984). The difference between mantle and crustal conductivities can mostly be ascribed to the density difference involved (since κ and C_P of the mantle and crust are similar). The resulting difference in conductivities has a major effect on the calculated geotherm through the lithosphere, as can be seen by the changes in slope at

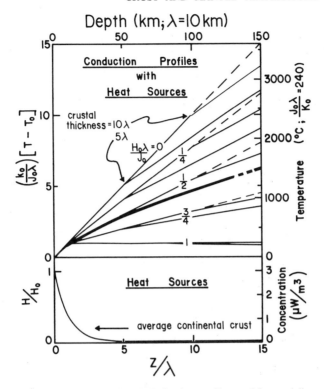

Figure 7 Steady-state, one-dimensional conduction profiles used for modeling geotherms through the lithosphere. The surface temperature and heat flux are specified as T_0 (0°C) and J_0, and the thermal conductivity of the crust is taken to be constant ($k_c = k_0$; subscripts 0, c, and m indicate surface, crust and mantle values, respectively). Dimensionalized scales are shown on the right and at the top. For both continental and oceanic crust, $J_0 \lambda / k_0 \cong 240$ is appropriate with $k_0 \cong 2.5$ W K^{-1} m^{-1} (Sclater et al 1980, England & Thompson 1984), but the contribution of fluid circulation to heat transfer through the crust is ignored (see Lachenbruch & Sass 1977, Lister 1980). The heat production in the crust is assumed to vary with depth (z) as $H = H_0 \exp(-z/\lambda)$, where λ is about 10 km. This distribution of heat production with depth has been shown to lead to the relationship that is observed at the surface between heat flow and heat production for continental crust (Lachenbruch 1970). The average crustal heat production determines the value of H_0, which is near 0 and about 3 μW m^{-3} for oceanic and continental crust, respectively (e.g. O'Connell & Hager 1980). Five groups of geotherms are shown, with the relative amount of crustal heat production contributing to the surface flux (i.e. the ratio $H_0 \lambda / J_0$) ranging between 0 and 1. For a given value of $H_0 \lambda / J_0$, cases for three crustal thicknesses are shown [z_c/λ being 5, 10, or 15 (the latter is dashed)]. In the mantle portion of the lithosphere ($z > z_c$), the thermal conductivity is taken to be $k_m = 1.35 k_0$ (as described in the text), and the heat production is assumed to be constant [$H(z > z_c) = H_0 \exp(-z_c/\lambda)$], in accord with the low values determined geochemically (e.g. O'Connell & Hager 1980). The preferred steady-state geotherm for continental lithosphere is given by the solid bold curve; for $z \gtrsim 13\lambda$, the curve is dashed to suggest that a steady conductive model may no longer be appropriate at such great depths.

depths of 5λ and 10λ (two possible values for the depth at the base of the crust) for the geotherms in Figure 7.

A representative continental geotherm at present (bold curve) is characterized by a crustal thickness of about 50 km, a decay length for radioactive heat production (λ) of $\cong 10$ km, and a ratio of crustal heat production to surface heat flux close to 0.5 (e.g. Sclater et al 1980, England & Thompson 1984). Downward curvature of the geotherm at a depth $z \sim \lambda$ is directly attributable to the fact that most of the heat production in this model is concentrated within a distance λ from the surface. Temperatures of 1100–1500°C are expected at a depth of 100–150 km, which is probably the maximum to which the steady-state conduction model is appropriate. The highest geotherm temperatures in the continental crust are predicted to be around 700–800°C according to this model. Such values are close to, but below, the melting temperature expected for deep crustal rocks (e.g. Winkler 1976, Bailey & Macdonald 1976).

Heat production in the oceanic crust is minor in comparison with that of the continental crust. Thus, to produce the heat flux characteristic of oceanic basins, temperatures of 1500°C are expected by about 65 km depth. These high temperatures cannot be extrapolated along a conduction geotherm to greater depths in the mantle. One reason is that this geotherm would exceed the melting temperature of the mantle beyond 100–150 km depth, and thus it would produce a massive layer of melt that is not observed at present (Figure 1). Moreover, we have already noted that steady-state conduction geotherms apply only at depths shallower than δ_{TBL}.

Cooling of the Oceanic Lithosphere: Time-Dependent Conduction

One of the most successful geophysical applications of the conductive heat transfer model has been the derivation of the age-dependent geotherm through the oceanic lithosphere. This model correctly predicts the bathymetry and surface heat flux as functions of age for oceanic plates, particularly once the effects of hydrothermal circulation in the crust are accounted for (e.g. Sclater et al 1980, Lister 1977, 1980). Although not discussed here, an analogous time-dependent modeling has been applied to continental geotherms by Pollack and colleagues (see Pollack 1982).

In contrast with the steady-state lithosphere geotherms presented above, the initial conditions describing the mantle temperatures beneath the ridges must be specified in order to solve the time-dependent heat-conduction equation that yields the oceanic geotherms. In the upwelling regions of the upper mantle, the temperature decreases adiabatically with increasing height, but the temperature at which melting begins (solidus) typically decreases more rapidly. Thus, as solid mantle rock approaches the surface, it begins to melt, and the effect of the latent heat of fusion is to cool the rock more rapidly upon further ascent. The dynamics of adiabatically upwelling, partially molten regions in the mantle have been the subject of considerable recent interest (Ahern & Turcotte 1979, Turcotte 1982, McKenzie 1984, Ribe 1985), and we use the results of these models for deriving the temperature distribution beneath ridges.

Figure 8 illustrates the initial temperatures beneath a ridge and the subsequent conductive cooling of the oceanic plate with age. Unlike the steady-state geotherms, the curvature evident in the geotherms for the suboceanic lithosphere is a reflection of the depth δ_{TBL} to which cooling has reached. In other words, Figure 8 illustrates the thickening of the thermal boundary layer with time ($\delta_{\text{TBL}} \sim \sqrt{\kappa t}$).

The present average age of oceanic lithosphere is 60 Myr, so temperatures exceeding $1200°C$ occur at depths of 100 km along the average oceanic geotherm. Whereas young oceanic lithosphere is hotter than the continental lithosphere at depth, geotherms for oceanic lithosphere that is older than about 70 Myr are comparable to the steady-state continental geotherm shown in Figure 7 (cf. Pollack & Chapman 1977). This is in accord with the conclusion of Sclater et al (1980) that the thermal states of old, "equilibrium" oceanic and continental lithosphere are essentially indistinguishable. In both cases, cooling has become negligible, and the temperature distribution of the old lithosphere is determined by a near-steady balance between outward heat flux, internal heat production, and heat input from below.

Uplift and Erosion of Crust: Combined Effects of Advection and Conduction

Up to this point, conduction and advection have been treated as separate mechanisms of vertical heat transfer. An important hybrid situation arises in the case of uplift and erosion of the crust (or downdropping and sedimentation). Although the local mechanism of heat transfer is still conductive (if we ignore the possible effects of fluid circulation), the effect of uplift is to provide an extra, advective contribution to the outward flow of heat. That is, conduction in a moving frame of reference (moving at the uplift velocity) corresponds to a combination of advection and conduction (e.g. Turcotte & Schubert 1982). For simplicity, we specifically refer only to the case of uplift, but the effects of downdropping are analogous.

The effect of imposing an uplift velocity u on the preferred steady-state continental geotherm (Figure 7) is shown in Figure 9. For a given heat flux observed at the surface (J_0), the temperature at depth decreases significantly with increasing uplift velocity (increasing Peclet number; see Equation 4).

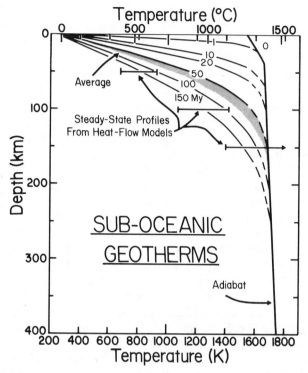

Figure 8 Geotherms through the oceanic lithosphere as a function of age. One-dimensional, conductive cooling of the zero-age adiabat is assumed, where the thermal diffusivity and surface temperature are 10^{-6} m^2 s^{-1} and 0°C, respectively. The initial adiabat (surface value of 1550 K) is taken from previous studies of the temperature associated with diapiric upwelling, partial melting, and magma migration (e.g. Basaltic Volcanism Study Project 1981, Turcotte 1982). The ages of successive geotherms are given in millions of years (to 150 Myr), with the shaded band showing the geotherm for oceanic lithosphere of average age (60 Myr). The present results for old ocean basins agree with Sclater et al's (1980) models for the corresponding temperatures under "equilibrium" oceanic crust. The latter are derived from modeling the observed heat flow with age, and the resulting temperatures lie within the brackets shown at 50, 100, and 150 km depth. For temperatures exceeding 90% of the initial adiabatic value, creep deformation may become significant: The transition to the corresponding portions of the geotherms, which are dashed, can be associated with the depth to the base of the lithosphere (see Pollack & Chapman 1977, Turcotte & Schubert 1982).

In other words if the temperature at depth is specified, the surface heat flux increases as u increases. The present illustration is for a steady-state situation, and uplift of the crust is imposed by an unspecified mechanism at great depth. More detailed, time-dependent models may be required for specific situations, however (e.g. Jaupart et al 1985).

As has been noted by others (England 1978a,b, 1979, England &

Richardson 1977, 1980), the thermal effect of uplift can be substantial. Uplift rates of 0.03–0.5 mm yr^{-1} have been recorded in many locations, and rates as high as a few millimeters per year are occasionally observed (e.g. Naeser et al 1983, Zeitler 1985, Carpena 1985, Jaupart et al 1985). From Figure 9 it is clear that even though these velocities are much smaller than plate tectonic velocities (or characteristic mantle convection velocities), advection is so much more efficient than conduction that the geotherm can be significantly perturbed. Corrections of up to 500°C for the temperature at

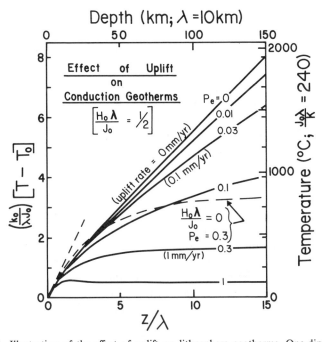

Figure 9 Illustration of the effect of uplift on lithosphere geotherms. One-dimensional, steady-state conduction profiles with exponentially decaying heat sources are assumed, as in Figure 7. The uplift velocity u is parameterized in terms of the nondimensional Peclet number Pe $= u\lambda/\kappa$. (The decay length for heat production and the thermal diffusivity are taken as $\lambda = 10$ km and $\kappa = 10^{-6}$ m^2 s^{-1}, respectively.) The solid curves correspond to the preferred geotherm for the continental lithosphere (bold curve in Figure 7), except that the difference between crust and mantle properties is ignored in this case. [The crustal contribution to the conductive surface heat flux $(H_0\lambda/J_0)$ is 50%, but the conductivity k_0 is taken to be constant.] For comparison, the dashed curves are for Pe = 0 and 0.3, with $H_0\lambda/J_0 = 0$. The temperature axis is dimensionalized (right-hand scale) under the assumption that the surface temperature and surface heat flux are known $(J_0\lambda/k_0 = 240$, as in Figure 7). Thus, increasing u or Pe results in a lower temperature at depth. Alternatively, for a given temperature profile the surface heat flux J_0 increases with increasing uplift velocity. The effect of uplift on conductive geotherms can be large because values of u between 0.1 and 1.0 mm yr^{-1} are not uncommon.

the base of the crust may well be required because of uplift in certain locations. Unfortunately, little quantitative information is generally available on uplift rates. Thus, Figure 9 demonstrates the importance of quantitative geomorphological and geochronological observations, combined with theoretical models of uplift, in order to characterize the thermal state of the continental lithosphere.

England and coworkers (England & Richardson 1977, England 1978a, 1979, England & Thompson 1984) have pointed out that the physical situation illustrated in Figure 9 is of considerable importance in metamorphic petrology. As rocks must be brought up from depth in order to be sampled, there is some overprinting of the initial equilibration conditions by the "anomalous" geotherm associated with uplift. In other words, in order to evaluate the degree to which the original pressures and temperatures of equilibration have been modified by temperature distributions associated with bringing the rock to the surface, the complete uplift process must be considered up to the point at which mineral compositions and textures are kinetically frozen in. The same problem also arises in the detailed interpretation of thermobarometry measurements made on xenoliths erupted from the mantle (Green & Gueguen 1974).

4. GEODYNAMICS

We have used concepts of heat transfer to interpret experimental data and to derive the broad temperature distribution through the crust and mantle (Figure 1); also, quantitative evaluations of heat transfer mechanisms allow more detailed steady-state and time-dependent geotherms to be estimated for the lithosphere (Figures 7, 8, and 9). The same concepts can be applied to studying the long-term thermal evolution of the Earth's interior. In particular, we wish to know whether the Earth is heating or cooling, as well as the time scale on which thermal evolution occurs. In fact, we shall find that the interior temperature has probably not changed much in the last 4 Gyr.

Archean Geotherms

For many, a compelling argument that temperatures in the crust and mantle were higher in Archean time (some 2.5–4.5 Gyr ago) is that the total heat production was undoubtedly much higher then than it is at present. For example, geochemical estimates of the current concentration in the Earth of uranium, thorium, and potassium, the dominant radioactive heat-producing elements, can be extrapolated to the concentrations at any previous time in the past because the decay constants of each of the radioactive isotopes is precisely known. A typical estimate for the current radioactive heat production ($H \sim 2 \times 10^{-8}$ W m^{-3} for the bulk Earth) would correspond to about one half the observed surface flux if the Earth were in steady state. This extrapolates to values between about 4 and 8 times larger at the origin of the planet 4.5 Gyr ago, with the main uncertainty being in the K/U ratio (e.g. O'Connell & Hager 1980, Verhoogen 1980, Davies 1980; cf. Sleep 1979). The resulting estimate for the total radioactive production over the age of the Earth is approximately 10^{31} J, which is sufficient to heat the whole planet by about 1700 K.

In addition, several sources of "primordial" heat may also have contributed to the heat production in the early Earth. These include accretional energy and heat from differentiation, both of which are difficult to estimate precisely. The former consists of the impact energy, or shock heating, that is retained in the planet as it grows by the accumulation of smaller planetesimals and meteoritic fragments. The latter includes the gravitational energy and heats of reaction that are released as the interior of the planet unmixes, primarily as the metal and silicate constituents separate to form the core and mantle, respectively. Current estimates suggest that the Earth heated sufficiently *either* by accretional processes or by core formation to begin melting by the time it had grown to about one half its final radius (Kaula 1979, Davies 1982, 1985). This conclusion is consistent with the observation that Mercury, Mars, the Moon, and meteorite parent bodies (planets considerably smaller than the Earth) have undergone differentiation and extensive internal melting (Basaltic Volcanism Study Project 1981).

For comparison, we note that Flasar & Birch's (1973) estimate of the amount of heat released upon core formation (10^{31} J) is the same as the total radioactive heat thought to have been produced throughout Earth history. Averaged over the age of the Earth, this value would imply a heat production from differentiation alone of 7×10^{13} W, or 1.7 times the total heat flux out of the surface at present. Thus, there can be little doubt that heat production was 2 to 5 times higher than at present 2.5 Gyr ago, and 5 to 10 times higher than at present 4.5 Gyr ago.

Does this imply that interior temperatures (the geotherm throughout the mantle and crust) were higher in the Archean than at present? Simple models of the thermal evolution of the Earth suggest that the answer is yes. Before describing these models, however, we summarize geological observations that document the cooling of the planetary interior with time.

The discovery of komatiites by Viljoen & Viljoen (1969) provides a significant datum regarding mantle temperatures in the Archean. These magnesium-rich, peridotitic rocks are highly refractory, but they exhibit quench textures indicating that they were erupted onto the surface as melts. Melting experiments have demonstrated that komatiites must have been erupted at temperatures of at least 1900 K and may have formed at 150–200 km depth and at temperatures exceeding 2050 K (Green 1975, Bickle 1978, Takahashi & Scarfe 1985). Komatiites were formed only during the Archean [see Basaltic Volcanism Study Project (1981) on this point], and since present melts are erupted at temperatures of 1500 K or less, this implies that temperatures some 400 K hotter than at present occurred in the early mantle.

In contrast, several independent observations prove that the lower crust and upper mantle were not uniformly hotter 3 to 4 Gyr ago than at present. For example, Archean metamorphic rocks containing high-pressure mineral assemblages indicate that temperatures in the lower crust and crustal thicknesses (i.e. the pressures achieved) were similar to what they are now (e.g. Bickle 1978, England 1979, Boak & Dymek 1982). More recently, the discovery of diamonds that are as much as 3.5 Gyr old has demonstrated, in conjunction with thermobarometry estimates from inclusions in the diamonds, that temperatures of only 1200–1500 K existed at depths of 150–200 km in the Archean mantle (Richardson et al 1984, Boyd et al 1985, Nickel & Green 1985).

Whereas the occurrence of komatiites implies significantly higher mantle temperatures, the other petrological observations suggest that temperatures indistinguishable from the present-day geotherm (Figure 1) existed in the Archean crust and upper mantle. These observations are not contradictory, however, because they tell us more about the horizontal variation of temperature than about the average temperature with depth (geotherm) in the Archean. Referring to Figure 5, it is clear that if the hot regions of the convecting mantle (the upwelling plumes) were 400–600 K hotter in the past, as indicated by the komatiites, the average interior temperature T_i would only have been about 250 K higher. Thus, a temperature difference between upwelling and downwelling regions of about 1600 K in the Archean (compared with a value of about 1200 K at present) translates to a geotherm through the crust and upper mantle that is within a few hundred kelvins of that shown in Figure 1 (cf. Sleep 1979).

The conclusion that horizontal temperature variations were larger 3 to 4 Gyr ago than at present has been advocated in a number of recent studies (e.g. Bickle 1978, Jarvis & Campbell 1983, Richter 1985). More specifically, regionalized models accounting for differences between the heat transfer through the oceanic lithosphere (cooling thermal boundary layer) and through the continental lithosphere (conductive lid over the convecting mantle), as well as for the higher heat production at that time, yield Archean crust and mantle temperatures that span the range indicated by the petrological observations (Bickle 1978, Richter 1984, 1985). An important assumption in this discussion, however, is that plate tectonics was active in the Archean. That the mantle was convecting is highly plausible based on the evidence that high temperatures (sufficient to begin melting) existed within the Earth before it was fully formed (cf. Figures 2 and 3). But the key issue is whether or not midocean ridges and subduction zones appeared at the surface. Without subduction, and the resultant downwelling plumes entering the mantle at 0°C, it is difficult to rationalize the existence of relatively low temperatures (i.e. present-day values) in the upper mantle and deep crust. If the convecting mantle were covered by an unbroken lid, making the Earth a one-plate planet without subduction, uniformly high internal temperatures would likely be required to explain the occurrence of komatiites. Thus, current models for the thermal state of the Archean mantle suggest that our planet has been, and still is, qualitatively different from Venus and Mars, which are thought to be convecting internally without plate tectonic activity at the surface (Basaltic Volcanism Study Project 1981, Phillips & Malin 1983, Kaula 1984).

Thermal Evolution of the Earth and Structure of the Mantle

Most thermal history models for the Earth's interior are constructed from an internal energy balance applied either to the whole mantle or to the upper and lower mantle separately. [Daly (1980) discusses this approach.] Since the interior temperature T_i is uniform outside the thermal boundary layers (Figure 5), this energy balance gives the rate of change of the temperature if the internal heat sources (H) and the heat flow (J) out through the top and in through the bottom boundaries are known:

$$\rho C_p V \frac{dT_i}{dt} = HV + (AJ)_{In} - (AJ)_{Out}. \tag{13}$$

The areas of the bottom and top surfaces (A_{In} and A_{Out}, respectively) differ for spherical geometry (the surface area of the core is one fourth the area at the Earth's surface), and V is the volume of the region considered (whole mantle or a part of the mantle). Also, recall that the heat flow across a given boundary depends on the thickness δ_{TBL} and temperature difference ΔT_{TBL} across the adjacent thermal boundary layer (Figure 5, Equation 8).

A time scale for the thermal evolution of the Earth can be derived from the energy balance. For example, in response to a change in the amount of internal heating relative to the cooling rate $\{[HV + (AJ)_{In}]/(AJ)_{Out}\}$, the internal temperature changes significantly on a time scale

$$\tau = \frac{\rho C_p V T_i}{(AJ)_{Out}} \tag{14}$$

according to Equation 13 (cf. O'Connell & Hager 1980). Each factor on the right-hand side of Equation 14 can be evaluated, at least approximately, for the present state of the Earth, which results in a thermal time constant of about 10^{10} yr. Such a large value (considerably longer than the age of the Earth) suggests that the thermal state of the planetary interior can only adjust itself very slowly. We shall see, however, that thermal evolution models can lead to much shorter response times.

All thermal history calculations assume that the top and bottom boundary layers have the same thickness, and that this thickness is given by the expression for steady convection [$\delta_{TBL} = D/(2Nu)$; the factor of two is explained after Equation 9]. As noted in Section 3, the Nusselt and Rayleigh numbers are related, and this relation is given by

$$Nu \simeq cRa^{\beta},\tag{15}$$

where c and β are constants that depend on specific characteristics of the flow, such as the planform boundary conditions and mode of heating (e.g. Frick et al 1983, Roberts 1979, McKenzie et al 1974). For example, we have $Nu = 2.24(Ra/Ra_{cr})^{0.319}$ for two-dimensional convection of an incompressible, isoviscous fluid with planar, traction-free boundaries (see Schubert & Anderson 1985; $Ra_{cr} = 27\pi^4/4$ is the critical Rayleigh number). In general, β lies in the range $0.2 \lesssim \beta \lesssim 0.33$, but because it is an exponent, thermal models of the Earth are sensitive to its value. Thus, Sharpe & Peltier (1979) report that reducing β from 0.33 to 0.30 changes their thermal history models significantly: Their calculated present-day geotherms go from being below the melting point to exceeding the liquidus, and the mean mantle viscosity decreases by an order of magnitude. The conclusion is that thermal evolution models can only indicate broad trends for the Earth's interior.

Plate velocities and values of heat flow observed at the surface of the Earth are reproduced by models of isoviscous, cellular convection (e.g. Jarvis & Peltier 1982). This suggests that the presence of very stiff material in the Earth's lithosphere only influences the heat transfer indirectly by controlling the size and shape of lithospheric plates, rather than controlling δ_{TBL}. With this reasoning, Tozer (1965, 1972) proposed that temperature-dependent rheology can be incorporated in thermal history calculations by using Equation 15, with the viscosity evaluated at the interior temperature T_i (see Equations 2 and 3). Lei (1985) has tested this idea for the convection driven by heating a strip of finite length at the bottom of a viscous half-space. He shows by asymptotic analysis that for a given large viscosity ratio, the heat transfer is given closely by the relationship for isoviscous convection, provided that the convection is sufficiently vigorous. All published models for the Earth use Tozer's approach, although it must be modified for planets without subduction (e.g. Schubert 1979, Morris & Canright 1984).

The essential effects of temperature-dependent rheology are to make the interior state self regulating and to lead to much shorter thermal response times than the 10^{10} yr derived above. The self-regulation means that cool bodies convect less vigorously, and therefore cool more slowly, than hot planets. Noting that all thermal histories must eventually converge (all bodies ultimately cool to $T = 0$ K regardless of the nature of the rheology), the importance of the temperature dependence is best seen for bodies starting from different initial temperatures and evolving by whole-mantle convection. For example, Schubert et al (1979) find that planets without heat sources and differing solely in their initial temperatures of 3300 and 1800 K (ratio of 1.8) cool to 1800 and 1700 K, respectively (ratio of 1.1), after only 10^9 yr. The influence of the rheology is most dramatic in the first billion years of cooling, and subsequent temperatures are essentially the same. Similarly, internal heat production is included in the calculations of Schubert et al (1980), who show that a model with a cold start (300 K) is essentially indistinguishable from models with a hot start (3300 K) after 2 to 3 Gyr.

The reason that the time scale of Equation 14 is shortened when convection by a fluid with temperature-dependent viscosity is considered can be seen as follows. According to Equation 8 (but substituting for the thermal conductivity $k = \kappa \rho C_P$), the outward heat flux is approximately $J_{\text{Out}} \cong \kappa \rho C_P T_i / \delta_{\text{TBL}}$ (recall that $\Delta T_{\text{TBL}} \cong T_i$; see Figure 5). With the volume $V \cong A_{\text{Out}} D$, Equation 14 becomes

$$\tau \cong \frac{D \delta_{\text{TBL}}}{\kappa} \cong \frac{D^2}{\kappa \text{Nu}} \tag{16}$$

(cf. Equations 9 and 11; Daly 1980). Thus, when the interior is hot, the thermal response time is shortened because the Nusselt number is enhanced by the effect of temperature on the viscosity and hence on the Rayleigh number: These interdependencies are described by Equations 15, 2, and 3, respectively. In particular, the existence of high internal temperatures early in Earth history can lead to intense convection and a relatively small value of τ.

It is evident that temperature-dependent rheology has two major consequences for the thermal evolution of a planet undergoing whole-mantle convection. First, the thermal response time can be as short as 10^9 yr (in particular, for planets cooling from a hot start), and the present thermal state no longer reflects the initial state. Second, bodies that have some internal heat generation and start hot, as is thought to be the case for the

Earth, cool by only a few hundred kelvins over the period from 1 to 4.5 Gyr after origin (e.g. Schubert et al 1980). This modest amount of cooling is in accord with the petrological observations on Archean rocks summarized above.

Are we to conclude that the present thermal state of the mantle contains no information about the internal temperatures early in Earth history? The answer is "not necessarily": A significant portion of the current heat content of the planet may be of primordial origin. In particular, if the mantle is layered, global heat transfer becomes considerably less efficient and the thermal response time of the interior can be comparable to (or longer than) the age of the Earth. That is, the effect of strongly temperature-dependent viscosity (enhanced heat transfer leading to a response time much less than 10^{10} yr) can be effectively counteracted by the occurrence of layering in the mantle.

The significant issue is whether or not the lower mantle is of the same bulk composition as the upper mantle (Figure 10; Jeanloz & Richter 1979). If not, only a few percent density increase due to a change in composition is required in order to keep the lower mantle from being swept up into the upper mantle as they both convect (see Olson 1984). Some of the most direct evidence that the mantle may be layered comes from interpreting the seismologically determined density and wave velocities of the lower mantle in terms of laboratory measurements on silicate perovskite and oxides (Jeanloz 1986). Thus, Knittle et al (1985) find that the lower mantle is about 2–3% denser than the corresponding high-pressure assemblage of upper-mantle composition. This result may not be conclusive, however, because seismological observations indicate that subducted slabs apparently penetrate into the lower mantle (Creager & Jordan 1984). We do not detail the arguments any further, but consider that the possibility of mantle layering, or partial layering (Jeanloz 1986), is an open issue at present (cf. Zindler & Hart 1986).

The consequence of layering is that heat must be conducted across the interface where the composition changes: There is no vertical motion across the compositional discontinuity, and hence no vertical advection at this level (Figure 10). Therefore, if the upper mantle and lower mantle differ in composition, thermal boundary layers are present at the 670–km seismological discontinuity defining the bottom of the transition zone. The effect is to increase the geotherm temperature in the lower mantle by some 500–1000 K (Jeanloz & Richter 1979). The two cases (with and without layering of the mantle) are illustrated in Figure 1 by the hot and cold geotherms, respectively. Evidently, the uncertainty about the presence or absence of layering results in one of the largest uncertainties in current estimates of the geotherm through the crust and mantle.

Another way of considering the effects of layering is to note the diminished efficiency of global heat transfer. In layered convection, the temperature of the upper mantle is similar to that in whole-mantle convection throughout Earth history; the lower mantle cools more slowly, however, and layered convection removes only about one third the heat that is removed by whole-mantle convection over 4.5 Gyr (Spohn &

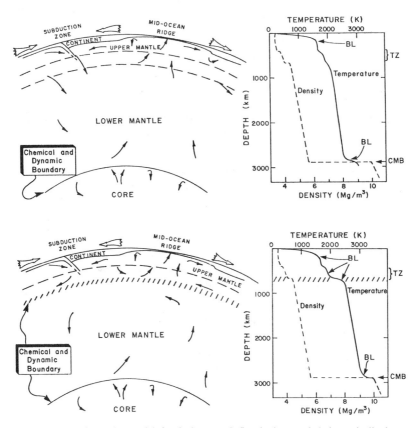

Figure 10 Two alternative models for the large-scale flow in the mantle (schematically shown on the left), with the resulting profiles of average temperature as a function of depth given on the right (from Jeanloz 1986). In the upper model, convection stirs the mantle homogeneously, and thermal boundary layers (BL) occur only at the top and bottom of the mantle. The presence of a barrier to convection (dynamic barrier) at the top of the lower mantle results in an additional set of thermal boundary layers and in higher temperatures in the lower mantle (lower model; Jeanloz & Richter 1979). Observed plate tectonic motions are schematically illustrated in the left-hand figures, and the average density as a function of depth (Dziewonski & Anderson 1981) and the locations of the transition zone (TZ) and core-mantle boundary (CMB) are given on the right.

Schubert 1982). Qualitatively, the upper mantle thermally insulates the lower mantle in the layered case. Thus, layering alone approximately doubles the thermal response time of the mantle relative to the unlayered case, and any differences in the properties of the two layers (e.g. viscosities of the upper and lower mantle) can further modify the response time (McKenzie & Richter 1981). The implications for estimates of the heat production within the Earth are especially severe, because the relationship between observed heat flux and inferred heat generation is critically dependent on the efficiency of heat transfer.

We have summarized the reasons for concluding that heat production within the planet has decreased since the early Archean. That the heat flux at the surface and the internal heat production are not equal through Earth history is simply a consequence of the finite time that is required for heat to be transported out of the deep interior (cf. Daly 1980). Thus, the ratio of heat loss to heat production at a given time—the Urey ratio (see Zindler & Hart, 1986)—deviates from 1, and the interior temperature is not steady in time (Equation 13).

Calculations of the Urey ratio by McKenzie & Richter (1981) illustrate the effect of layering (Figure 11). They consider both a single layer and a

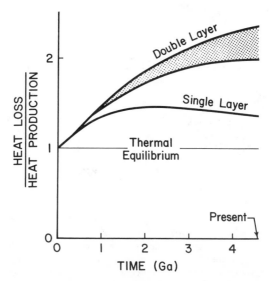

Figure 11   Time evolution of the ratio of heat loss through the top surface to the internal heat production for isoviscous models simulating convection in the mantle (McKenzie & Richter 1981). Two cases are shown relative to the steady-state value of 1 (which indicates an equilibrium between heat loss and heat production): uniform mantle convection (single layer), and convection in a layered mantle (double layer). Stippling indicates the range of results derived for viscosity ratios between upper and lower layers of 1 to 1/1000.

double layer of isoviscous fluid, with radioactive internal heat sources decaying by about a factor of 5 over 4.5 Gyr and with an initial condition assumed to be steady state. Thus, as heat production decreases, there is initially a time lag before the effect is seen at the surface as a decreasing heat flux: For both the double-layer and single-layer cases (corresponding to layered and unlayered mantles, respectively), the ratio of heat loss to heat production increases at first. Because this ratio is assumed to start at 1, the whole subsequent thermal history involves cooling (heat loss exceeds heat production). Eventually, the heat loss at the surface begins to follow the decaying heat production, and the Urey ratio achieves a maximum (Figure 11). The time required for this to happen is a direct reflection of the efficiency of global heat transfer. For a single layer, this time is about 2 Gyr, but doubly layered systems are still evolving *away* from steady state after 4.5 Gyr.

Determining the bulk composition of the lower mantle (the region comprising the single largest fraction of the Earth) is clearly of geochemical interest for evaluating the composition of the planetary interior. The constitution of the lower mantle has considerably broader implications, however. As we have shown, models of the current temperature distribution through the mantle, as well as the thermal evolution of the planet, are critically dependent on whether the mantle is of uniform composition or not. If the mantle is not layered, it is expected to have a thermal response time that is short compared with the age of the Earth. In this case, the present temperature distribution through the mantle would retain no memory of the earliest conditions inside the planet. With mantle layering, however, insulation of the lower mantle by the upper mantle can lead to a long thermal response time. Considering the evidence that the mantle may be layered, the present temperature distribution of the deep interior may still reflect the original thermal state, and studies of the internal temperatures may yield unique insights into the formation and earliest differentiation of the Earth.

ACKNOWLEDGMENTS

We thank F. Bishop, M. Bukowinski, and J. Verhoogen for critical comments on an early version of the manuscript. Also, we benefited from discussions with M. Bukowinski, D. Heinz, A. Kirkpatrick, E. Knittle, C. Mead, K. C. McNally, L. E. Weiss, and Q. Williams. This work is supported by the National Science Foundation and NASA.

Literature Cited

Ahern, J. L., Turcotte, D. L. 1979. Magma migration beneath an ocean ridge. *Earth Planet. Sci. Lett.* 45:115–22

Akaogi, M., Akimoto, S. 1979. High-pressure phase equilibria in a garnet lherzolite, with special reference to Mg^{2+}–Fe^{2+} partitioning among constituent minerals. *Phys. Earth Planet. Inter.* 19 : 31–51

Akimoto, S., Akaogi, M., Kawada, K., Nishizawa, O. 1976. Mineralogic distribution of iron in the upper half of the transition zone in the Earth's mantle. In *The Geophysics of the Pacific Ocean Basin and Its Margin*, ed. G. H. Sutton, M. H. Manghnani, R. Moberly, 1 : 399–405. Washington, DC: Am. Geophys. Union

Anderssen, R. S., DeVane, J. F., Gustafson, S. A., Winch, D. E. 1979. The qualitative character of the global electrical conductivity of the Earth. *Phys. Earth Planet. Inter.* 20 : p15–p21

Aoki, K. 1984. Petrology of materials derived from the upper mantle. In *Materials Science of the Earth's Interior*, ed. I. Sunagawa, pp. 415–44. Tokyo: Terra Sci. Publ. Co.

Bailey, D. K., Macdonald, R., eds. 1976. *The Evolution of the Crystalline Rocks.* New York: Academic. 484 pp.

Banks, R. J. 1972. The overall conductivity distribution of the Earth. *J. Geomagn. Geoelectr.* 24 : 337–51

Basaltic Volcanism Study Project. 1981. *Basaltic Volcanism on the Terrestrial Planets.* New York: Pergamon. 1286 pp.

Bickle, M. J. 1978. Heat loss from the Earth: a constraint on Archean tectonics from the relation between geothermal gradients and the rate of plate production. *Earth Planet. Sci. Lett.* 40 : 301–15

Birch, F. 1952. Elasticity and constitution of the Earth's interior. *J. Geophys. Res.* 57 : 227–86

Birch, F. 1964. Density and composition of mantle and core. *J. Geophys. Res.* 69 : 4377–88

Birch, F. 1968. Thermal expansion at high pressures. *J. Geophys. Res.* 73 : 817–19

Birch, F. 1972. The melting relation of iron and temperatures in the Earth's core. *Geophys. J. R. Astron. Soc.* 29 : 373–87

Boak, J. L., Dymek, R. F. 1982. Metamorphism of the ca. 3800 Ma supracrustal rocks at Isua, West Greenland: implications for early Archean crustal evolution. *Earth Planet. Sci. Lett.* 59 : 155–76

Boyd, F. R. 1973. The pyroxene geotherm. *Geochim. Cosmochim. Acta* 37 : 2533–46

Boyd, F. R., Gurney, J. J., Richardson, S. H. 1985. Evidence for a 150–200 km thick Archean lithosphere from diamond inclusion thermobarometry. *Nature* 315 : 387–89

Brett, R. 1976. The current status of speculations on the composition of the core of the Earth. *Rev. Geophys. Space Phys.* 14 : 375–83

Brown, J. M., McQueen, R. G. 1980. Melting of iron under core conditions. *Geophys.*

Res. Lett. 7 : 533–36

Brown, J. M., McQueen, R. G. 1982. The equation of state for iron and the Earth's core. In *High-Pressure Research in Geophysics*, ed. A. Akimoto, M. H. Manghnani, 1 : 611–23. Tokyo: Cent. for Acad. Publ.

Brown, J. M., Ahrens, T. J., Shampine, D. L. 1984. Hugoniot data for pyrrhotite and the Earth's core. *J. Geophys. Res.* 89 : 6041–48

Carpena, J. 1985. Tectonic interpretation of an inverse gradient of zircon fission-track ages with respect to altitude: alpine thermal history of the Gran Paradiso basement. *Contrib. Mineral. Petrol.* 90 : 74–82

Carslaw, H. S., Jaeger, J. C. 1959. *Conduction of Heat in Solids.* Oxford, Engl: Oxford Univ. Press. 510 pp.

Christian, J. W. 1975. *The Theory of Transformations in Metals and Alloys, Part 1.* New York: Pergamon. 586 pp.

Clark, S. P. Jr. 1957. Radiative transfer in the Earth's mantle. *Trans. Am. Geophys. Union* 38 : 931–38

Creager, K. C., Jordan, T. H. 1984. Slab penetration into the lower mantle. *J. Geophys. Res.* 89 : 3031–49

Daly, S. F. 1980. Convection with decaying heat sources: constant viscosity. *Geophys. J. R. Astron. Soc.* 61 : 519–47

Davies, G. F. 1980. Thermal histories of convective Earth models and constraints on radiogenic heat production in the Earth. *J. Geophys. Res.* 85 : 2517–30

Davies, G. F. 1982. Ultimate strength of solids and formation of planetary cores. *Geophys. Res. Lett.* 9 : 1267–70

Davies, G. F. 1985. Heat deposition and retention in a solid planet growing by impacts. *Icarus.* In press

Dawson, J. B. 1980. *Kimberlites and Their Xenoliths.* New York: Springer-Verlag. 252 pp.

Dziewonski, A. M. 1984. Mapping the lower mantle: determination of lateral heterogeneity in *P* velocity up to degree and order 6. *J. Geophys. Res.* 89 : 5929–52

Dziewonski, A. M., Anderson, D. L. 1981. Preliminary reference Earth model. *Phys. Earth Planet. Inter.* 25 : 297–356

England, P. C. 1978a. Some thermal considerations of the Alpine metamorphism—past, present and future. *Tectonophysics* 46 : 21–40

England, P. C. 1978b. The effect of erosion on paleoclimatic and topographic corrections to heat flow. *Earth Planet. Sci. Lett.* 399 : 427–34

England, P. C. 1979. Continental geotherms during the Archean. *Nature* 277 : 556–58

England, P. C., Richardson, S. W. 1977. The influence of erosion upon the mineral facies of rocks from different metamorphic environments. *J. Geol. Soc. London* 134: 201–13

England, P. C., Richardson, S. W. 1980. Erosion and the age dependence of continental heat flow. *Geophys. J. R. Astron. Soc.* 62:421–37

England, P. C., Thompson, A. B. 1984. Pressure-temperature-time paths of regional metamorphism. I. *J. Petrol.* 25: 894–928

Finnerty, A. A., Boyd, F. R. 1984. Evaluation of thermobarometers for garnet peridotites. *Geochim. Cosmochim. Acta* 48:15–27

Flasar, F. M., Birch, F. 1973. Energetics of core formation: correction. *J. Geophys. Res.* 78:6101–3

Frick, H., Busse, F. H., Clever, R. M. 1983. Steady three-dimensional convection at high Prandtl numbers. *J. Fluid Mech.* 127: 141–53

Green, D. H. 1975. Genesis of Archean peridotitic magmas and constraints on Archean geothermal gradients and tectonics. *Geology* 3:15–18

Green, H. W. II, Gueguen, Y. 1974. Origin of kimberlite pipes by diapiric upwelling in the upper mantle. *Nature* 249:617–20

Gubbins, G., Masters, G. 1979. Driving mechanisms for the Earth's dynamo. *Adv. Geophys.* 21:1–50

Hager, B. H., Clayton, R. W., Richards, M. A., Comer, R. P., Dziewonski, A. M. 1985. Lower mantle heterogeneity, dynamic topography and the geoid. *Nature* 313: 541–45

Hood, L. L., Herbert, F., Sonett, C. P. 1982. The deep lunar electrical conductivity profile: structural and thermal inferences. *J. Geophys. Res.* 87:5311–26

Jacobs, J. A. 1975. *The Earth's Core.* New York: Academic. 253 pp.

Jarvis, G. T., Campbell, I. H. 1983. Archean komatiites and geotherms: solution to an apparent contradiction. *Geophys. Res. Lett.* 10:1133–36

Jarvis, G. T., McKenzie, D. P. 1980. Convection in a compressible fluid with infinite Prandtl number. *J. Fluid Mech.* 96: 515–83

Jarvis, G. T., Peltier, W. R. 1982. Mantle convection as a boundary layer phenomenon. *Geophys. J. R. Astron. Soc.* 68:389–427

Jaupart, C., Francheteau, J., Shen, X.-J. 1985. On the thermal structure of the southern Tibetan crust. *Geophys. J. R. Astron. Soc.* 81:131–55

Jeanloz, R. 1983. The Earth's core. *Sci. Am.* 249:56–65

Jeanloz, R. 1986. High-pressure chemistry of the Earth's mantle and core. In *Mantle Convection*, ed. W. R. Peltier. New York: Gordon & Breach. In press

Jeanloz, R., Heinz, D. L. 1984. Experiments at high temperature and pressure: laser heating through the diamond cell. *J. Phys. (Paris)* 45(C8): 83–92

Jeanloz, R., Richter, F. M. 1979. Convection, composition, and the thermal state of the lower mantle. *J. Geophys. Res.* 84:5497–5504

Jeanloz, R., Thompson, A. B. 1983. Phase transitions and mantle discontinuities. *Rev. Geophys. Space Phys.* 21:51–74

Kaula, W. M. 1979. Thermal evolution of Earth and Moon growing by planetesimal impacts. *J. Geophys. Res.* 84:999–1008

Kaula, W. M. 1984. Tectonic contrasts between Venus and the Earth. *Geophys. Res. Lett.* 11:35–37

Kennedy, C. S., Kennedy, G. C. 1976. The equilibrium boundary between graphite and diamond. *J. Geophys. Res.* 81:2467–70

Kirby, S. H. 1983. Rheology of the lithosphere. *Rev. Geophys. Space Phys.* 21: 1458–87

Knittle, E., Jeanloz, R., Smith, G. L. 1985. The thermal expansion of silicate perovskite and stratification of the Earth's mantle. *Nature.* In press

Lachenbruch, A. H. 1970. Crustal temperature and heat production: implications of the linear heat flow relation. *J. Geophys. Res.* 75:3291–3300

Lachenbruch, A. H., Sass, J. H. 1977. Heat flow in the United States and the thermal regime of the crust. In *The Earth's Crust*, ed. J. G. Heacock, pp. 626–75. Washington, DC: Am. Geophys. Union

Lay, T., Helmberger, D. V. 1983. A lower mantle *S*-wave triplication and the shear velocity structure of *D''*. *Geophys. J. R. Astron. Soc.* 75:799–837

Lei, U. 1985. *The effects of strongly temperature-dependent viscosity on free convection from a horizontal hot strip.* PhD thesis. Dept. Mech. Eng., Univ. Calif., Berkeley. 155 pp.

Lister, C. R. B. 1977. Estimators for heat flow and deep rock properties based on boundary layer theory. *Tectonophysics* 41:157–71

Lister, C. R. B. 1980. Heat flow and hydrothermal circulation. *Ann. Rev. Earth Planet. Sci.* 8:95–117

MacGregor, I. D. 1974. The system MgO-Al_2O_3-SiO_2: solubility of Al_2O_3 in enstatite for spinel and garnet peridotite compositions. *Am. Mineral.* 59:110–19

McKenzie, D. P. 1984. The generation and compaction of partial melts. *J. Petrol.* 92: 713–65

McKenzie, D. P., Richter, F. M. 1981. Parameterized thermal convection in a layered region and the thermal history of the Earth. *J. Geophys. Res.* 86:11667–80

McKenzie, D. P., Roberts, J. M., Weiss, N. D. 1974. Convection in the Earth's mantle: towards a numerical solution. *J. Fluid Mech.* 62:465–538

Mercier, J.-C., Carter, N. L. 1975. Pyroxene geotherms. *J. Geophys. Res.* 80:3349–62

Merrill, R. T., McElhinny, M. W. 1983. *The Earth's Magnetic Field.* New York: Academic. 401 pp.

Morris, S., Canright, D. 1984. A boundary-layer analysis of Bénard convection in a fluid of strongly temperature-dependent viscosity. *Phys. Earth Planet. Inter.* 36:355–73

Naeser, C. W., Bryant, B., Crittenden, M. D. Jr., Sorensen, M. L. 1983. Fission-track ages of apatite in the Wasatch Mountains, Utah: an uplift study. *Geol. Soc. Am. Mem.* 157:29–36

Navrotsky, A., Akaogi, M. 1984. α-β-γ phase relations in Fe_2SiO_4–Mg_2SiO_4 and Co_2SiO_4–Mg_2SiO_4: calculation from thermochemical data and geophysical applications. *J. Geophys. Res.* 89:10135–40

Nickel, K. G., Green, D. H. 1985. Empirical geothermobarometry for garnet peridotites and implications for the nature of the lithosphere, kimberlites and diamonds. *Earth Planet. Sci. Lett.* 73:158–70

O'Connell, R. J., Hager, B. H. 1980. On the thermal state of the Earth. In *Physics of the Earth's Interior*, ed. A. M. Dziewonski, E. Boschi, pp. 270–317. New York: Elsevier

Olson, P. 1984. An experimental approach to thermal convection in a two-layered mantle. *J. Geophys. Res.* 89:11293–11301

Oxburgh, E. R. 1980. Heat flow and magma genesis. In *Physics of Magmatic Processes*, ed. R. B. Hargraves, pp. 161–99. Princeton, NJ: Princeton Univ. Press

Peltier, W. R. 1981. Ice age geodynamics. *Ann. Rev. Earth Planet. Sci.* 9:199–225

Phillips, R. J., Malin, M. C. 1983. The interior of Venus and tectonic implications. In *Venus*, ed. D. M. Hunten, L. Colin, T. M. Donahue, V. I. Moroz, pp. 159–214. Tucson: Univ. Ariz. Press

Poirier, J. P. 1985. *Creep of Crystals.* New York: Cambridge Univ. Press. 260 pp.

Pollack, H. N. 1982. The heat flow from the continents. *Ann. Rev. Earth Planet. Sci.* 10:459–81

Pollack, H. N., Chapman, D. S. 1977. On the regional variation of heat flow, geotherms and lithospheric thickness. *Tectonophysics* 38:279–96

Ribe, N. M. 1985. The generation and composition of partial melts in the Earth's mantle. *Earth Planet Sci. Lett.* 73:361–76

Richardson, S. H., Gurney, J. J., Erlank, A. J., Harris, J. W. 1984. Origin of diamonds in old enriched mantle. *Nature* 310:198–202

Richter, F. M. 1984. Regionalized models for the thermal evolution of the Earth. *Earth Planet. Sci. Lett.* 68:471–84

Richter, F. M. 1985. Models for the Archaean thermal regime. *Earth Planet. Sci. Lett.* 73:350–60

Ringwood, A. E. 1975. *Composition and Petrology of the Earth's Mantle.* New York: McGraw-Hill. 618 pp.

Roberts, G. O. 1979. Fast viscous Bénard convection. *Geophys. Astrophys. Fluid Dyn.* 12:235–72

Schubert, G. 1979. Subsolidus convection in the mantles of terrestrial planets. *Ann. Rev. Earth Planet. Sci.* 7:289–342

Schubert, G., Anderson, C. 1985. Finite element calculations of very high Rayleigh number thermal convection. *Geophys. J. R. Astron. Soc.* 80:575–601

Schubert, G., Cassen, P., Young, R. E. 1979. Subsolidus convective cooling histories of terrestrial planets. *Icarus* 38:192–211

Schubert, G., Stevenson, D., Cassen, P. 1980. Whole planet cooling and the radiogenic heat source contents of the Earth and Moon. *J. Geophys. Res.* 85:2531–38

Sclater, J. G., Jaupart, C., Galson, D. 1980. The heat flow through oceanic and continental crust and the heat loss of the Earth. *Rev. Geophys. Space Phys.* 18:269–311

Shankland, T. J., Duba, A. G., Woronow, A. 1974. Pressure shifts of optical absorption bands in iron-bearing garnet, spinel, olivine, pyroxene, and periclase. *J. Geophys. Res.* 79:3273–82

Shankland, T. J., Nitsan, U., Duba, A. G. 1979. Optical absorption and radiative heat transport in olivine at high temperature. *J. Geophys. Res.* 84:1603–10

Shankland, T. J., O'Connell, R. J., Waff, H. S. 1981. Geophysical constraints on partial melt in the upper mantle. *Rev. Geophys. Space Phys.* 19:394–406

Sharpe, H. N., Peltier, W. R. 1979. A thermal history model for the Earth with parameterized convection. *Geophys. J. R. Astron. Soc.* 59:171–205

Sleep, N. H. 1979. Thermal history and degassing of the Earth: some simple calculations. *J. Geol.* 87:671–86

Sonett, C. P. 1982. Electromagnetic induction in the Moon. *Rev. Geophys. Space Phys.* 20:411–55

Spiliopoulos, S., Stacey, F. D. 1984. The Earth's thermal profile: is there a mid-mantle thermal boundary layer? *J. Geodyn.* 1:61–77

Spohn, T., Schubert, G. 1982. Modes of mantle convection and the removal of heat from the Earth's interior. *J. Geophys. Res.* 87:4682–96

Stevenson, D. J. 1981. Models of the Earth's core. *Science* 214:611–19

Stevenson, D. J., Turner, J. S. 1979. Fluid models of mantle convection. In *The Earth: Its Origin, Structure and Evolution*, ed. M. W. McElhinny, pp. 227–63. New York: Academic

Stolper, E., Walker, D., Hager, B. H., Hays, J. F. 1981. Melt segregation from partially molten source regions: the importance of melt density and source region size. *J. Geophys. Res.* 86:6261–71

Takahashi, E., Scarfe, C. M. 1985. Melting of peridotite to 14 GPa and the genesis of komatiite. *Nature* 315:566–68

Taylor, S. R. 1982. *Planetary Science: A Lunar Perspective*. Houston: Lunar Planet. Inst. 481 pp.

Tozer, D. C. 1965. Heat transfer and convection currents. *Philos. Trans. R. Soc. London Ser. A* 258:252–71

Tozer, D. C. 1972. The present thermal state of the terrestrial planets. *Phys. Earth Planet Inter.* 6:182–97

Turcotte, D. L. 1982. Magma migration. *Ann. Rev. Earth Planet. Sci.* 10:397–408

Turcotte, D. L., Schubert, G. 1982. *Geodynamics*. New York: Wiley. 450 pp.

Verhoogen, J. 1980. *Energetics of the Earth*. Washington, DC: Natl. Acad. Sci. 139 pp.

Viljoen, M. J., Viljoen, R. P. 1969. The geology and geochemistry of the lower ultramafic unit of the Onverwacht Group and a proposed new class of igneous rock. *Geol. Soc. S. Afr. Upper Mantle Project Spec. Publ.* 2:221–44

Wakatsuki, M., Ichinose, K. 1982. A wedge-type cubic anvil high-pressure apparatus and its application to material synthesis research. In *High-Pressure Research in Geophysics*, ed. S. Akimoto, M. H. Manghnani, pp. 13–26. Tokyo: Cent. for Acad. Publ.

Weertman, J., Weertman, J. R. 1975. High temperature creep of rock and mantle viscosity. *Ann. Rev. Earth Planet. Sci.* 3:293–315

Weidner, D. J. 1985. A mineral physics test of a pyrolite mantle. *Geophys. Res. Lett.* 12:417–20

Weidner, D. J., Sawamoto, H., Sasaki, S., Kumazawa, M. 1984. Single-crystal elastic properties of the spinel phase of Mg_2SiO_4. *J. Geophys. Res.* 89:7852–60

Winkler, H. G. F. 1976. *Petrogenesis of Metamorphic Rocks*. New York: Springer-Verlag. 334 pp. 4th ed.

Woodhouse, J. H., Dziewonski, A. M. 1984. Mapping the upper mantle: three-dimensional modeling of Earth structure by inversion of seismic waveforms. *J. Geophys. Res.* 89:5953–86

Yoder, H. S. 1976. *Generation of Basaltic Magma*. Washington, DC: Natl. Acad. Sci. 265 pp.

Zeitler, P. K. 1985. Cooling history of the NW Himalaya, Parkistan. *Tectonics* 4:127–51

Zindler, A., Hart, S. 1986. Chemical geodynamics. *Ann. Rev. Earth Planet. Sci.* 14:493–571

Ann. Rev. Earth Planet. Sci. 1986. 14:417–54

PETROGENESIS OF ANDESITES

Timothy L. Grove and Rosamond J. Kinzler

Department of Earth, Atmospheric, and Planetary Sciences, Massachusetts
Institute of Technology, Cambridge, Massachusetts 02139

INTRODUCTION

Andesites are most commonly found at convergent plate margins, where
they constitute part of the erupted lavas of volcanic arcs that parallel
subduction zones. They are also found less frequently as eruptive products
in divergent plate boundaries, in association with midocean ridge basaltic
volcanism. Next to basalt, andesite is the most common volcanic rock type
on Earth. Andesites provide information on the complex tectonic and
magmatic processes that occur in subduction zones. An understanding of
the processes that lead to island-arc and convergent margin volcanic
activity can provide information on the relative roles played by cold subducted oceanic lithosphere, overlying mantle, and continental or oceanic
crust. Moreover, subduction zone magmatism is the process through
which continental crust has been extracted from the mantle. The parent
magmas of andesites, the differentiation paths followed en route to the
surface, and the magmatic cumulate residues left in the crust determine
the contribution of island-arc magmatism to continental crustal growth.

This review discusses the petrogenesis of andesites in the context of their
relation to specific rock series and to the magmatic processes that control
the development of the series. The review advocates andesites as derivative
magmas produced by crustal-level fractional crystallization of primary
basaltic magma derived by partial melting of mantle peridotite that has
been fluxed by a component released from the underlying subducted
oceanic slab [see Gill (1981) and Thorpe (1982) for reviews]. By fractional
crystallization we mean the continued removal of crystals from melt as they

417

form, which leads to a continuous variation in the composition of the residual liquid, called the liquid line of descent. Much of this review involves calculations modeling this process. The eruption of lavas that are following this liquid line of descent during different stages of the process leads to the development of rock series, which allows the process of fractional crystallization to be studied.

Several interrelated factors influence the liquid line of descent followed during fractional crystallization: the bulk composition of the primary (parental) magma, the fractionating phase assemblages, and the open-system magmatic processes (contamination, mixing). These factors are discussed with reference to rock suites that exemplify one of these controls, and relevant experimental phase equilibrium, mineralogical, geochemical, and isotopic data are used to determine the processes that led to the development of a rock series. Bulk composition of the parent melt is a key factor that can determine whether the liquid line of descent leads to the development of calc-alkaline or tholeiitic andesites. An important control is exercised by the proportions of silicate phases that crystallize early from the basaltic parent melt. The tholeiitic trend and its associated oceanic andesites or icelandites are produced by fractional crystallization of basalt at low pressures. The crystallization sequence is olivine, followed by plagioclase, and then by augite. Plagioclase dominates the assemblage, and the three-phase crystallization continues to a reaction point where olivine + liquid react to augite, plagioclase, and pigeonite. Total iron in the liquid increases throughout this crystallization process, and a dramatic increase in iron and a mild decrease in SiO_2 occur at the reaction point.

The calc-alkaline trend and its associated andesites develop when olivine, augite, and/or amphibole are the dominant crystallizing phases early in the differentiation history. These phases precipitate under conditions of moderate pressure and water undersaturation in the middle to upper crust and lead to a moderate increase in total iron, an increase in SiO_2, and a decrease in MgO in derivative liquids. Liquids derived by such a process can evolve to calc-alkaline andesites by continued crystallization involving plagioclase, orthopyroxene, pigeonite, augite, amphibole, and/or magnetite. The assimilation of a crustal component by a fractionating basaltic melt and the mixing of basaltic liquids with siliceous residual liquids produced by either fractionation or fractionation aided by assimilation are additional processes that assist in the production of calc-alkaline andesites. Magnetite, a phase that has been proposed to play a key role in the development of calc-alkaline andesites, is an important crystallizing phase in the later evolutionary stages that lead to andesites in both tholeiitic and calc-alkaline series.

Any discussion of andesites encounters problems in the definition of the term andesite and in the distinction between the two nonalkaline differentiation series that contain andesites—the tholeiitic and calc-alkaline series. Andesites, defined here as hypersthene-normative volcanic rocks with 53 to 63% SiO_2, constitute the intermediate part of tholeiitic and calc-alkaline rock series. Because rock series are under consideration, the distinction of andesite from its basalt and dacite neighbors within a series is not a key issue. However, the problem of distinguishing calc-alkaline from tholeiitic rock series is not as easily resolved. Throughout the history of naming these series and distinguishing one from another, it has been realized that clear-cut distinctions do not exist and that subtle differences separate a continuum of rock series from calc-alkaline to tholeiitic (Yoder & Tilley 1962, Kuno 1968, Jakes & White 1972, Miyashiro 1974). As a consequence, the classification schemes for distinguishing calc-alkaline vs tholeiitic trends do not provide clear distinctions between the trends, but they do supply criteria for separating or contrasting magma series within the continuum.

The existence of a tholeiitic trend was proposed by Fenner (1931, 1948) and Wager & Deer (1939) to be one characterized by iron enrichment with increasing differentiation. Rock series of this type are found at the ocean ridges and form in shallow crustal-level basaltic magma reservoirs in extensional environments. The calc-alkaline trend is characterized by slight iron enrichment or iron depletion with increasing differentiation. Rock series of this type are found in island-arc and continental convergent margin volcanoes.

The models for the origin of calc-alkaline andesites have been modified as more data on the compositional variations of rock series have become available. In the pre–plate tectonics era, when most of the known examples of rock series were from continental environments, field-based petrologists argued for an origin of andesites from parent basaltic melts in which contamination by melted silicic crust played an important part (Larsen et al 1938, Tilley 1950, Waters 1955). During this era, experimental petrologists, led by N. L. Bowen, argued that fractional crystallization of parent basalt was the dominant factor that controlled the development of calc-alkaline rock series. The problem became more complex when it was recognized that tholeiitic and calc-alkaline trends developed from roughly similar olivine-normative basaltic parents. If fractionation was responsible for producing these two trends, the processes that led to the calc-alkaline and tholeiitic trends must be different. Osborn (1959) proposed that the difference was caused by the appearance of an iron oxide phase early (in the case of the calc-alkaline trend) or late (in the case of the tholeiitic trend). But the magnetite hypothesis did not resolve the problem, because many calc-alkaline series basalts and basaltic andesites do not contain magnetite as an early phenocryst phase.

Following the acceptance of plate-tectonic theory, experimental petrologists turned away from the notion that andesites were part of rock series and considered the hypothesis that andesites could be primary melts produced by subduction-related processes [see Ringwood (1974) and Boettcher (1973) for reviews]. High pressure experimental investigations (Nicholls & Ringwood 1972, 1973, Mysen et al 1974) demonstrated that partial melting of the hydrated peridotite or eclogite source materials proposed to generate primary andesite magmas produced melts that were basaltic in composition. In many island-arc settings, basaltic lavas are often found in close association with andesite. Although the basalts need not be the parent of the andesite lavas, a parent-derivative relation warrants testing.

PARENT MAGMAS AND ROCK SERIES

Four magma series of the calc-alkaline type and three of the tholeiitic type have been chosen for discussion. These rock series are displayed on two of the diagrams that have been used in classifying calc-alkaline and tholeiitic lavas: the AFM diagram (Figures 1, 2) of Wager & Deer (1939), and the Miyashiro (1974) discrimination diagram of FeO^*/MgO vs SiO_2 (Figures 3, 4). Also shown on these diagrams are the boundaries defined by Irvine & Baragar (1971) and Miyashiro (1974) for distinguishing calc-alkaline and tholeiitic series. On both of the discrimination diagrams the series trends merge at their basaltic ends. Some overlap results because the different trends and their derivatives are generated from similar olivine-normative parent magmas. There are also significant compositional differences among the basalts chosen for this discussion. The mafic compositions range from midocean ridge tholeiites through high-alumina basalts to basanitoids. These differences in basaltic parent composition arise as a combination of the effects of segregation depth and degree of melting of the mantle source peridotite and subsequent differentiation upon separation from source. We show in what follows how the differences in parent melt composition can influence the differentiation trend that is followed and the andesite that is produced. We also show that roughly similar olivine-normative parent basalt compositions can produce derivative magmas that follow either a tholeiitic or calc-alkaline differentiation trend. The tholeiitic trends develop by fractional crystallization at low pressures. The calc-alkaline series develop by fractional crystallization at moderate crustal-level pres-

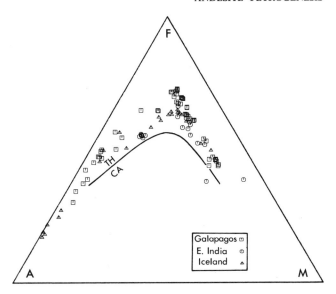

Figure 1   Tholeiitic lava series projected on the AFM diagram (A = $Na_2O + K_2O$, F = FeO $+ 0.9 \times Fe_2O_3$, M = MgO). Curve separates the fields of tholeiitic (TH) and calc-alkaline (CA) suites, as defined by Irvine & Baragar (1971). The Galapagos data set consists of compositions of glasses and rocks from the Galapagos spreading center (Clague & Bunch 1976, Byerly et al 1976, Byerly 1980, Clague et al 1981). The East Indian Ocean data set is composed of whole-rock compositions from the Ninetyeast Ridge (Frey et al 1977, Ludden et al 1980). The Iceland tholeiitic trend is defined by rocks from Thingmuli (Carmichael 1964).

sures, and in continental environments where open-system magmatic processes lead to contamination of basaltic magma by silicic crustal material.

If Figures 1 and 2 or 3 and 4 were combined, they would show a continuum of trends. The separation of tholeiitic from calc-alkaline trends is made here on the basis of tectonic setting. The calc-alkaline examples are chosen from island arcs and continental convergent margins. The tholeiitic lavas are from midocean ridge extensional environments. The tholeiitic trend that shows the most extreme iron enrichment on the AFM diagram is that of the Galapagos data set. In the following section, we show that this trend is produced by near-surface fractional crystallization of olivine-normative basalt. The other tholeiitic trends differ subtly from the Galapagos trend. These differences have arisen because differentiation occurred at elevated pressure. The calc-alkaline trend that shows the highest iron enrichment, the Witu Island trend, is produced primarily by fractional crystallization at low crustal-level pressures and low water partial pressure (p_{H_2O}). The Antilles trend shows less iron enrichment on the

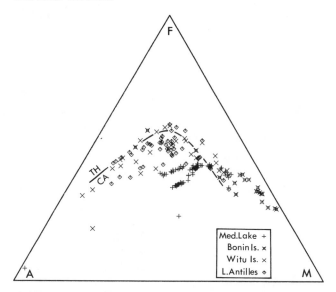

Figure 2   Calc-alkaline lava series projected on the AFM diagram (see Figure 1 caption). Bonin suite is whole-rock data from Dallwitz (1968) and Shiraki & Kuroda (1977), the Witu Island suite consists of whole-rock compositions from Johnson & Arculus (1978), and the Lesser Antilles suite is a compilation of whole-rock and averaged whole-rock analyses from Sigurdsson et al (1973), Arculus (1976), Brown et al (1977), and Dostal et al (1983).

AFM diagram and is produced by fractional crystallization at moderate crustal level pressures and higher p_{H_2O}. The trend that shows less iron enrichment than all others, the Medicine Lake series, is one in which open-system contamination by a silicic crustal component has played an important role.

BULK COMPOSITION AND FRACTIONAL CRYSTALLIZATION CONTROLS

Phase Relations in Natural Systems

This discussion uses phase diagrams to predict crystallization paths. The reader should consult Grove & Baker (1984) for discussion on the limitations of the use of projected phase equilibrium data in natural basalt systems. First, the phase relations that have been determined for natural basalt systems on a variety of bulk compositions, and under different physical conditions (P_{total}, f_{O_2}, and p_{H_2O}) are presented graphically. Second, the information on phase proportions and compositions from the experimental data are used as constraints for calculating crystallization paths. At 1 atm the existing experimental data allow nearly complete character-

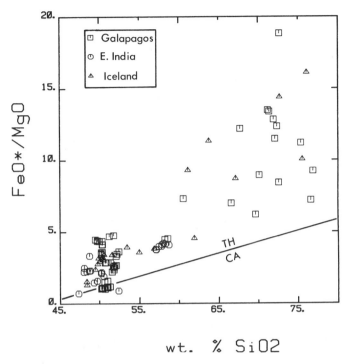

wt. % SiO2

Figure 3 Tholeiitic lava series projected on the FeO*/MgO vs SiO$_2$ variation diagram (FeO* = FeO + 0.9 × Fe$_2$O$_3$). The tholeiitic (TH) vs calc-alkaline (CA) dividing line is from Miyashiro (1974). Data sources for lava suites are listed in Figure 1 caption.

ization of the phase assemblage that coexists with liquid. At elevated pressures, the experimental information currently available can be used as a guide for estimating phase proportions.

One-atm Phase Relations

Results of 1-atm melting experiments performed at the quartz-fayalite-magnetite (QFM) buffer on lavas from Medicine Lake Highland, California (Grove et al 1982, 1983), on basalts from the Oceanographer Fracture Zone (Walker et al 1979), and on basalts from the FAMOUS area of the Mid-Atlantic Ridge (Grove & Bryan 1983) were used to construct the diagram shown in Figure 5. The CMAS projection scheme of O'Hara (1968) was modified and used to display the compositions of experimentally produced liquids saturated with plagioclase and Fe, Mg silicates. Analyses are recalculated from weight-percent oxides to the oxygen-normalized molar mineral components—olivine (oliv), clinopyroxene (cpx), silica (qtz),

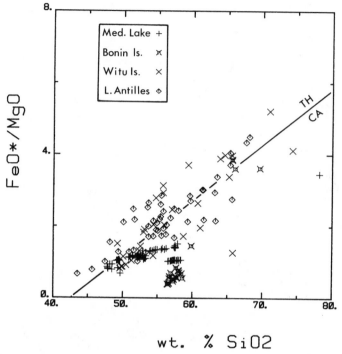

wt. % SiO2

Figure 4   Calc-alkaline lava series projected on the FeO*/MgO vs SiO$_2$ variation diagram (see Figure 3 caption). Data sources for lava suites are listed in Figure 2 caption.

plagioclase (plag), orthoclase, and spinel (see Figure 5 caption). This six-component system is simplified by projection from orthoclase and spinel into the tetrahedron oliv-cpx-qtz-plag. (In effect, the orthoclase and spinel components are calculated, then discarded, and the oliv, cpx, plag, and qtz components are renormalized.) A further simplification is made by projecting from plagioclase onto the pseudo-ternary oliv-cpx-qtz (Figure 5), and this pseudo-ternary diagram can be used qualitatively to infer the crystallization paths of plagioclase-saturated liquids. The 1-atm experiments on Medicine Lake, Oceanographer Fracture Zone, and FAMOUS basalts locate the olivine-augite(aug)-plagioclase cotectic and the augite-pigeonite(pig)-plagioclase cotectic. Experiments on Medicine Lake basaltic andesites, andesites, and a Shasta andesite determine the olivine-pigeonite-plagioclase, olivine-orthopyroxene(opx)-plagioclase, and orthopyroxene-pigeonite-plagioclase reaction curves. Grove et al (1982, 1983) and Grove & Bryan (1983) discuss the uncertainties and details of locating the phase boundaries and evaluate errors that limit the utility of this projection scheme.

The topology of the plagioclase-saturated projected phase diagram is similar to that of the system forsterite-diopside-silica (Kushiro 1972a, Longhi & Boudreau 1980) in several respects (Figure 6). With the exception of the protoenstatite field, which is absent in the plagioclase-saturated pseudo-ternary and present in forsterite(Fo)-diopside(Di)-qtz, the primary phase volumes have similar configurations. In both, the olivine-augite boundary is cotectic (mostly) and the olivine-pigeonite and olivine-orthopyroxene boundaries are reaction curves. The most important difference of the projection versus the Fo-Di-qtz ternary is that the projected liquids contain a variable plagioclase component that is removed by projection onto the plane oliv-cpx-qtz. The projection also suppresses the important effects of Fe-Mg and Ca-Na variation in crystalline phases and coexisting liquids, and it masks the effects of plagioclase and magnetite crystallization on the liquid line of descent. A further limitation of the diagram is that the low-Ca pyroxene phase volumes are only correct for liquids with $Mg/(Mg + Fe) < 0.75$ and low K_2O and TiO_2 contents. Longhi & Boudreau (1979) have determined that liquids saturated with augite, olivine, and plagioclase with $Mg/(Mg + Fe) > 0.75$ will reach a reaction point that involves orthopyroxene, rather than pigeonite.

Low-Pressure Liquid Lines of Descent Followed by Magmas of Contrasting Bulk Composition

The projected oliv-cpx-qtz pseudoternary diagram (Figure 5) contains two reaction points (A and B) terminal to the coexistence of olivine and liquid. These two reaction points require that the olivine-pigeonite-plagioclase reaction curve have a thermal divide (TD). The reaction coefficients for points A and B have been calculated for the pseudo-quarternary components by using a Schreinemakers analysis (Zen 1966) to obtain the stoichiometric coefficients for each reaction; these coefficients are (in weight proportions)

point A: 1 liquid + 0.18 oliv = 0.58 pig + 0.23 aug + 0.37 plag

point B: 1 liquid + 0.38 oliv = 0.37 opx + 0.58 pig + 0.43 plag.

As a first example, consider the low-pressure crystallization path characteristic of liquids with different bulk compositions. One is an olivine tholeiite (T; see Figure 7), and the second is a boninitic andesite (B). Under equilibrium conditions (not shown in Figure 7), T precipitates olivine + plagioclase as liquidus phases, and the residual liquid composition moves away from the oliv corner until it reaches the aug-oliv-plag cotectic. Here augite begins to crystallize, and the liquid follows the cotectic to point A. At reaction point A, liquid and olivine react to form augite + pigeonite +

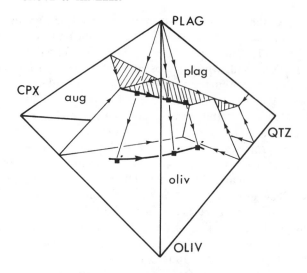

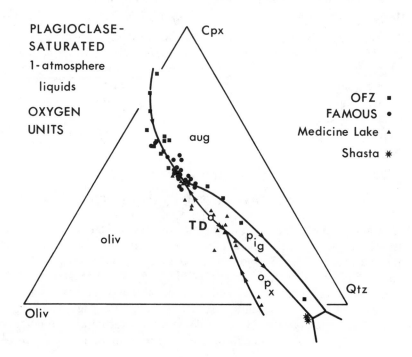

plagioclase. Any olivine-normative composition that lies to the left of a line oliv-TD crystallizes completely at point A to an assemblage of olivine-plagioclase-augite-pigeonite. During fractional crystallization of liquid T, all phases are removed from reaction with liquid continually. When point A is reached, the liquid moves away from A along the pigeonite-augite-plagioclase curve. Continued fractionation drives residual liquids toward the pigeonite-augite-plagioclase-quartz point. The path outlined in Figure 7 represents such a fractional crystallization path.

The lavas of the Galapagos spreading center are shown in the pseudo-ternary projection (Figure 8) as an example of the tholeiitic (iron-enrichment) trend. Grove & Baker (1984) modeled the crystallization path followed by the Galapagos suite using the results of 1-atm phase equilibrium experiments. The Galapagos samples include typical midocean ridge tholeiites, FeTi-rich ferrobasalts, andesites, and rhyodacites. The

---

Figure 5   (*Top*) A perspective drawing of part of the pseudo-quaternary system olivine(oliv)-clinopyroxene(cpx)-silica(qtz)-plagioclase(plag). The diagram shows some of the cotectics and reaction curves in the pseudo-quarternary and ternary subsystems. In the ternary subsystem oliv-cpx-plag, the oliv-cpx, plag-cpx, and oliv-plag cotectics are shown as light lines. In the pseudo-quaternary these cotectic lines extend as surfaces, and their intersection forms the oliv-cpx-plag boundary curve (dark line). The ruled surfaces bound the plagioclase primary phase volume. No attempt is made in this schematic drawing to show the complex low-Ca pyroxene phase relations. Experimentally produced liquids that plot on the oliv-cpx-plag boundary are represented by three squares. The projection of these liquids from pseudo-quaternary space to the oliv-cpx-qtz pseudo-ternary is accomplished by projecting on a line drawn from plagioclase through the liquid composition to the oliv-cpx-qtz base. In practice, this is done by renormalizing the oliv-cpx-qtz components to unity. These projected liquid compositions (shown by squares with an apostrophe) and the projected oliv-cpx-plag boundary are shown on the oliv-cpx-qtz pseudo-ternary. This plagioclase-saturated pseudo-ternary can be used to predict crystallization paths for plagioclase-saturated liquids, and the oliv-plag and cpx-plag surfaces (shown as ruled surfaces) are also projected onto the pseudo-ternary plane. (*Bottom*) The pseudo-ternary system oliv-cpx-qtz, constructed using plagioclase-saturated liquids. The squares and circles are glass compositions from experiments on midocean ridge basalts from the Oceanographer Fracture Zone (Walker et al 1979) and FAMOUS (Grove & Bryan 1983) regions, respectively. Triangles are glass compositions from experiments on Medicine Lake Highland lavas (Grove et al 1982), and stars are glass compositions from experiments on a Shasta andesite (T. L. Grove, unpublished data, 1979). This projection and all subsequent projections use oxygen units. The oxygen-based components are calculated by converting wt% oxides into mol% following the method described in Figure 3 of Grove et al (1982). The mole units of the projection are then weighted by the number of oxygens in each mineral component. With respect to SiO_2, olivine is weighted by a factor of 2, clinopyroxene by a factor of 3, and plagioclase by a factor of 4. These weighted units are renormalized and projected from plagioclase. Such a renormalization is useful for visualizing volume relations among phases and coexisting liquids. The abbreviations used to label primary phase regions are oliv (olivine), aug (augite), pig (pigeonite), and opx (orthopyroxene). TD marks a thermal divide on the reaction boundary.

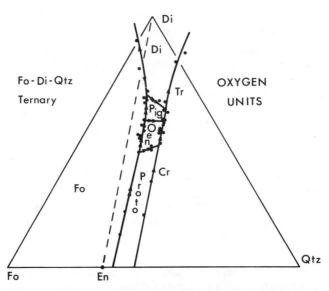

Figure 6 The forsterite(Fo)-diopside(Di)-SiO$_2$(qtz) ternary determined by Kushiro (1972a) and Longhi & Boudreau (1980) at 1 atm. The abbreviations used to label primary phase volumes are Fo (forsterite), Proto (proto-enstatite), Oen (orthoenstatite), Pig (pigeonite), Di (diopside), Tr (tridymite), and Cr (cristobalite).

glass data reported by Clague & Bunch (1976), Byerly et al (1976), Byerly (1980), and Clague et al (1981) from the region around 85°W and 95°W are projected onto the pseudo-ternary oliv-cpx-qtz in Figure 8, along with the cotectics and reaction curves from Figure 5. The typical midocean ridge tholeiites cluster near the oliv-aug-plag cotectic and contain olivine and plagioclase ± augite phenocrysts. Ferrobasalts cluster around point A and contain plagioclase, augite, and pigeonite(± magnetite) phenocrysts; these glass compositions record the olivine reaction relation. The andesites and rhyodacites plot on the pig-aug-plag cotectic near the qtz apex and are found in association with these phases and magnetite.

Iron enrichment that occurs during development of the tholeiitic series is a consequence of two combined effects. First, plagioclase dominates the low-pressure fractionation assemblage (experimentally determined cumulative proportions are 0.75 plag + 0.25 oliv and 0.58 plag + 0.25 aug + 0.17 oliv), which causes residual liquids to become depleted in the plagioclase component and increases the proportions of Fe, Mg silicate, and total iron in the melt. Second, further iron enrichment occurs at the reaction point A. The addition of an SiO$_2$-poor, FeO- and MgO-rich olivine component and the removal of augite and pigeonite that have higher Mg/Fe ratios than the dissolving olivine cause iron enrichment accompanied by a modest decline

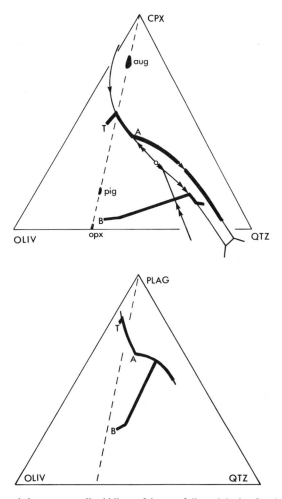

Figure 7   Schematic low-pressure liquid lines of descent followed during fractional crystallization of lavas series, with parent melts of contrasting bulk composition projected on the
pseudo-ternaries oliv-cpx-qtz and oliv-plag-qtz. One bulk composition (T) and its liquid line of
descent are traced from the Galapagos data set (Figure 8). The second series is from the
boninite (B) data set (Figure 9). The experimentally determined 1-atm cotectics are shown for
reference. Curves on the oliv-cpx-qtz projection are the plagioclase-saturated boundaries,
similar to those in Figure 5. The oliv-plag-qtz projection shows the clinopyroxene-saturated
oliv-aug-plag and pig-aug-plag boundaries determined at 1 atm. The oliv-opx reaction curve is
not shown, but the reaction line is crossed at the kink in the B trend.

in silica content in the residual liquid. Once olivine is exhausted from the assemblage, the liquid moves away from point A and follows the augite-pigeonite-plagioclase cotectic on a silica enrichment trend; eventual appearance of titanomagnetite causes iron depletion. Both effects mentioned above will be sensitive to variations of f_{O_2}, but for natural basalt compositions the sensitivity lies at f_{O_2} values that are more oxidizing than the NNO (nickel–nickel oxide) buffer. The 1-atm crystallization experiments performed by Grove et al (1982) on a high-alumina tholeiite

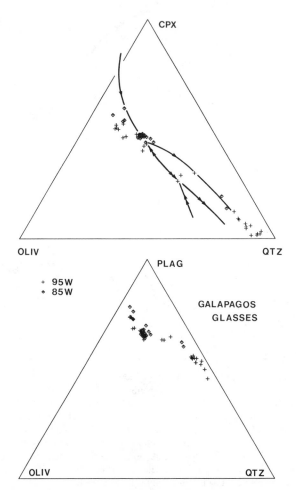

Figure 8   Compositions of glasses from 85°W and 95°W on the Galapagos spreading center projected into the oliv-cpx-qtz and oliv-plag-qtz pseudo-ternaries. One-atm cotectics are taken from Figures 5 and 7.

(performed at an f_{O_2} one order of magnitude more oxidizing than the NNO buffer) show development of an iron enrichment trend (13 wt% FeO at oliv-aug-plag saturation; see run 79-35g-41). Although magnetite saturation occurs at a higher temperature in these runs, magnetite appears late in the crystallizing assemblage (after augite), and Fe enrichment has occurred. At three orders of magnitude above NNO, spinel is a near-liquidus phase (see Table 3 of Grove et al 1982), and the FeO contents in the augite-saturated residual liquids are similar (11 wt%) to that of the starting composition.

The second example—a boninite (B; Figure 7)—has the low-pressure crystallization sequence olivine, then orthopyroxene, and finally plagioclase. Equilibrium crystallization (not shown in Figure 7) begins with olivine as the liquidus phase, and the residual liquid composition moves away from B along a line drawn from the oliv corner until it reaches the oliv-opx reaction curve. When the reaction curve is reached, olivine and liquid react and orthopyroxene precipitates, the proportion of olivine in the solid assemblage decreases, and the liquid moves down the reaction curve until all olivine is consumed. When olivine is depleted from the solid assemblage by reaction with liquid to orthopyroxene, the liquid enters the ortho-pyroxene primary phase volume, orthopyroxene crystallizes, and the residual liquid composition moves away from orthopyroxene. The liquid reaches the orthopyroxene + plagioclase primary phase volume, and finally the pigeonite-orthopyroxene-plagioclase boundary. Detailed information on the SiO_2-enriched part of this differentiation trend is not available. The olivine, orthopyroxene, and plagioclase phase appearances have been determined experimentally at 1 atm by Howard & Stolper (1981).

During fractional crystallization (shown in Figure 7), the liquid line of descent again begins with olivine crystallization. When the liquid reaches the oliv-opx reaction curve, olivine is removed from reaction with liquid, and continued fractional crystallization drives the liquid away from opx through the opx phase volume to the opx-plag boundary. Orthopyroxene and plagioclase crystallize, and the residual liquid moves to the opx-pig-plag reaction curve, and finally to silica-enriched residual compositions.

The lavas of the Bonin Islands in the Mariana arc are shown in the pseudo-ternary projection (Figure 9) as an example of a calc-alkaline trend that is produced by low-pressure fractional crystallization. The boninite series includes Mg-rich andesites, dacites, and some highly porphyritic cumulate rocks. The analyses plotted in Figure 9 are whole-rock data from Dallwitz (1968) and Shiraki & Kuroda (1977). The magnesian andesites contain phenocrysts of clinoenstatite and orthopyroxene as well as less abundant olivine (which is mantled by clinoenstatite). The groundmass of these andesites is glassy and contains microlites of pyroxene. These lava compositions are highly depleted in diopside and plagioclase components, and they plot well away from any low-pressure saturation boundary involving plagioclase. The 1-atm experiments performed on a parental boninitic melt show a 100°C temperature interval of orthopyroxene crystallization (Howard & Stolper 1981). Silicic andesites and dacites plot near the low-pressure plag–low-Ca pyroxene boundaries and contain phenocrysts of orthopyroxene and plagioclase, as well as augite and magnetite.

Like many of the calc-alkaline series, the boninites qualify as a calc-alkaline trend on one projection (the Miyashiro plot of FeO*/MgO vs

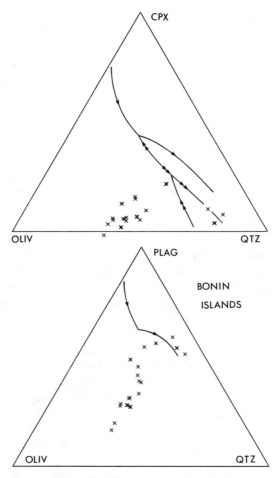

Figure 9 Compositions of lavas of the Bonin Islands in the Mariana arc projected into the oliv-cpx-qtz and oliv-plag-qtz pseudo-ternaries. One-atm cotectics are taken from Figures 5 and 7.

SiO_2; see Figure 4) but overlap with tholeiitic lavas on the other projection (the AFM plot; see Figure 2). In Figure 4 the boninite series shows increasing FeO^*/MgO with increasing SiO_2, and total FeO is constant or increases slightly, then decreases slightly as SiO_2 increases; hence, these lavas show a calc-alkaline trend. The FeO trend is generated during the large temperature interval over which a low-Ca pyroxene (either clino-enstatite, protoenstatite, or orthopyroxene) is the sole crystallizing phase. Magnetite does not appear until late in the crystallization sequence, after plagioclase.

The discussion above traced the low-pressure crystallization paths of two different initial bulk compositions. One of the composition trends corresponds to a tholeiitic trend, and the other corresponds to a rare and unusual type of calc-alkaline trend. The important control on the development of these two contrasting rock series is the bulk composition of the parent melt. The typical tholeiitic basalt follows an iron-enrichment trend because plagioclase is a dominant early crystallizing phase, and residual liquids are depleted in plagioclase component and enriched in FeO. The calc-alkaline trend displayed by the boninite series starts from a parent melt that is highly depleted in plagioclase component, and the variation in composition of the low-Ca pyroxene that crystallizes early controls the variation in the FeO content of the residual liquid. The boninite series is a comparatively rare one, and the parent to the series is also rare and unusual among island-arc parent magmas. Most of the parents to calc-alkaline series are olivine tholeiites, subalkaline tholeiites, or high-alumina basalts. These parent basalts do not resemble the boninite parent but instead more closely resemble the parents of the tholeiitic trend. In the following sections we show that differentiation at elevated pressure is responsible for producing the calc-alkaline trend commonly found in convergent margin volcanic arcs. Furthermore, the boninite trend is not typical of the calc-alkaline trends that provided the basis for the heated Bowen-Fenner controversy of the twenties and thirties. The calc-alkaline lavas known at that time were from continental convergent margins (Mt. Lassen, California, and Katmai, Alaska), and it is probable that these suites develop their iron-depletion trend as a consequence of contamination of a basaltic melt by a silicic crustal component. This issue is discussed in a following section.

Effects of Elevated Pressure on Phase Relations

Experimental results obtained at elevated pressures on simple and natural systems can be used to predict the effects of increasing pressure on the positions of the cotectics and reaction curves. The effects of increasing pressure under anhydrous conditions in the system Fo-Di-SiO_2 (Kushiro

1969) and in natural basalts (Kushiro 1974, Bender et al 1978, Stolper 1980, Takahashi & Kushiro 1983) are to shrink the olivine primary phase volume and to expand the orthopyroxene and augite phase volumes (Figure 10). The opposite effects occur under conditions of H_2O saturation; the olivine

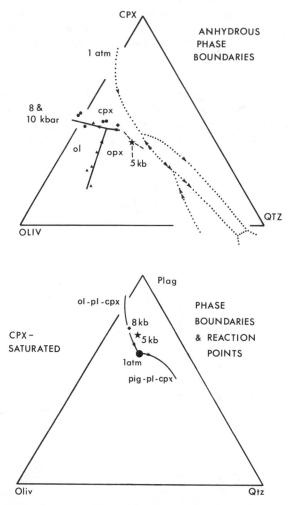

Figure 10 Effects of pressure on the phase volumes in the oliv-cpx-qtz and oliv-plag-qtz pseudo-ternaries. The 1-atm saturation surfaces are shown for reference. Experimentally produced liquids locate the position of anhydrous phase boundaries at 5, 8, and 10 kbar. Triangles locate the olivine-orthopyroxene surface (Stolper 1980), and circles locate the oliv-cpx-plag boundary (Thompson 1974, 1975) at 10 kbar. The diamond is the composition of a liquid saturated with oliv+cpx+plag±pig at 8 kbar (Kushiro 1973). The star locates the projected position of a liquid saturated with oliv+cpx+plag+opx at 5 kbar (Takahashi & Kushiro 1983).

phase volume expands at the expense of the pyroxene volumes (Kushiro 1972b). The effects of H_2O-undersaturated conditions on phase relations can be addressed qualitatively by using the experimental results of Spulber & Rutherford (1983) on a midocean ridge basalt and a Kilauea tholeiite (Figure 11). Compared with the 1-atm conditions, the olivine field has now

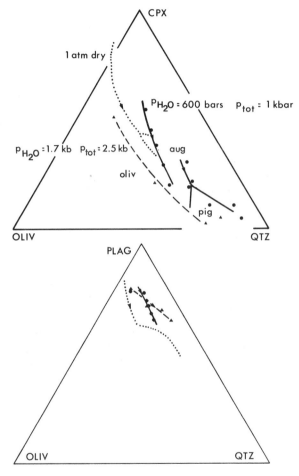

Figure 11   Experimentally produced liquids from experiments of Spulber & Rutherford (1983), performed at $P_{total} = 1$ kbar and $p_{H_2O} = 600$ to 700 bars (circles) and $P_{total} = 2.5$ kbar and $p_{H_2O} = 1.7$ kbar (triangles). On the upper diagram, experiments on a MORB from the Galapagos outline an oliv-aug-plag boundary. Experiments on a second composition, a Kilauea tholeiite, outline oliv-aug-plag and pig-aug-plag boundaries. Differences in the bulk compositions of the MORB and Kilauea tholeiite starting materials cause a shift of the pseudo-ternary projected positions. On the lower diagram, the experiments on the MORB at 1 kbar and 2.5 kbar at $p_{H_2O} < P_{total}$ saturated with oliv-aug-plag are projected on the oliv-plag-qtz pseudo-ternary, and these show the contraction of the plagioclase phase volume.

expanded, reaction point A has moved to a much higher qtz value, and the oliv-aug-plag boundary has moved to higher plagioclase contents. The effect of increasing total pressure on plagioclase stability in natural systems is not well known. The plagioclase phase volume contracts with increasing pressure (Kushiro 1974, Takahashi & Kushiro 1983), but plagioclase may be present as a crystallizing phase over a pressure interval that approaches 20 kbar (Baker & Eggler 1983). The effect of H_2O addition is to destabilize plagioclase and to increase the anorthite content of the liquidus plagioclase (Yoder 1968, Eggler 1972, Sekine et al 1979).

High-Pressure Liquid Lines of Descent

Grove & Baker (1984) have suggested that an important effect on the crystallization path followed by a basaltic parent melt is the proportion of plagioclase relative to Fe/Mg-bearing phases that crystallize early during the differentiation history of a basalt parent melt. Presnall et al (1979) found that in the system CaO-MgO-SiO_2-Al_2O_3, increased pressure shrinks the plagioclase phase volume and moves the forsterite-diopside-anorthite(An) piercing point toward the An apex. Thus, multiply saturated liquids are richer in An at higher pressures, and the plagioclase primary phase volume shrinks. Kushiro (1973) and Takahashi & Kushiro (1983) report natural basalt liquids saturated with oliv + aug + plag + pig at 8 kbar and oliv + aug + plag + opx at 5 kbar, respectively. These liquids contain a higher proportion of plagioclase than do similar 1-atm multiply saturated liquids in natural basalt compositions (Figure 10). The addition of water as a component aids pressure in decreasing the size of the plagioclase phase volume. The experiments of Spulber & Rutherford (1983) show a marked contraction of the plagioclase volume at water-undersaturated conditions (Figure 11) with increasing pressure. The proportion of plagioclase to olivine + augite that crystallizes at the oliv-aug-plag boundary at elevated water pressure is also reduced. As a result, the liquids that crystallize on the high-pressure oliv-aug-plag boundary are richer in plag component, since the proportion of plagioclase in the crystallizing assemblage is decreased. A schematic high-pressure liquid line of descent that is consistent with the high-pressure phase relations summarized above is shown in Figure 12. Olivine is the liquidus phase, and augite precipitates early and is accompanied by crystallization of an Al-rich, Si-poor calcic plagioclase. The three-phase crystallization assemblage contains nearly equal proportions (by weight) of oliv + aug + plag and moves the residual liquid toward SiO_2-rich and hypersthene-normative compositions. Continued crystallization at elevated pressure on the oliv-aug-plag cotectic moves the liquid beyond the 1-atm reaction point A. This liquid is a basaltic andesite, and if it separates from its oliv + plag + aug residue and moves to a shallow magma

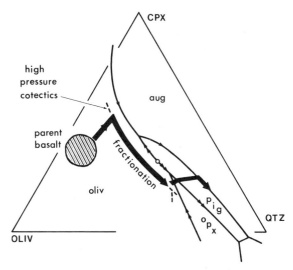

Figure 12   Proposed fractionation scheme that allows a tholeiitic basalt parent melt to follow a calc-alkaline trend through early crystallization of olivine + augite + calcic plagioclase at elevated pressure. The position of the high-pressure cotectic (dashed) is the inferred shift caused by increased P_{total} at water-undersaturated conditions. The liquid derived by high-pressure fractionation is separated to a shallow magma reservoir and evolves through low-Ca pyroxene + augite + plagioclase fractionation to silica-rich residual liquids.

reservoir, continued fractionation involves only oliv + plag, which drives the liquid toward andesitic compositions and to the oliv-pig-plag or oliv-opx-plag reaction curve. Removal of olivine from reaction with the melt allows the liquid to move into a low-Ca pyroxene + plagioclase phase volume, to intersect the aug-pig-plag cotectic, and finally to reach the aug-pig-plag-qtz point. Therefore, this process of polybaric fractional crystallization produces calc-alkaline andesites from olivine tholeiitic basalt parent magmas.

CALC-ALKALINE ANDESITES PRODUCED BY FRACTIONATION INVOLVING ANHYDROUS SILICATES

Grove & Baker (1984) proposed that one important control on the development of a calc-alkaline lava series from an olivine tholeiite parent was fractionation at elevated pressure, which would involve higher proportions of olivine and augite and a lesser proportion of plagioclase during the early stages of differentiation. Grove & Baker (1984) used the lavas of the Witu Islands (Johnson & Arculus 1978) as an example of a calc-alkaline trend in an island-arc setting, which was produced by fractionation at elevated pressure. When plotted in the oliv-cpx-qtz projection (Figure 13), the general trend of these lavas with advancing fractionation is one of increasing qtz component. When viewed in the oliv-plag-qtz projection, the general trend during the early part of the differentiation history is one of increasing plagioclase component, a distinctively different trend of calc-alkaline series lavas that is opposite the early trend followed by tholeiitic series (Figure 8). With advancing fractionation, iron increases gently, then

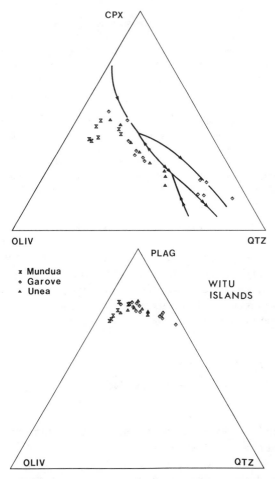

Figure 13 Compositions of phenocryst-poor lavas from the Witu Islands projected onto the oliv-cpx-qtz and oliv-plag-qtz pseudo-ternaries. One-atm cotectics are taken from Figures 5 and 7.

decreases (Figure 12 of Grove & Baker 1984), and on the AFM diagram (Figure 2) and the FeO^*/MgO vs SiO_2 plot (Figure 4) this lava series straddles the calc-alkaline and tholeiitic dividing lines. In the development of this calc-alkaline series, olivine and augite dominate the early crystallization assemblage, and plagioclase is calcic and precipitates in reduced proportions. Magnetite is an important mineral in the development of the chemical trends during the later evolutionary stages of the Witu calc-alkaline lavas.

CALC-ALKALINE ANDESITES PRODUCED BY FRACTIONATION INVOLVING AMPHIBOLE

In some calc-alkaline series amphibole plays an important role in the early fractionation history of basaltic parent melts. Amphibole has been documented as an early participant in the crystallization assemblage in rock series from the Lesser Antilles and the Aleutian arcs. The Aleutian series is described in papers by Kay et al (1982), Conrad & Kay (1984), and Kay & Kay (1985). The Lesser Antilles rocks are discussed in papers by Lewis (1973), Sigurdsson et al (1973), Arculus (1976), Brown et al (1977), Arculus & Wills (1980), and Dostal et al (1983). In both associations, the crystallization sequence has olivine as a liquidus phase, followed by augite, then amphibole, and finally plagioclase. This crystallization sequence is a result of the combined effects of increased pressure and an elevated water content of the parent magma. Experimental studies of Cawthorn & O'Hara (1976), Cawthorn et al (1973), Holloway & Burnham (1972), and Helz (1973, 1976) show that with increasing pressure at $P_{fluid} = P_{total}$, plagioclase is destabilized while the amphibole field expands to progressively higher temperatures. As an example, the water-undersaturated phase relations of a Kilauea tholeiite (Holloway & Burnham 1972) are reproduced in Figure 14. The appropriate crystallization sequence is stable above 3 kbar for the Kilauea tholeiite composition. This fractionation process involving amphibole is similar to the case described previously, which involved only anhydrous silicates. In both, the effect of fractionation at elevated pressure and at water-undersaturated conditions is to destabilize plagioclase. Furthermore, in both cases the proportion of plagioclase crystallization early in the differentiation history is reduced, and the derivative andesitic liquids have the plagioclase-rich bulk compositions that are diagnostic of fractionation at elevated total pressure and water content. The fractionation process involving early appearance of amphibole differs in having occurred at higher pressure (Figure 14), where amphibole appears before plagioclase. The Witu anhydrous trend resulted from fractionation at lower pressure.

As an example of a lava series where amphibole played an important role early in the differentiation history, we have chosen the lavas of the Lesser Antilles. Chemical analyses (Sigurdsson et al 1973, Arculus 1976, Brown et al 1977, Dostal et al 1983) of the lavas of the island arc and petrologic characterization of phenocryst assemblages and cognate magmatic inclusions in the lavas (Lewis 1973, Arculus & Wills 1980) provide sufficient information to characterize a fractionation assemblage for the lava series. The major problem in trying to model the lava series in the Lesser Antilles is the choice of a suitable parent. Chemical analyses of the lavas are plotted in Figure 15 on the pseudo-ternary along with the projected position of amphiboles and clinopyroxenes from xenoliths found in the lavas. The highly undersaturated lavas (ranging from basanitoid through basanites, alkali-olivine basalts, and subalkaline basalts) that plot to the left of the olivine-clinopyroxene sideline have only been found on Grenada. The trace, rare earth element, and major element compositions of these undersaturated lavas prohibit a parent-derivative relation between them and the calc-alkaline lavas (Sigurdsson et al 1973, Arculus 1976). In the pseudo-ternary projection (Figure 15), the lavas that project into silica-

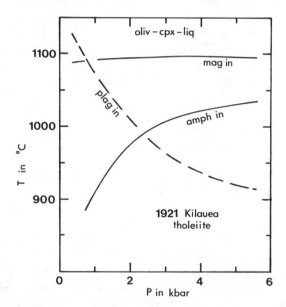

Figure 14 Pressure-temperature diagram at $p_{H_2O} < P_{total}$ determined on a 1921 Kilauea tholeiite (Holloway & Burnham 1972). Experiments were carried out at the nickel-nickel oxide buffer in an H_2O (40% molar)-CO_2 (60% molar) vapor. Lines show the effect of pressure on phase appearance sequence. Amphibole (amph) is stabilized to higher temperatures by increasing pressure.

undersaturated space cannot produce calc-alkaline derivatives by any fractionation process involving any combination of the phases olivine, amphibole, and clinopyroxene. Lavas that lie on the quartz-rich side of the oliv-cpx line in Figure 15 could fractionate to calc-alkaline derivatives.

Using the data available in the literature, we tried fractionation calculations involving olivine, augite, and amphibole with all available nepheline-normative compositions. The fractionation models (see Grove & Baker 1984) used phase assemblages and compositions that were con-

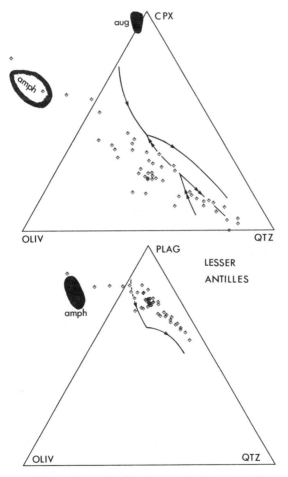

Figure 15 Compositions of Lesser Antilles lavas projected onto the oliv-cpx-qtz and oliv-plag-qtz pseudo-ternaries. Dark areas on the projections are the fields for amphibole and clinopyroxene from magmatic inclusions found in the lavas. One-atm cotectics are taken from Figures 5 and 7.

strained by the compositional information and paragenetic sequence recorded in the magmatic inclusions found in the lavas. The most primitive lava that produced a successful fractionation path to the andesites and dacites present in the Lesser Antilles suite was an average of several subalkaline high-alumina basalts from St. Kitts, which has a bulk composition that is in equilibrium with olivine, augite, amphibole, plagioclase, and magnetite (Arculus & Wills 1980). Fractional crystallization of these five phases (Table 1) produced an andesitic residual liquid. At this point in the model, the fractionation assemblage was changed. The plutonic blocks record reaction relations terminal to the coexistence of liquid with olivine and augite (Lewis 1973). The common textural manifestation of these reactions is the overgrowth of olivine by rims of orthopyroxene and the overgrowth/reaction rimming of augite by amphibole. These reaction relations have been recognized in experimental studies of amphibole stability in basalts and andesites (Helz 1973, 1976, Holloway & Burnham 1972). The compositions of Fe-Mg silicates in the Lesser Antilles plutonic blocks that mark the disappearance of olivine and amphibole (Arculus & Wills 1980) correspond to those calculated by the fractionation model. Further crystallization involves amphibole, plagioclase, orthopyroxene, and magnetite (Table 1), and it generates the dacitic derivative lavas, which are the most evolved lavas found in the arc.

CALC-ALKALINE ANDESITES PRODUCED BY FRACTIONATION ACCOMPANIED BY CONTAMINATION

In studies of Hakone volcano, Japan (Kuno 1950), the San Juan volcanic field, Colorado (Larsen et al 1938), Paricutin volcano, Mexico (Wilcox 1954), and Medicine Lake volcano, California, (Anderson 1941) petrographic and field evidence substantiated the operation of mixing of mafic and silicic magmas in generating magmas of intermediate composition. Isotopic variation within andesitic lavas from continental environments (James et al 1976, Briqueu & Lancelot 1979, DePaolo 1981) sometimes shows that the mantle-derived ratios of $^{87}Sr/^{86}Sr$, $^{143}Nd/^{144}Nd$, and $^{18}O/^{16}O$ have been modified with a crustal signature by assimilation. In these lavas, open-system behavior involving the incorporation of a crustal component by a mantle-derived melt is in operation. In other calc-alkaline volcanic systems, basaltic parent magma from a deeper part of the magmatic reservoir replenishes and mixes with fractionated silicic melt. In these systems the petrographic and geochemical signature of magma mixing should support mixing of two genetically related liquids, and isotopic systems should remain undisturbed.

Medicine Lake volcano provides an example of a calc-alkaline system where assimilation of a crustal component by basaltic parent magma has played an important role. Glassy lavas (T. L. Grove et al, unpublished data) from single, compositionally zoned lava flows are plotted in the oliv-cpx-qtz pseudo-ternary in Figure 16. These flows range from high-alumina basalt to andesite and constitute only the early part of the compositional spectrum from high-alumina basalt (HAB) to basaltic andesite (BA), andesite, dacite, and rhyolite that is present at this volcano (see Figures 10 and 11 of Grove & Baker 1984). Grove et al (1982) concluded from the elevated abundances of incompatible elements in the differentiated lavas and from the general

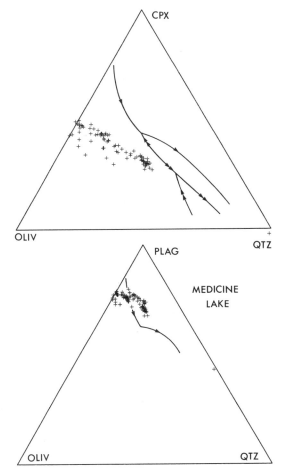

Figure 16 Compositions of Medicine Lake Highland lavas projected onto the oliv-cpx-qtz and oliv-plag-qtz pseudo-ternaries. One-atm cotectics are taken from Figures 5 and 7.

Table 1 Fractionation calculations for Lesser Antilles lavas

					wt% oxides				
Liquids									
Stage one	SiO$_2$	Al$_2$O$_3$	TiO$_2$	FeO	MgO	CaO	Na$_2$O	K$_2$O
Initial composition:[a]	48.23	20.30	1.01	9.56	6.21	12.40	2.12	0.21
Intermediate composition:[b]	55.68	18.37	0.99	8.27	4.05	8.57	3.50	0.58
Calculated composition:	55.34	18.37	0.83	8.38	4.21	9.25	3.23	0.40
Stage two								
Initial composition:[b]	55.68	18.37	0.99	8.27	4.05	8.57	3.50	0.58
Evolved composition:[c]	66.36	16.61	0.47	4.59	1.48	5.46	3.54	1.49
Calculated composition:	66.33	16.31	0.11	4.49	1.40	6.15	4.26	0.95

Equilibrium phases	Oliv Fo	Aug Mg#	Amph Mg#	Plag An	Mag	Opx En
Stage one						
Compositions predicted to be in equilibrium with range of liquids:	0.78–0.73	0.82–0.77	0.75–0.69	0.87–0.76		
Proportions removed in calculation:	0.06	0.16	0.165	0.54	0.075	
Variation of F (fraction of liquid remaining):			1.0–0.42			
Stage two	Oliv Fo	Aug Mg#	Amph Mg#	Plag An	Mag	Opx En
Compositions predicted to be in equilibrium with range of liquids:			0.69–0.60	0.73–0.63		0.73–0.65
Proportions removed in calculation:	0.00	0.00	0.39	0.47	0.09	0.05
Variation of F:			0.42–0.23			
	Oliv Fo	Aug Mg#	Amph Mg#	Plag An		Opx En
Compositions observed in inclusions and lavas:	0.82–0.64	0.83–0.70	0.86–0.38	1.0–0.36		0.77–0.43

[a] Average of 9 analyses of 48–50 wt% SiO_2 lavas from St. Kitts (Brown et al 1977).
[b] Average of 8 analyses of andesites from St. Vincents (Dostal et al 1983).
[c] Average of 60 analyses of dacites from Grenada, Dominica, and St. Kitts (Brown et al 1977).

trend of increasing $^{87}Sr/^{86}Sr$ ratios with increasing SiO_2 content that assimilation of a radiogenic crustal component accompanied fractional crystallization. Textural evidence in some intermediate lavas indicates that magma mixing of basaltic and more silicic lavas also operated to produce some intermediate lavas. These compositionally zoned flows record the contamination of parent HAB by a silicic crustal component. The HAB lavas contain forsteritic olivine (Fo_{90-84}) and calcic plagioclase (An_{86-80}) as sparse microphenocrysts, and the andesites contain oliv-plag phenocrysts and xenocrysts of quartz, sodic andesine, and Fe-rich orthopyroxene. Additionally, xenoliths of partly fused crustal material are present in the contaminated andesites.

These compositionally zoned flows record a process of combined assimilation and fractional crystallization (AFC). An AFC model that produces a close match to the Medicine Lake compositionally zoned lava flows requires that 1–1.4 g of assimilation accompany fractionation of 1 g of basalt. This large ratio of assimilation to fractionation violates simple heat-budget calculations. In these calculations it is assumed that basalt undergoes fractional crystallization to supply both the heat required to raise the crust to its solidus and the latent heat necessary for melting (Bowen 1928, Chap. 10). The heat required for melting crust depends on the temperature differential between basalt and crust and on the latent heats of basalt and crust. Estimates from thermochemical data (Hon & Weill 1982) indicate that latent heats of melting for basalt and granite are 100 and 50 cal g^{-1}, respectively. If the temperature of the basalt is assumed to be 1200°C and the solidus of the crustal assimilant is 1000°C, the amount of material that can be assimilated per gram of basaltic liquid crystallized at upper-crustal levels (6 km, $T = 200°C$), is approximately 3 g of basalt to melt 1 g of wall rock (assuming a specific heat of 0.33 cal g^{-1} °C^{-1}). This discrepancy can be reconciled in the context of geologic evidence. The proportion of contaminated andesite in these flows is less than 5% of the total erupted volume. Hence, the magma system contained a large volume of basalt that could supply heat to melt the assimilant. The large volume of basalt crystallizes and provides heat to drive the process, and the melted assimilant dilutes only a small fraction of the total mass of basalt in the magma reservoir. The resulting AFC-produced andesitic lavas could be boundary-layer liquids that formed at the margins, collected near the top of the chamber, and were sampled during an eruptive cycle of the system.

The addition of a silicic component by a continuous assimilation process, by mixing an evolved silicic melt produced by fractionation of basalt, or by AFC can have a dramatic effect on the resulting liquid line of descent. In fact, these two processes move liquids to reaction curves and phase volumes that would never be reached by fractional crystallization alone. The employment of the process of assimilation as a mechanism for producing the calc-alkaline series has had a long history. A calculation similar to the one presented above was carried out for lavas from Paricutin volcano, Mexico (Wilcox 1954, Bryan et al 1969), where it was demonstrated that either magnetite fractionation or assimilation could cause the observed compositional variation. The choice in favor of assimilation at Paricutin and in many field-based studies of calc-alkaline andesites in continental settings (Waters 1955, Tilley 1950) was made on field evidence in the form of granitic xenoliths in the andesite lavas. It is clear from the discussion above that a combination of field, petrologic, and isotopic evidence is required to quantify the assimilation process.

A variant of the process described above involves the remixing of an evolved melt produced by fractionation with a new pulse of basaltic liquid. Such a process would seem a logical consequence of a long-lived volcanic system, where the same magmatic plumbing system is used for each eruptive cycle. Sakuyama (1979, 1981) has made the case for this process at Shirouma-Oike and Myoko volcanoes in Japan. In such a system, one would expect to find volcanic products that record the fractionation path followed by basalt to the evolved dacitic liquids, as well as the recombined mixes of mafic basalt and evolved dacite. Sakuyama (1981) names these two trends the N and R types, respectively, and suggests that they correspond to Kuno's pigeonitic and hypersthenic trends. In Figure 17 we show the pigeonite and hypersthene trends based on Kuno's original data (Kuno 1950, 1968). Kuno devised his classification scheme for these lavas based on their groundmass low-Ca pyroxene. Note that the groundmass pyroxene is one that would be predicted by the 1-atm phase relations to be the first pyroxene to crystallize after olivine and plagioclase (Figure 17). Kuno identified the hypersthenic trend as one produced by a contamination/mixing process, and its trend on the pseudo-ternary is consistent with such a process. Hypersthene becomes the groundmass pyroxene because mixing moves the bulk composition to an SiO_2-rich part of the plagioclase-olivine primary phase volume next to the orthopyroxene-plagioclase phase volume. The pigeonitic series parallels a trend that would be followed by fractional crystallization of an olivine-normative tholeiitic parent at crustal-level pressures, similar to that of the Witu Island suite.

CONCLUSIONS

This review has emphasized the role of andesites as part of the continuum of a differentiation series. In several well-characterized geologic examples we have (a) shown that either calc-alkaline or tholeiitic derivatives can be generated from similar olivine-normative parent basalts, (b) documented

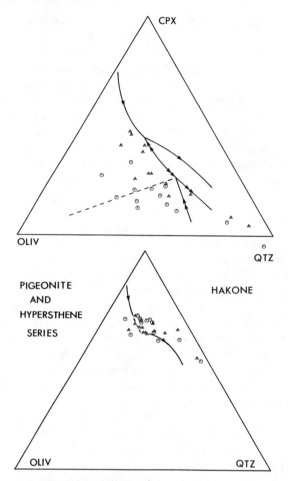

Figure 17   Compositions of rocks of the pigeonite and hypersthene series from Hakone volcano, Japan, projected onto the oliv-cpx-qtz and oliv-plag-qtz pseudo-ternaries (Kuno 1950, 1968). Pigeonite series lavas are plotted as triangles. Hypersthene series lavas are plotted as octagons. One-atm cotectics are taken from Figures 5 and 7. Dashed line separates lavas that would crystallize groundmass pigeonite and hypersthene during low-pressure crystallization. Lavas that project toward the oliv-cpx sideline would crystallize pigeonite, whereas lavas that project toward the oliv-qtz sideline would crystallize hypersthene.

the divergence of crystallization paths and derivative lavas produced by differentiation of compositionally distinct parent melts at low pressure, and (c) demonstrated the important role played by the open-system magmatic processes of crustal contamination and magma mixing in generating andesitic derivatives. Important controls that determine whether an olivine-normative tholeiitic basalt parent will follow the path to a calc-alkaline vs a tholeiitic derivative andesite are the pressure of differentiation and the H_2O content of the parent melt. Low-pressure differentiation produces the tholeiitic derivatives, and elevated pressure produces calc-alkaline derivatives.

In the preceding discussions the processes of differentiation of parental basaltic magma at increased pressure and water content and of contamination of basalt magma by assimilation of silicic crust have been held responsible for generating the diversity of differentiation trends and the variety in derivative andesites. Figures 18 and 19 show that the calculated fractional crystallization paths shift systematically in response to these controls. Differentiation at low pressures produces the tholeiitic series, and derivative andesites are enriched in normative ferromagnesian components and depleted in normative plagioclase component. Differentiation at increased pressure and elevated water content reduces the proportion of plagioclase in the early crystallizing assemblage and systematically lowers the iron enrichment in differentiation paths followed on the AFM and

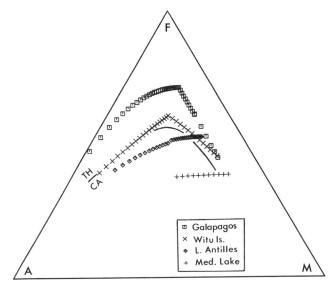

Figure 18   This AFM diagram contains the calculated fractional crystallization paths for the lava series discussed in this review. The model liquid line of descent for the Galapagos tholeiitic trend is found in Table 1 of Grove & Baker (1984). The calculated liquid line of descent for the Witu Islands calc-alkaline suite is found in Table 2 of Grove & Baker (1984). The fractionation model for the Lesser Antilles is found in Table 1 of the present review. The Medicine Lake trend is calculated assuming 20 wt% fractionation of 0.65-g plagioclase and 0.35-g olivine accompanied by assimilation of 1.5 g of silicic crust (see Table 3 of Grove & Baker 1984).

FeO*/MgO vs SiO_2 projections. The Witu Island trend is generated at lower pressures and lower magmatic water contents, and the Lesser Antilles trend is produced at higher crustal-level pressures and higher water contents. The dramatic iron depletion trend found in the Medicine Lake suite results from dilution of normative ferromagnesian components by assimilation of silicic continental crust. The two effects cannot be separated on these commonly used discrimination diagrams, but other geochemical and isotopic criteria can distinguish the processes. Our preferred discrimination test for fractionation vs contamination is to use incompatible trace element abundances in parent and derivative lavas to test fractionation

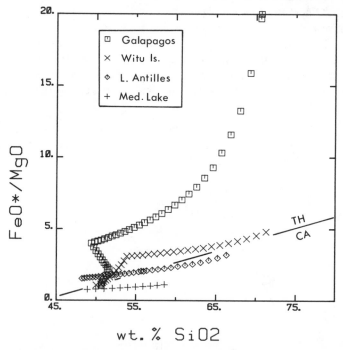

Figure 19 This FeO*/MgO vs SiO_2 diagram contains the calculated liquid lines of descent for the lava series discussed in this review (see Figure 18 for data sources). The kinks in the Galapagos model correspond to changes in the crystallization assemblage. The initial increase in FeO*/MgO and SiO_2 occurs during oliv + plag + aug crystallization. At the reaction point, fractional crystallization by zoning overgrowth generates an increase in FeO*/MgO and a decrease in SiO_2. The final stage of fractional crystallization of aug + pig + plag + mag produces an increase in FeO*/MgO and SiO_2. Both calc-alkaline fractionation trends (Witu Islands and Lesser Antilles) contain kinks where olivine disappears from the fractionation assemblage. The Medicine Lake trend is calculated from high-alumina basalt to silicic andesite. Figures 18 and 19 show systematic lowering of iron enrichment of fractionation paths in response to increased pressure and water content of parent magma and to contamination by assimilation of a silicic crustal component.

models. Isotopic differences between parent and derivative are also diagnostic.

The magmatic differentiation trends described here are found in different tectonic settings. The tholeiitic trend forms at divergent plate margins, specifically midocean ridges, where the crust is thin and parent basalts reach the surface having experienced little differentiation at crustal pressures. When differentiation occurs, it is at low pressures within the thin oceanic lithosphere. The low-pressure calc-alkaline trend [called island-arc tholeiite by Jakes & Gill (1970)] is found in young island arcs, where subduction rates are fast and the overlying crust is thin. The New Britain, South Sandwich, and Tonga-Kermadec arcs are examples [see Gill (1981) and Thorpe (1982) for summaries]. The higher pressure, higher water content calc-alkaline trend in the Lesser Antilles is developed in an arc where the crust is thicker and subduction rates are slower. The crustal contamination trend is found at continental convergent margins. The island-arc differentiation story is not always simple. Kay & Kay (1985) have recognized the development of both lower and higher pressure calc-alkaline trends in the Aleutian arc, and the Japan arc contains examples of all three trends. Finally, there are, no doubt, differentiation trends in existence that have not been discussed here, or even anticipated. Situations where tholeiitic parent magmas encounter and interact with unusual crustal material or unusually O_2-rich environments can be imagined.

A final consequence of this discussion of andesite petrogenesis is that the component added to developing continental crust must be on average basaltic. The fractional crystallization models proposed to explain the development of island-arc andesitic lavas generate mafic cumulates at midcrustal depths and gabbroic cumulates at shallower levels, and they concentrate andesite and more evolved differentiates at the shallowest levels. Such mafic cumulates are the consequence of the substantial amounts of fractional crystallization required to produce andesitic lavas.

ACKNOWLEDGMENTS

Research for this paper was supported through NSF Grants OCE-8315394 and EAR-8407829. The authors thank M. B. Baker, F. A. Frey, J. Donnelly-Nolan, and T. Sisson for stimulating discussions on the problem of andesite petrogenesis.

Literature Cited

Anderson, C. A. 1941. Volcanoes of the Medicine Lake Highland, California. *Univ. Calif. Dep. Geol. Sci. Bull.* 25:347–442

Arculus, R. J. 1976. Geology and geochemistry of the alkali basalt-andesite association of Grenada, Lesser Antilles island arc. *Geol. Soc. Am. Bull.* 87:612–24

Arculus, R. J., Wills, K. J. A. 1980. The petrology of plutonic blocks and inclusions from the Lesser Antilles island arc. *J. Petrol.* 21:743–99

Baker, D. R., Eggler, D. H. 1983. Fractionation paths of Atka (Aleutians) high-alumina basalts: constraints from phase relations. *J. Volcanol. Geotherm. Res.* 18: 387–404

Bender, J. F., Hodges, F. N., Bence, A. E. 1978. Petrogenesis of basalts from the project FAMOUS area: experimental study from 0 to 15 kbars. *Earth Planet. Sci. Lett.* 41: 277–302

Boettcher, A. L. 1973. Volcanism and orogenic belts—the origin of andesites. *Tectonophysics* 17: 223–40

Bowen, N. L. 1928. *The Evolution of the Igneous Rocks.* Princeton, NJ: Princeton Univ. Press. 332 pp.

Briqueu, L., Lancelot, J. R. 1979. Rb-Sr systematics and crustal contamination models for calc-alkaline igneous rocks. *Earth Planet. Sci. Lett.* 43: 385–96

Brown, G. M., Holland, J. G., Sigurdsson, H., Tomblin, J. F., Arculus, R. J. 1977. Geochemistry of the Lesser Antilles volcanic island arc. *Geochim. Cosmochim. Acta* 41: 785–801

Bryan, W. B., Finger, L. W., Chayes, F. 1969. Estimating proportions in petrographic mixing equations by least squares approximation. *Science* 163: 926–27

Byerly, G. R. 1980. The nature of differentiation trends in some volcanic rocks from the Galapagos spreading center. *J. Geophys. Res.* 85: 3797–3810

Byerly, G. R., Melson, W. G., Vogt, P. R. 1976. Rhyodacites, andesites, ferro-basalts and ocean tholeiites from the Galapagos spreading center. *Earth Planet. Sci. Lett.* 30: 215–21

Carmichael, I. S. E. 1964. The petrology of Thingmuli, a Tertiary volcano in eastern Iceland. *J. Petrol.* 5: 435–60

Cawthorn, R. G., O'Hara, M. J. 1976. Amphibole fractionation in calc-alkaline magma genesis. *Am. J. Sci.* 276: 309–29

Cawthorn, R. G., Curran, E. B., Arculus, R. J. 1973. A petrogenetic model for the origin of the calc-alkaline suite of Grenada, Lesser Antilles. *J. Petrol.* 14: 327–37

Clague, D. A., Bunch, T. E. 1976. Formation of ferrobasalt at east Pacific mid-ocean spreading centers. *J. Geophys. Res.* 81: 4247–56

Clague, D. A., Frey, F. A., Thompson, G., Rindge, S. 1981. Minor and trace element geochemistry of volcanic rocks dredged from the Galapagos spreading center: role of crystal fractionation and mantle heterogeneity. *J. Geophys. Res.* 86: 9469–82

Conrad, W. K., Kay, R. K. 1984. Ultramafic and mafic inclusions from Adak Island: crystallization history and implications for the nature of primary magmas and crustal evolution in the Aleutian arc. *J. Petrol.* 25: 88–125

Dallwitz, W. B. 1968. Chemical composition and genesis of clinoenstatite-bearing volcanic rocks from Cape Vogel, Papua: a discussion. *Proc. Int. Geol. Congr., 23rd, Prague,* 2: 229–42

DePaolo, D. J. 1981. Trace element and isotopic effects of combined wall rock assimilation and fractional crystallization. *Earth Planet. Sci. Lett.* 53: 189–202

Dostal, J., Dupuy, C., Carron, J. P., le Guen de Kerneizon, M., Maury, R. C. 1983. Partition coefficients of trace elements: application to volcanic rocks of St. Vincent, West Indies. *Geochim. Cosmochim. Acta* 47: 525–33

Eggler, D. H. 1972. Water-saturated and undersaturated melting relations in a Paricutin andesite and an estimate of water content in the natural magma. *Contrib. Mineral. Petrol.* 34: 261–71

Fenner, C. N. 1931. The residual liquids of crystallizing magmas. *Mineral. Mag.* 22: 539–60

Fenner, C. N. 1948. Immiscibility of igneous magmas. *Am. J. Sci.* 246: 465–502

Frey, F. A., Dickey, J. S., Thompson, G., Bryan, W. B. 1977. Eastern Indian Ocean DSDP sites: correlations between petrography, geochemistry and tectonic setting. In *A Synthesis of Deep Sea Drilling in the Indian Ocean,* ed. J. R. Heirtzler, J. G. Sclater, pp. 189–257. Washington, DC: US Gov. Print. Off.

Gill, J. B. 1981. *Orogenic Andesites and Plate Tectonics.* New York: Springer-Verlag. 390 pp.

Grove, T. L., Baker, M. B. 1984. Phase equilibrium controls on the tholeiitic versus calc-alkaline differentiation trends. *J. Geophys. Res.* 89: 3253–74

Grove, T. L., Bryan, W. B. 1983. Fractionation of pyroxene-phyric MORB at low pressure: an experimental study. *Contrib. Mineral. Petrol.* 84: 293–309

Grove, T. L., Gerlach, D. C., Sando, T. W. 1982. Origin of calc-alkaline series lavas at Medicine Lake volcano by fractionation, assimilation and mixing. *Contrib. Mineral. Petrol.* 80: 160–82

Grove, T. L., Gerlach, D. C., Sando, T. W., Baker, M. B. 1983. Origin of calc-alkaline series lavas at Medicine Lake volcano by fractionation, assimilation and mixing: corrections and clarifications. *Contrib. Mineral. Petrol.* 82: 407–8

Helz, R. T. 1973. Phase relations of basalts in their melting range at $p_{H_2O} = 5$ kb as a function of oxygen fugacity. *J. Petrol.* 14: 249–302

Helz, R. T. 1976. Phase relations of basalts in their melting range at $p_{H_2O} = 5$ kb. Part II. Melt compositions. *J. Petrol.* 17: 139–93

Holloway, J. R., Burnham, C. W. 1972. Melting relations of basalt with equilibrium water pressure less than total pressure. *J. Petrol.* 13:1–29

Hon, R., Weill, D. F. 1982. Heat balance of basaltic intrusion vs granitic fusion in the lower crust. *Eos, Trans. Am. Geophys. Union* 63:470 (Abstr.)

Howard, A. H., Stolper, E. 1981. Experimental crystallization of boninites from the Mariana Trench. *Eos, Trans. Am. Geophys. Union* 62:1091 (Abstr.)

Irvine, T. N., Baragar, W. R. A. 1971. A guide to the chemical classification of the common volcanic rocks. *Can. J. Earth Sci.* 8:523–48

Jakes, P., Gill, J. B. 1970. Rare earth elements and the island arc tholeiitic series. *Earth Planet. Sci. Lett.* 9:17–28

Jakes, P., White, A. J. R. 1972. Major and trace element abundances in volcanic rocks of orogenic areas. *Geol. Soc. Am. Bull.* 83:29–40

James, D. E., Brooks, C., Cuyubamba, A. 1976. Andean Cenozoic volcanism: magma genesis in the light of strontium isotopic composition and trace element geochemistry. *Geol. Soc. Am. Bull.* 87:592–600

Johnson, R. W., Arculus, R. J. 1978. Volcanic rocks of the Witu Islands, Papua New Guinea: the origin of magmas above the deepest part of the New Britain Benioff zone. *Bull. Volcanol.* 41:609–55

Kay, S. M., Kay, R. W. 1985. Aleutian tholeiitic and calc-alkaline magma series. I: the mafic phenocrysts. *Contrib. Mineral. Petrol.* 90:276–90

Kay, S. M., Kay, R. W., Citron, G. P. 1982. Tectonic controls of Aleutian arc tholeiitic and calc-alkaline magmatism. *J. Geophys. Res.* 87:4051–72

Kuno, H. 1950. Petrology of Hakone volcano and the adjacent areas, Japan. *Geol. Soc. Am. Bull.* 61:957–1020

Kuno, H. 1968. Differentiation of basalt magmas. In *Basalts*, ed. H. H. Hess, A. Poldevàart, 2:623–88. New York: Interscience

Kushiro, I. 1969. The system forsterite-diopside-silica with and without water at high pressures. *Am. J. Sci.* 267A:269–94

Kushiro, I. 1972a. Determination of liquidus relations in synthetic silicate systems with electron probe analysis: the system forsterite-diopside-silica at 1-atmosphere. *Am. Mineral.* 57:1260–71

Kushiro, I. 1972b. Effect of water on the composition of magmas formed at high pressures. *J. Petrol.* 13:311–34

Kushiro, I. 1973. Origin of some magmas in oceanic and circumoceanic regions. *Tectonophysics* 17:211–22

Kushiro, I. 1974. On the nature of silicate melt and its significance in magma genesis: regularities in the shift of the liquidus boundaries involving olivine, pyroxene, and silica minerals. *Am. J. Sci.* 275:411–31

Larsen, E. S., Irving, J., Gonyer, F. A., Larsen, E. S. III. 1938. Petrologic results of a study of the minerals from the tertiary volcanic rocks of the San Juan region, Colorado. *Am. Mineral.* 23:227–57, 417–29

Lewis, J. F. 1973. Petrology of the ejected plutonic blocks of the Soufrière volcano, St. Vincent, West Indies. *J. Petrol.* 14:81–112

Longhi, J., Boudreau, A. E. 1979. Pyroxene liquidus fields in basaltic liquids at low pressure. *Lunar Planet. Sci. X*, pp. 739–41 (Abstr.)

Longhi, J., Boudreau, A. E. 1980. The orthoenstatite liquidus field in the system forsterite-diopside-silica at one atmosphere. *Am. Mineral.* 65:563–73

Ludden, J. N., Thompson, G., Bryan, W. B., Frey, F. A. 1980. The origin of lavas from the Ninetyeast Ridge, eastern Indian Ocean: an evaluation of fractional crystallization models. *J. Geophys. Res.* 85:4405–20

Miyashiro, A. 1974. Volcanic rock series in island arcs and active continental margins. *Am. J. Sci.* 274:321–55

Mysen, B. O., Kushiro, I., Nicholls, I. A., Ringwood, A. E. 1974. A possible mantle origin for andesite magmas: discussion and replies. *Earth Planet. Sci. Lett.* 21:221–29

Nicholls, I. A., Ringwood, A. E. 1972. Production of silica-saturated magmas in island arcs. *Earth Planet. Sci. Lett.* 17:243–46

Nicholls, I. A., Ringwood, A. E. 1973. Effects of water on olivine stability in tholeiite and the production of silica-saturated magmas in the island-arc environment. *J. Geol.* 81:285–300

O'Hara, M. J. 1968. The bearing of phase equilibria studies in synthetic and natural systems on the origin and evolution of basic and ultrabasic rocks. *Earth Sci. Rev.* 4:60–133

Osborn, E. F. 1959. Role of oxygen pressure in the crystallization and differentiation of basaltic magma. *Am. J. Sci.* 257:609–47

Presnall, D. C., Dixon, J. R., O'Donnell, T. H., Dixon, S. A. 1979. Generation of mid-ocean ridge tholeiites. *J. Petrol.* 20:3–35

Ringwood, A. E. 1974. The petrological evolution of island arc systems. *J. Geol. Soc. London* 130:183–204

Sakuyama, M. 1979. Evidence of magma mixing: petrological study of Shiroumaoike calc-alkaline andesite volcano, Japan. *J. Volcanol. Geotherm. Res.* 5:179–208

454   GROVE & KINZLER

Sakuyama, M. 1981. Petrological study of the Myoko and Kurohime volcanoes, Japan: crystallization sequence and evidence for magma mixing. *J. Petrol.* 22: 553–83

Sekine, T., Katsura, T., Aramaki, S. 1979. Water saturated phase relations of some andesites with application to the estimation of the initial temperature and water pressure at the time of eruption. *Geochim. Cosmochim. Acta* 43: 1367–76

Shiraki, K., Kuroda, N. 1977. The boninite revisited. *J. Geol. Soc. Jpn.* 86: 34–50

Sigurdsson, H., Tomblin, J. F., Brown, G. M., Holland, J. G., Arculus, R. J. 1973. Strongly undersaturated magmas in the Lesser Antilles island arc. *Earth Planet. Sci. Lett.* 18: 285–95

Spulber, S. D., Rutherford, M. J. 1983. The origin of rhyolite and plagiogranite in oceanic crust: an experimental study. *J. Petrol.* 24: 1–25

Stolper, E. 1980. A phase diagram for mid-ocean ridge basalts: preliminary results and implications for petrogenesis. *Contrib. Mineral. Petrol.* 74: 13–27

Takahashi, E., Kushiro, I. 1983. Melting of a dry peridotite at high pressures and basalt magma genesis. *Am. Mineral.* 68: 859–79

Thompson, R. N. 1974. Primary basalts and magma genesis, I, Skye, north-west Scotland. *Contrib. Mineral. Petrol.* 45: 317–41

Thompson, R. N. 1975. Primary basalts and magma genesis, II, Snake River Plain, Idaho, U.S.A. *Contrib. Mineral. Petrol.* 52: 157–64

Thorpe, R. S., ed. 1982. *Andesites: Orogenic Andesites and Related Rocks.* New York: Wiley. 724 pp.

Tilley, C. E. 1950. Some aspects of magmatic evolution. *J. Geol. Soc. London* 106: 37–61

Wager, L. R., Deer, W. A. 1939. The petrology of the Skaergaard intrusion, Kangerlassuaq, East Greenland. *Medd. Groenl.* 105(4): 1–352

Walker, D., Shibata, T., De Long, S. F. 1979. Abyssal tholeiites from the Oceanographer Fracture Zone. II. Phase equilibrium and mixing. *Contrib. Mineral. Petrol.* 70: 111–25

Waters, A. C. 1955. Volcanic rocks and the tectonic cycle. *Geol. Soc. Am. Spec. Pap.* 62: 703–22

Wilcox, R. E. 1954. Petrology of Paricutin Volcano, Mexico. *US Geol. Surv. Bull.* 965C, pp. 281–353

Yoder, H. S. Jr. 1968. Albite-anorthite-quartz-water at 5 kb. *Carnegie Inst. Washington Yearb.* 66: 477–78

Yoder, H. S. Jr., Tilley, C. E. 1962. Origin of basalt magmas: an experimental study of natural and synthetic rock systems. *J. Petrol.* 3: 342–532

Zen, E. 1966. Construction of pressure-temperature diagrams for multicomponent systems after the method of Schreinemakers—a geometric approach. *US Geol. Surv. Bull. 1225.* 56 pp.

Ann. Rev. Earth Planet. Sci. 1986. 14 : 455–92

GEOLOGIC SIGNIFICANCE OF PALEOZOIC AND MESOZOIC RADIOLARIAN CHERT[1]

David L. Jones and Benita Murchey

US Geological Survey, Menlo Park, California 94025

INTRODUCTION

Bedded rocks containing abundant skeletons of radiolarians are an important member of a group of highly siliceous, nonclastic sedimentary rocks generally referred to as "chert." As a general class of sedimentary rocks, this group remains poorly known in comparison to the other broad classes of sedimentary rocks, including carbonates and clastics. Yet these rocks are becoming increasingly important for both economic and scientific reasons, and they are currently the subject of intensive study in many parts of the world.

The major impetus that sparked the current high level of interest in cherty rocks was the discovery in the early 1970s (Pessagno & Newport 1972) that the siliceous skeletons of microfossils entombed within such rocks could be liberated from their siliceous matrix by leaching with hydrofluoric acid. The resulting freed specimens could then be studied using standard micropaleontologic techniques, with the result that a useful biostratigraphic zonation for rocks ranging in age from middle Paleozoic through the Mesozoic was quickly achieved through broad international cooperative studies. This development now permits dating of rocks that hitherto could not be dated directly and provides a wealth of geochronometric, paleoecologic, and sedimentologic data that are critical for understanding the geologic and tectonic histories of continental margins.

In this brief report, we summarize the results of several of the major lines of research concerning bedded radiolarian chert and emphasize important

---

[1] The US Government has the right to retain a nonexclusive, royalty-free license in and to any copyright covering this paper.

problems that remain to be solved, particularly those that have significant geologic or tectonic implications.

LITHOLOGIC ATTRIBUTES OF RADIOLARIAN CHERT: COMPOSITION

Chert is defined as a microcrystalline or cryptocrystalline sedimentary rock consisting dominantly of quartz crystals less than 30 μm in diameter (Bates & Jackson 1980). In this report we are concerned with radiolarian-bearing bedded chert and related siliceous mudstones and claystones formed from marine sediments. The primary silica source for the quartz in these rocks is generally considered to be the biogenic opal tests of silica-secreting marine organisms. Biogenic opal ($SiO_2 \cdot nH_2O$) is chemically unstable. It is eventually dissolved and reprecipitated, first as opal-CT (a more ordered form of opal) and then as quartz (Jones & Segnet 1972, Kastner et al 1977, Hein et al 1978). The rate of this well-documented process is controlled by temperature and chemical environment. For all practical purposes, the Paleozoic and Mesozoic rocks we discuss in this manuscript are diagenetically mature because their initial biogenic silica has been converted entirely to quartz.

Binomial Classification for Fine-Grained Siliceous Sedimentary Rocks

Fine-grained siliceous rocks such as chert and siliceous mudstone are products of a wide variety of depositional environments and diagenetic histories that are reflected by differences in rock type, biogenic and detrital components, and mineralogy. References in the literature to these rocks are commonly both inconsistent and incomplete. For instance, many references to bedded chert make no distinction between radiolarian chert and sponge spicule chert. Conversely, the term "radiolarite," while emphasizing the biogenic component, may refer to conchoidally fracturing chert, siliceous claystone, or radiolarian-rich limestone. Commonly, no distinction is made between primary chert formed directly by silica cementation of unlithified sediments and secondary chert formed by replacement of limestone. Even though both primary bedded chert and secondary replacement chert may be radiolarian-rich, their depositional and diagenetic histories are distinct. While chert nodules in limestones are easy to recognize as being of replacement origin, completely silicified limestone beds may not be so readily identified. Criteria for distinguishing primary bedded chert from silica-replaced bedded chert need to be established.

We propose a simple binomial system for classifying fine-grained siliceous sedimentary rocks in order to facilitate comparisons and analyses. Our descriptive scheme utilizes a binomial classification system in which the rock type is a noun and the principal grain type is the primary modifier of the rock type. The classification system is meant to be applied only to fine-grained siliceous rocks formed from marine or freshwater sediments. Our discussion emphasizes those siliceous rocks in which radiolarians may be abundant. A more comprehensive manuscript discussing a wider spectrum of fine-grained siliceous rocks is in preparation.

The distinction of rock types is critical for the development of a useful data base for basin analyses and stratigraphic comparisons. In particular, discrimination of depositional environments and of the possible genetic relationship of strata-bound mineral resources to siliceous deposits requires careful recognition of different rock types.

Rock Type

By definition, the classification system is limited to sedimentary rocks that are fine grained and siliceous. These rocks can be plotted on a ternary diagram with SiO_2, carbonate, and clay end members (Figure 1). The rock-type name is a field or petrographic designation determined by the physical properties of the rock and the composition of the rock matrix. In the ternary diagram, rock-type fields include chert (vitreous, conchoidal fracture, very high silica-to-clay ratio), clay-rich chert or porcellanite (earthy appearance,

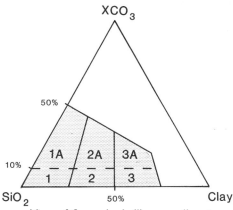

Figure 1 Matrix compositions of fine-grained siliceous sedimentary rocks shown on a ternary diagram with SiO_2, clay, and carbonate end members: (1) chert; (2) clay-rich or argillaceous chert (porcellanite); (3) siliceous claystone, mudstone, argillite, or shale; (A) modifier indicating type of carbonate present.

well indurated, may or may not fracture conchoidally, intermediate silica-to-clay ratio), and siliceous claystone (high clay content). A significant carbonate component (10–49%) should be reflected as a modifier to the rock type (e.g. dolomitic chert, calcareous siliceous claystone).

Grains

In the binomial classification system, the most abundant grain type is the principal modifier of the rock type and immediately precedes the rock-type name. Grains may be either biogenic or detrital. Less abundant grain types can be included as secondary modifiers preceding the principal modifier. For instance, chert in which the grains are composed of 60% radiolarian grains and 40% quartz silt grains is called silty (quartz) radiolarian chert, or quartzitic radiolarian chert. The former descriptive phrase is preferable because it clarifies the detrital nature of the quartz. Conversely, chert with 40% radiolarian grains and 60% quartz silt grains is radiolarian-bearing silty (quartz) chert. Because clay content is reflected in the rock-type name, it is not used as a modifier in the binomial classification system. Many fine-grained rocks contain no recognizable grains at all. The absence of granular silt in these rocks need not be noted in the binomial name, but the absence of fossils must be. For example, chert with no fossils and no granular silt is called unfossiliferous chert.

Radiolarians, sponge spicules, and diatoms are the principal sources for the silica with which their host rocks are cemented. Any other types of microfossils or small fragments of megafossils may be components of siliceous rocks, although they are usually not the most abundant fossil unless the rock is a silicified carbonate. Carbonaceous material is another form of organic detritus that is present in significant quantities in some fine-grained siliceous rocks. When fossils are not the most abundant grain type in the rock, their presence should still be noted as a secondary modifier to the binomial name [e.g. radiolarian-bearing tuffaceous (vitric) chert].

Nonbiogenic detrital grains are a significant component of many fine-grained siliceous rocks. By definition, the silt-sized detrital component of the rock must be less than 33% of the total rock or else the rock must be classified as a mudstone or a siltstone (Folk 1968). As previously discussed, we have specifically excluded clay as a modifier because we include clay abundance in the name of the rock type. Detrital grains are as compositionally variable in fine-grained siliceous rocks as in other sedimentary rocks, although grains of obvious metamorphic origin are rare. Detrital megacrystalline quartz and chert are particularly common. When the detritus is volcanogenic, it may be useful to use the adjective "tuffaceous" rather than silty [e.g. tuffaceous (lithic) clay-rich chert, tuffaceous (vitric) siliceous claystone].

Quantitative Analyses

In detailed analyses, quantitative determinations of the grain-to-matrix ratio and the relative proportions of the types of biogenic and detrital components to one another for individual samples are useful. In Paleozoic and Mesozoic fine-grained siliceous rocks, the three most abundant grain types are radiolarians, sponge spicules, and mineral detritus. For general stratigraphic comparisons, the relative ratios of these three groups can be plotted in fields of a ternary diagram.

Obviously, the composition of nonbiogenic detrital grains provides important clues concerning provenance. When silt is abundant, the ratios of detrital components can be plotted on a quartz-feldspar-lithic (QFL) ternary diagram. We suggest that detrital chert be plotted with the lithic group, as suggested by Folk (1968) for sandstones. In addition, the ratios of lithic grain types can also be plotted on a ternary diagram with sedimentary, igneous, and metamorphic end members. Each of the three lithic groups can be further subdivided by composition.

As with all other sedimentary rocks, the biogenic and detrital grains in fine-grained siliceous rocks can provide information about the depositional environment in which the sediments were deposited. In the future, much more attention must be paid to the relative abundances of various particle types as they relate to paleoenvironment and basin history.

Geochemistry

Geochemical analysis of bedded radiolarian chert also promises to be a useful tool for deciphering the depositional history of individual chert sequences and individual basins. The elements in radiolarian chert have four potential sources: hydrothermal, detrital, biogenic, and hydrogenous. Hydrogenous components are those precipitated or adsorbed from seawater. One of the goals of geochemists is to discriminate the relative influence of these different elemental sources on the rock composition. In order to do so, geochemists have compared major-element, trace-element, and rare earth abundances in radiolarian chert with element abundance patterns in marine sediments of known depositional environments. The inherent assumption in these types of studies is that the geochemistry of ancient marine sediments was similar to the geochemistry of modern marine sediments. Recent papers by Barrett (1981), Crerar et al (1982), Steinberg et al (1983), Karl (1984), and Pinto-Auso & Harper (1986) try to interpret depositional environments for Mesozoic chert based upon comparisons of the rock chemistry with the geochemistry of marine sediments from known environments.

Geochemical analyses, combined with sedimentological, stratigraphic, and petrographic analyses, will undoubtedly be widely used in future studies of radiolarian chert-bearing sequences and basins. The discrimination of systematic geochemical changes vertically or laterally in radiolarian-bearing sequences may reveal previously unrecognized sedimentary patterns. Ultimately, such studies will facilitate more accurate reconstructions of the geologic histories of accreted terranes.

Silica Diagenesis

Whereas whole-rock geochemistry provides information about the depositional history of chert, the study of SiO_2 polymorphs provides information about the diagenetic history, thermal history, tectonic history, and age of bedded chert.

Biogenic silica from opal-secreting organisms such as radiolarians, diatoms, and sponge spicules is the principal silica source for the quartz cement in bedded chert sequences. Biogenic silica, or opal-A, is an unstable amorphous form of hydrous silica. With increasing time and temperature, opal-A undergoes a transformation to quartz, with an intermediate stage as disordered low-temperature cristobalite/tridymite (opal-CT) (Jones & Segnit 1971). Solution and reprecipitation occur at both stages of the transformation of biogenic opal to quartz.

The rate of silica diagenesis is controlled by many factors, including temperature, pore-water chemistry, lithology, and time (Jones & Segnit 1971, 1972, Mizutani 1977, Kastner et al 1977, Hein et al 1978, Pisciotto 1981). In oceanic sediments, the transformation of biogenic silica to quartz requires an estimated 40 to 100 Myr (Siever 1979, Riech & von Rad 1979). Tectonic and sedimentary processes tend to increase this rate for on-land sequences. The Miocene Monterey Formation of California is a siliceous sequence that has partly converted to quartz, especially in basins that have been deeply buried (Isaacs et al 1983, Pisciotto 1981). On-land Paleozoic to middle Cretaceous radiolarian chert sequences of the world are virtually all diagenetically mature, and many have been altered by the effects of postdiagenetic metamorphism. For three Mesozoic radiolarian siliceous shale sequences, Mizutani (1983) used $^{87}Rb/^{87}Sr$ isochrons to calculate that diagenesis was completed in less than 35 Myr.

Textures, varieties, and grain sizes of quartz and chalcedony are related to the diagenetic and metamorphic histories of these minerals. As previously mentioned, microcrystalline and cryptocrystalline quartz may have a variety of textures ranging from very fine interlocking crystals forming mosaic patterns to equigranular subrounded crystals forming a curdled, cottage cheese-like pattern (see Folk & McBride 1978, McBride & Folk 1977, 1979). In addition, fibrous SiO_2 (chalcedony) has many varieties distinguished by their extinction patterns. The texture, variety, and grain size of quartz and chalcedony are functions of their chemical (Keene 1983, Folk & Pittman 1971) and thermal (Murata & Norman 1976) diagenetic histories and their metamorphic histories.

Silica diagenesis is a destructive process for radiolarians. The vast majority of radiolarians are lost in the process of dissolution of biogenic opal in the water column. Of those radiolarians that are deposited on the seafloor, many more are destroyed in the process of transformation from opal-A to opal-CT to quartz. Metamorphic recrystallization of chert further destroys or deforms the remainder. Therefore, well-preserved radiolarians in Paleozoic and Mesozoic siliceous sequences are the exception, rather than the rule. Radiolarians can be extracted from siliceous rocks by hydrofluoric acid etching when (*a*) the quartz grains composing the radiolarian fossil are larger than the quartz grains composing the rock matrix and (*b*) the grain boundaries between radiolarian fossils and the rock matrix are distinct.

In diagenetically immature sequences, nonuniform diagenesis can produce interlayered zones with different physical properties dependent on the diagenetic stage. When stress is applied to these interlayered rocks, each zone will exhibit a deformational style related to its diagenetic state. Folding styles, stylolites, dilation breccias, fractures, and sedimentary dikes are all structural features that can be related to deformation of beds during specific diagenetic stages. One may be able to infer the timing of deformation in older sequences by observing their deformation styles and comparing them to deformation styles in diagenetically immature sequences (Snyder & Brueckner 1983, Snyder et al 1983).

Summary of Compositional Analyses

The composition of fine-grained siliceous sedimentary rocks can be expressed in terms of rock type, total grain abundance, relative grain-type abundance, whole-rock geochemistry, and SiO_2 polymorph composition. Within these categories, systematic rock comparisons can discriminate differences or similarities within and between rock sequences. Systematically combining two or more of these methods for studying rock composition would more than double the utility of each individual method.

LITHOLOGIC ATTRIBUTES: PHYSICAL CHARACTERISTICS

The physical characteristics of bedded chert sequences are the product of both the depositional and diagenetic histories of the rocks. Radiolarian chert displays a wide variety of bedding styles, sedimentary and diagenetic features, and rock colors. Until fairly recently, scant attention had been paid to these differences. However, interest in the sedimentology of siliceous sediments has been spurred by the new developments in radiolarian biostratigraphy and by the economic potential of petroliferous Cenozoic diatomaceous rocks.

Bedding

The terms "bedded chert" or "radiolarian chert" commonly denote rock sequences that include both chert and siliceous claystone or mudstone. Radiolarian chert is famous for a bedding style characterized by rhythmically alternating thin beds of chert and shale, otherwise known as ribbon chert or ribbon bedding. Within this somewhat broad definition, ribbon bedding can be variable in terms of bed thicknesses, chert-to-shale ratios, and lateral variations in bed thickness. Radiolarian-rich sequences dominated by siliceous mudstone, shale, or chert may lack rhythmically alternating bedding entirely. Bedding styles can vary considerably within a single radiolarian chert sequence or laterally within a basin. For example, Baltuck (1983) shows 12 bedding styles for interbedded chert, mudstone, and shale over 1-m intervals at various localities in radiolarian-rich sequences in Greece. She also shows an additional style of bedded chert in which chert partially replaces a radiolarian-bearing limestone sequence.

In the past, several different processes were proposed as the single cause for the rhythmic bedding of radiolarian chert. McBride & Folk (1979) summarized four hypotheses for the origin of ribbon bedding: "(1) diagenetic segregation of silica from initially sub-homogeneous siliceous mud; (2) episodes of rapid and slow production of radiolaria in surface waters during a constant rate of mud deposition; (3) episodes of current deposition of radiolarian silt during constant mud deposition; (4) episodes of current deposition of mud during constant radiolaria sedimentation." As with most long-standing controversies, evidence exists supporting all points of view. We agree with the conclusions made by several workers in the last few years (e.g. Steinberg et al 1983, Baltuck 1983, Jenkyns & Winterer 1982) that the bedding in radiolarian chert sequences cannot be related to a single process. Sedimentary processes contributing to bedding style can be highly variable, and diagenetic processes can strongly overprint primary bedding characteristics.

Because the geographic distribution of different styles of radiolarian chert bedding in any given basin has never been adequately described or studied, the environmental significance of various bedding styles is not known. Clearly, the field of radiolarian chert sedimentology is in its infancy.

Sedimentary Features

Primary sedimentary structures in bedded chert include structures commonly associated with rapid deposition by currents or turbidites, such as

Bouma sequences, grading, starved current ripples, current orientation of sponge spicules and radiolarians, flute casts, and structures commonly associated with slow deposition (e.g. millimeter-scale varvelike laminae and radiolarian-studded micronodules). Surface trace fossils and burrows are common in some bedded chert sequences and rare in others. The relative contributions of turbidites, contourites, current winnowing, and planktonic productivity to the accumulation of siliceous sedimentary sequences should vary from basin to basin. So, too, should the relative proportions of silt, clay, sponge spicules, and radiolarians be expected to vary. In North America, sedimentary structures associated with current deposition are more common in bedded chert sequences containing silty chert and silty spiculitic chert than in sequences characterized by radiolarian chert and radiolarian siliceous shale. Although the correlation between sedimentary structures and lithologies may be, in part, an artifact of grain size, we believe that the correlation is also related to depositional modes.

Diagenetic Features

Diagenetic features can be related to variations in the rate of silica diagenesis, dilation resulting from silica diagenesis, and syndiagenetic fluid movement through lithifying sediments.

The rates and uniformity of diagenetic processes strongly affect bedding style. Relatively uniform diagenesis along horizontal surfaces seems to result in fairly regular, flat beds. The compositional homogeneity of the original sediments and the permeability of sediments along particular horizons are both important parameters controlling the uniformity of diagenesis. Therefore, discrete sedimentary layers such as turbidites are likely to produce discrete beds or bedding couplets (chert and shale). In our opinion, relatively homogeneous radiolarian clays may also be able to produce discrete chert and shale beds through diagenesis. Evidence for this mode of bed formation can be found in bedding bifurcations and pinch-outs seemingly unrelated to sedimentary features or structural boudinage. Secondary evidence for diagenetically produced bedding comes from symmetrical geochemical "grading" and symmetrical radiolarian abundance grading perpendicular to individual chert beds (Mizutani & Shibata 1983, Sano 1983, Steinberg et al 1983, Murchey 1984).

Laterally irregular bedding is a common feature caused by variations in the rate of silica diagenesis. When local areas along a horizontal plane are preferentially lithified very early in the diagenetic history of the sediment sequence, these areas (usually centimeter scale) are more resistant to compaction than surrounding sediments. The resulting beds may have pinch and swell structures or pronounced knoblike structures. In the Tethys, pinch and swell bedding in chert is common in a transition zone between bedded chert and overlying pelagic limestone. Bosellini &

Winterer (1975) concluded that the pinch and swell bedding was related to the lowering of the calcite compensation depth in the Tethys, and that the sediments that formed the pinch and swell chert beds originally contained both biogenic silica and carbonate. Differential rates of lithification were related to local differential rates of carbonate dissolution and silica diagenesis. Knobby beds, which are not necessarily related to a transition from bedded chert to limestone, are clearly diagenetic phenomena. The interior portion of the knob is commonly very silica rich and radiolarian rich, while laterally the bed may be an argillaceous chert (e.g. Murchey 1984).

Differential compaction and fluid-escape phenomena can result in a spectrum of structures from knobby bedding to diapirs and dikes. Diapiric structures and chert dikes are relatively common features in some bedded chert sequences. Although most published descriptions of these structures refer to radiolarian-poor or nonradiolarian sequences (e.g. Steinitz 1970, Snyder et al 1983), diapirs and chert dikes also occur in radiolarian-rich sequences as well. The smallest diapirlike structures are the "Monroe" structures of Folk & McBride (1976), which are hemispherical mounds (generally 2–6 cm high, 15 cm diameter) with a central nipplelike protuberance. The central nipple suggests that "Monroes" are fluid-escape structures and not simply knobs resulting from early lithification. Somewhat longer, thin chert diapirs (30 cm high, 2–3 cm diameter) intruding several chert and shale beds were observed by B. Murchey (unpublished data) in calcareous radiolarian chert of northern Alaska. Snyder et al (1983) illustrate small-scale chert dikes and argillite dikes intrusive into chert beds for Paleozoic siliceous rocks from Nevada.

Diagenetic processes capable of forming breccias in siliceous sedimentary sequences include (a) dilation caused by water release and associated volume loss occurring during the transformation from opal ($SiO_2 \cdot H_2O$) to quartz (SiO_2), and (b) disruption caused by movement of pressured fluids through lithifying sediments (Snyder et al 1983, Steinitz 1970, Paris et al 1985). The dilation process is likely to increase fluid pressure as well as fluid volume; and therefore both processes are related. The movement of heated, high-pressure hydrothermal fluids associated with volcanism appears to be a common cause of brecciation and veining in chert sequences (Paris et al 1985, Crerar et al 1982, Murchey 1984). Sedimentary loading may also cause overpressuring, fluid migration, and brecciation along well-defined horizons in chert sequences (Steinitz 1970, Murchey 1984). Because the origin of brecciation in chert sequences is controversial (e.g. McBride & Folk 1977, 1979), the criteria for differentiating sedimentary breccias from purely diagenetic breccias in chert sequences need to be carefully established. Paris et al (1985) showed the need for cautious interpretations when they documented the diagenetic formation of a chert pebble conglomerate in Precambrian chert of South Africa.

Color

Alpine geologists have long differentiated variations in chert colors with descriptive rock names such as diasporiti (= jasperite, red chert) or lydite (black chert). Likewise, Japanese geologists have also developed distinctive names for particular associations of chert colors. For instance, a special Japanese term is applied to cherry red chert with white veins (Imoto 1984). The significance of the rock color of radiolarian cherts was first discussed in some detail by Thurston (1972), who tried to relate rock color to physical properties such as radiolarian preservation. McBride & Thomson (1970) and Folk & McBride (1978), recognizing the importance of rock color, proposed that rock color descriptions be systematized and standardized by employing Munsell color numbers and names. Particular colors can be diagnostic of particular stratigraphic intervals or distinctive textural styles.

Because chert color is related to the presence or absence of carbonaceous organic material and to the oxidation state of iron and other metals present in the rock, systematic color variations in chert sequences provide information about local depositional environments, as well as about worldwide oceanographic conditions through time. Bedded chert associated with basaltic basement is commonly red and green, whereas bedded chert associated with continental margins is commonly dark gray and black. In North America, Ordovician to Lower Mississippian radiolarian and spiculitic chert sequences are predominantly black and gray, although red chert is locally conspicuous. Upper Mississippian to Permian chert sequences are predominantly gray, green, and red, with red being especially common in Pennsylvanian and Permian sequences. Red Permian chert and argillite is a circum-Pacific phenomenon apparently related to worldwide oceanographic conditions. A notable exception is the dark gray and black Permian Rex Chert of the Phosphoria Formation in Utah and Idaho, which is a spiculitic chert and siliceous argillite associated with carbonate and phosphorite rocks. In North America, Triassic chert is dominantly dark gray and black, but Jurassic and Cretaceous chert is dominantly maroon, red, and green. With the exception of red chert and argillite associated with latest Paleozoic sequences, rock color seems to correlate more closely with depositional environment than with geologic age.

Summary

Temporal and spatial variations in the physical characteristics of chert sequences need to be compared not only in small-scale studies of individual basins but also in regional studies of large oceanic systems. Once again, we emphasize the necessity for systematizing comparisons and for integrating geologic disciplines. By so doing, geologists will learn which lithologic attributes of bedded chert have the most paleoceanographic and tectonic significance.

PALEOCEANOGRAPHIC SIGNIFICANCE

Depositional Models

Radiolarians are single-celled planktonic organisms that appear in abundance as early as the Ordovician, although rare forms that may be radiolarians are reported from the Cambrian. Only the silica-secreting radiolarians are ever preserved as fossils. From the Ordovician to the early Late Cretaceous, radiolarians were important rock-formers. Their demise as major rock-formers appears to be related to the late Mesozoic rise of calcareous planktonic foraminifers, coccolithophorids, and silica-secreting, photosynthetic diatoms. Ordovician to upper Mesozoic pelagic rocks are dominated by radiolarians, but most Late Cretaceous and Cenozoic pelagic rocks are dominated by foraminifers, coccolithophorids, or diatoms.

Although all radiolarian chert formed from marine sediments, the specific settings where deposition occurred are a matter of conjecture and dispute. Early views held that chert formed far from continental margins in deep ocean basins, and therefore it represents abyssal sediment deposited directly on oceanic crust. This view was supported by the joint occurrence, particularly in the Mediterranean region, of pillow basalt, ultramafic rocks, and red chert forming the famous "Steinmann trinity," thought by some to represent typical oceanic crust. It was also supported by the discovery of radiolarian oozes on the deep-sea floor. Results of deep ocean drilling dispelled this interpretation, however, as countless drill holes failed to substantiate the presence of chert at the basalt-sediment interface within the modern oceans. This discovery forced reinterpretation of chert-forming environments and led to several alternate suggestions, ranging from continental margin to lagoonal (e.g. Folk & McBride 1978, Jenkyns & Winterer 1982).

Because more than 98% of the unstable opal tests of silica-secreting radiolarians dissolve in the water column and on the seafloor prior to burial, rapid production of radiolarians is now considered by many to be a requirement for the eventual formation of bedded radiolarian chert. In the modern oceans, high planktonic productivity is associated with nutrient-rich waters produced by upwelling. Areas with upwelling in the present oceans include east-west trending zones at very high latitudes and along the equator, north-south trending zones along western continental margins, and localized zones in small basins such as the Gulf of California and the Sea of Okhotsk.

Recently, the idea that radiolarian cherts represent high productivity in marginal basins has become very popular because of the present-day association between diatomaceous sediments and upwelling in local marginal basins of the modern oceans (Jenkyns & Winterer 1982, Hein & Karl 1983). The role of equatorial high productivity in the formation of radiolarian chert may be underestimated. All Mesozoic cherts from the Mediterranean region (= Tethys) formed at low latitudes because that region lay along the Mesozoic equator. Many circum-Pacific radiolarian chert-bearing terranes have apparently been displaced northward or southward from low latitudes (Grommé 1984, Murchey 1984, Karl 1984, Hillhouse & Grommé 1983, Mizutani & Hattori 1983, Hirooka et al 1983, Ozawa & Kanmera 1983). In addition, Hein & Parrish (1986) have found some correlation between present-day worldwide distribution patterns of bedded chert (both radiolarian and spiculitic chert) and low paleolatitudes, although they did not make necessary corrections for the displaced Pacific deposits.

The depositional environment in which radiolarian-rich siliceous sediments accumulated must have been away from sources of terrigenous detritus (in order to prevent dilution by clastics) and near or below the calcite compensation depth, the depth at which calcite dissolves in ocean water (in order to prevent dilution by carbonate). The calcite compensation depth is known to be elevated in regions of high upwelling, and before the Mesozoic explosion of planktonic calcareous organisms, the calcite compensation depth in the oceans may have been relatively shallow.

The radiolarian-rich siliceous sediments that formed Ordovician to lower Upper Cretaceous radiolarian chert probably accumulated in equatorial and high-latitude east-west upwelling belts and in continental-margin upwelling zones. Radiolarians lived in the widespread inland seas, but they did not accumulate in sufficient quantity to form bedded radiolarian chert sequences. In intracratonic basins, radiolarians are preserved in phosphatic, cherty, or calcareous nodules or in scattered cherty beds in carbonate or shale sequences. In North America, all bedded radiolarian chert sequences lie seaward of the ancient continental shelf edges.

In contrast, bedded sponge spicule chert, which is generally not adequately distinguished from bedded radiolarian chert, is locally common in sediment-starved intracratonic basins. A good example of bedded spiculitic chert in an intracratonic basin is the Rex Chert member of the Phosphoria Formation of Idaho and Utah. In the oceans surrounding the cratons, siliceous sponges on the flanks of carbonate platforms and on drowned platforms added to the accumulating siliceous sediments along continental margins, rifting margins, and island arcs (Imoto 1983, Murchey et al 1986). Therefore, all gradations between bedded sponge spicule chert and bedded radiolarian chert are preserved in the rock record. The demise of siliceous sponges as major components of rock-forming siliceous sediments predates the decline of radiolarians as rock-formers. Siliceous sponges ceased to be important rock-formers in the circum-Pacific region by the end of the Triassic, and in the Mediterranean region by the end of the Jurassic.

Paleobiogeography of Radiolarians

Establishing radiolarian distribution patterns and correlating these with tectonic and paleoceanographic settings is an exciting new field for Paleozoic and Mesozoic radiolarian workers. In the modern oceans, oceanography controls the relative abundances and geographic distribution of radiolarian taxa. For fossil radiolarian assemblages in accreted sequences, the locations of depositional sites and the oceanographic parameters controlling radiolarian taxa are all interpretive.

In 1983, we recognized that Permian radiolarian faunas in the Cordillera of North America can be divided into three coeval assemblage groups that are geographically separate (Murchey & Jones 1983, Murchey et al 1983a, 1986). One group is characterized by high diversity and by a high number of multi-rayed and discoidal radiolarians. This group is found in argillaceous chert and siliceous argillite with abundant sponge spicules and silt of siliciclastic and/or volcaniclastic origin. The second group is characterized by low diversity and by a high number of bilaterally symmetrical radiolarians. This group is found in sponge spicule–poor radiolarian chert that is areally associated with ophiolites. Both groups are characteristic of faunas south of the Denali fault in Alaska, with the first group lying inboard of the second. The third faunal group is characterized by very low diversity and is dominated by spheroidal forms. The host rock may contain few to abundant sponge spicules and silt. This group occurs north of the Denali fault and is particularly characteristic of the northern Brooks Range. Whereas upwelling, depth, or water temperature may have controlled the difference in radiolarian populations of the Permian faunas south of the Denali fault, the most important oceanographic parameter controlling the Permian radiolarian populations north of the Denali fault seems to have been paleolatitude.

Mesozoic faunas also show regional differences in the taxonomic composition of coeval radiolarian assemblages. The interpretation of such differences is complicated by the effects of major latitudinal translation of terranes in the circum-Pacific region during the Mesozoic and Cenozoic.

On the other hand, the latitudinal translation may be in large part responsible for the juxtaposition of different but coeval radiolarian assemblages. Murchey (1984) and Murchey & Jones (1984) utilized faunal comparisons, paleomagnetic data, and paleoceanography to argue that specific oceanic terranes in the Franciscan Complex of California record a history of low-latitude deposition of radiolarian-rich sediments in the Mesozoic Pacific. Pessagno & Blome (1986) have interpreted the radiolarian biostratigraphic sequence of Jurassic rocks in eastern Oregon as indicating northward movement from a low-latitude site to a high-latitude boreal site during the history of deposition. They cite specific criteria for distinguishing low-latitude from boreal-latitude faunas, including relative abundances of specific genera.

Temporal Distribution

Recently, Hein & Parrish (1986) have compiled a list of published references to chert around the world. They attempted to analyze chert abundance (as measured by the number of occurrences of any size) through time and in relation to paleogeography. They found the greatest number of deposits in post-Triassic rocks and sediments. They also found a correlation between predicted upwelling and chert localities. This type of compilation is very useful, but analyses of this kind of information have their pitfalls. First, older rock sequences will be proportionally less well represented in the rock record than younger rock sequences because of the cumulative effects of destruction by burial, erosion, metamorphism, and volcanism. Second, most cherts in the circum-Pacific region are in terranes that have moved from their depositional site. The amount of movement may be minor or may be very large both latitudinally and longitudinally. Third, some stratigraphic intervals are easier to date than others because faunas characterizing those intervals are particularly distinctive or robust. Fourth, radiolarians, sponge spicules, and diatoms are different organisms, and they do not necessarily have the same paleogeographic requirements. Therefore, bedded chert formed from one type of siliceous organism is not necessarily the paleogeographic equivalent of bedded chert formed from another siliceous organism.

Summary

Geologists are a long way from accurately reconstructing the paleoceanography of Paleozoic and Mesozoic oceans, particularly for the deeper, more siliceous basins. However, multidisciplinary studies using megafossil and microfossil faunal distribution data, paleomagnetic data, and sedimentologic data ultimately will illuminate parts of the ancient picture.

LITHOLOGIC ASSOCIATIONS AND TECTONIC SETTINGS OF CHERT DEPOSITION

Recognition that radiolarian chert forms only in regions of high productivity generated by upwelling of cold nutrient-laden waters is an important advance in understanding the paleoecological significance of chert. Nevertheless, this generalization is too broad to explain the specific depositional setting of any particular chert deposit. A case in point is seen in regard to cherts associated with mafic and ultramafic igneous rocks generally referred to as ophiolite. Some of these rocks are accepted as products of seafloor spreading but not necessarily of spreading at midoceanic ridges. By definition, chert is an integral part of ophiolites, and few such assemblages lack pelagic deposits. Does this mean that all ophiolites formed in regions of upwelling, either along continental margins or near the equator? Such an interpretation, if true, would have profound implications for the tectonic history of the oceans and for processes of Earth dynamics. An alternate interpretation is that certain tectonic settings are themselves conducive to generating local upwelling conditions in settings that are neither along a continental margin nor near the equator. To test these kinds of ideas, we have studied the lithologic associations in which chert occur and from these have attempted to establish the tectonic setting in which the chert formed. These studies indicate that a broad spectrum of environments can be recognized, as indicated by the following six major associations (Murchey et al 1983b).

Ophiolitic Chert Association

The classic ophiolitic suite of serpentinite, gabbro, sheeted diabase dikes, and pillow basalt is commonly capped by red radiolarian chert. Metalliferous deposits, including umbers and manganiferous chert lacking radiolarians, are common at or near the basalt-sediment interface. Small pods of chert may also be intercalated with basalt near the top of the pillowed sequence. Typical chert sequences are 50–100 m thick, maroon to dark red in color, and contain abundant radiolarians and few sponge spicules. Clay content is variable, but diagenetically enhanced ribbons tend to be well developed. Where stratigraphic sequences are preserved, the chert may be overlain by a variety of sedimentary rocks, including pelagic limestone (Mediterranean area), turbiditic sandstone and shale (California), or volcaniclastic rocks (southern Alaska).

In North America, deposits of this type are known to range in age from Late Devonian to Jurassic. The depositional setting is clearly oceanic in the sense that continentally derived terrigenous material is scant or lacking throughout stratigraphic sequences that may span tens of millions of years.

The amount of separation from continental margins, however, need not exceed 500 to 1000 km, and deposition in wide back-arc basins cannot be ruled out.

Ophiolitic rocks that represent oceanic crust are rare in western North America : Most so-called ophiolites belong to the seamount association (see below), which appears to reflect midplate volcanism rather than volcanism along an active spreading center. Nevertheless, rocks of the ophiolitic association, as here defined, form a minor but tectonically significant component within the accreted terranes of western North America.

Seamount Association

The depositional intercalation of thick piles of pillowed and massive basalt with thin lenses and layers of chert, cherty tuff, and basaltic aquagene tuff characterize another oceanic association that we attribute to midplate volcanism similar to that which formed Mesozoic and younger seamounts in the Pacific Ocean (e.g. Batiza & Vanko 1984). Volcanic products tend to be tholeiitic-to-alkalic basalts with moderate contents of TiO_2, Zr, and light rare earth elements (REEs). Mafic rocks commonly can be observed to intrude sedimentary assemblages of chert, argillite, and other deep marine sedimentary rocks that formed the substrate for the volcanic edifice. Some of the volcanic piles extended into shallow water, since they are capped with reefal carbonates, conglomerates, or other types of shallow-water deposits. Many of the cherts in this association are tuffaceous, are red to dark brown in color, and occur with volcanic mudstone or aquagene tuff. Most of the chert deposits are very thin, ranging from a few centimeters to a few meters, and they rarely show a long age span.

Island-Arc Chert Association

Varying admixtures of biogenic silica (either radiolarians or sponge spicules) and silicic volcanic ash characterize the island-arc chert association. The lithic spectrum ranges from bedded silicic tuff lacking biogenic silica to tuffaceous cherts containing feldspar crystals and siliceous fossils in a matrix of silicified smectitic clays. Volcanic shards are sometimes preserved, and olistostromal blocks of limestone are locally abundant. The chert is commonly green to gray-green and usually is well laminated. Because of the highly particulate nature of the rock, recognizing the presence of radiolarians in the field by hand-lens observation may be difficult.

Subsidence Association

A vertical succession from shallow-water fossiliferous carbonate or clastic rock upward to spiculitic chert and then to radiolarian chert characterizes a subsidence-induced lithologic change. Argillite or pelagic limestone commonly lies at the transition zone between shallow deposits and bedded chert. An upward shift in the calcium carbonate dissolution depth could also produce a change from carbonate-rich to silica-dominated deposition, but such a shift would not explain the upward transformation from spiculitic chert to radiolarian chert, which we believe to be depth controlled (see below).

The subsided platform itself may represent a foundered continental block, a cooled volcanic arc, or a seamount or plateau that sank as a result of cooling and contraction of oceanic crust.

A general rise in sea level could produce a similar lithologic change from shallow- to deep-water deposits. We see no evidence in the geologic record, however, that would substantiate that worldwide rises in sea level were ever sufficient to permit widespread and contemporaneous deposition of radiolarian chert–forming sediments on existing continental platforms. Thus, continental blocks that exhibit this subsidence curve must have had local and individual histories that do not seem to be tied to global events controlling sea level.

Continental Margin Chert Association

Chert interbedded with terrigenous detritus indicates deposition within range of continentally derived sedimentary material (mainly quartzose sand and silt). Cherts typically are silty, black, laminated, and may have thick silty interbeds. Toward the source area, cherts grade into argillite and then to shale. Where clastic deposition in the source area is replaced by carbonate deposition, a similar transition to offshore chert deposits can be demonstrated. In proximal facies, cherty limestone is interbedded with spiculitic chert, which gradually replaces limestone laterally and grades to radiolarian-rich chert in distal facies. This lateral transformation parallels the upward change from spiculitic chert to radiolarian chert observed in the subsidence assemblage described above. Spiculitic cherts are commonly black and well laminated, and they generally exhibit strong preferential alignment of long axes of spicules due to current activity. Occasionally, broken or abraded silicified shelly material occurs with the spicules, which attests to the presence of nearby shallow-water environments. Shelly material in radiolarian-dominated chert is rare.

Mélange Chert Association

Interbedded silty black argillite, minor graywacke, basaltic tuff with or without pillow basalt, cherty tuff, and radiolarian chert form the disrupted matrix of Mesozoic mélange terranes along the Pacific Coast of North America and elsewhere in the circum-Pacific region. The matrix chert differs in significant ways from blocks of nonmatrix chert that may be incorporated in the mélange. Careful sampling is required so that blocks and matrix are not confused. In most outcrops this matrix displays evidence of soft sediment slumping and mixing, which is in part responsible for the disrupted nature of the mélange itself (Aalto & Murphy 1984, Cowan 1985). The mélange chert is usually light greenish grey to brownish red, thin bedded, and radiolarian rich. The mélange chert association, which is often associated with a subduction setting (e.g. Cowan 1985), requires deposition in relatively deep water (as evidenced by the presence of chert and the lack of limestone in the matrix) along a continental margin where minor basaltic volcanism was prevalent and contemporaneous with clastic and pelagic deposition. Because active volcanism is not characteristic of modern oceanic trenches, this association appears to be related to some type of rift or leaky transform system, perhaps analogous to the present Gulf of California.

Summary

From the foregoing descriptions, it is apparent that radiolarian chert has formed in a number of different geologic settings, and that no single environmental interpretation applies to all chert occurrences. Major differences are seen in comparison with Cenozoic diatomaceous deposits, which are closely linked to upwelling along continental margins or to cold polar seas (where the calcite compensation depth occurs at or near sea level). Because the best Paleozoic and Mesozoic analogues for Cenozoic diatomaceous sequences are generally sponge spicule dominated, no known radiolarian cherts appear to duplicate precisely either of these two settings; thus it remains risky to apply models of more recent sedimentary systems to ancient siliceous deposits. The association of rock types offers the best clue as to the depositional setting, but more information is needed concerning the nature of the faunas, bedding characteristics, and geochemical parameters in order to specify more accurately the paleogeographic setting of any particular chert deposit.

GEOLOGIC OCCURRENCE AND TECTONIC SIGNIFICANCE OF RADIOLARIAN CHERT IN WESTERN NORTH AMERICA

Radiolarian chert is widespread in the Cordillera of western North America and has received intensive study in local areas during the past 10 years. Documented ages of chert range from Ordovician to early Late Cretaceous; although older cherts are reported, they are not known to be radiolarian bearing. No Phanerozoic radiolarian cherts are present in that part of North America underlain by thick Precambrian crust (the cratonal part of the continent). Nor are bedded radiolarian cherts abundant in the marginal miogeoclinal deposits that rim the craton throughout much of the west, although in a few places such as the Selwyn Basin of Canada, displaced lower Paleozoic distal miogeoclinal strata are cherty. The radiolarian content of these cherts, however, has not been determined, and they may be mainly of spiculitic origin.

Most deposits of radiolarian chert occur in terranes of noncontinental origin that have been added to North America during an intensive episode of continental growth that occurred during the Mesozoic. A few terranes, such as in western and central Nevada, were added to the continental margin during an earlier period of accretion that occurred in late Paleozoic time. These accreted terranes consist of oceanic island arcs, scraps of seamounts and remnants of oceanic plateaus, fragments of oceanic basins and continental rises, and foundered continental blocks of uncertain cratonal affinity. Radiolarian cherts incorporated in many of these terranes provide a wealth of sedimentological and biostratigraphic data that are crucial to deciphering the geologic history of the terranes themselves, as well as to elucidating the tectonic process of crustal growth through accretion of allochthonous terranes.

The following consists of a brief description of some of the more important chert deposits that occur within the Cordillera. The purpose is to illustrate the kinds of significant information that can be derived from study of the chert itself. This discussion should not be construed as constituting a comprehensive catalogue of chert occurrences in western North America.

Marin Headlands, California

Perhaps the best known Cordilleran example of the ophiolitic chert association is the Marin Headlands terrane, which is part of the Franciscan Complex of the California Coast Ranges. The stratigraphic sequence of this terrane consists of a lower unit of pillow basalt, a middle unit of chert, and an upper clastic unit of graywacke and shale. Recent detailed studies of the chert succession by Murchey (1984) have shown that the chert deposit occurred from Early Jurassic (Pliensbachian) to Cretaceous (late Albian or early Cenomanian) and represents a time span of about 100 Myr. This is one of the longest durations known for continuous chert deposition and is remarkable in that it records pelagic deposition isolated from continentally derived clastic sources for much of this time. Only the last few centimeters in the section are clay rich, and these are directly overlain by terrigenous deep-sea fan deposits of mid-Cretaceous age. Geochemical studies by Karl (1984) show that the lower part of the section lacks a recognizable clastic component, whereas higher in the section, clastic material can be recognized based on an increase in the ratio $Al/(Al + Fe + Mn)$. The increase in alumina with respect to the other metals is thought to signify an approaching continental source, and this is borne out by the fact that the overlying clastic rocks contain appreciable amounts of quartz and feldspar (Jayko & Blake 1984).

Based on the known distribution of Mesozoic radiolarians on oceanic plates, Murchey (1984) suggested that the cherts of the Marin Headlands terrane formed in a midoceanic setting along the Jurassic paleoequator.

Chulitna Terrane

Another example of the ophiolitic chert association occurs in the Chulitna terrane of south-central Alaska (Jones et al 1980). The stratigraphic sequence of this terrane is much more complex than that of Marin Headlands, as it contains four separate chert horizons that are widely separated in time and that record distinct tectonic and depositional events.

The basal chert horizon consists of several tens of meters of red radiolarian-rich chert intercalated with and overlying pillow basalt. Associated with the basaltic rocks are large bodies of serpentinite and minor amounts of gabbro. The chert contains a well-preserved Upper Devonian radiolarian fauna, including the distinctive and diagnostic form *Holoeciscus*, as well as rare conodonts. The upper part of this chert unit is green cherty argillite and chert of Mississippian age. Clasts of this rock occur in overlying volcaniclastic conglomerates and related finer-grained volcanogenic deposits of Permian age. These rocks record a period of arc-related volcanism that ceased by Early Permian time, and they were succeeded by spiculitic and radiolarian-bearing chert of Permian age. This subsidence may have been induced by cooling of the underlying volcanic arc.

Fossiliferous limestone of Permian and Early Triassic ages overlies the Permian chert and records a return to shallow-water conditions. The uppermost bed of the cherty Lower Triassic limestone, however, consists of a few centimeters of phosphatic chert with abundant conodonts and bone fragments suggesting slow deposition in water depths close to the calcite compensation depth. These Lower Triassic rocks are unconformably overlain by Upper Triassic nonmarine redbeds and basalt that contain thin beds of shallow marine fossiliferous carbonate rocks. Subsidence is recorded by an upward change to shallow marine sandstone and shale of latest Triassic and Early Jurassic age, and finally to radiolarian chert of Late Jurassic age. The Triassic redbeds contain abundant detritus derived from the basal Upper Devonian ophiolite assemblage, as well as pebbles of polycrystalline quartz derived from a continental source; thus they provide evidence for a major Late Triassic orogenic event involving collision of an oceanic assemblage and a continental margin, followed by rifting, subsidence, and northward displacement. Evidence for the northward displacement is seen in the Triassic faunas, which are of tropical nature and now occur far north of their cratonally linked coeval counterparts (Nichols & Silberling 1979).

Angayucham Terrane

Parts of northern and east-central Alaska contain enormous thrust sheets composed of complex assemblages of pillow basalt, diabase, gabbro, and minor ultramafic rocks. Associated sedimentary rocks are mainly chert and argillite, with minor amounts of volcaniclastic rock. Chert, which ranges in age from Late Devonian to Early Jurassic, occurs in three different settings, which are now structurally bounded and closely juxtaposed as follows:

1. Thick piles of isoclinally folded strata that are not associated with pillow lavas nor intruded by sills and dikes of diabase and gabbro;
2. Similar kinds of chert and argillite, but intruded by large sills and dikes of diabase and gabbro; and
3. Small pods, lenses, and thin beds of chert interstratified with pillow basalt or associated with tuffaceous sedimentary rocks.

Based on field associations, intrusive relations, and chemical characteristics (Barker et al 1986), this oceanic mafic assemblage appears to be a product of off-ridge volcanic activity, and the pillow piles seem to represent seamounts, rather than primary oceanic crust. The major differences that separate this seamount association from the ophiolitic association are less abundant ultramafic rocks (including harzburgite and cumulate ultramafic rocks), no sheeted dikes or sills, abundant evidence for multiple intrusion of mafic rocks into older deep-water sedimentary sequences, and local evidence that the basaltic edifices grew upward into shallow-water environments.

Coast Range Ophiolite

The Coast Range ophiolite, which is widely distributed throughout the California Coast Ranges, is a composite entity (Blake & Jones 1981, Shervais & Kimbrough 1985) that actually contains three separate chert-bearing associations, none of which seem to represent primary oceanic crust. The best-studied part is in the southern Coast Ranges, where pillow lava (island-arc tholeiite according to Shervais & Kimbrough 1985) is capped by a few tens of meters of radiolarian-rich tuffaceous chert and minor silicic limestone. Radiolarian faunas from these strata have been studied extensively (Pessagno 1977a,b, Pessagno & Blome 1980, Hopson et al 1981, Pessagno et al 1984), and they serve as the local biostratigraphic standard of reference for Upper Jurassic western North American siliceous deposits.

The cherts themselves contain abundant volcanogenic material, including crystals of plagioclase, hornblende, and magnetite, as well as shardlike and pumaceous fragments; these deposits grade into water-laid tuffs and other coarser volcaniclastic material (Hopson et al 1981). Both mixing of wind-blown volcanic material from an active volcanic arc with pelagic organisms and deposition in a deep-water environment beyond the reach of continentally derived clastic material are clearly indicated.

Arc-related tuffaceous cherts like those of the southern Coast Ranges are not present in the northern Coast Ranges, although Shervais & Kimbrough (1985) report that basaltic rocks near Paskenta are chemically similar to the island-arc tholeiites of the southern Coast Ranges. The Paskenta rocks, however, are mainly breccias with large clasts of mafic and ultramafic plutonic rocks that are thought to have formed along fault scarps in a transform fault zone. Locally present at the base of the breccias are a few meters of red argillite occurring with crudely pillowed to massive basalt. A few thin layers within the argillite are radiolarian rich; although these fossils are poorly preserved and difficult to identify, they appear to be about the same age as those from the base of the tuffaceous cherts of the southern Coast Ranges. On the basis of associated rocks, this cherty argillite unit resembles the ophiolitic assemblage.

Structurally below this breccia unit is another tectonic assemblage consisting of serpentinite mélange containing large blocks of basalt with intercalated beds of radiolarian chert up to tens of meters in thickness. Radiolarians from these cherts are at least in part younger than those from the structurally higher red cherty argillite, and the volcanic rocks are different in that they are subalkalic basalts geochemically akin to ocean-island tholeiites (Shervais & Kimbrough 1985). These differences indicate that the Coast Range ophiolite represents a number of different assemblages that formed in very different environments that were subsequently tectonically juxtaposed.

Golconda Terrane

An extensive, tectonically complex assemblage of chert, argillite, volcaniclastics, limestone turbidites, quartzose turbidites, and minor pillowed greenstone constitutes the Golconda terrane of northwestern Nevada. Radiolarian cherts range in age from Late Devonian to Late Permian (Murchey et al 1983b, Miller et al 1984) and provide most of the useful biostratigraphical data from the terrane. These rocks were strongly folded in latest Permian time and unconformably overlain by Early Triassic volcanic and sedimentary rocks of the Koipato Group (Silberling &

Roberts 1962) and by younger Triassic strata of the Auld Lang Syne Group (Burke & Silberling 1973).

The Golconda terrane contains several different chert associations, which are at least in part coeval. The oldest (Late Devonian to Early Mississippian) is an ophiolitic or seamount assemblage consisting of pillowed greenstone and red, manganiferous, iron-rich chert and jasper. This assemblage is apparently overlain locally by black, finely laminated chert or by cherty volcaniclastic rocks and possible andesitic flows (Miller et al 1984) of earliest Mississippian age.

Another chert assemblage, which forms the dominant part of the Golconda terrane, consists of chert, argillite, and interbedded turbidites of either clastic or carbonate composition. Dated cherts range in age from Mississippian to Late Permian. These rocks occur both structurally below the ophiolitic assemblage, as in the Tobin Range (Stewart et al 1977), or structurally above, as in the Independence Range (Miller et al 1984). The regional distribution of these differing chert facies has not yet been determined, and their original paleogeographic relations are also unknown. The entire Golconda terrane was deformed during the Sonoma orogeny and was emplaced on the continental margin of North America either during or subsequent to this deformation.

Various plate tectonic models have been proposed to explain the tectonic history of the Golconda terrane. Earlier interpretations centered on a back-arc setting for the Golconda terrane, with subsequent east-directed back-arc thrusting leading to deformation and emplacement of the allochthon on the continental margin (Silberling 1973, Burchfiel & Davis 1972, 1975, Miller et al 1984). More recent interpretations have stressed an open-ocean setting for the Golconda terrane, with formation of an accretionary prism in front of an east-facing island arc. Final emplacement of the allochthon is explained simply by a collision between the North American margin and the island arc as the thick continental plate entered the west-dipping subduction zone.

Neither of these models are compelling in their present, oversimplified form. However, they are testable, and with sufficient detailed analyses of the chert-bearing parts of the terrane, the tectonic setting of the Golconda terrane may ultimately be elucidated.

CHERT-BEARING TERRANES IN OTHER PARTS OF THE CIRCUM-PACIFIC REGION

Much of the circum-Pacific rim is characterized by accreted terranes of oceanic character in which radiolarian chert forms a conspicuous component. In only a few places, however, have the cherts been studied in sufficient detail to establish their age, faunal and lithic characteristics, and structural relations. Dated radiolarian cherts are known from Japan, the Philippines, Indonesia, New Zealand, Antarctica, Chile, Costa Rica, and Mexico. Space does not permit discussion of all of these localities, except to note that when thoroughly analyzed, these deposits will provide a wealth of important sedimentological, structural, and paleontological data needed for reconstructing the geologic and tectonic history of the ancient Pacific Ocean.

Radiolarian cherts from Japan are among the best known as a result of the detailed studies of Mizutani & Shibata (1983), Imoto (1983, 1984), Yao (1983), Taira et al (1980), Sano (1983) and many other Japanese workers. The bulk of Japanese radiolarian cherts are of Permian, Triassic, Jurassic, and Cretaceous ages. Most of these deposits are associated with accreted terranes that have moved northward during the Mesozoic from low paleolatitudes (Mizutani & Hattori 1983, Hirooka et al 1983). On the other hand, radiolarian cherts from both New Zealand and Chile are associated with tropical faunas (Ozawa & Kanmera 1983) that appear to have been displaced southward from low paleolatitudes. Because of these large-scale dislocations that occurred during the Mesozoic, establishing the original deposition and paleobiogeographic setting of many of these scattered radiolarian faunas is now extremely difficult.

STRUCTURAL RELATIONS IN RADIOLARIAN CHERT-BEARING ASSEMBLAGES

Because radiolarian cherts typically were deposited in oceanic to pericratonal environments, their incorporation into continental landmasses has involved minimum tectonic displacements on the order of tens to hundreds of kilometers. As a consequence, most radiolarian chert assemblages are highly folded, disrupted, and cut by innumerable faults. The Marin Headlands terrane, for example, contains 10 separate thrust sheets, each a few hundred meters thick, composed of the same assemblage of basalt, chert, and graywacke. According to detailed geologic mapping by Wahrhaftig (1984), the average horizontal spacing between imbricate thrust faults is about half a kilometer.

Intact stratigraphic sequences are rarely observable within these deformed chert terranes, and normal biostratigraphic methods that involve detailed measured sections to establish faunal successions usually cannot be applied. As a result of this structural complexity, the development of a useful biostratigraphic time scale based on radiolarians has been very difficult to achieve. Only in the past few years have the broad outlines of radiolarian distribution through time been sufficiently understood so that meaningful biogeochronometric ages could be determined. The relatively rare but very important joint occurrence of radiolarians and conodonts (whose relative ages were established in shallow-water carbonate deposits) in single samples has permitted calibration of many radiolarian faunas that otherwise could not have been dated (Holdsworth & Jones 1980a).

Unraveling the structural history and reconstructing the depositional environments as well as the paleogeographic settings of many of these chert-bearing terranes are some of the most difficult but important geologic problems yet to be solved.

DATING RADIOLARIAN CHERT

Radiolarian Biostratigraphy

The development of a technique for extracting radiolarians from siliceous rocks by hydrofluoric acid etching (Pessagno & Newport 1972) created great interest in radiolarian biostratigraphy. The technique provided a means for dating previously updated rock units all over the world. Since that time, the number of workers in the field of radiolarian biostratigraphy has increased exponentially.

Opal-secreting radiolarians, the only type of radiolarians that are preserved as fossils, first appear in the fossil record in Cambrian(?) (Anita Harris, oral communications, 1984) and Ordovician rocks. The first bedded radiolarian chert sequences are Ordovician. Opal-secreting radiolarians still inhabit the oceans, although they have been eclipsed by other planktonic organisms as important rock-formers. Since their initial appearance in the fossil record, radiolarians have undergone periods of rapid diversification followed by extinctions and periods of relatively low diversity. Figures 2a and 2b illustrate the relative abundance and diversity of important radiolarian morphotypes from the early Paleozoic to the Cretaceous. The morphological lineages shown in these two figures are polyphyletic, especially those rows encompassing both Paleozoic and Mesozoic time.

Paleozoic faunas are characterized by radiolarians with an internal bar or spheroid structure. The earliest radiolarian populations were dominated by spheroidal radiolarians with an internal structure of connecting rods and with multiple simple spines (see Figure 2a, Ordovician radiolarian). By the Silurian, fragile ancestral forms of important Paleozoic lineages with bilateral symmetry (*Paleoscenidium* and *Ceratoikiscum* lineages) appear in the rock record (Figure 2b, top row).

Rapid diversification of radiolarians occurred during the Late Devonian and Early Mississippian (Figures 2a,b). Bilaterally symmetrical forms became increasingly diverse and complex. One lineage of the bilaterally

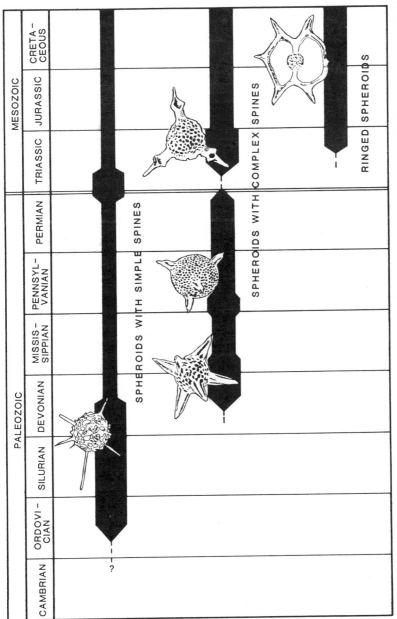

Figure 2a The relative abundance and diversity of spheroidal radiolarians during the Paleozoic and Mesozoic. Radiolarians are subdivided into three polyphyletic groups based on the complexity of their primary spines and the presence or absence of a ring.

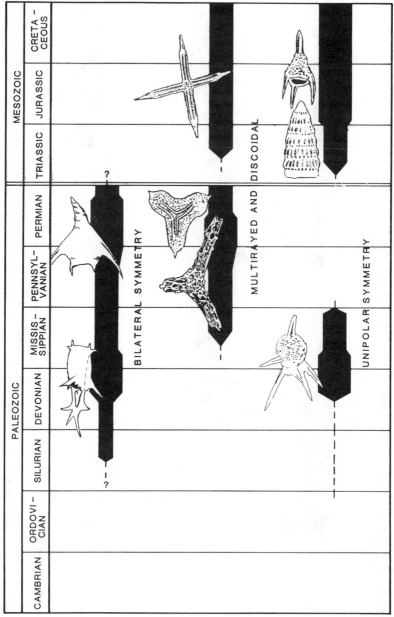

Figure 2b   The relative abundance and diversity of nonspheroidal radiolarians during the Paleozoic and Mesozoic. Radiolarians are subdivided into three polyphyletic groups on the basis of shape and symmetry.

symmetrical forms (superfamily Albaillellaria) provides the assemblage names of Upper Devonian–to–Permian radiolarian assemblages in the zonation schemes of Holdsworth & Jones (1980a,b) and Ishiga et al (1982). Paleozoic genera characterized by spheroids with a hole or pylome surrounded by three or four "feet" (pylentonemiids) were important during this period (Figure 2b, bottom row; Deflandre 1963, Holdsworth & Jones 1980b). Beginning in the Devonian, the major (primary) spines of spheroidal species were commonly tribladed or grooved in cross section (Figure 2a, middle row). The spines of many Late Devonian and Early Mississippian spheroidal species were robust and complexly shaped.

The rapid diversification of the Late Devonian and Early Mississippian was followed by a loss in diversity and the extinction or impoverishment of some genera (Figures 2a,b). However, a new and stratigraphically important Paleozoic lineage of multi-rayed forms, the Latentifistulidea, first appeared during the late Early Mississippian (Figure 2b, middle row; Nazarov & Ormiston 1983, Ormiston & Lane 1976, Holdsworth & Jones 1980b). New species and genera of bilaterally symmetrical forms (Albaillellaria) and multi-rayed forms (Latentifistulidea) evolved during the Late Mississippian, Pennsylvanian, and Permian. Radiolarians with pylomes surrounded by three or four "feet" apparently disappeared from the fossil record near the end of the Mississippian, although nassellarian homeomorphs reappeared in the Mesozoic. The Late Pennsylvanian and Early Permian were periods of diversification, particularly for low-latitude faunas. Late Permian faunas were generally lower in diversity than Early Permian faunas.

The Permian-Triassic boundary marks the greatest extinction in the history of radiolarians. All morphologies except simple spheroidal forms and some simple spicular forms disappeared from the record (Figures 2a,b). The earliest Triassic radiolarian assemblages were characterized by spheroidal spumellarian radiolarians with simple primary spines, circular in cross section (D. L. Jones, unpublished data). The massive extinction was followed by rapid diversification during the Middle and Late Triassic. An important new Mesozoic and Cenozoic lineage of radiolarians with external uniaxial radial symmetry, the nassellarians, first appeared in the Triassic (Figure 2b, bottom row). Also, new lineages of multi-rayed, discoidal, and spongy morphotypes appeared (Figure 2b, middle row). While they are externally similar to the Paleozoic multi-rayed forms (the Latentifistulidea), their internal test structure is quite distinct. In addition, a new lineage of Saturn-like spheroids with discoidal rings (Parasaturnalidae) evolved during the Triassic (Figure 2a, bottom row; e.g. Kozur & Mostler 1972). Late Triassic spheroidal radiolarians also include genera with very robust, complex spines (Figure 2a, middle row; e.g.

DeWever et al 1979, Blome 1984). Some of these spines are homeomorphs of Late Devonian and Early Mississippian spines. The great burst of diversification that followed the Permian-Triassic extinctions subsided at the end of the Triassic, but the major new lineages that appeared during the Triassic continue, in some form, today.

The Jurassic and Cretaceous assemblages were composed of radiolarians belonging to the major lineages established in the Triassic. Radiolarians with uniaxial radial symmetry (nassellarians) were abundant and became increasingly ordered and complex. For this reason, nassellarians are important as key index fossils for the Jurassic and Cretaceous (Pessagno 1977a,b). Multi-rayed forms also show a pattern of increasing organization and complexity in structure during the Jurassic and Cretaceous (Baumgartner 1980).

In summary, periods of low diversity (the early Paleozoic and the Early Triassic) were dominated by spheroids with simple primary spines. Periods of high diversity (Late Devonian and Early Mississippian, Early Permian, Late Triassic, and Jurassic) were characterized by faunas with elongated unipolar radiolarians, with spheroidal radiolarians possessing robust, complex spines, and with multi-rayed or discoidal radiolarians (Mississippian and younger).

Paleozoic and Mesozoic radiolarian biostratigraphy is still a new field, and many (if not most) species remain undescribed. Further refinement of zonation schemes will require detailed collections from many sequences. At present, faunal correlations from one region to another or from one terrane to another are complicated by environmentally controlled assemblage differences (e.g. Pessagno et al 1984). The discrimination between those radiolarians most useful as stratigraphic index markers and those radiolarians most useful as environmental markers will take many years of painstaking comparisons of faunal assemblages, distribution patterns, sedimentary associations, basin histories, and paleolatitudinal histories. Tools for determining chronometric ages of chert sequences, such as isotopic dating and magnetostratigraphic dating, will be valuable for establishing good correlation between regions.

Magnetostratigraphic Dating

Channell & Baumgartner (1985) documented magnetozones corresponding to oceanic magnetic anomalies M15 to M22 in Jurassic siliceous limestones and cherts in the Mediterranean region; this correspondence permits correlation of magnetozones with previously established radiolarian zones (Baumgartner 1984). The magnetozones recognized in the radiolarian-bearing sequences permit the correlation of these sequences with ammonite-bearing sequences, and hence with ammonite-based stage boundaries, for which the same magnetozones have also been recognized. Similar applications of magnetostratigraphy to radiolarian-bearing sequences around the world would improve intraregional correlations dramatically.

Isotopic Dating

Isotopic age dating of pre-Cenozoic radiolarian-bearing rocks is a new field of study that promises to provide valuable information concerning timing of deposition and duration of diagenesis. Good Rb-Sr isochrons have been obtained from Mesozoic radiolarian chert and radiolarian siliceous mudstone (Shibata & Mizutani 1980, 1982, Mizutani 1983, Chyi et al 1984). In the studies by Shibata & Mizutani, the whole-rock Rb-Sr isochron ages for radiolarian siliceous mudstone were compared with their biostratigraphic ages. Shibata & Mizutani considered the difference between the isochron and biostratigraphic ages to represent the duration of chemical diagenesis of the siliceous sediments. In other words, when biogenic opal-A is converted to quartz, the system becomes chemically closed with respect to rubidium and strontium, and the isochron clock begins ticking. The calculated values for the duration of chemical diagenesis ranged from 0–12 to 17–35 Myr for samples from three Mesozoic formations of different ages. These values are considerably less than the 40–100 Myr required for formation of quartz in siliceous sediments in the deep sea (Siever 1979, Riech & von Rad 1979). Mizutani concluded that elevated temperatures during diagenesis (estimated 38–56°C) were responsible for the fast diagenetic rates in the Japanese sequences. Shibata & Mizutani also obtained K-Ar whole-rock ages for the same formations, but these were diversified and not very satisfactory.

Chyi et al (1984) obtained a whole-rock Rb-Sr isochron age of 158 ± 5 Myr for radiolarian chert associated with a manganese deposit in the Franciscan Complex of California. The age agrees with an age of approximately 160 Myr calculated from Th-Pb data. Chyi et al considered that the initial $^{87}Sr/^{86}Sr$ value was derived wholly from seawater and not partly contributed by continental detritus. The assumption of a lack of a detrital component is supported by $^{143}Nd/^{144}Nd$ values in the chert and by an isochron value for initial $^{87}Sr/^{86}Sr$ close to that of Jurassic seawater as determined by the $^{87}Sr/^{86}Sr$ seawater curve (Burke et al 1982). The isotopic age is in apparent agreement with a Late Jurassic radiolarian biostratigraphic age for the rocks. Chyi et al concluded that the Rb-Sr age represents an absolute age for the time of deposition of the siliceous sediments, especially since the radiolarian faunas are Jurassic.

The utility of isotopic dating of siliceous sedimentary rocks needs to be tested on more rock sequences, but it may prove to be very useful. Whether this method of absolute dating ultimately provides more information concerning the rate and timing of diagenetic events than the timing of deposition remains to be discovered. To our knowledge, other methods of isotopic dating, such as correlating $^{87}Sr/^{86}Sr$ values with the $^{87}Sr/^{86}Sr$ seawater curve for Phanerozoic time (Burke et al 1982), have not been applied to Paleozoic or Mesozoic chert sequences.

Summary of Dating Techniques

Radiolarian biostratigraphy is the easiest method for dating radiolarian-bearing rocks. The radiolarian zonations will be improved by further stratigraphic studies and by interdisciplinary studies correlating radiolarian zones to magnetozones and to isotopic age determinations. Isotopic age determinations may be giving diagenetic ages in most or all cases. If so, they may at least provide a minimum stratigraphic age for deposition.

MINERALS

On-land stratabound mineral deposits in radiolarian-bearing chert sequences are easily accessible settings for studying oceanic mineral-forming processes. Therefore, they are interesting from a scientific point of view as well as from an economic perspective. The distribution of stratabound mineral deposits in radiolarian-bearing siliceous rocks is controlled by two factors: (a) the volcanic-hydrothermal history of the rocks, and (b) the depositional and early diagenetic history of the rocks.

Some manganese, barite, metal sulfides, and metalliferous shale deposits in radiolarian-bearing siliceous rock sequences apparently formed as the result of oceanic hydrothermal activity. Modern analogues can be found in the hot springs of oceanic ridge systems such as the Mid-Atlantic Ridge, the East-Pacific Rise, and the Gorda Ridge (Scott et al 1974, Lonsdale 1977). Oceanic hot-spring systems may also occur in other tectonic settings with active volcanism, such as oceanic islands, island arcs, and back-arc basins (Snyder 1978).

In contrast, sedimentary and diagenetic processes directly related to depositional environment and only indirectly related to tectonic setting account for phosphate deposits and possibly many barite deposits in radiolarian-bearing siliceous rocks.

CONCLUSIONS

Siliceous sedimentary rocks, particularly those varieties containing abundant skeletons of radiolarians, are important because they are now yielding significant new information concerning paleoceanography, paleoecology, evolution of planktonic forms of life, and the tectonic evolution of continental margins during the Paleozoic and Mesozoic. Many cherts record deposition in oceanic settings, and thus they uniquely provide a historical record of events of major importance that are not preserved in any other rock system. The development of a radiolarian biostratigraphic time scale with adequate resolving power has opened this record to detailed examination. This development has had a profound effect on our understanding of geologic history and of the dynamic processes that shape the Earth. Much credit is owed those pioneers who perfected the extraction method and proved that radiolarians are biostratigraphically useful.

In this brief report, we have outlined the composition and character of cherts, proposed a classification system that will facilitate exchange of information regarding various kinds of siliceous sedimentary rocks, and described some of the biologic, geologic, and tectonic ramifications of chert investigations. Many problems remain to be resolved, some of the more important of which are listed below:

1. The basic descriptive taxonomy of radiolarians is primitive; only Cenozoic and upper Mesozoic faunas are adequately described, although a good start has been made on the lower Mesozoic forms. Paleozoic faunas are 95% undescribed, and until massive taxonomic studies are published, the biostratigraphic utility of these forms remain available only to a few specialists. Because of this situation, documentation of some important geologic and tectonic interpretations based on chert ages is difficult.

2. Most chert occurrences are known only through reconnaissance studies, and few have been examined in sufficient detail to establish the full range of ages, lithologies, and faunas present. Consequently, we have only scratched the surface in our studies to date, and many localities will require additional sampling. Chemical parameters are virtually unknown except for a few places. Quantitative studies of biologic components are only now getting underway. They should yield important data on sedimentary facies despite the formidable problems presented by selective leaching, diagenesis, and metamorphism that have altered original compositions.

3. Basin analysis of cherty deposits is a new field that is emerging now that biostratigraphic control is available. Such analyses will include rock composition and distribution; age of deposition; character of the faunas; facies relations with noncherty clastic, volcanic, and carbonate rocks; transport direction of sedimentary materials; nature of the basement, if present; and postdepositional structural history. No chert basin in the North American Cordillera has yet been adequately studied, although initial efforts are underway. Such studies are extremely labor intensive owing to the detailed nature of the sampling, measuring, and mapping

required, and few students have received the requisite training needed for independent investigations.

4. Paleobiogeographic analyses of radiolarian faunas are in their infancy. More comparative studies are required from broadly separated areas and particularly from places whose paleogeographic locations through time are adequately known. Combined paleontologic and paleomagnetic studies will be required to establish the original latitudinal settings of many displaced chert assemblages. Preliminary data suggest that broad geographical faunal differences as well as locally controlled facies differences existed throughout the later Paleozoic and Mesozoic. Separating the local influences from broad regional patterns is difficult, however, because of the extreme degree of disruption that has occurred within chert assemblages.

From the above list, it is apparent that many exciting and challenging problems await solution for those interested in the history, character, and significance of fine-grained siliceous rocks composed dominantly of radiolarian skeletons.

Literature Cited

Aalto, K. R., Murphy, J. M. 1984. Guide to Franciscan geology of the Crescent City area, California, with a road log from Eureka to Crescent City. Unpublished report for Geological Society of America, Penrose Conference on Structural Styles and Deformation Fabrics of Accretionary Complexes, Eureka, Calif., April 30–May 4, 1984. 70 pp.

Baltuck, M. 1983. Some sedimentary and diagenetic signatures in the formation of bedded radiolarite. In *Siliceous Deposits in the Pacific Region*, ed. A. Iijima, J. R. Hein, R. Siever, pp. 299–315. Amsterdam: Elsevier

Barker, F., Jones, D. L., Boudahn, J. R., Coney, P. 1986. Ocean plateau-seamount origin for basaltic rocks of the Angayucham terrane, north-central Alaska. *Can. J. Earth Sci.* In press

Barrett, T. J. 1981. Chemistry and mineralogy of Jurassic bedded chert overlying ophiolites in the north Apennines, Italy. *Chem. Geol.* 34:289–317

Bates, R. L., Jackson, J. A., eds. 1980. *Glossary of Geology*. Falls Church, Va: Am. Geol. Inst.

Batiza, R., Vanko, P. 1984. Petrology of young Pacific seamounts. *J. Geophys. Res.* 89:1235–60

Baumgartner, P. O. 1980. Late Jurassic Hagiastridae and Patulibracchiidae (Radi-

olaria) from the Argolis Peninsula (Peloponnesus, Greece). *Micropaleontology* 26:274–322, pls. 1–12

Baumgartner, P. O. 1984. Middle Jurassic–Early Cretaceous low-latitude radiolarian zonation based on unitary associations and age of Tethyan radiolarities. *Ecol. Geol. Helv.* 77(3):729–837

Blake, M. C. Jr., Jones, D. L. 1981. The Franciscan assemblage and related rocks in northern California: a reinterpretation. In *Geotectonic Development of California*, ed. W. G. Ernst, pp. 308–28. Englewood Cliffs, NJ: Prentice-Hall

Blome, C. D. 1984. Upper Triassic Radiolaria and radiolarian zonation from western North America. *Bull. Am. Paleontol.* 85 (318):88 pp.

Bosellini, A., Winterer, E. L. 1975. Pelagic limestone and radiolarite of the Tethyan Mesozoic: a genetic model. *Geology* 3:279–82

Burchfiel, B. C., Davis, G. A. 1972. Structural framework and evolution of the southern part of the Cordilleran orogen, western United States. *Am. J. Sci.* 272:92–118

Burchfiel, B. C., Davis, G. A. 1975. Nature and controls of Cordilleran orogenesis, western United States: extension of an earlier synthesis. *Am. J. Sci.* 275A:363–96

Burke, D. B., Silberling, N. J. 1973. The Auld Lang Syne Group of Late Triassic and

Jurassic age, north-central Nevada. *US Geol. Surv. Bull. 1394E.* 14 pp.

Burke, W. H., Denison, R. E., Hetherington, E. A., Koepnick, R. B., Nelson, H. F., Otto, J. B. 1982. Variation of seawater $^{87}Sr/^{86}Sr$ throughout Phanerozoic time. *Geology* 10: 516–19

Channell, J. E. T., Baumgartner, P. O. 1985. Correlation of Late Jurassic radiolarian and magnetic stratigraphy in northern Italian sections. *Eos, Trans. Am. Geophys. Union* 66: 259 (Abstr.)

Chyi, M. S., Crerar, D. A., Carlson, R. W., Stallard, R. F. 1984. Hydrothermal Mn-deposits of the Franciscan Assemblage, II. Isotope and trace element geochemistry, and implications for hydrothermal convection at spreading centers. *Earth Planet. Sci. Lett.* 71: 31–45

Cowan, D. S. 1985. Structural styles in Mesozoic and Cenozoic mélanges in the western Cordillera of North America. *Geol. Soc. Am. Bull.* 96: 451–62

Crerar, D. A., Namson, J., Chyi, M. S., Williams, L., Feigenson, M. D. 1982. Manganiferous cherts of the Franciscan Assemblage, I. General geology, ancient and modern analogues, and implications for hydrothermal convection at oceanic spreading centers. *Econ. Geol.* 77: 519–40

Deflandre, G. 1963. Pylentonema, nouveau genre de Radiolaire du Viseen: Sphaerellaire ou Nassellaire? *C. R. Acad. Sci., Paris* 257: 3981–84

DeWever, P., Sanfilippo, A., Riedel, W. R., Gruber, B. 1979. Triassic radiolarians from Greece, Sicily, and Turkey. *Micropaleontology* 25: 75–110

Folk, R. L. 1968. *Petrology of Sedimentary Rocks.* Austin, Tex: Hemphill's. Reprinted in 1974.

Folk, R. L., McBride, E. F. 1976. The Caballos Novaculite revisited. Part I. Origin of Novaculite members. *J. Sediment. Petrol.* 46: 659–69

Folk, R. L., McBride, E. F. 1978. Radiolarites and their relation to subjacent "oceanic crust" in Liguria, Italy. *J. Sediment. Petrol.* 48: 1069–1101

Folk, R. L., Pittman, J. S. 1971. Length-slow chalcedony—a new testament for vanished evaporites. *J. Sediment. Petrol.* 41: 1045–58

Grommé, S. 1984. Paleomagnetism of Franciscan Basalt, Marin County, California, revisited. In *Franciscan Geology of Northern California*, ed. M. C. Blake Jr., pp. 113–20. Los Angeles: Soc. Econ. Paleontol. Mineral., Pac. Sect.

Hein, J. R., Karl, S. M. 1983. Comparisons between open-ocean and continental margin chert sequences. In *Siliceous Deposits in the Pacific Region*, ed. A. Iijima, J. R. Hein,

R. Siever, pp. 25–43. Amsterdam: Elsevier

Hein, J. R., Parrish, J. 1986. Distribution of silica in space and time. In *Sedimentary Siliceous Rock Hosted Ores and Petroleum*, ed. J. R. Hein. In press

Hein, J. R., Scholl, D. W., Barron, J. A., Jones, M. G., Miller, J. 1978. Diagenesis of late Cenozoic diatomaceous deposits and formation of the bottom simulating reflector in the southern Bering Sea. *Sedimentology* 25: 155–81

Hillhouse, J. W., Grommé, C. S. 1984. Northward displacement of Wrangellia: new paleomagnetic evidence from Alaska. *J. Geophys. Res.* 89: 4461–77

Hirooka, K., Uchiyama, S., Date, T., Kanai, H. 1983. Paleomagnetic evidence of accretion and tectonism of the Hida or the Circum-Hida terranes, central Japan. *Proc. Circum-Pac. Terrane Conf.*, ed. D. G. Howell, D. L. Jones, A. Cox, A. Nur, 18: 115–17. Stanford, Calif: Stanford Univ. Publ., Geol. Sci.

Holdsworth, B. K., Jones, D. L. 1980a. Preliminary radiolarian zonation for Late Devonian through Permian time. *Geology* 8: 281–85

Holdsworth, B. K., Jones, D. L. 1980b. A provisional radiolaria biostratigraphy, Late Devonian through Late Permian. *US Geol. Surv. Open-File Rep. 80–876.* 32 pp.

Hopson, C. A., Mattinson, J. M., Pessagno, E. A. 1981. Coast Range ophiolite, western California. In *Geotectonic Development of California*, ed. W. G. Ernst, pp. 232–48. Englewood Cliffs, NJ: Prentice-Hall

Imoto, N. 1983. Sedimentary structures of Permian-Triassic cherts in the Tamba district, southwest Japan. In *Siliceous Deposits in the Pacific Region*, ed. A. Iijima, J. R. Hein, R. Siever, pp. 377–94. Amsterdam: Elsevier

Imoto, N. 1984. Late Paleozoic and Mesozoic cherts in the Tamba belt, southwest Japan. *Bull Kyoto Univ. Educ., Ser. B* 65: 15–71

Isaacs, C. M., Pisciotto, K. A., Garrison, R. E. 1983. Facies and diagenesis of the Miocene Monterey Formation, California: a summary. In *Siliceous Deposits in the Pacific Region*, ed. A. Iijima, J. R. Hein, R. Siever, pp. 247–82. Amsterdam: Elsevier

Ishiga, H., Kito, T., Imoto, N. 1982. Permian radiolarian biostratigraphy. *News Osaka Micropaleontol. Spec. Vol.* 5: 17–25

Jayko, A. S., Blake, M. C. Jr. 1984. Sedimentary petrology of graywacke in the Franciscan Complex in the northern San Francisco Bay region, California. In *Franciscan Geology of Northern California*, ed. M. C. Blake Jr., pp. 121–34. Los Angeles: Soc. Econ. Paleontol. Mineral. Pac. Sect.

Jenkyns, H. C., Winterer, E. L. 1982. Palae-

oceanography of Mesozoic ribbon radio-larities. *Earth Planet. Sci. Lett.* 60:351–75
Jones, D. L., Silberling, N. J., Csejtey, B. Jr., Nelson, W. H., Blome, C. D. 1980. Age and structural significance of ophiolite and adjoining rocks in the upper Chulitna district, south-central Alaska. *US Geol. Surv. Prof. Pap. 1121-A.* 21 pp.
Jones, J. B., Segnit, E. R. 1971. Nomenclature and constituent phases. Part I. The nature of opal. *J. Geol. Soc. Aust.* 18:57–68
Jones, J. B., Segnit, E. R. 1972. Genesis of cristobalite and tridymite at low tempera-tures. *J. Geol. Soc. Aust.* 18:419–22
Karl, S. M. 1984. Sedimentologic, diagenetic, and geochemical analysis of upper Meso-zoic ribbon cherts from the Franciscan assemblage at the Marin Headlands, Cali-fornia. In *Franciscan Geology of Northern California*, ed. M. C. Blake Jr., pp. 71–88. Los Angeles: Soc. Econ. Paleontol. Miner-al. Pac. Sect.
Kastner, M., Keene, J. B., Gieskes, J. M. 1977. Chemical controls on the rate of opal-A to opal-CT transformation—an experi-mental study. Part I of Diagenesis of siliceous oozes. *Geochim. Cosmochim. Acta* 41:1041–59
Keene, J. B. 1983. Chalcedonic quartz and occurrence of quartzine (length-slow chal-cedony) in pelagic sediments. *Sedimentol-ogy* 30:449–54
Kozur, H., Mostler, H. 1972. Beitrage zur Erforschung der mesozoischen Radiola-rien. Teil I: Revision der Oberfamilie Coccodiscacea Haeckel, 1862, emend. und Beschreibung ihrer triassischen Vertreter. *Geol. Palaont. Mitt. Innsbruck* 2:1–60
Lonsdale, P. 1977. Deep-tow observations at the Mounds abyssal hydrothermal field, Galapagos rift. *Earth Planet. Sci. Lett.* 36:92–110
McBride, E. F., Folk, R. L. 1977. The Caballos Novaculite revisited. Part II. Chert and shale members and synthesis. *J. Sediment. Petrol.* 47:1261–86
McBride, E. F., Folk, R. L. 1979. Features and origin of Italian Jurassic radiolarites deposited on continental crust. *J. Sedi-ment. Petrol.* 49:837–68
McBride, E. F., Thomson, A. 1970. The Caballos Novaculite, Marathon region, Texas. *Geol. Soc. Am. Spec. Pap. 122.* 129 pp.
Miller, E. L., Holdsworth, B. K., Whiteford, W. B., Rodgers, D. 1984. Stratigraphy and structure of the Schoonover sequence, northeastern Nevada: implications for Paleozoic plate-margin tectonics. *Geol. Soc. Am. Bull.* 95:1063–76
Mizutani, S. 1977. Progressive ordering of cristobalitic silica in the early stage of diagenesis. *Contrib. Mineral. Petrol.* 61:

129–40
Mizutani, S. 1983. Duration of chemical diagenesis. *J. Earth Sci. Nagoya Univ.* 31:17–35
Mizutani, S., Hattori, I. 1983. Hida and Mino: Tectonostratigraphic terranes in central Japan. In *Accretion Tectonics in the Circum-Pacific Regions*, ed. M. Hashi-moto, S. Uyeda, pp. 169–78. Tokyo:Terra Sci. Publ.
Mizutani, S., Shibata, K. 1983. Diagenesis of Jurassic siliceous shale in central Japan. In *Siliceous Deposits in the Pacific Region*, ed. A. Iijima, J. R. Hein, R. Siever, pp. 283–98. Amsterdam: Elsevier
Murata, K. J., Norman, M. B. 1976. An index of crystallinity of quartz. *Am. J. Sci.* 276:1120–30
Murchey, B. 1984. Biostratigraphy and litho-stratigraphy of chert in the Franciscan Complex, Marin Headlands, California. In *Franciscan Geology of Northern California*, ed. M. C. Blake Jr., pp. 51–70. Los Angeles: Soc. Econ. Paleontol. Mineral. Pac. Sect.
Murchey, B., Jones, D. L. 1983. Tectonic and paleobiologic significance of Permian radiolarian distribution in circum-Pacific region. *Am. Assoc. Pet. Geol., Ann. Meet., Abstr. with Programs*, p. 134
Murchey, B., Jones, D. L. 1984. Age and significance of chert in the Franciscan Complex in the San Francisco Bay region. In *Franciscan Geology of Northern Cali-fornia*, ed. M. C. Blake Jr., pp. 23–30. Los Angeles: Soc. Econ. Paleontol. Mineral. Pac. Sect.
Murchey, B., Jones, D. L., Blome, C. D. 1983a. Comparison of Permian and lower Mesozoic radiolarian chert in western ac-creted terranes. *Geol. Soc. Am., Cordilleran and Rocky Mt. Sects., Abstr. with Pro-grams*, p. 371
Murchey, B., Jones, D. L., Holdsworth, B. K. 1983b. Distribution, age, and depositional environments of radiolarian chert in west-ern North America. In *Siliceous Deposits in the Pacific Region*, ed. A. Iijima, J. R. Hein, R. Siever, pp. 109–26. Amsterdam: Elsevier
Murchey, B., Jones, D. L., Blome, C. 1986. The environmental and tectonic signifi-cance of two coeval Permian radiolarian-sponge associations in eastern Oregon. *US Geol. Surv. Prof. Pap.* In press
Nazarov, B. B., Ormiston, A. R. 1983. A new superfamily of stauraxon polycystine Radiolaria from the late Paleozoic of the Soviet Union and North America. *Senckenbergiana Lethaea* 64:363–379
Nichols, K. M., Silberling, N. J. 1979. Early Triassic (Smithian) ammonites of paleo-equatorial affinity from the Chulitna ter-rane, south-central Alaska. *US Geol. Surv. Prof. Pap. 1121-B.* 5 pp., 3 pls.

Ormiston, A. R., Lane, H. 1976. A unique radiolarian fauna from the Sycamore Limestone (Mississippian) and its biostratigraphic significance. *Paleontogr. Am.* 154:158–80

Ozawa, T., Kanmera, K. 1983. Tectonic terranes of late Paleozoic rocks and their accretionary history in the circum-Pacific region viewed from Fusulinacean paleobiogeography. *Proc. Circum-Pac. Terrane Conf.*, ed. D. G. Howell, D. L. Jones, A. Cox, A. Nur, 18:158–60. Stanford, Calif: Stanford Univ. Publ., Geol. Sci.

Paris, I., Stanistreet, I. G., Hughes, M. J. 1985. Cherts of the Barberton Greenstone belt interpreted as products of submarine exhalative activity. *J. Geol.* 93:111–29

Pessagno, E. A. Jr. 1977a. Upper Jurassic Radiolaria and radiolarian biostratigraphy of the California Coast Ranges. *Micropaleontology* 23:56–113, pl. 1–12

Pessagno, E. A. Jr. 1977b. Lower Cretaceous radiolarian biostratigraphy of the Great Valley Sequence and Franciscan Complex, California Coast Ranges. *Cushman Found. Foram. Res. Spec. Publ.* 15:1–86, pl. 1–12

Pessagno, E. A. Jr., Blome, C. D. 1980. Upper Triassic and Jurassic Pantenelliinae from California, Oregon, and British Columbia. *Micropaleontology* 26:225–73, pl. 1–11

Pessagno, E. A. Jr., Blome, C. D. 1986. Tectonogenesis of Mesozoic rocks in the Blue Mountains province of eastern Oregon and western Idaho. *US Geol. Surv. Prof. Pap.* In press

Pessagno, E. A. Jr., Newport, R. L. 1972. A technique for extracting Radiolaria from radiolarian cherts. *Micropaleontology* 18:231–34, pl. 1

Pessagno, E. A. Jr., Blome, C. D., Longoria, J. F. 1984. A revised radiolarian zonation for the Upper Jurassic of western North America. *Bull. Am. Paleontol.* 18:51 pp.

Pinto-Auso, M., Harper, G. D. 1986. Pelagic/hemipelagic sedimentation and metallogenesis in a Late Jurassic backarc basin, northwestern California (USA). *Earth Planet. Sci. Lett.* In press

Pisciotto, K. A. 1981. Diagenetic trends in the siliceous facies of the Monterey Shale in the Santa Maria region, California. *Sedimentology* 28:547–71

Riech, V., von Rad, U. 1979. Silica diagenesis in the Atlantic Ocean: diagenetic potential and transformations. In *Deep Drilling Results in the Atlantic Ocean: Continental Margins and Paleoenvironments, Maurice Ewing Ser.*, ed. M. Talwani, W. W. Hay, W. B. F. Ryan, 3:315–40. Washington, DC: Am. Geophys. Union

Sano, H. 1983. Bedded cherts associated with greenstones in the Sawadani and Shimantogawa Groups, southwest Japan. In *Sili-*

ceous Deposits in the Pacific Region, ed. A. Iijima, J. R. Hein, R. Siever, pp. 427–40. Amsterdam: Elsevier

Scott, M. B., et al. 1974. Rapidly accumulating manganese deposit from the median valley of the Mid-Atlantic Ridge. *Geophys. Res. Lett.* 1:355–58

Shervais, J. W., Kimbrough, D. L. 1985. Geochemical evidence for the tectonic setting of the Coast Range ophiolite: a composite island arc–oceanic crust terrane in western California. *Geology* 13:35–38

Shibata, K., Mizutani, S. 1980. Isotopic ages of siliceous shale from Hida-Kanayama, central Japan. *Geochem. J.* 14:235–42

Shibata, K., Mizutani, S. 1982. Isotopic ages of Jurassic siliceous shale and Triassic bedded chert in Unuma, central Japan. *Geochem. J.* 16:213–23

Siever, R. 1979. Plate-tectonic controls on diagenesis. *J. Geol.* 87:127–55

Silberling, N. J. 1973. Geologic events during Permian-Triassic time along the Pacific margin of the United States. In *The Permian and Triassic Systems and Their Mutual Boundary*, ed. A. Logan, L. V. Hills, pp. 345–62. Calgary: Alberta Soc. Pet. Geol.

Silberling, N. J., Roberts, R. J. 1962. Pre-Tertiary stratigraphy and structure of northwestern Nevada. *Geol. Soc. Am. Spec. Pap.* 72. 58 pp.

Snyder, W. S. 1978. Manganese deposited by submarine hot springs in chert-greenstone complexes, western United States. *Geology* 6:741–44

Snyder, W. S., Brueckner, H. K. 1983. Tectonic evolution of the Golconda allochthon, Nevada: problems and perspectives. In *Pre-Jurassic Rocks in Western North America Suspect Terranes*, ed. C. H. Stevens, pp. 103–23. Los Angeles: Soc. Econ. Paleontol. Mineral., Pac. Sect.

Snyder, W. S., Brueckner, H. K., Schweikert, R. A. 1983. Deformational styles in the Monterey and other siliceous sedimentary rocks. In *Symposium Volume on Monterey Oil Fields*, ed. C. M. Isaacs, R. Garrison, pp. 151–70. Los Angeles: Soc. Econ. Paleontol. Mineral., Pac. Sect.

Steinberg, M., Bonnot-Courtois, C., Tlig, S. 1983. Geochemical contribution to the understanding of bedded chert. In *Siliceous Deposits in the Pacific Region*, ed. A. Iijima, J. R. Hein, R. Siever, pp. 193–210. Amsterdam: Elsevier

Steinitz, G. 1970. Chert "dike" structures in Senonian chert beds, southern Negev, Israel. *J. Sediment. Petrol.* 40:1241–54

Stewart, J. H., MacMillan, J. R., Nichols, K. M., Stevens, C. 1977. Deep-water upper Paleozoic rocks in north-central Nevada: a study of the type area of the Havallah

Formation. In *Paleozoic Paleogeography of the Western United States, Paleogeogr. Symp. 1*, ed. J. H. Stewart, C. H. Stevens, A. E. Fritsche, pp. 337–47. Los Angeles: Soc. Econ. Paleontol. Mineral., Pac. Sect.

Taira, A., Okamura, M., Katto, J., Tashiro, M., Saito, Y., et al. 1980. Lithofacies and geologic age relationship within mélange zones of northern Shimanto Belt (Cretaceous), Kochi Prefecture, Japan. In *Geology and Paleontology of the Shimanto Belt: Selected Papers in Honor of Prof. Jiro Katto*, ed. A. Taira, M. Tashiro, pp. 197–213. Kochi, Jpn: Rinyakosaikai (In Japanese; English summary, p. 214)

Thurston, D. 1972. Studies on bedded cherts. *Contrib. Mineral. Petrol.* 36:329–34

Wahrhaftig, C. 1984. Structure of the Marin Headlands block, California: a progress report. In *Franciscan Geology of Northern California*, ed. M. C. Blake Jr., pp. 31–50. Los Angeles: Soc. Econ. Paleontol. Mineral., Pac. Sect.

Yao, A. 1983. Late Paleozoic and Mesozoic radiolarians from southwest Japan. In *Siliceous Deposits in the Pacific Region*, ed. A. Iijima, J. R. Hein, R. Siever, pp. 361–76. Amsterdam: Elsevier

Ann. Rev. Earth Planet. Sci. 1986. 14: 493–571

CHEMICAL GEODYNAMICS

Alan Zindler

Lamont-Doherty Geological Observatory and Department of Geological Sciences, Columbia University, Palisades, New York 10964

Stan Hart

Center for Geoalchemy, Department of Earth, Atmospheric, and Planetary Sciences, Masachusetts Institute of Technology, Cambridge, Massachusetts 02139

1. INTRODUCTION

Chemical geodynamics, developed by Allegre (1982) as an integrated study of the chemical and physical structure and evolution of the solid Earth, is a field of inquiry that has evolved from a marriage of mantle geochemistry and geophysics. By its very nature, geophysics can only characterize the present state of the Earth; geochemistry, on the other hand, supplies the necessary historical or time-averaging power but is inherently weak in terms of providing three-dimensional information. The two approaches are thus necessary and complementary to each other, and our prognosis is that this relatively young marriage will surely flourish in the years to come.

The evidence for a chemically heterogeneous mantle is, today, un-equivocal. However, the nature, development, and scale of this hetero-geneity in the mantle remains problematic. In this synthesis, we discuss the various lines of observation and argument that have been used to approach this problem and to work toward the development of quantitative geo-dynamic models. We first present the geochemical evidence, which is organized in terms of three principal "boundary conditions": (*a*) scale lengths inferred for mantle heterogeneities, (*b*) correlation and inter-relationships of various mantle isotopic tracers (Sr, Nd, Pb, and He), and (*c*) initial chemical composition of the Earth. We then consider the interplay between geophysics and geochemistry and attempt to present an unbiased assessment of the current state of the art.

493

0084–6597/86/0515–0493$02.00

2. MANTLE CHEMICAL HETEROGENEITY: A MATTER OF SCALE

The Role of Chemical Diffusion

Estimates for the scale length over which demonstrable isotopic hetero-geneities exist range from mineralogical (cm) scale to scale lengths comparable to mantle dimensions (> 1000 km). The short end of this range can be evaluated by use of existing data on diffusion of cations in relevant mantle phases. For the purposes of this discussion, it is not necessary to review all available diffusion data in detail. We note only that cation diffusion in olivine, diopside, and basalt melt is observed to follow a compensation law (Hart 1980, Sneeringer et al 1984), and this allows several brief generalizations to be made. Compensation (the existence of a well-defined relationship between the preexponential factor and the activation energy in the Arrhenius-type description of diffusion) implies a tempera-ture at which individual element Arrhenius lines intersect. For olivine, diopside, and basalt melt, the crossover temperatures are 1360, 1300, and 1370°C and the diffusion coefficients are $\sim 4 \times 10^{-11}$, 3×10^{-14}, and 2×10^{-7} cm^2 s^{-1}, respectively.

By virtue of the compensated nature of diffusion in olivine, diopside, and basalt melt, we need not separately consider the differences in diffusion rate as a function of species, pressure, or diffusion direction (Hart 1980); at the temperature of particular interest to isotopic homogenization between mantle phases (1300–1400°C), diffusion in each of these phases may be characterized by a single diffusion coefficient. Hofmann & Hart (1978) used 10^{-13} cm^2 s^{-1} for solid phases and 10^{-7} cm^2 s^{-1} for melt phases; the more comprehensive set of diffusion data now available supports the choice of these values, and thus the following conclusions of Hofmann & Hart (1978) remain valid:

1. In the presence of a melt phase, mantle minerals will achieve isotopic equilibrium on time scales of 10^5 yr (at 1350°C).
2. Maintenance of isotopic disequilibrium between minerals for a period of 1 b.y. requires a melt-free system where the host phases are not in contact (i.e. are separated by low-solubility barrier phases such as olivine) and where the temperature does not exceed 1100–1200°C.
3. In a mantle containing 5% basaltic partial melt at 1370°C, and for trace elements with bulk crystal/liquid partition coefficients < 0.1, the maximum scale length ($\sim \sqrt{Dt}$) over which these elements can be diffusively transported is $\sim 500\sqrt{t}$ (in meters), where t is in billions of years (Hofmann & Magaritz 1977). Thus, over the age of the Earth,

diffusion will not eradicate isotopic heterogeneities that are developed on larger than a kilometer scale, even in the presence of a melt.

This latter calculation can be used to constrain the minimum dimension of a heterogeneity that can survive a melting event in a mantle segment with a *homogeneous* solidus temperature. For a mantle diapir rising with an ascent velocity of 5 cm yr^{-1}, it will take ~ 0.3 m.y. to generate 5% partial melt (for 180 cal g^{-1} heat of fusion, and 2°C km^{-1} difference between the melting curve and adiabat). If isotopic heterogeneities existed in this diapir prior to the onset of melting, they will be diffusively erased over scale lengths of some 10 m, even if we assume no convective mixing of the melt. Since most basaltic eruptions almost certainly integrate much larger volumes of mantle than this, we conclude that the minimum size of mantle heterogeneities that can be visualized by studying basalts is set by the volumes and dynamics of melt segregation and transport rather than by diffusion considerations, at least for a source with a homogeneous solidus temperature.

In contrast, we can consider a diapir with relatively fertile zones that have lower solidus temperatures than the surrounding depleted matrix. At the onset of melting, the melt will be restricted to the fertile zones, and it is conceivable that very small-scale heterogeneities may be sampled. Thus, when heterogeneities are observed in basalts, we cannot be sure whether they document kilometer- or meter-scale heterogeneity in the source mantle.

Ultramafic Rocks

Scale lengths of mantle isotopic heterogeneities may be investigated by direct observations of mantle-derived rocks. Two types of material are accessible for this kind of study: (*a*) ultramafic xenoliths erupted in alkaline volcanic rocks and kimberlites; and (*b*) tectonically exposed ultramafic rocks, such as alpine ultramafics and high-temperature peridotites. Xenolith material, by virtue of its small size (< 30 cm), is useful for studying only very small scale lengths. In contrast, high-temperature peridotites may have outcrop scale lengths of up to 35 km (e.g. Ronda massif, Spain).

Numerous early studies reported isotopic disequilibrium between minerals in anhydrous ultramafic xenoliths (e.g. Stueber & Ikramuddin 1974, Dasch & Green 1975); it now appears that much of this observed heterogeneity was a result of alteration and analytical artifacts (Jagoutz et al 1980, Zindler & Jagoutz 1986, Richardson et al 1985).

Ultramafic xenoliths that contain hydrous (metasomatic?) phases often display Sr and Nd isotopic disequilibria between minerals such as phlogopite, amphibole, and clinopyroxene (Menzies & Murthy 1980b, Kramers et al 1983, Richardson et al 1985). However, it is not yet clear whether these metasomatic phases are long-term residents of mantle environments or whether they simply represent young events in some way related to the early stages of the volcanic event in which the xenoliths were erupted. As discussed by Zindler & Jagoutz (1986), intermineral isotopic disequilibria do not appear to be a common long-lived feature of the upper mantle, although internodule isotopic variations are extreme (Figure 1).

In contrast, larger-scale isotopic variations have been clearly documented in orogenic lherzolite massifs. In a pioneering study, Polve & Allegre (1980) analyzed diopside separates from various parts of the Lherz, Lanzo, and Beni Bousera massifs and observed ranges in $^{87}Sr/^{86}Sr$ of 0.7020–0.7032, 0.7020–0.7030, and 0.7020–0.7033, respectively. Results of a sys-

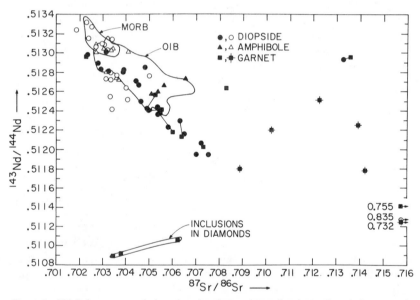

Figure 1   Nd-Sr isotope correlation plot, showing analyses of various minerals from mantle xenoliths and megacrysts. Localities are divided either as African kimberlites (solid symbols) or other localities, usually with alkali basalt hosts (open symbols). The xenolith data of Richardson et al (1985) have first been age corrected back to 90 Myr using mineral Sm/Nd, then corrected ahead to present using average bulk rock Sm/Nd ~ 0.10; solid lines connect coexisting garnet-cpx pairs. Other African kimberlite data are plotted as present-day values, as insufficient data are available to make age corrections (though these will shift points only very slightly on this plot—generally < 0.00006 in $^{143}Nd/^{144}Nd$). Data references are as follows: Basu & Tatsumoto (1979, 1980), Cohen et al (1984), Jagoutz et al (1980), Kramers et al (1981, 1983), Menzies & Murthy (1980a,b), Menzies & Wass (1983), Richardson et al (1984, 1985), Stosch et al (1980), Zindler & Jagoutz (1986).

tematic scale-length study of the Ronda massif (Reisberg & Zindler 1985, 1986; see Figure 2), using Sr and Nd isotopes in peridotite clino-pyroxene separates, show significant isotopic variation on scale lengths of 10–20 m (0.7031–0.7034 for Sr and 0.51255–0.51285 for Nd) in agreement with the work of Polve & Allegre (1980). Reisberg & Zindler (1985, 1986) have further shown that mean ^{143}Nd/^{144}Nd ratios for 5–10 km^2 terrains (as documented by Cr-diopside separates from river sediments) vary from 0.51325 to 0.51300 for terrains separated by less than 10 km. The total range in ^{143}Nd/^{144}Nd measured at Ronda over a scale length of ~ 30 km is on the order of 95% of the total variation observed in oceanic basalts (Figure 3). Results from St. Paul's Rocks (in the equatorial mid-Atlantic) and Zabargad Island [in the Red Sea (Figures 2, 3)] also provide qualitative support for the above conclusions (Roden et al 1984, Brueckner et al 1986).

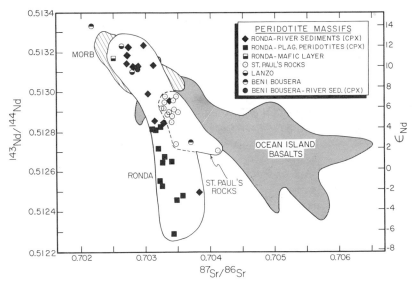

Figure 2 ^{87}Sr/^{86}Sr vs ^{143}Nd/^{144}Nd variation diagram for peridotite massifs. While all localities, with the exception of St. Paul's Rocks, contain some samples that fall within the depleted MORB (mid-oceanic ridge basalts) field (^{143}Nd/^{144}Nd > 0.5132), extreme heterogeneity occurs within these massifs. The enriched portions of the Ronda, Beni Bousera, and Zabargad massifs tend to be EM I type (see discussion in Section 3), a result that tends to support an intramantle metasomatic origin for this component (Menzies 1983). Scale lengths for the observed isotopic heterogeneities are discussed in the text and are shown in Figure 3. Data sources are as follows: Ronda (Reisberg & Zindler 1986, Zindler et al 1983); St. Paul's Rocks (Roden et al 1984); Zabargad (Brueckner et al 1986); Lanzo and Beni Bousera (Richard & Allegre 1980).

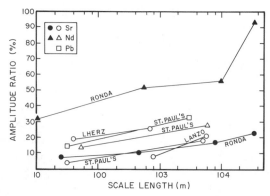

Figure 3   Isotopic heterogeneity versus scale length for ultramafic rocks from three orogenic massifs (Lherz, Ronda, and Lanzo) and one oceanic massif (St. Paul's Rocks). Isotopic variability is expressed as a percentage of the total isotopic range observed in all oceanic basalts (MORB and OIB; see discussion later in text). Two end points plotted represent the scale length of the maximum observed variation in each massif (*right*) and the largest variation observed at the smallest scale length (*left*). St. Paul's samples are whole rocks; Lherz, Ronda, and Lanzo samples are diopsides. Data are from Polve & Allegre (1980), Roden et al (1984), and Reisberg & Zindler (1985, 1986).

Basalts

Inferring domain sizes of mantle heterogeneities based on observations of erupted basalts is a very complex issue, though this avenue is the principal source of the proof that mantle geochemical heterogeneities do in fact exist. The basic problem is one of trying to elucidate three-dimensional domain sizes from "planform" observations. First, we do not truly know the degree of partial melting represented by any given basaltic melt, and estimates vary widely, from values < 1% to values as high as 20–30%. Second, we have very poor control on the depth of origin of basaltic magmas. The problem is worse, however, than simply establishing a depth of segregation, as source materials may represent diapirs that have ascended from deep within the mantle (e.g. the "plume" model of Morgan 1971 or Schilling 1973). Furthermore, if the uprising plumes or diapirs undergo any solid-state mixing during their ascent, the final basaltic melt product may represent some confused average of large vertical sections of mantle. Thus, it is obvious that basalts from two volcanoes on Hawaii, for example, separated by 10 km of horizontal distance, need not be giving us interpretable information regarding isotopic mantle variations on 10-km scale lengths.

Despite these problems, we believe that useful constraints may be defined by working at the extreme ends of the scale-length spectrum. For example, significant isotopic heterogeneity has been observed between samples of

MORB from a single dredge haul (leRoex et al 1983) and within the eruptives of a single small seamount (10-km base diameter, 42-km^3 volume) (Zindler et al 1984). In the seamount case, it seems unlikely that this variability can be ascribed to several major mantle plumes coming from disparate depths. Zindler et al (1984) used these data to argue for a mantle that is heterogeneous on a kilometer or smaller scale but homogeneous on a 20–100 km scale.

At the other extreme (large scale lengths), there appear to be large areas of the Earth that, in planform, show coherent and characteristic isotopic signatures (Dupre & Allegre 1983). One of these, termed the Dupal anomaly (Hart 1984b), is more or less continuously present worldwide between the equator and 60°S latitude. Figure 4 shows a planform of this anomaly, which is contoured using several different isotopic criteria. It is difficult to conceive of the isotopic delineation of such a megascale anomaly by melting of mantle that is anything but heterogeneous on a large ($>$ 1000 km) scale. These results, taken together with the aforementioned seamount data, enable us to conclude that mantle heterogeneities must exist *at least* on both small and very large scales.

In light of these results, we consider it useful to establish a measure of the amplitude of isotopic variability and then to relate this measure to the planform scale length. As an amplitude measure, we take the extreme values of $^{87}Sr/^{86}Sr$, $^{143}Nd/^{144}Nd$, and $^{206}Pb/^{204}Pb$ measured on any oceanic basalt [both ocean island basalt (OIB) and midocean ridge basalt (MORB); see Figure 14]. The isotopic "variability" (amplitude ratio, or AR) of a given basalt suite (island, volcano, section of ridge, etc) is then expressed as a percentage of this total possible variation. Amplitude ratios for selected basalt suites are shown as a function of sampling scale length in Figure 5. Also shown for comparison is the field of "first-order" observations derived from ultramafic massifs (taken from Figure 3). It is apparent from this plot that the maximum attainable amplitude ratio increases with scale length, whereas relatively small amplitude ratios can be observed at all scale lengths.

The important feature of this plot is the *upper* bound of the wedge-shaped field, which convincingly shows that large amplitude ratios occur only for large scale lengths. Note that the data from orogenic massifs suggest an upper bound that is shifted toward considerably smaller scale lengths. This difference indicates that the magmatic processes involved in basalt production are capable of considerable averaging on a 10-km scale.

This analysis suggests several important implications with respect to the nature of mantle heterogeneities. While heterogeneities of relatively small isotopic amplitude may exist on small scales (meters to kilometers), as documented by the seamount studies, it does not appear possible to explain

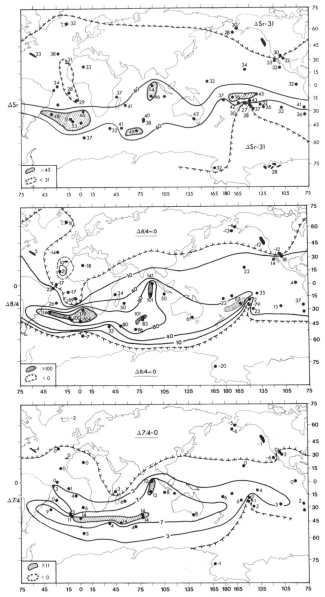

Figure 4 World maps showing the distribution of three isotopic criteria used to outline the very large-scale anomalous region that encircles the globe between the equator and 60°S. This region is termed the Dupal anomaly (Hart 1984b). The three anomaly criteria are ΔSr (upper figure: absolute $^{87}Sr/^{86}Sr$ value, third & fourth significant figures only; that is, ΔSr for 0.70350 = 35), Δ8/4 [middle figure: vertical deviation in $^{208}Pb/^{204}Pb$ from a Northern Hemisphere reference line (NHRL on Figures 9A,C)], and Δ7/4 (lower figure: vertical deviation in $^{207}Pb/^{204}Pb$ relative to the NHRL in Figures 9B,D).

the whole observed range of oceanic mantle heterogeneity with small-scale (< 10 km) heterogeneities alone. To accomplish this with the preferential melting model proposed by Sleep (1984) would require that melting processes in certain broad regions of the oceanic mantle can selectively tap only the more enriched portions of small heterogeneities, whereas in other regions the melting process is unable to be so selective, thereby averaging in a larger fraction of more depleted matrix. Since it is difficult to envision a mechanism for such large-scale modulation of the melting process, one could argue that the small-scale heterogeneities (blobs) are themselves variable in composition, with large volumes of the mantle typified by blobs

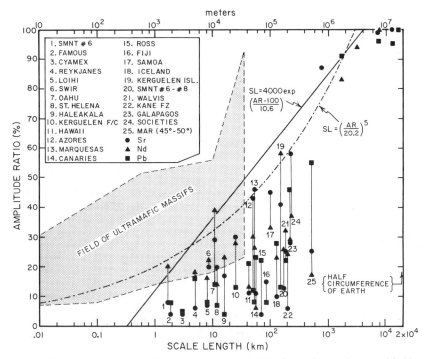

Figure 5 Amplitude ratio (*AR*) versus scale length (*SL*) for various volcanoes, islands, island groups, and ridge segments. Vertical lines connect the *AR* values observed for Sr, Nd, and Pb at each locality. Points for Sr, Nd, and Pb plotted at 100% *AR* are the normalizing values, given for Nd by Kane–Walvis (*SL* = 7780 km), for Sr by Kane–Societies (*SL* = 12,400 km), and for Pb by Mid-Indian Ocean ridge–Tubuai (16,700 km). Other points plotted in upper-right quadrant are for other various island pairs that provide large *AR* for minimal scale lengths (Nd: Walvis–Mid-Atlantic Ridge, 94%, 3030 km; Sr: Tahaa–Juan de Fuca, 99%, 7100 km; Tahaa–Tubuai, 87%, 770 km; Pb: Walvis–Tubuai, 95%, 13,500 km; St. Helena–Mid-Indian Ocean ridge, 96%, 7600 km; Walvis–St. Helena, 91%, 1700 km). Arbitrary exponential and power-law curves are shown as possible upper-bound fits to the data. Also shown is the outline of the field for ultramafic massifs, taken from Figure 3. Data references may be found in Zindler et al (1982), Hart (1984b), and White (1985).

that are more enriched than those in other regions. For example, in this model, all of the mantle underlying the Dupal anomaly region (Figure 4) would contain blobs of a more enriched character than the blobs in mantle lying outside the Dupal region. In essence, this model implies that the isotopic *contrast* between blobs and their depleted mantle matrix is systematically larger in Dupal mantle regions than elsewhere. The upper bound in Figure 5 would then be providing a measure of the degree of isotopic contrast between small-scale blobs and their matrix.

Alternative interpretations derived from the data of Figures 4 and 5 are equally possible. The simplest of these would view the oceanic mantle as isotopically heterogeneous not only on small scales, but also on the largest scales possible (thousands of kilometers), with the extent of heterogeneity increasing with scale length. In the context of a plume model, this could be accomplished by allowing the lower mantle to carry the enriched isotopic signatures, with sub-Dupal lower mantle being substantially more "enriched" than other regions, and then allowing this material to mix with (be "polluted" by) the depleted and homogeneous upper mantle (Dupre & Allegre 1983). The isotopic contrast would then be between the upper and lower mantles, with Dupal regions showing the largest contrast because of the more enriched nature of the lower-mantle component. Yet another view (Anderson 1985) suggests that basaltic magmas generated in the transition region could rise into, and become contaminated by, a variably enriched shallow mantle. The enhanced enrichment of Dupal-type shallow mantle may document extensive Pangeatic subduction of altered oceanic crust and sediment into the sub-Dupal mantle (Anderson 1982a). In this model, the isotopic contrast is between the upper and lower parts of the *upper* mantle, with Dupal uppermost mantle being more enriched than ambient upper mantle.

These various interpretations (models) for the systematics depicted in Figures 4 and 5 lead to very different views of convective processes in the mantle. The first model would be consistent with whole-mantle convection, the second would require a two-layered mantle, with limited exchange between the layers over geologic time and relatively inefficient convective mixing of the lower mantle, and the third requires a three-layer mantle with a depleted transition region overlain by a relatively stagnant, enriched upper layer.

3.   ISOTOPIC CHARACTERIZATION OF GEOCHEMICALLY DISTINCT MANTLE COMPONENTS

In recent years, isotope geochemists have devoted considerable attention to investigating the relationships between various isotope systems in oceanic basalts and their implications for mantle chemical structure. There are now seven decay schemes for which a significant body of analytical data exist for oceanic basalts: ^{87}Rb-^{87}Sr, ^{147}Sm-^{143}Nd, ^{238}U-^{206}Pb, ^{235}U-^{207}Pb, ^{232}Th-^{208}Pb, ^{176}Lu-^{176}Hf, and (^{238}U + ^{235}U + ^{232}Th)-^{4}He. Over the past decade, our understanding has evolved from thinking in terms of linear correlations and two-component mixing for two-isotope systems (e.g. the Sr-Nd isotope correlation; DePaolo & Wasserburg 1976a, O'Nions et al 1977, Allegre et al 1979), to thinking in terms of at least three components, which may define an approximate planar or two-dimensional mixing array in three-, four-, or five-dimensional space (e.g. ^{87}Sr/^{86}Sr, ^{143}Nd/^{144}Nd, ^{206}Pb/^{204}Pb, ^{207}Pb/^{204}Pb, ^{208}Pb/^{204}Pb; Zindler et al 1982). Allegre & Turcotte (1985) and White (1985) have recently proposed that there are five types of sources from which all oceanic basalts may be produced by variable mixing relationships.

The present status of oceanic isotopic data is summarized in a series of two-dimensional variation diagrams (Figures 6–12). Hafnium isotopic data are not represented here because, in general, the overall high degree of correlation between Hf and Nd isotope ratios permits accurate inference of the behavior of ^{176}Hf/^{177}Hf based on ^{143}Nd/^{144}Nd (e.g. Patchett & Tatsumoto 1980). Brief consideration of the data arrays in Figures 6–12 clearly demonstrates the inadequacy of simple two-component mixing models to explain isotopic variations in oceanic basalts [as originally pointed out by Sun & Hanson (1975) based on Sr-Pb systematics]. Even the Sr-Nd variation diagram (Figure 6), until recently thought to represent a linear array (the "mantle array" of, for example, DePaolo & Wasserburg 1977), requires a substantially more complex interpretation. This is an important, first-order conclusion based on all currently available data and is not subject to interpretation. Thus, we must consider that any model for the evolution of the mantle-crust system that presumes only two geochemically distinct components within the mantle is necessarily inaccurate.

Until recently many geochemists felt that the apparent differences in the relative positions of various localities in the Sr-Nd and Pb-Pb variation diagrams (Figures 6, 9) documented a decoupling of the U-Th-Pb system from Rb-Sr and Sm-Nd. Zindler et al (1982) proposed that there was a coherence in these variations that became apparent when they were viewed as different two-dimensional projections of an approximately planar three-dimensional surface. Delineation of this three-dimensional mantle plane is shown in Figure 13, where every data point represents the averaged isotopic values for an island or ridge system. It should be noted that this plane is defined by averaged data, and that many localities show intraisland isotopic variations that do not lie in the plane (e.g. Stille et al 1983, Staudigel et al 1984). There are also several localities for which data have recently been reported that, even when averaged, do not lie on the plane [e.g.

Koolau series, Oahu (Stille et al 1983); Walvis Ridge (Richardson et al 1982); Samoa (White & Hofmann 1982, White 1985)]. While for many localities, the mantle plane remains a useful representation, an accumulating body of evidence suggests that it too is an oversimplification.

The interrelationships displayed by the various isotope ratios in Figures 6–12 delineate a number of important constraints regarding the definition of isotopically distinct mantle components:

1. The "depleted" isotopic character of N-type MORB (high $^{143}Nd/^{144}Nd$, low $^{87}Sr/^{86}Sr$ and $^{206}Pb/^{204}Pb$) requires the existence of a depleted end-member MORB-mantle component (DMM) that is clearly identifiable along ridge segments well removed from hotspot volcanic activity.

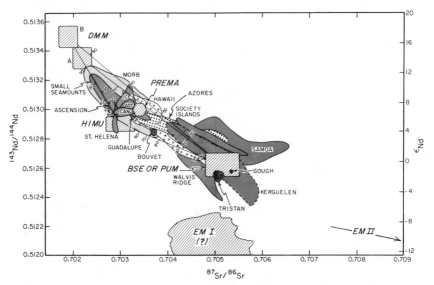

Figure 6 $^{143}Nd/^{144}Nd$ vs $^{87}Sr/^{86}Sr$ variation diagram for oceanic basalt suites. [Data sources may be found in Zindler et al (1982), Hart (1984b), and White (1985).] Proposed mantle components are shown as cross-hatched boxes and blobs labeled DMM (A and B), HIMU, PREMA, BSE or PUM, EM I, and EM II (see discussion in text). Mixing lines between DMM (A and B), HIMU, and BSE or PUM, with tick marks indicating percentages of one end member, are constructed using the following source compositions for Sr and Nd, respectively (in ppm): DMM (1)—6, 0.33; DMM (2)—12, 0.65; HIMU—120, 6.5; BSE—18.4, 1.0. The HIMU composition assumes a magmatic origin for this component; had we assumed it to be a high-μ peridotite (where $\mu = {}^{238}U/{}^{204}Pb$), with lower concentrations of Sr and Nd but a similar Sr/Nd ratio, the shapes of the mixing curves would not change, only the positions of the tick marks (see discussion in text). The two values given for DMM delimit the possible concentration range for this component and result in two sets of tick marks on each mixing line that goes to DMM A or DMM B. Note that none of the mixing lines are highly curved because of the similarity in Sr/Nd ratios in all components.

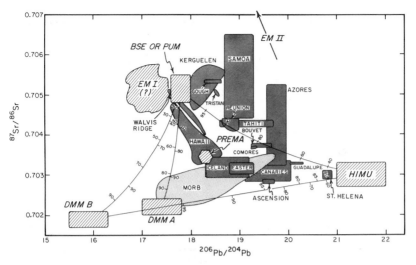

Figure 7 $^{87}Sr/^{86}Sr$ vs $^{206}Pb/^{204}Pb$ variation diagram for oceanic basalt suites. Data sources, components, and mixing lines are as in Figure 6. Pb concentrations are as follows (in ppm): DMM (1)—0.02; DMM (2)—0.04; HIMU—0.4; BSE—0.135. Islands lying outside the DMM-HIMU-BSE mixing triangle document the existence of EM II and are geographically situated within the Dupal anomaly. The Koolau series of Oahu (Stille et al 1983) and the Walvis Ridge (Richardson et al 1982) document the existence of EM I.

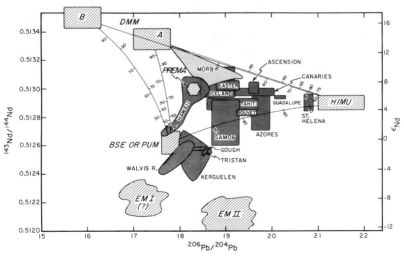

Figure 8 $^{143}Nd/^{144}Nd$ vs $^{206}Pb/^{204}Pb$ variation diagram for oceanic basalt suites. Data sources, components, mixing lines, and other considerations are as in Figures 6 and 7.

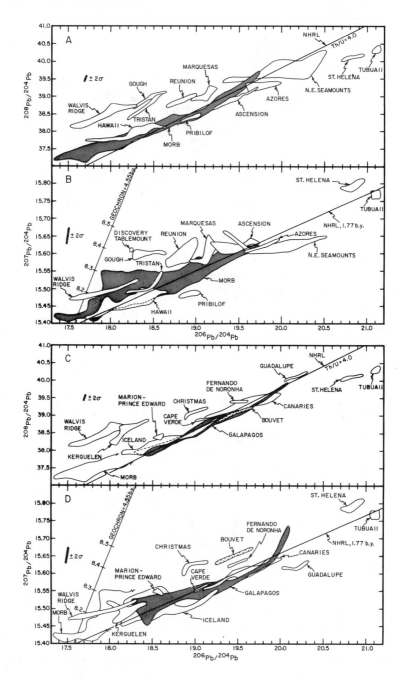

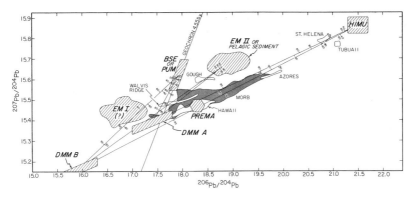

Figure 9 (*opposite and above*)   (*A–D*) $^{208}Pb/^{204}Pb$–$^{207}Pb/^{204}Pb$–$^{206}Pb/^{204}Pb$ isotope variation diagrams for a variety of oceanic basalts. Two panels are shown for each plot in the interest of clarity. A 4.55-b.y. geochron is shown, with single-stage μ-values (tick marks), along with the Northern Hemisphere reference line (NHRL) of Hart (1984b). This NHRL, when interpreted as a secondary isochron, has a slope of 1.77 b.y. and a second-stage Th/U (wt. ratio) of 4.0. Islands lying significantly above the NHRL are part of the Dupal anomaly belt. Data sources may be found in Hart (1984b). Typical 2σ analytical precision is also shown ($\pm 0.05\%$ amu^{-1}). (*above*) In this panel, DMM (A and B), HIMU, BSE, EM I, and EM II component compositions are indicated on a $^{207}Pb/^{204}Pb$ vs $^{206}Pb/^{204}Pb$ diagram, with MORB and selected island fields from panels *B* and *D* shown for reference. Parameters and considerations for mixing lines are as in Figures 6 and 7.

2. The very high $^{206}Pb/^{204}Pb$ ratios observed at St. Helena and Tubuaii (Figure 9), coupled with the low $^{87}Sr/^{86}Sr$ and intermediate $^{143}Nd/^{144}Nd$ ratios at these sites, suggest a mantle component that is markedly enriched in U and Th relative to Pb (HIMU) without an associated increase in Rb/Sr.

3. Trends for Walvis Ridge, Kerguelen, and Samoa in Figure 6, which extend beyond presumed bulk silicate Earth (BSE) values for $^{143}Nd/^{144}Nd$ and/or $^{87}Sr/^{86}Sr$, require the existence of at least two "enriched" mantle components (EM I and EM II).

4. The high $^{3}He/^{4}He$ ratios that are found at Iceland and Hawaii demand the involvement of relatively undegassed mantle, or a mantle segment that was enriched in volatiles during the Archean (see Figure 16). Although this component is often described as "primitive" or undifferentiated mantle, the Sr, Nd, and Pb characteristics of Iceland and Hawaii (Figures 10–12) are not consistent with such an interpretation.

5. The high frequency of Nd and Sr isotopic compositions at about 0.5130 and 0.7033, respectively (Figures 6, 14, 15), suggests that the existence of a mantle component with this isotopic character (at least when scale lengths for melting volumes are considered) is more likely than constant

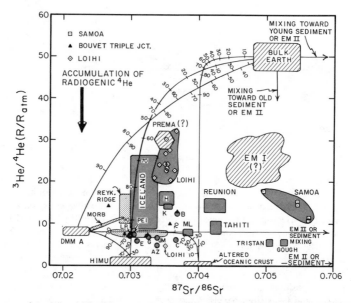

Figure 10   ³He/⁴He vs ⁸⁷Sr/⁸⁶Sr variation diagram of oceanic basalt suites. DMM, HIMU, BSE (labeled Bulk Earth), PREMA, EM I and EM II components are shown as crosshatched fields. Mixing lines and considerations are as in Figure 6; Helium concentrations for the various components are as follows (std cm³ g⁻¹) DMM (1)—6.5 × 10⁻⁶; DMM (2)—1.3 × 10⁻⁵; HIMU—2.7 × 10⁻⁵; EM II (for "young" sediment or EM II case)—5.3 × 10⁻⁶. Sr concentrations are given in Figure 6 caption. Mixing lines are also shown to an altered crust component presumed to have a He concentration of 1.5 × 10⁻⁶ std cm³ g⁻¹. In general, we have assumed fairly low He concentrations in those components presumed to have low ³He/⁴He ratios. However, this assumption may be highly variable, since it is strongly dependent on the age of these components because of the rapid growth of ⁴He in materials with high U/⁴He ratios. This situation is qualitatively depicted by contrasting mixing trajectories shown for "young" and "old" EM II or sediment mixing. Because of the potential for pre- or posteruptive accumulation of radiogenic ⁴He (as indicated by the black arrow), ³He/⁴He ratios measured at ocean islands (particularly those in the waning stages of their volcanic cycles) must be considered as minimum values for mantle sources. The large number of components, together with the potential for radiogenic ⁴He accumulation, precludes the unique definition of mixing scenarios for individual localities. Localities given by abbreviation are as follows: LP—La Palma, Canary Islands; PEI—Prince Edward Island; E—Easter Island; G—Galapagos; AZ—Azores; JM—Jan Mayen; B—Bouvet; C—Comores; H—Hualalai, Hawaii; K—Kilauea, Hawaii; ML—Mauna Loa, Hawaii. Data sources for ³He/⁴He shown in this diagram (as well as for Figures 11 and 12) are as follows: Poreda et al (1980), Rison (1981), Kurz & Jenkins (1981), Rison & Craig (1982), Craig & Rison (1982), Kurz (1982), Kurz et al (1982, 1983, 1985), Allegre et al (1983a), Condomines et al (1983).

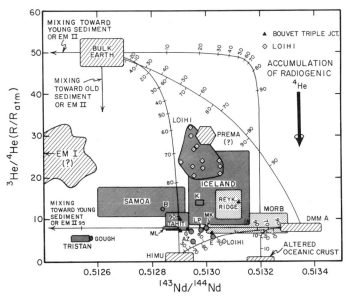

Figure 11 ³He/⁴He vs ¹⁴³Nd/¹⁴⁴Nd variation diagram for oceanic basalt suites. Components, considerations, mixing lines, locality abbreviations, and data sources are as in Figure 10 (except that MK stands for Mauna Kea). He and Nd concentrations for mixing lines are given in captions to Figures 10 and 6, respectively.

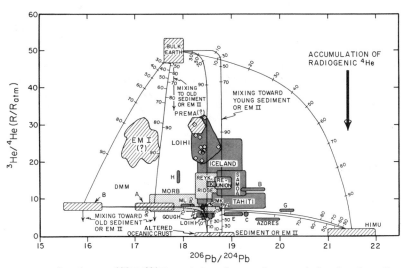

Figure 12 ³He/⁴He vs ²⁰⁶Pb/²⁰⁴Pb variation diagram for oceanic basalt suites. Components, considerations, mixing lines, locality abbreviations, and data sources are as in Figure 10 (except that T stands for Tristan). He and Pb concentrations for mixing lines are given in captions to Figures 10 and 7, respectively.

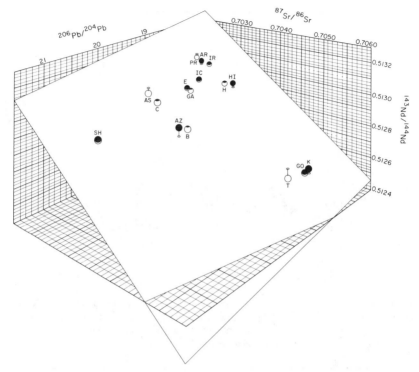

Figure 13   Three-dimensional plot of average $^{206}Pb/^{204}Pb$, $^{143}Nd/^{144}Nd$, and $^{87}Sr/^{86}Sr$ for basalts from oceanic ridges and islands. The best-fit plane is shown. Closed and open symbols, whose size approximates the uncertainty of a single analysis, lie above and below the "mantle plane," respectively. Lines connecting data points to the plane are parallel to the $^{143}Nd/^{144}Nd$ axis and indicate those points that do not intersect the plane. Localities are as follows: AR— Mid-Atlantic Ridge; PR—East Pacific Rise; IR—Indian Ocean ridges; C—Canary Islands; AZ—Azores; AS—Ascension; SH—St. Helena; GO—Gough; T—Tristan da Cunha; B— Bouvet; E—Easter Island; HI—Hawaiian Islands; H—Island of Hawaii; GA—Galapagos; K—Kerguelen; IC—Iceland. Data sources are given in Zindler et al (1982).

mixing proportions between spatially distant components that are far removed from these isotopic values. We refer to this component as PREMA (*pre*valent *ma*ntle).

Mantle He Signals: Potential Problems

Because of its dominantly primordial origins, ^{3}He offers tremendous potential as a tracer for geochemically distinct mantle components and reservoirs. However, the interpretation of mantle $^{3}He/^{4}He$ signals may not be as straightforward as is often presumed, and thus it merits our further consideration here.

$^3He/^4He$ variations in oceanic basalts are displayed in histogram fashion in Figure 14 (along with Sr, Nd, and Pb isotope ratios for reference). In contrast to Sr, Nd, and Pb, MORB $^3He/^4He$ values occupy a relatively small portion of the OIB array (1–32 Ra ; Ra denotes the atmospheric value of $^3He/^4He$) and lie totally within this array, rather than defining one end of it (as is the case for the other isotope ratios). Basalts from Loihi Seamount, a young submarine volcano located on the southeast flank of Kilauea on the

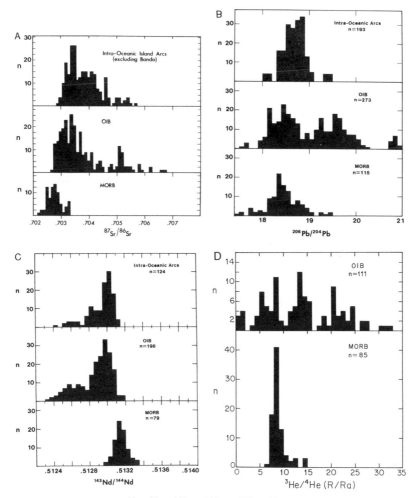

Figure 14   Histograms of $^{87}Sr/^{86}Sr$, $^{143}Nd/^{144}Nd$, $^{206}Pb/^{204}Pb$, and $^3He/^4He$ in MORB, oceanic island basalts, and intraoceanic arcs (adapted from Morris & Hart 1983). Data sources for the He data in (D) are those given in Figure 10 caption plus Ozima & Zashu (1983).

Island of Hawaii, span nearly the entire known range of oceanic ^{3}He/^{4}He values (4–32 Ra; Kurz et al 1983) while displaying only minor variations in ^{87}Sr/^{86}Sr (from 0.70335 to 0.70370; Staudigel et al 1984). Some of the lowest, or least primitive, ^{3}He/^{4}He ratios found on oceanic islands come from Tristan da Cunha and Gough (~ 5 Ra; Kurz et al 1982); these areas are known for their Sr and Nd isotope ratios, which approximately coincide with many estimates for the bulk silicate earth (BSE). Recently, Graham et al (1984, 1986) have measured ^{3}He/^{4}He ratios of ~ 1 Ra in alkali basalts from several small seamounts adjacent to the East Pacific Rise at about 12°N.

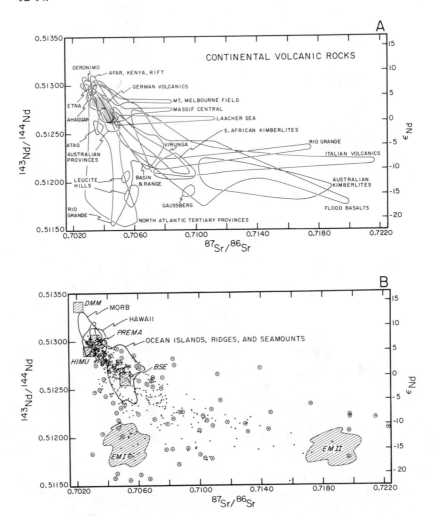

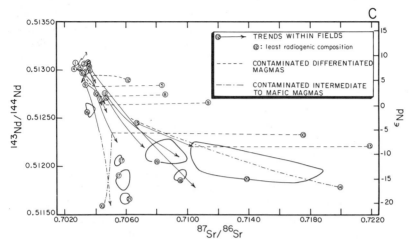

Figure 15   Comparison of more than 500 individual Sr and Nd isotope ratios for continental volcanic rocks taken from Worner et al (1986). Data sources can be found in Worner et al (1986). Panel *A* shows fields for individual provinces (except that all flood basalts are grouped into a single field). Note that fields converge toward PREMA in the upper left-hand corner of the diagram. Individual "branches" of the combined arrays diverge toward EM I–type compositions at low $^{87}Sr/^{86}Sr$, and EM II–type compositions at high $^{87}Sr/^{86}Sr$. These mixing trends are similar to those observed in oceanic basalts, except that the continental basalts extend to much more extreme values.

Panel *B* shows individual data points, with fields for MORB, Hawaii, and ocean islands (OIB) shown for reference. The paucity of data between the two lower branches of the array ($^{87}Sr/^{86}Sr \sim 0.7065$, $^{143}Nd/^{144}Nd < 0.5124$) suggests that EM I– and EM II–type mixing processes are mutually exclusive. Worner et al (1986) propose [in accord with prior suggestions by Menzies (1983) and Hawkesworth et al (1984)] that these enriched components exist within the subcontinental source mantle, since selective assimilation of only upper or lower crustal materials, without mixing between them, seems unlikely (although, without question, many of these rocks have been contaminated by crustal components; such data points are circled where specifically discussed in the original publication).

Panel *C* shows depleted and enriched end members for individual fields. Numbers correspond to the following localities: (1) North Atlantic Tertiary provinces; (2) Afar and Kenya Rift; (3) Geronimo volcanic field; (4) Australian provinces; (5) Mt. Melbourne, Antarctica; (6) Etna; (7) Ahaggar; (8) Massif Central; (9) Central European provinces; (10) Rio Grande Rift; (11) Virunga volcanic field; (12) Basin and Range, western United States; (13) Italian provinces; (14) continental flood basalts; (15) Oslo Rift; (16) Ataq; (17) Leucite Hills; (18) Gaussberg; (19) South African kimberlites; (20) Marie Byrd Land; (21) Smokey Butte. Depleted end-member compositions for ten of the fields fall precisely at PREMA. This, together with the fact that none of the more than 500 analyses fall within the N-type or depleted end of the MORB field (Panel *B*), suggests PREMA rather than DMM as a depleted mixing end member (Carlson 1984, Worner et al 1986). Horizontal arrays extending to high values of $^{87}Sr/^{86}Sr$ are interpreted by Worner et al (1986) as documenting assimilation of high-Sr crust by differentiated magmas with very low Sr contents (due to extensive plagioclase fractionation).

Collectively, these and other observations have led to the widely held view (e.g. Anderson 1985) that the U-Th-He system has been "decoupled" from the U-Th-Pb, Sm-Nd, Rb-Sr, and Lu-Hf systems (i.e. that the U-Th-He system, by virtue of the gaseous daughter product, has been subject to perturbations that have not affected the other systems). It is the nature, location, and timing of these perturbations that are of interest to us if we are to accurately interpret the He isotope signal in oceanic basalts.

The petrogenesis of mantle magmas typically results in an increase of U/Pb, Rb/Sr, Nd/Sm, and probably U/Pb (see Watson et al 1986) ratios in derived basalts relative to presumed mantle sources. These increases, however, are modest and rarely, if ever, approach an order of magnitude (e.g. Zindler et al 1979). In contrast, U/He ratios in OIB magmas are higher than estimated mantle source values by up to six orders of magnitude (if we assume single-stage source evolution; see Figure 16). The high degree of incompatibility expected for both U and He in candidate mantle phases makes it appear highly unlikely that such fractionations can occur during magma generation (see Zindler & Hart 1986). Several authors (e.g. O'Nions & Oxburgh 1983, Allegre et al 1986) have suggested or implied that rapid

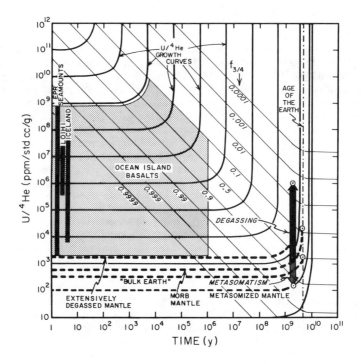

Figure 16 $U/^4He$ [ppm/(std cm^3/g)] vs time (in years). Diagram is designed to aid in the assessment of the evolution of $^3He/^4He$ ratios in systems with a given $U/^4He$ ratio. Heavy solid lines, labeled "$U/^4He$ growth curves," document closed-system changes in $U/^4He$ due to growth of radiogenic ^4He from U and Th decay. Contours are shown for values of $f_{3/4}$ ranging from 0.9999 to 0.0001 [thin solid lines with slopes of ~ -1; $f_{3/4} = (^3He/^4He)_p/(^3He/^4He)_t$, where $(^3He/^4He)_p$ is the present-day He isotope ratio in a system that has been closed since time t]. The $^{232}Th/^{238}U$ ratio is taken as 3.0; use of values of 3.5 or 4.0 would shift the $f_{3/4}$ contours slightly to the left without qualitatively changing the appearance or the implications of the diagram.

The inner two heavy dashed lines toward the base of the diagram indicate closed-system evolution paths for the MORB source reservoir and BSE over the past 4.4×10^9 and 4.55×10^9 yr, respectively (following R. Hart et al 1985). U concentrations are from Table 3, and initial $^3He/^4He$ ratios are taken as 100 Ra, although use of 200 Ra would not significantly shift the positions of the paths because of the log-log scale. Present-day $^3He/^4He$ ratios in the MORB source and BSE reservoirs are assumed to be 8 and 50 Ra, respectively; again, use of 30 Ra or 80 Ra for BSE would not significantly change the diagram. The resulting present-day $U/^4He$ ratios in the MORB source and BSE reservoirs are ~ 630 and ~ 310, respectively. Note that these values are substantially lower than $U/^4He$ ratios measured in basalts from Iceland (Condomines et al 1983, Kurz et al 1985), Loihi (Kurz et al 1983), and several small seamounts adjacent to the East Pacific Rise at $\sim 12°N$ (Graham et al 1984, 1986), a result which suggests large amounts of pre- or posteruptive degassing.

The outer two heavy dashed lines show hypothetical closed-system evolution paths for extensively degassed and metasomatized mantle segments over the past 2×10^9 yr. These mantle segments are presumed to have evolved from the MORB reservoir at this time. The metasomatized reservoir, considered a possible analog for EM I, has incorporated a He-rich, U-poor fluid component that has resulted in a factor of 10 reduction in its $U/^4He$ ratio. This has the effect of "freezing-in" its 2-b.y. $^3He/^4He$ ratio, as indicated by the downward traverse across $f_{3/4}$ contours from ~ 0.70 to ~ 0.93. The extensively degassed reservoir is considered to be a possible analog for HIMU, originating as either a metasomatic residuum or a degassed magma. This increase in $U/^4He$ shifts the material above the $f_{3/4} = 0.01$ contour, leading to a present-day $^3He/^4He$ ratio less than the atmospheric value. Note that even this extensively degassed mantle material will have a relatively low $U/^4He$ ratio today owing to the 2-b.y. aging period.

Heavy solid bars at the left side of the diagram show ranges for ocean island and seamount basalts from sources mentioned above. (Note that these bars are for reference only, and that their positions have no time significance.) The shaded field labeled "ocean island basalts" is drawn to include these bars and is arbitrarily terminated at 10^6 yr and $f_{3/4} = 0.1$ (i.e. factor of ten closed-system reduction in $^3He/^4He$). This field documents the potential for $^3He/^4He$ reduction in real systems on time scales that may be appropriate for high-level magma evolution. "Worst-case" scenarios involve the preeruptive loss of He from a magma [via continuous diffusive loss (CDL) or exhalative loss (EL) models], which results in an increase in $U/^4He$ from ambient mantle values (heavy dashed lines) into the shaded field. Subsequent "aging" of the magma results in $^3He/^4He$ reductions, as indicated by the $f_{3/4}$ contours. For example, a mantle melt that is degassed 2×10^5 yr ago to a $U/^4He$ ratio of $\sim 2 \times 10^7$ will have suffered about a 50% reduction of its original mantle $^3He/^4He$ ratio by the present day; this magma will then have a present-day $U/^4He$ ratio of 10^7. As another example, a magma that degasses 5×10^4 yr ago to a $U/^4He$ ratio of $\sim 10^9$ will suffer a 90% reduction in $^3He/^4He$ by today, at which time it will have a $U/^4He$ ratio of $\sim 10^8$. The observed disparity between mantle and basalt $U/^4He$ ratios clearly documents the potential for $^3He/^4He$ reduction during high-level magma evolution.

solid-state diffusion of He relative to U may explain large U/He variations; this, however, now seems unlikely based on measured diffusion rates for He in olivine (Hart 1984a). We conclude, therefore, that U/He ratios document magma degassing, which may occur prior to or during eruption. The important question in the present context is whether or not significant amounts of ^{4}He can accumulate subsequent to this degassing and prior to sampling and analysis (see also the comprehensive discussion in Kurz & Jenkins 1981).

By and large, ^{3}He/^{4}He ratios are determined on "zero-age" rocks or in fluid (or melt) inclusions in phenocrysts, so that in-situ production of ^{4}He is not presumed to be significant. Even in rocks that are not "zero age," the extent of the problem can be assessed by measuring U, Th, and He concentrations and correcting ^{3}He/^{4}He ratios for in-situ accumulation of ^{4}He. However, ^{3}He/^{4}He ratios measured in volcanic rocks are typically equated with mantle source values, and thus it is implicity assumed that ^{4}He accumulation during mantle or crustal magma storage is not important. We suggest that in the face of abundant evidence for preeruptive degassing of magmas, this may be a dangerous presumption.

The change in ^{3}He/^{4}He with time in a given system can be assessed by defining a factor $f_{3/4}$ that represents the fractional change in ^{3}He/^{4}He with time:

$$f_{3/4} = (^3\text{He}/^4\text{He})_f/(^3\text{He}/^4\text{He})_o = 1 - [2.178 \times 10^{-13}(\text{U}/^4\text{He})t],$$

where U is in ppm total U, ^{4}He is in std cm^3 g^{-1}, t is in years, and the subscripts f and o refer to final and initial values respectively.

There will be essentially no change in the ^{3}He/^{4}He ratio of a system (that is, $f_{3/4} \geq 0.95$ when the product $(\text{U}/^4\text{He})t \leq 2.3 \times 10^{11}$). U/^{4}He ratios (in many cases estimated from K contents using a K/U value of 1.27×10^4 from Jochum et al 1983) in the range 5×10^5 to $> 10^8$ are not uncommon in basalts from ocean islands and seamounts (Graham et al 1984, 1986, Condomines et al 1983, Kurz et al 1983, 1985), and "zero age" ($f_{3/4} \geq 0.95$) in such cases may stipulate times since degassing of < 2000 to $500,000$ yr. These relationships are shown in Figure 16, where U/^{4}He [ppm/(std cm^3/g)] is plotted versus time and contoured for values of $f_{3/4}$ ranging from 0.9999 to 0.0001. Thus, for a given present-day U/^{4}He ratio at the left side of the diagram, changes in ^{3}He/^{4}He back in time can be traced. For example, a rock with U/^{4}He $= 10^7$ and ^{3}He/^{4}He $= 5$ Ra today would have had a ^{3}He/^{4}He ratio of 30 Ra 383,000 yr ago (had it or the predecessor magma existed as a closed system over that time interval). Based on U/^{4}He ratios shown in Figure 16 for Hawaii, Iceland, and several small seamounts near the East Pacific Rise, it is clear that accumulation of radiogenic ^{4}He during magma transit and storage may result in perturbation of indigenous mantle He isotope signatures. Extensive preeruptive degassing of magmas will likely result in correlation of U/^{4}He ratios with factors that document increasing magma differentiation. The relationships in Figure 17 document a tendency for cogenetic suites to display an inverse correlation of U/^{4}He with Mg#. A number of factors, however, preclude a strictly quantitative analysis of this kind of diagram (Zindler & Hart 1986), although the convincing, and in some cases, striking negative correlations that are observed support the contention that preeruptive degassing may be an important phenomenon.

MODELING OF HE LOSS VIA DEGASSING   We have modeled the effects of preeruptive He loss on ^{3}He/^{4}He ratios using two end-member models: (a)

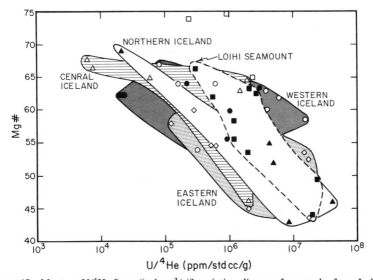

Figure 17   Mg# vs U/^{4}He [ppm/(std cm^3/g)] variation diagram for samples from Iceland (Condomines et al 1983, Kurz et al 1985) and Loihi (Kurz et al 1983). Mg# (atomic) calculated with Fe$_2$O$_3$/FeO = 0.2. In many cases, U had to be estimated from K contents by taking advantage of the relative constancy of K/U in basalts (Jochum et al 1983). Data for Iceland are broken down into northern (closed triangles), central (open triangles), eastern (diamonds), and western (open and closed circles) regions (see Kurz et al 1985). The Kurz et al data were obtained by crushing only, and so they may tend to underestimate He contents. Loihi data include closed and open squares, but the open squares are not included in the Loihi field because they have clearly accumulated olivine and therefore have anomalously high Mg#s. Note that all fields, to greater and lesser degrees, display negative correlations between the two plotted parameters. We interpret this as strong evidence that degassing of magmas, resulting in increased U/^{4}He, proceeds in conjunction with crystal fractionation, which causes reduction of the Mg#.

continuous diffusive loss from a magma chamber (CDL model), and (b) exhalative loss via partitioning into an evolved CO_2 gas phase (EL model). The general behavior of the CDL model (described in detail in Zindler & Hart 1986) is illustrated in Figure 18, which shows the rapid loss of initial ^3He and ^4He, the buildup of radiogenic $^4He*$, and a crossover of $^4He_0$ and $^4He*$ at time $t \sim 2.3$ m.y. The total ^4He ($^4He_0 + {}^4He*$) becomes constant for

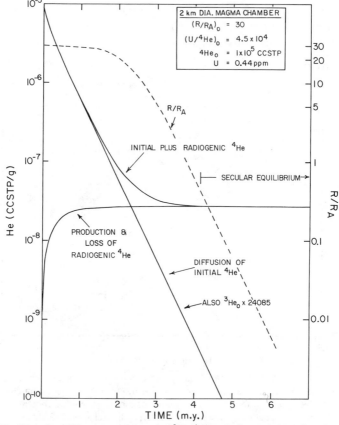

Figure 18 Variation of He concentrations and $^3He/^4He$ ratio (log scales) as a function of time for the CDL model. Model parameters are given on the figure; a He diffusion coefficient of 7.5×10^{-5} cm^2 s^{-1} was used (see text). Curves show the diffusion loss of initial ^3He and ^4He (approximately log-linear), the in-situ production and concurrent diffusional loss of radiogenic ^4He, and the sum of initial plus radiogenic ^4He. This latter quantity approaches steady state (secular equilibrium) for times > 4 m.y. Also shown (dashed curve) is the evolution of $^3He/^4He$ ratios (right-side axis) as a function of time. For the chosen parameters, the $^3He/^4He$ ratio decreases very rapidly for times > 2 m.y. Governing equations are derived in Zindler & Hart (1986).

$t > 4$ m.y., which represents a state of secular equilibrium between ^{4}He* loss and ^{4}He* production. The decrease of ^{3}He/^{4}He with time (also shown in Figure 18) is related to the fact that ^{4}He loss is being balanced by ^{4}He production, whereas ^{3}He is simply lost. For the particular parameters chosen for Figure 18, the ^{3}He/^{4}He ratio drops by 10% in $\sim$ 1.5 m.y., and by 50% in $\sim$ 2.3 m.y.

Ultimately, however, the evolutionary history of magma chambers depends on thermal budget considerations. By noting that thermal diffusivities are $\sim$ 100 times higher than He diffusivities (at 1200°C), we see that it may be difficult to keep a magma chamber molten long enough for He diffusion to cause significant reduction in ^{3}He/^{4}He. For example, a 2-km static magma chamber will cool below its solidus in $\sim$ 20,000 yr [for ΔH (fusion) = 150 cal g^{-1}, thermal diffusivity = 0.006 cm^2 s^{-1}, and ΔT (magma–wall rock) = 150°C]. The only conceivable way of avoiding such a short lifetime is to intrude, or refill, a magma chamber that has been in prior existence long enough to substantially preheat the environment so that ΔT between wall rock and magma approaches zero.

Figure 19 gives solidification times (in millions of years) as a function of ΔT for spherical magma chambers of various diameters (0.5–10 km) that have been refilled for the last time [see Zindler & Hart (1986) for details]. Curves represent the time required by the CDL model to decrease ^{3}He/^{4}He by 10% as a function of various initial U/^{4}He ratios (10^3–3×10^6). Clearly, the effect on ^{3}He/^{4}He is a sensitive function of the initial U/^{4}He ratio. For magmas that start at mantle values ($\leq 10^3$), changes in ^{3}He/^{4}He will likely not exceed 10%, even with a magma chamber life of several million years and $\Delta T = 10$°C. However, for magmas with high initial U/^{4}He values (perhaps resulting from degassing at depth, as discussed below), changes in ^{3}He/^{4}He due to evolution in such a system may be dramatic.

The EL model assumes that He is lost from a magma by partitioning into, and subsequent loss of, a CO_2-rich fluid or gas phase once CO_2 solubility is exceeded at a given pressure. Experimental work has shown that basaltic magmas may dissolve large quantities of CO_2 at depth (3–7 wt%; Mysen et al 1975). In contrast, the average CO_2 content in Mid-Atlantic Ridge basalts is on the order of 0.13 wt% (Delaney et al 1978), a value that is in accord with saturation limits predicted for MORB (at pressures of $\sim$ 250–400 bars) by Khitarov & Kadik (1973). Greenland et al (1985) have recently constructed a mass-balance model for several volatile species in Kilauea magmas. They estimate that the magmas arrive at the summit magma chamber (2–5 km depth) laden with 0.32 wt% of both CO_2 and H_2O. More than 90% of this CO_2 is lost by preeruptive degassing, $\sim$ 6% is lost during eruption, and less than $\sim$ 3% is retained in the magma. (In contrast, H_2O

solubility is not exceeded within the magma chamber, and most of the H_2O is outgassed at eruption.)

These observations suggest that preeruptive degassing of CO_2 from magmas is not only important, but that it dominates the CO_2 flux from the mantle to the atmosphere, with eruptive degassing accounting for a relatively minor component of this flux. During the exsolution of a CO_2-rich gas phase from a magma, He is expected to behave as an infinitely dilute trace element and partition according to Henry's law between the silicate melt and gas or fluid phases. We have modeled this process using the partitioning data of Kurz & Jenkins (1981) together with temperature corrections discussed by Kurz (1982).

In order to predict the change in $U/^4He$ in a system that has outgassed CO_2 prior to eruption, we need to know the amount of CO_2 that has been

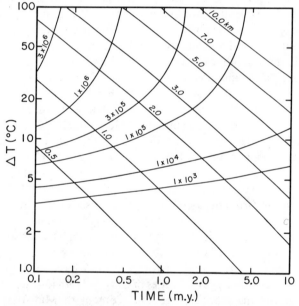

Figure 19   Solidification time (in millions of years) versus temperature contrast [$\Delta T(^\circ C)$] between a refilled spherical magma chamber and uniform wall-rock temperature subsequent to refilling for the last time. Magma chambers of various diameters (in km) are shown as solid diagonal lines. Curves represent the time required by the CDL model to decrease the $^3He/^4He$ ratio by 10% as a function of various initial $U/^4He$ ratios [ppm (std cm^3 g^{-1})] ranging from 10^3–3×10^6, of magma chamber diameters, and of values of ΔT. Cooling models calculated for conductivity of 2.5×10^{-3} cal s^{-1} cm^{-1} deg^{-1}, specific heat of 0.36 cal g^{-1} $^\circ C^{-1}$ (basalt at 1200°C), and density of 2.9 g cm^{-3} (Birch et al 1942); latent heat taken as 110 cal g^{-1} (Yoder 1976). The nonconvecting fixed melting-point formulation of Flemings (1974) was used for the thermal model; the fact that basalt has a melting interval of $\sim 100°C$ was accounted for to first order by using an "effective" latent heat of 146 cal g^{-1}.

lost and the depth of outgassing. Figure 20 shows the relationship between the change in the $U/^4He$ ratio ($f_{U/He}$) and F(the wt% of CO_2 that has been outgassed) for values of D ranging from 1.35×10^{-3} to 6.77×10^{-3} corresponding to pressures of 300–1500 bars. [D is the liquid/gas partition coefficient (moles He per gram of liquid/moles He per gram of CO_2) and derives directly from K, the Henry's law constant for He solubility (Kurz & Jenkins 1981).] The curves are calculated using a Rayleigh model that presumes that gas bubbles are rapidly removed from the system as they are formed, so that the total gas is not maintained in equilibrium with the residual liquid system.

Unfortunately, the CO_2 budget of the mantle is ill constrained, so that primary magma CO_2 contents are difficult to estimate. We can, nevertheless, compare increases in $U/^4He$ in magmas relative to the mantle (see Figure 16) with predicted CO_2 losses (Figure 20) and measured CO_2

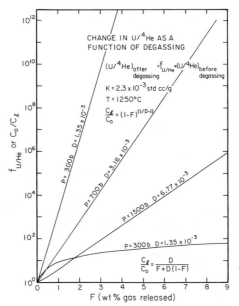

Figure 20   Changes in $U/^4He$ are shown as a function of F, the wt% of CO_2 gas released from a magma. Here $f_{U/He}$ is the ratio of $U/^4He$ in the magma after degassing to that before degassing. Curves for 300, 700, and 1500 b (bars) are shown for a Rayleigh degassing model $[C_l/C_0 = (1-F)^{(1/D)-1}]$; a 300-b curve for an equilibrium degassing model is also shown for reference $\{C_l/C_0 = D/[F+D(1-F)]\}$. In these equations, the subscript "1" refers to the final magma and "0" refers to the magma prior to degassing. The Rayleigh model essentially assumes that as CO_2 exsolves from the magma, it is rapidly removed from the system (that is, it either escapes or is segregated at the top of a body of magma). Variation of D with pressure is discussed in Zindler & Hart (1986). Note that, for a given F or wt% CO_2 released, He is lost much more efficiently at lower pressures.

solubilities in magmas. MORBs, for example, have $U/^4He$ ratios on the order of 6×10^3, or about a factor of 10 higher than a model mantle source ($f_{U/He} = 10$). If we assume that this outgassing occurs in the upper kilometer of the oceanic crust ($P \sim 300$–700 bars), we would predict the loss of something between 0.30 and 0.75 wt% CO_2, compared with a value of 0.13 wt% CO_2 retained in the erupted lava (Delaney et al 1978). Such amounts of CO_2 are likely to exceed solubility limits in basalt at pressures less than 1 kbar, and it is likely, therefore, that He loss from MORBs occurs via more extensive CO_2 outgassing at greater depths, or that MORBs arrive at shallow levels supersaturated with CO_2.

To produce large values of $f_{U/He}$ by outgassing CO_2, OIB magmas must arrive at relatively shallow levels supersaturated with CO_2. Because of this requirement, the time scales permitted for radiogenic ^4He accumulation will be on the order of magma residence times in high-level magma chambers. Such time scales may range from very short times in high-throughput environments such as midoceanic ridges, to several million years for large magma chambers in stagnating systems, such as might occur during the waning stages of volcanism at an ocean island.

The potential for producing changes in $^3He/^4He$ in the context of the EL model is still significantly greater than for the CDL model because the He loss may occur over much shorter time spans and is not strongly coupled to heat loss. That is, in the CDL model, He diffusion from the magma chamber occurs as the magma cools, and effective evolution times at high values of U/He will be limited. In the EL model, however, significant outgassing of a high-level magma may occur with very little associated cooling, allowing more time for evolution of the high U/He system. For example, with the CDL model, a magma chamber diameter of 1 km, and a magma with $(U/^4He)_0 = 5 \times 10^3$ and $(U/^4He)_f = 4.2 \times 10^6$, reduction of $^3He/^4He$ over 7×10^6 yr is on the order of 6.7% (with an initial $^3He/^4He$ value of 30 Ra). This is a minimum value for the CDL model, however, because outgassing of CO_2, even at depth, will substantially increase the potential for change in $^3He/^4He$ by diffusive loss of He from a magma chamber. By comparison, magma that attains a $U/^4He$ ratio of 4.2×10^6 by degassing will suffer more than 50% reduction of $^3He/^4He$ over 7×10^5 yr, or a 6.7% reduction in less than 10^5 yr (see Figure 16).

The arguments presented here are necessarily qualitative because of the lack of constraints on magma residence times, CO_2 budgets, and depths for degassing. We cannot, therefore, make specific statements regarding changes in $^3He/^4He$ ratios measured at individual localities. We can, however, draw several general conclusions, which may be used as guidelines for the interpretation of $^3He/^4He$ data until such time as a more quantitative treatment is feasible:

1. Low $U/^4He$ ratios in MORBs preclude any significant preeruptive reduction of $^3He/^4He$ ratios.
2. Low $U/^4He$ ratios in some OIBs ($U/^4He < 10^5$) are subject to a maximum of a few percent reduction of $^3He/^4He$ over time scales ranging up to 10^6 yr.
3. OIBs with $U/^4He$ between 10^6 and $\sim 5 \times 10^8$ may be subject to significant reduction of $^3He/^4He$ ($\geq 10\%$), with preeruptive aging over time periods ranging from 5.6×10^5 to 1.0×10^3 yr.
4. Where $U/^4He$ ratios of magmas are not available because $^3He/^4He$ ratios have been measured in fluid inclusions in phenocrysts, the potential for preeruptive $^3He/^4He$ reduction may be considered to increase with decreasing magma throughput in the system and with the degree of differentiation of the host magma.
5. Where $^3He/^4He$ values may have suffered preeruptive reduction, measured ratios should be considered as minimum values for the mantle source.

The Proverbial Bulk Silicate Earth (BSE)

The question of whether or not there exists today some primitive, undifferentiated segment of the Earth's mantle is central to the development of a viable and comprehensive model for the geochemical evolution of the Earth. The concept of such a primitive mantle segment has become quite common in the modern literature, although the terminology used to describe it has been varied. Essentially, the idea is that during accretion, core formation, and associated degassing, the silicate Earth achieves a fairly homogeneous state as a result of high heat flow and rapid convection. Those elements that do not enter the core or the atmosphere will then be represented in their initial, or primordial, abundance ratios in the silicate portion of the Earth. The elusive, primitive mantle segment (BSE) is then one that is supposed to have survived from this time as a closed system.

Documentation of the existence of such material today has been sought via two separate lines of reasoning: (a) that some oceanic and continental basalts are observed to have Nd isotope ratios similar to chondrites and to the presumed BSE (e.g. DePaolo & Wasserburg 1976a,b, O'Nions et al 1977, Wasserburg & DePaolo 1979, Jacobsen & Wasserburg 1979, DePaolo 1981); and (b) that some oceanic and continental basalts have $^3He/^4He$ ratios that are significantly higher than the atmospheric value (e.g. Lupton & Craig 1975, Kurz et al 1982, 1983). Although tacitly accepted throughout much of the literature, these two lines of reasoning do not represent unique interpretations of the available data, nor, in fact, are they altogether consistent with each other [since the highest observed ^{3}He/^{4}He ratios do not represent samples or locales with chondritic ^{143}Nd/^{144}Nd, or single-stage Pb that lies on the geochron (see Figures 11, 12)].

As fate would have it, the first series of high-precision measurements of ^{143}Nd/^{144}Nd in young basalts ranged from $\varepsilon_{Nd} \sim 13$ [where $\varepsilon_{Nd} = 10^4 \, (^{143}$Nd/^{144}Nd$_{measured} - \,^{143}$Nd/^{144}Nd$_{chondrites}) \, (^{143}$Nd/^{144}Nd$_{chondrites})^{-1}$, and ^{143}Nd/^{144}Nd$_{chondrites} = 0.51262$] in MORBs to $\varepsilon_{Nd} \sim 0 \pm 2$ for continental flood basalts and the island of Tristan da Cunha (Richard et al 1976, DePaolo & Wasserburg 1976a,b, O'Nions et al 1977). The observed correlation of ^{143}Nd/^{144}Nd with ^{87}Sr/^{86}Sr gave rise to the "mantle array" (early version of Figure 6) and was presumed to permit estimation of an average ^{87}Sr/^{86}Sr value for BSE (0.7045–0.7050). Implicit in this line of reasoning, and explicit in many of the models that followed (e.g. Jacobsen & Wasserburg 1979, DePaolo 1980, O'Nions et al 1979, Allegre et al 1980), was the idea first discussed by Schilling (1973) that isotopic variations in mantle-derived basalts result from mixing between depleted and undepleted (or primitive) mantle reservoirs. The data, however, which were taken to justify the notion of primitive undifferentiated mantle, would always have been equally well or better explained by mixing between a MORB source component and an enriched component with $\varepsilon_{Nd} < 0$ (Anderson 1982b). As more data have been acquired, the need for enriched mantle material has become more evident, and consequently compelling arguments for an undifferentiated mantle reservoir have been diminished.

Even when we explain away chondritic ^{143}Nd/^{144}Nd ratios as mixing phenomena, we must deal with the source for ^{3}He in the mantle. Despite the problems concerning production of ^{4}He, the range of ^{3}He/^{4}He ratios measured in mantle-derived basalts (Figure 14) clearly demonstrates that there are at least more and less "primitive", if not primordial, materials in the suboceanic mantle. However, as noted earlier, when we take ^{3}He/^{4}He ratios at face value, there is little evidence to suggest that high ^{3}He/^{4}He ratios occur in rocks that have ε_{Nd} values close to zero (see Figure 11). Furthermore, high ^{3}He/^{4}He ratios need not indicate the existence of primitive mantle; they could just as well document intramantle metasomatism of a previously depleted mantle segment, which resulted in a relative stabilization of high ^{3}He/^{4}He ratios at some time in the past (Figure 16; Rison & Craig 1982, Anderson 1985).

Based on the above discussion, we deem it reasonable to at least examine the consequences of a mantle that no longer contains any primitive undifferentiated components. Such a scenario could evolve in two different ways. First, enriched and depleted segments could evolve via continuous differentiation of a homogeneous mantle that had a chondritic Sm/Nd value subsequent to accretion. Second, borrowing from Armstrong (1981), we might consider that separation of the core did not result in homogenization of the silicate portion of the Earth. Certainly, with separation of the core and atmosphere, extensive differentiation of the silicate Earth may occur in conjunction with a substantial upward flux of incompatible elements. This likely resulted in at least the transient production of a granitic or alkali basaltic crust. Whether or not the silicate portion of the Earth was ever homogenized subsequent to this time depends on how effective early mantle convection was with regard to destruction of these earliest differentiates. If they survived to be reworked and cratonized during the Archean, then even mantle that has existed as a closed system for the past 4.4 b.y. may have a depleted Nd isotopic signature due to the separation of these early differentiates with low Sm/Nd ratios.

Even if "primordial" mantle material has survived unfractionated since accretion or core formation, we contend that no geochemical observation to date demands this survival. However, neither do the data demand that no such "primordial" component survives. Accordingly, we shall consider both possibilities in our subsequent evaluation of geochemical mantle models.

The Source of Midocean Ridge Basalts (MORBs)

The observation that MORBs that are well removed from any hotspot volcanism display unique "depleted" isotopic characteristics that are not observed elsewhere in oceanic or continental volcanic domains (see Figures 6–12, 15) documents the existence of a depleted mantle segment, which constitutes an important volumetric fraction of the MORB source reservoir. We refer to this depleted segment as end-member-depleted MORB mantle (DMM) and suggest that MORBs with more "enriched" isotopic signatures result from the inclusion of relatively enriched components in a DMM matrix.

Recent modeling of midocean ridge isotopic variations (Cohen & O'Nions 1982a, leRoex et al 1983, Allegre et al 1984) and studies of seamounts situated near the East Pacific Rise (Batiza & Vanko 1984, Zindler et al 1984, Graham et al 1984, 1986) all support this view of the MORB source mantle. In fact, this is one of the more notable successes of recent geochemical work on the mantle, and we embrace it without reservation in our subsequent discussion of mantle chemical dynamics.

The question of whether the MORB source mantle is heterogeneous on a large or global scale (100–10,000 km) is a bit more difficult to approach. The scale of melting and the rate of magma production at a given ridge segment will likely affect the chemical influence of the dispersed enriched components if their solidi are lower than that of DMM (Zindler et al 1984, Sleep

1984). The chemical influence of the dispersed components will also be diminished by processing through large magma chambers, which may be present at faster spreading ridges (Allegre et al 1984). Within this context, we might expect that some large-scale geochemical variations between MORBs may result from variable styles of magma processing at ridges with variable rates of magma throughput.

Recent work by Dupre & Allegre (1983) on MORBs from the Indian Ocean has shown that these rocks, even when far removed from hotspots, tend to be isotopically distinct from MORBs of other latitudes and oceans (Hart 1984b). For present purposes, the critical question with regard to these data is whether they document distinct differences in the isotopic character of the DMM component, the enriched components, or both, or whether they simply represent an increased abundance of enriched components. While none of these explanations can be categorically excluded, the latter is consistent with the large amounts of subduction known to have characterized the Southern Hemisphere during the Phanerozoic (Anderson 1982a) and the hypothesized origins for enriched mantle components (EMs), which are discussed subsequently.

With regard to the location of the MORB reservoir, we know that MORB source material is present beneath all active ridges. Essentially, all other volcanic materials (e.g. ocean islands, continental basalts, intra-oceanic arcs, etc) known to have feeder systems that extend to deeper mantle levels than at least the melt segregation region under midocean ridges (≥ 100 km) tend to have OIB-type isotopic signatures that are distinct from N-type MORB or DMM (Figures 14, 15). Taken at face value, this may mean that the MORB source mantle is restricted to the uppermost hundred or several hundred kilometers of the upper mantle. Such a suggestion is perhaps reasonable if DMM comprises relatively low-density depleted peridotite. In the context of the present discussion, we note that the only rocks at the surface of the Earth, other than normal MORBs, that are known to have DMM isotopic signatures are a subset of ultramafic xenoliths and orogenic lherzolites (Figures 1, 2). This suggests that DMM is peridotitic, as opposed to eclogitic or "piclogitic" (the latter having been suggested by Anderson 1983, Anderson & Bass 1985), and documents the existence of at least some DMM in the uppermost mantle beneath continents, particularly the relatively young mobile zones in which the xenolith localities and orogenic peridotites occur (e.g. the Rio Grande Rift, the Rhine Graben, East Africa, and Alpine zones of the western Mediterranean).

Recent work with rare gases has shown that DMM is characterized by (a) $^3He/^4He$ ratios of 8.0 ± 1.0 Ra (e.g., Kurz et al 1982; Figure 14); (b) $^{40}Ar/^{36}Ar$ ratios that are probably in excess of 25,000 (Allegre et al 1983a, Sarda et al 1985); and (c) occasionally observed excesses of ^{129}Xe, formed by the ancient decay of now-extinct ^{129}I (Staudacher & Allegre 1982, Allegre et al 1983a). The xenon and argon data, taken together, strongly support a catastrophic degassing of proto-DMM during the first $\sim 10^8$ yr of Earth history (Staudacher & Allegre 1982, R. Hart et al 1986, Sarda et al 1985). $^3He/^4He$ ratios of 8 Ra, although substantially lower than "planetary" or solar values of ~ 100 to 200 Ra (Jeffrey & Anders 1970, Craig & Lupton 1978), are still markedly enriched in primordial ^3He compared with the atmosphere, and these values suggest that the early degassing was more efficient for Ar and Xe than for He (e.g. R. Hart et al 1985, Sarda et al 1985); this is consistent with the enhanced solubility of the lighter noble gases in silicate melts (e.g. Ozima & Podosek 1983).

When the He isotope data for MORBs are compared with the data for other isotopic systems displayed in Figures 10–12 and 14, it is apparent that MORB $^3He/^4He$ ratios cover a relatively small fraction of the total oceanic basalt range, and that in contrast to all the other systems, MORBs fall *within* the oceanic range rather than defining one end of it. The latter observation may relate to the rapid accumulation of radiogenic ^4He in degassed and/or recycled materials and the influence of such processes and/or materials at some ocean islands. The relatively rapid diffusion of He in the presence of a small amount of melt ($D = 7 \times 10^{-5} cm^2 s^{-1}$ at 1200°C, extrapolated from Kurz & Jenkins 1981) may contribute to the smoothing of $^3He/^4He$ variations in the MORB reservoir during magmatic processing, if heterogeneities are on the order of 1 km or less.

A Prevalent Mantle Composition (PREMA)

Isotopic compositions of ocean islands and enriched MORBs, which are intermediate between DMM, BSE, HIMU, and EM components in Figures 6–9, have traditionally been interpreted as mixtures of these or other end-member components. Following the reasoning of Morris & Hart (1983) and Carlson (1984), we suggest the existence of a distinct mantle component with this isotopic character ("PREMA" for *pre*valent *ma*ntle). The position of this component is shown in Figures 6–12 and 15 and is seen to coincide with (a) fields for many ocean islands, including Iceland and Hawaii; (b) the most depleted representatives of many continental basalt suites (Figure 15); and (c) fields for many intraoceanic island arcs (see Figure 14). The coincidence of this component with Iceland and Hawaii, and with their respective high $^3He/^4He$ ratios (Figures 10–12), suggests that at least some portion of PREMA is relatively gas rich.

If PREMA represents a mixture of DMM, HIMU, and EM (with or without BSE) on a scale of 100 m or less, then the process of sampling during melting may necessarily result in a fairly restricted range of compositions, as we have already discussed for the MORB mantle. Such a result could not be easily distinguished from a mantle segment with PREMA characteristics that is homogeneous down to a meter or centimeter scale. On the other hand, if distinct components that compose PREMA are separated by large distances in the mantle (> 100 km) or reside within separate convective regimes, then the PREMA composition must result from a very reproducible mixing process that acts over large distances. We believe that the ramifications of both of these scenarios should be considered, and that the current geochemical data base is consistent with either of them.

The origin of a PREMA component may be envisioned in several different ways. First, as discussed earlier in Section 2, we might imagine that significant differentiation of the silicate portion of the Earth occurred contemporaneously with core segregation (e.g. Anderson 1979, Armstrong 1981). If something on the order of 50% of the present-day continental crust (or the trace-element equivalent) were segregated at this time, then PREMA today might represent the most primitive remaining mantle, having essentially survived unscathed since the earliest days of Earth history. In this context, the high $^3He/^4He$ ratios at Iceland and Hawaii would not be surprising. A variation of this scenario might suggest that, over the first part of Earth history (~ 1–2×10^9 yr), the entire mantle mixed convectively, which resulted in a PREMA-type depletion of the whole mantle. Subsequent splitting into upper and lower mantle convective regimes would then preserve PREMA in the lower mantle today, while the upper mantle would continue to be depleted by crust extraction, evolving toward DMM.

Alternatively, PREMA may evolve via the quasi-continuous separation of a crustal component from the mantle over time. We can hypothesize that the creation of new continental crust at volcanic arcs results in the complementary production of enriched and depleted mantle segments (EM, HIMU, and DMM). The EM, HIMU, and DMM components may then remix into the convecting mantle with time constants of 1–2×10^9 yr, so that at any given time, the mantle contains "quasi-steady-state" amounts of these materials. The ambient remixed mantle material will then represent PREMA. It would then follow that (a) PREMA is the chemical complement of the continental crust, (b) PREMA is approximately the mean composition of the mantle, and (c) that DMM + EM I + EM II + HIMU = PREMA.

These scenarios allow significant latitude for the location of PREMA in the mantle today, providing that it does represent a distinct component. If PREMA is close to the average mantle composition and is complementary to the continental crust, then it may occupy much of the mantle and be overlain by a relatively thin layer dominated by low-density DMM (see

Figure 26). Alternatively, PREMA may reside totally within the lower mantle, with the upper mantle comprising a mixture of DMM, EMs, and HIMU. Volcanic rocks from arcs, continents, and islands with PREMA signatures are generally thought to be derived from depths in excess of 100 km, so that either of the above scenarios would be feasible.

A High U/Pb Mantle Component (HIMU)

Isotopic evidence for the existence of the HIMU component derives largely from consideration of the Pb isotope systematics displayed in Figure 9, where oceanic basalts appear to mix from the depleted MORB zone near the geochron out toward the islands of St. Helena and Tubuaii. The extreme enrichments of ^{206}Pb and ^{208}Pb observed at these islands suggest an enrichment in $U + Th$ relative to Pb on the order of 1.5–2.0×10^9 yr ago (Tatsumoto 1978). These enrichments are not paralleled by an increase in Rb/Sr, and thus the islands are characterized by relatively low ^{87}Sr/^{86}Sr values (~ 0.7029). This unradiogenic Sr, which characterizes HIMU, suggests the possibility that HIMU has evolved from DMM some time in the past.

Chase (1981) and Hofmann & White (1982) suggested that HIMU might evolve from subduction of ancient altered oceanic crust, which is significantly enriched in U by interaction with seawater. At face value, it would appear difficult to provide U enrichments by seawater alteration without introducing a high ^{87}Sr/^{86}Sr signature. However, if the oceanic crust underwent alteration a few billion years ago, when the seawater Sr isotope curve was much closer to the mantle Sr evolution curve, a significant seawater Sr imprint might have been avoidable. It may be, however, that the fractionation of U from Th in the alteration cycle is sufficient to preclude the model.

Zindler et al (1982) proposed that HIMU might result from the subduction of ancient oceanic crust from which the alteration component had been removed by subduction-related processes. This proposal was made largely to avoid the problem of Th/U fractionations during alteration. The viability of this model hinges on the likelihood of a primary magmatic increase in U/Pb and Th/Pb during the generation of MORB melts. Relatively low μ values in hand-picked MORB glasses (~ 9.4), however, now seem to suggest that significant enrichments of U relative to Pb do not occur during MORB magmatism. However, further processing of MORB through subduction zones may effect such enrichments, as evidenced by low-μ values in island-arc volcanics (e.g. Tatsumoto 1978).

Vollmer (1977), Vidal & Dosso (1978), and Allegre and co-workers (e.g. Allegre et al 1980, Allegre 1982) have proposed that HIMU may result from extraction of Pb from some portion of the mantle to the core over a significant portion of Earth history ($\sim 15\%$ of the core formed since 4.4×10^9 yr ago). A lack of constraints not only on the time scales for core formation, but also on the nature of the light elements in the core (S, Si, O), as well as the hypothesized chalcophilic tendencies for Pb at current lower-mantle pressures and temperatures, makes it difficult to evaluate this mechanism. For present purposes, we must allow that it remains a possibility.

Another mechanism that must be considered as a plausible means by which to produce HIMU is the mobilization of Pb (and Rb?) in a metasomatic fluid and its subsequent removal from some mantle segment. The behavior of Pb in the continental crust suggests its affinity for volatile-rich ore-bearing fluids and provides qualitative support for this hypothesis, although there is no available data with which to evaluate such behavior in the mantle. Mantle metasomatism might thus result in the production of a residual degassed HIMU component and a complementary gas-rich low-μ component.

Regardless of the mechanism by which HIMU is created, its ubiquitous presence in the interactive mantle is clearly documented by the mixing trends in Figure 9. We thus propose that HIMU is present as a dispersed component within the mantle.

Enriched Mantle Components (EMs)

The isotopic character of islands that lie outside the DMM, HIMU, and BSE mixing domains (outlined in Figures 7 and 8) necessitates the existence of EM components, whether or not we think a BSE component actually exists (Anderson 1982b, Cohen & O'Nions 1982b, Richardson et al 1982). Proposed modes of origin for EM components now include (a) injection into the mantle via subduction of continentally derived sediment or crust (Cohen & O'Nions 1982b), altered oceanic crust (Hofmann & White 1982), ocean island crust, or seamounts; (b) delamination of subcontinental lithosphere (McKenzie & O'Nions 1983); and (c) mantle metasomatism (e.g. Menzies & Murthy 1980a, Menzies 1983). It is significant that all of these mechanisms can operate in convergent margin tectonic settings.

EM components are characterized by low $^{143}Nd/^{144}Nd$, variable $^{87}Sr/^{86}Sr$, and high $^{207}Pb/^{206}Pb$ and $^{208}Pb/^{204}Pb$ (at a given value of $^{206}Pb/^{204}Pb$). We refer to the low and high $^{87}Sr/^{86}Sr$ EMs as EM I and EM II, respectively. Comparison of Figures 6 and 15 reveals that the tendency for the continental data array to "fan out" at low $^{143}Nd/^{144}Nd$ is mimicked, to a lesser degree, by the oceanic data array (Worner et al 1986). This behavior in the continental suites has been attributed both to the existence of enriched subcontinental mantle components and to the contamination of some mantle-derived magmas by upper and lower continental crustal materials, lying at high and low values of $^{87}Sr/^{86}Sr$, respectively (e.g. Carter et al 1978, Hawkesworth et al 1983, 1984, Worner et al 1986). The similarity between the continental and oceanic arrays suggests similar styles of enrichment for the subcontinental and suboceanic mantle (although enrichment of the continental mantle is clearly more extreme), as well as the possible involvement of the continental crust in the production of EMs.

A review of the isotopic character of ultramafic nodules shows that many of those that contain hydrous phases plot below the oceanic array (Figure 1). If these materials represent metasomatized DMM, then the effect has been to decrease Sm/Nd substantially while increasing Rb/Sr only moderately (Menzies 1983, Zindler & Jagoutz 1986). These nodules, as well as some portions of ultramafic massifs (Reisberg & Zindler 1985, 1986, Brueckner et al 1986; Figure 2), are displaced from the oceanic array toward measured compositions for lower continental crustal materials (e.g. Hamilton et al 1979, Stosch & Lugmair 1984). Although little is known of the Pb isotopic character of ultramafic nodules, the range observed for $^{206/204}$Pb in massifs encompasses the whole range of oceanic Pbs (Hamelin & Allegre 1985a). We suggested earlier that metasomatized materials might have low U/Pb ratios and, with time, unradiogenic Pb (similar to that thought to characterize the lower continental crust). It is thus difficult to distinguish between recycled lower continental crust and metasomatized mantle as candidates for the EM I component.

EM II, characterized by high ^{87}Sr/^{86}Sr and low ^{143}Nd/^{144}Nd, has a strong similarity with upper continental crust or continentally derived sediment (e.g. Hawkesworth & Vollmer 1979, Carlson 1984, Worner et al 1986). Subducted continental material has been strongly implicated in several island-arc volcanic suites [e.g. Banda arc (Morris 1984), Martinique (Davidson 1983), Lesser Antilles (White & Patchett 1984)], and the isotopic signatures of these "contaminated" arcs bear a strong kinship with the EM II clan of OIB. (In contrast, there are no EM I–type arcs.) There are, as well, some ultramafic nodules, predominantly from South African kimberlites, that display this EM II isotopic signature (Figure 1). Menzies (1983) has proposed that the high ^{87}Sr/^{86}Sr, or high Rb/Sr (EM II–type) character of these nodules is imprinted by incorporation of a hydrous metasomatizing agent, whereas EM I–type signatures are produced by CO_2-rich fluids. Experimentally determined solubilities for CO_2 and H_2O in mantle melts (e.g. Khitarov & Kadik 1973) document the very high solubility of H_2O, relative to CO_2, and make it likely that mantle-derived metasomatic fluids are dominated by CO_2. EM II–type metasomatism may thus require a source of crustal water (perhaps in a subduction environment) and derive its isotopic character indirectly from the continental crust.

As noted earlier, the oceanic islands exhibiting strong EM II signatures are almost exclusively restricted to the Southern Hemisphere and serve to delineate the Dupal anomaly. Enhanced rates of Pangeatic subduction into the Dupal mantle (Anderson 1982a) may account for this observed localization. Based on the high $\Delta 7/4$ (Hart 1984b) character of the EM components, they must have been isolated from other mantle components (DMM, PREMA, etc) for a long period of time (> 3 b.y.). Isolation in the continental crust or subcontinental lithosphere is one obvious solution, although long-term isolation within the mantle is not precluded by our present understanding of mantle dynamics.

Anderson (1982c, 1983, 1985) has proposed that the MORB reservoir is overlain by an "enriched" layer of material situated in the uppermost mantle. While this tends to contravene conventional geochemical wisdom, which places enrichments deeper with the mantle, it does not violate constraints that derive solely from analysis of the present geochemical data base. The question of a shallow versus deep enriched zone in the mantle essentially depends on one's view of the mechanism by which the enrichment is created. For example, if we accept that enrichments are related in some way to subduction, then rapid extraction and stabilization of the enriched components from the downgoing slab in the uppermost mantle will result in a shallow enriched zone, while entrainment of enriched components to great depth along with the bulk of the subducting slab will result in relative enrichment at depth.

Isotope Systematics in Oceanic Basalts: The Role of Mixing

Implicit in our discussion of component characterization is our contention that oceanic basalt isotope systematics reflect mixing. This interpretation is supported by studies that document mixing at individual localities (e.g. Langmuir et al 1978, Zindler et 1979), as well as the various lines of reasoning that suggest relatively small-scale heterogeneities within magma source regions. Early interpretations of basalt isotope systematics suggested two- or three-component mixing to explain the various arrays. Here, we identify a *minimum* of four distinct components (DMM, HIMU, EM I, EM II), and it is thus not possible to delineate unique "component recipes" for individual ocean islands or ridge segments. It is, however, useful to qualitatively discuss constraints on mixing that derive from simultaneous consideration of the two-dimensional variation diagrams (Figures 6–12).

Our estimated component isotopic compositions are schematically depicted in Figures 6–9. Mixing lines between DMM, HIMU, and BSE are shown for reference. (Element concentrations used for mixing calculations are given in the figure captions.) Essentially, these mixing lines lie on the mantle plane (Figure 13; Zindler et al 1982), and (as discussed earlier) those localities that lie outside the DMM-HIMU-BSE mixing domain (Kerguelen, Walvis Ridge, Tristan, Gough, Samoa, Azores, Reunion, Tahiti) demand the existence of the EM components. Having invoked the EM components, we no longer require BSE as a mixing end member, and all observed compositions can result from mixing between DMM, HIMU, EM I, and EM II. If this mixing hypothesis is accurate, then the relative positions of individual localities, with respect to the components, should remain constant from one diagram to the next. Comparison of Figures 6–9 provides general support for this hypothesis.

The estimated composition of PREMA is also shown in Figures 6–9. In general, it is intermediate between DMM and HIMU but is displaced slightly toward EM II. Insofar as PREMA does represent a distinct mantle component and not simply a highly reproducible mixture, then PREMA may represent the depleted end-member OIB component, obviating the need for DMM in the OIB source reservoir (Morris & Hart 1983). This is readily apparent from the position of PREMA in Figures 7 and 8, even though mixing lines to PREMA are not shown (i.e. they are not expected to be highly curved). In this context, only the isotopic composition of N-type MORB documents the existence of DMM.

Direct extrapolation of the mixing relationships discussed above to the He variation diagrams (Figures 10–12) is not straightforward, partly because of the degassing considerations discussed earlier. Intramantle decoupling of He and lithophile trace elements (e.g. during fluid-phase metasomatism) may also contribute to this difficulty. We have, nevertheless, estimated $^3He/^4He$ ratios in the various components to facilitate direct comparison of Figures 10–12 with Figures 6–9.

While the $^3He/^4He$ ratio in DMM is tightly constrained by the narrow range of $^3He/^4He$ observed in MORBs (see Figure 14), values for other components are ill constrained. The BSE $^3He/^4He$ ratio is taken to be 50 Ra, although a choice of 30 or 100 Ra would not significantly alter the relationships in Figures 10–12. If we assume a BSE U value of 20.8 ppb (Table 1) and a BSE $U/^4He$ ratio of ~ 310 (Figure 16), the BSE He concentration (for single-stage evolution) will be on the order of 7×10^{-5} std cm^3 g^{-1}. A similar analysis for DMM, with $U/^4He$ taken to be ~ 600 (Figure 16) and U to be 8.4 ppb (Table 3), yields a He concentration of $\sim 1.4 \times 10^{-5}$ std cm^3 g^{-1}. Using the 9.6 enrichment factor for MORB over the MORB source (Table 3), we predict He concentrations on the order of 1.3×10^{-4} for MORBs. This is about a factor of four higher than the highest He concentration measured in a MORB glass (Kurz & Jenkins 1981) and thus implies $\sim 80\%$ He loss from even the most He-rich MORB (equivalent to ~ 0.2 wt% CO_2 degassing; see Figure 20). These considerations suggest that BSE will be relatively gas rich compared with

DMM, and that mixing lines between them in Figures 10–12 will be convex upward.

Given that $^3He/^4He$ ratios at ocean islands can only safely be interpreted as minimum values, we cannot use interisland relationships in Figures 10–12 to document component $^3He/^4He$ ratios. For the case of HIMU, we lack even the first-order constraint that would derive from $^3He/^4He$ analyses of St. Helena or Tubuaii lavas. Even so, our preferred modes of production of HIMU, involving either magmatism and degassing or loss of a metasomatic fluid within the mantle ("intramantle degassing"), are expected to lead to enrichments of U over He in HIMU source regions. Thus, we propose, following Allegre & Turcotte (1985), that HIMU will have $^3He/^4He$ that is close to or less than the atmospheric value (see Figure 16 caption).

He concentrations in HIMU, as a function of U concentrations, can be gleaned from Figure 16. Based on a present-day $U/^4He$ ratio of ~ 2000 and an "age" of 2×10^9 yr, He will range between 7.5×10^{-6} and 7.5×10^{-5} std cm^3 g^{-1} for U concentrations of 15 and 150 ppb, respectively. These U concentrations are appropriate for metasomatically depleted peridotite and degassed basalt, respectively. The mixing lines from DMM and BSE to HIMU (Figures 10–12) are based on the lower of these U concentrations; taking the higher one, the BSE-HIMU mixing line will be nearly straight, and the DMM-HIMU mixing line will be slightly convex downward.

Because of its affinity with upper continental crust and/or sediment, EM II is presumed to have $^3He/^4He$ in the range of HIMU. EM I, on the other hand, which may represent ancient metasomatized mantle, is likely to be at least moderately enriched in ^3He (Rison & Craig 1982). He concentrations in EM I will be strongly dependent on the age of the material. He concentrations, if produced as suggested by the "metasomatism" arrow in Figure 16, will be on the order of 2.5×10^{-4} std cm^3 g^{-1} (assuming 20–50 ppb U in metasomatized peridotite).

PREMA is shown near the top of the Loihi field in Figures 10–12; this is because of the coincidence of high $^3He/^4He$ Hawaiian and Icelandic localities with the overall PREMA isotopic character. The high $^3He/^4He$ ratio for PREMA is by no means certain; in fact, PREMA-like intraoceanic arcs with $^3He/^4He$ ratios of ~ 6 Ra (Craig & Rison 1983) argue against such a hypothesis. The He isotopic character of PREMA is, then, somewhat problematic, as is the existence of PREMA itself.

Having placed the various components on the He variation diagrams (Figures 10–12), we see that the possibilities for generating individual island compositions are essentially endless, particularly in conjunction with possible recent accumulation of radiogenic ^4He. While it is tempting to call on mixing between BSE and either HIMU or DMM, perhaps in conjunction with late-stage radiogenic He accumulation, to produce

Hawaiian and Icelandic arrays, mixing between PREMA, HIMU, and the EMs will serve as well. EM II islands (e.g. Samoa, Societies, Reunion, and Tahiti), identified on the basis of Sr, Nd, and Pb systematics, may represent mixing between the Hawaii-Iceland array out toward EM II.

We propose the mixing relationships in Figures 10–12 as an interim working hypothesis. Our best guess is that these relationships are significantly obscured by recent accumulation of radiogenic ^{4}He in some or all of the OIB localities, perhaps in conjunction with the paucity of data for many localities. A more quantitative understanding of He isotope systematics as they relate to Pb, Nd, and Sr will become possible as more He data are obtained in conjunction with U and Th determinations, and as more quantitative models for degassing time scales and magnitudes at specific localities are developed.

4. THE CHEMICAL AND PHYSICAL STRUCTURE OF THE MANTLE

Cosmochemical Constraints on the Bulk Composition of the Earth

A precise knowledge of the composition of the bulk silicate earth (BSE) can be used to constrain the dynamics and evolution of the crust-mantle system. For example, knowing the U content of BSE can be used to accurately estimate a value for the Urey ratio (ratio of total heat flow from the Earth to the total internal radioactive heat production), which in turn may provide constraints on the nature of mantle convection (Richter 1984). In addition, the determination of how much mantle has been depleted in large-ion lithophile (LIL) elements by continental crust extraction (e.g. Jacobsen & Wasserburg 1979, DePaolo 1980, Goldstein et al 1982, Allegre et al 1983c) depends heavily on knowledge of BSE Sr and Nd contents. In fact, our inability to specify BSE Sr and Nd contents is presently the "weak link" in deciding whether the whole mantle or just the upper mantle (~ 670 km) has been depleted. Obviously these considerations also impact the question of whole-mantle versus layered-mantle convection.

Jagoutz et al (1979) presented a novel and very important approach to determining BSE chemistry by noting that terrestrial ultramafic rocks and meteorites formed arrays of opposite slope on plots of Mg/Si and Al/Si, and proposed that the BSE composition could be established from the intersection of these arrays. They concluded that the BSE composition was about 2.5 times that of the C1 chondrites for refractory incompatible elements such as Ca, Al, and the rare-earth elements (REE). This approach (Jagoutz et al 1979) appears to be very promising (see discussion in Hart & Zindler 1986), and we have pursued it further to see if BSE refractory element concentrations could be determined to better than 10%, which is the level required for providing useful constraints on mantle models such as those mentioned above.

First, let us set down the assumptions involved in this approach:

1. The ultramafic samples accessible to geochemists cannot be proven to sample more than the upper ∼ 150–200 km of the mantle, and thus the Jagoutz et al (1979) approach will define only the uppermost mantle chemistry. Extrapolation of this to BSE will involve additional assumptions regarding the mode of accretion of the Earth and subsequent intramantle chemical fractionations.

2. The Earth is presumed to have some chemical "kinship" with meteorites. In particular, the BSE is assumed to have similar ratios of those refractory (nonvolatile) elements that can be shown to be constant throughout all classes of meteorites and the Sun.

3. Various upper-mantle ultramafic rocks are presumed to be "related" to each other by simple petrogenetic processes. In particular, since all known ultramafic samples appear to have undergone at least some modification in the light REE (LREE), and none have present-day $^{143}Nd/^{144}Nd$ isotope ratios consistent with closed-system growth for 4.55 b.y., some extrapolation will be required to project observed residual chemistries back to predepletion values. The usual model that is adopted, and which we develop below, is that depletion of a "primitive" lherzolite occurs by extraction of a partial melt, which leads to residual lherzolites with lower Ca, Al, and REE, and higher Mg.

PRIMITIVE UPPER MANTLE (PUM) ESTIMATES   We have assembled a data set comprising averages of 7 classes of meteorites (E, H, L, LL, Cl, CM2, and CV-CO chondrites) and 33 individual lherzolites, including both nodules and orogenic lherzolite samples [see Hart & Zindler (1986) for details of data selection].

Figure 21A is a plot of our data on the Mg/Si-Al/Si plot used by Jagoutz et al (1979). The appearance of the plot is very similar to theirs, though only eight samples are common to both diagrams. We have plotted the position of Jagoutz et al's (1979) preferred composition of primitive upper mantle (JPUM) and have drawn the terrestrial and meteorite (exclusive of CM2 and CV-CO classes) arrays to intersect at this point. Note that the CM2 and CV-CO chondrites appear to have been fractionated away from Cl and ordinary chondrites by a process that did not affect these latter chondrite classes (Larimer 1979). As discussed in Hart & Zindler (1986), an "earth" composition estimated from the intersection of the terrestrial and carbonaceous chondrite arrays is clearly inappropriate and thus, the CM2 and CV-CO chondrite classes are not considered further here. Two immediate conclusions are obvious from Figure 21A: (a) the primitive upper mantle (PUM hereafter) is chemically distinct from known meteorite classes, and (b) Mg/Si and Al/Si of PUM may be defined to better than $\pm 5\%$ in the context of the assumptions outlined earlier. This is indeed a major step forward, thanks to Jagoutz et al (1979)! It was our hope that similar plots involving Ca and the REE would equally well define Ca/Si and REE/Si for PUM. Figure 21B shows a Mg/Si-Ca/Si plot, and again the terrestrial and meteorite arrays are well defined—however, a number of samples, including JPUM, fall to the right of the intersection, which indicates the failure of one of the assumptions stated above.

Let us digress briefly here to discuss chemical fractionation in chondrites. Concentrating on the meteorite classes exclusive of CM2 and CV-CO, we find that ratios involving Nd, Sm, Ca, and Al are constant to better than 6%, and part or all of this may be analytical. Ratios involving Yb show variations of 9–13%, clearly beyond analytical error. This is probably related to the relatively low condensation temperature (T_c) of Yb compared with those of the other REE (Boynton 1975). Ratios involving Mg show variations of 23–39% except for Yb/Mg, which is only 13%. Mg also has a comparatively low T_c, relative to Ca and Al (Grossman & Larimer 1974), and apparently follows Yb to some extent. Ratios involving Si are highly fractionated (factors of 1.6–1.7); again, T_c for Si is less than that for Mg (Grossman & Larimer 1974). Thus, four ratios are of particular importance in estimating terrestrial chemistry from meteorite analogues: Sm/Nd and Ca/Al (which are constant in all meteorite classes), and Nd/Ca and Nd/Al (which are constant for meteorite classes exclusive of CM2 and CV-CO).

Returning now to the lherzolite data, we note that the Ca/Al ratio ranges from 0.556–1.60 versus the constant value of 1.09 for meteorites (Ahrens & Von Michaelis 1969). Furthermore, all but one of the lherzolites lying to the right of the meteorite-terrestrial array intersection in Figure 21B have Ca/Al ratios > 1.20. The terrestrial array is commonly viewed as a "melt-depletion" array (Loubet et al 1975, Suen 1978, Frey & Suen 1985, O'Hara et al 1975). However, recognizing that all common mantle melts have approximately chondritic Ca/Al ratios [average MORB = 1.02; oceanic alkali basalts = 0.91; all basalt types (worldwide) = 0.88 (Manson 1967, Melson et al 1976)], we can easily show that these high Ca/Al ratios cannot be produced by melt extraction.

The simplest explanation of the high Ca/Al ratios is to postulate that these samples have high and unrepresentative clinopyroxene (cpx) contents, not related to any petrogenetic process but simply a function of local heterogeneity in mineral abundances, compounded by the fact that in the field, geologists and geochemists will go for the "pretty green" cpx-rich rocks more often than the barren harzburgites. We have therefore "corrected" the lherzolite data set for excess cpx by subtracting cpx until all

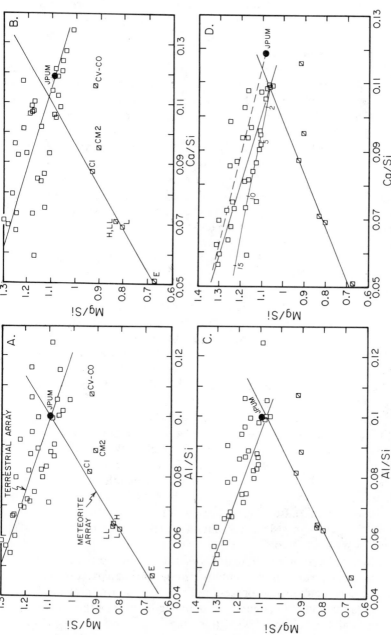

Figure 21   Mg/Si-Al/Si-Ca/Si relationships of meteorites and lherzolites. Panels *A* and *B* display data as published; panels *C* and *D* show lherzolites after subtraction of clinopyroxene (cpx) to force Ca/Al = 1.09 [=chondrites; see Hart & Zindler (1986) for complete discussion]. Solid circle labeled "JPUM" is the estimate of primitive mantle composition from Jagoutz et al (1979). Solid lines are eyeball fits to the data (forced to go through "JPUM" in *A* and *B*). Dashed line in *D* is the best-fit terrestrial line from *B* (included for reference). Solid line with the tick marks in *D* is trajectory of residual compositions resulting from melt removal; the percentage of melt removed is indicated by the numbers [see Hart and Zindler (1986) for description of this model]. CM2 and CV-CO chondrite data have been ignored in drawing the best-fit line to the meteorite array. Note that all concentration ratios are given as metal weight ratios.

Ca/Al ratios are ≤1.09 [see Hart & Zindler (1986) for details of this correction]. Comparisons of corrected and uncorrected data are shown in Figures 21*A* and 21*C*, and in Figures 21*B* and 21*D*. For the corrected data, only one sample (a serpentinized, recrystallized lherzolite from the Lizard peridotite) is anomalously high in Al/Si (Figure 21*C*), and all of the anomalously high Ca/Si samples have moved to lower Ca/Si ratios (Figure 21*D*), so they are now at or to the left of the meteorite-terrestrial array intersection. Based on *subjective* consideration of the data, we have placed the intersection (and its uncertainty) at Mg/Si = 1.06 ± 0.06, Al/Si = 0.100 ± 0.003, and Ca/Si = 0.109 ± 0.003. A similar plot, of Mg/Si versus Nd/Si (not shown), yields an estimate for Nd/Si ~ 0.054 ± 0.003. (Note that we will give REE/major element ratios in ppm/wt% throughout this discussion; true weight ratios can be derived simply by dividing by 10^4.) Only our Ca/Si value differs significantly from that of Jagoutz et al (1979).

Since most of the lherzolites in the data set have Sm/Nd > chondrites (> 0.325), it is clear that some depletion of LREE is typical. In order to understand how the REE and major elements are coupled during the

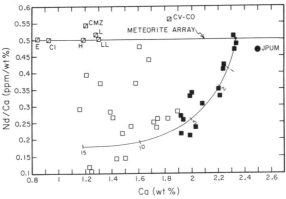

Figure 22 Nd/Ca versus Ca relationships for meteorites and lherzolites (cpx corrected). Discrepancy between our estimate for PUM (small solid circle) and that of Jagoutz et al (1979) (solid circle marked JPUM) reflects the cpx correction we have applied to the lherzolite data. The 15 most primitive lherzolites (Ca > 1.9%, Mg/Al < 14) are shown as solid squares; many of the low-Ca lherzolites (mostly xenoliths) have anomalously high Nd/Ca, perhaps due to metasomatic LREE enrichment (Frey & Green 1974). Note that many of the most metasomatized refractory xenoliths of Frey & Green (1974), not included in this data set, have Nd/Ca > chondrite. The intersection of the lherzolite array and the meteorite array at Nd/Ca (ppm/wt%) = 0.500 is taken to define Ca and Nd in PUM (= 2.34 ± 0.02 wt% Ca and 1.17 ± 0.02 ppm Nd). Al in PUM then equals 2.15%, since the chondrite Ca/Al ratio is 1.09. Solid curve is trajectory for residual lherzolite compositions derived from partial melt extraction model; the model predicts 20 ppm Nd and Sm/Nd = 0.24 for the 2% melt, and 13.7 ppm Nd and Sm/Nd = 0.26 for the 5% melt.

depletion process, we have formulated a simple depletion model to gain at least a first-order understanding of what depletion trajectories will look like on various compositional plots. Model depletion trajectories are shown in Figures 21–23, where it can be seen that the lherzolites that are least depleted (closest to PUM) have suffered < 0.5% melt extraction; typically, the average level of melt extraction for the more primitive lherzolites (black squares) is on the order of 2–3%. While this produces relatively insignificant depletions in major elements, it does significantly change REE concentrations.

We can now use all of the above considerations in concert to develop a best estimate for PUM. From a large variety of compositional plots (Hart & Zindler 1986), Figures 21–23 were selected to illustrate the approach. A summary of our estimate for PUM is given in Table 1, along with our admittedly subjective estimates of the uncertainties (which reflect only the application of the chosen data set, not the various assumptions concerning the development of the data set or its applicability to the actual upper mantle). Other refractory element abundances in PUM may be estimated

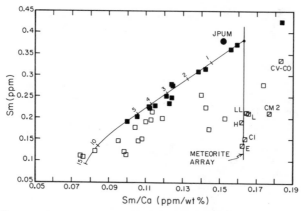

Figure 23   Sm versus Sm/Ca relationships for meteorites and lherzolites (cpx corrected). Best-fit array, as defined by the 15 most primitive lherzolites (see Figure 22 caption), intersects the meteorite array (Sm/Ca = 0.162) at Sm = 0.380±0.006 ppm (small solid circle). Note that the Jagoutz et al (1979) estimate for (JPUM) falls neither on the terrestrial nor on the meteorite arrays because of the cpx correction problem; note also that sample SC-1, the most primitive of Jagoutz et al., falls above and to the right of the intersection point and is therefore quite anomalous in Sm content. The Sm value derived from the intersection, coupled with the Nd value from Figure 22, leads to Sm/Nd = 0.325 (exactly chondritic), which provides an independent verification of these Sm and Nd estimates for PUM. The solid curve is the trajectory of residual compositions resulting from melt extraction and is seen to be an excellent fit to the most primitive lherzolites. As in the Nd/Ca-Ca plot (Figure 22), many of the more residual lherzolites (open squares; mostly xenoliths) have been apparently enriched in Sm relative to Ca, probably by metasomatic processes.

based on C1 abundances relative to Ca (Anders & Ebihara 1982). Also given in Table 1 is the JPUM estimate derived by Jagoutz et al (1979). The agreement for the REE is absurdly good. [We did not expect this, nor were we predisposed to the Jagoutz et al (1979) values.] The differences in major elements arise largely from our insistence that PUM must be chondritic in Ca/Al, which thereby forces a downward revision in the Jagoutz et al (1979) Ca estimate.

THE EARTH AS A C1 CHONDRITE   In addition to the above approach, we have derived a C1 chondrite–based Earth model (LOSIMAG C1) using average C1 values from Anders & Ebihara (1982) (Hart & Zindler 1986). We have assumed the following: (a) a near total loss or noncondensation of the

Table 1   Primitive upper mantle (PUM) composition

	This paper	Jagoutz et al (1979)
Si	21.5 ± 0.6 wt%	21.1%
Al	2.15 ± 0.02%	2.10%
Mg	22.8 ± 0.6%	23.1%
Ca	2.34 ± 0.02%	2.50%
Nd	1.17 ± 0.02 ppm	1.17 ppm[a]
Sm	0.380 ± 0.006 ppm	0.38 ppm
Yb	0.420 ± 0.02 ppm	0.42 ppm
SiO_2	46.0%	45.2
Al_2O_3	4.06%	3.97
MgO	37.8%	38.3
CaO	3.27%	3.50
Sr	19.6 ppm[b]	28 ppm
U	20.8 ppb[c]	26 ppb

[a] Derived using the Sm data and Sm/Nd chondrite = 0.325.

[b] Derived using Sr data for meteorite classes E, H, L, LL, and C1 (from Allegre et al 1983c). These Sr data were coupled with the Nd and Ca data for these meteorite classes to generate an average value for these five classes of meteorites of Sr/Nd = 16.7 and Sr/Ca = 8.40 (ppm/wt%) (see Hart & Zindler 1986).

[c] Derived using U concentration data for E, H, L, LL, and C1 chondrites, (from Chen & Tilton 1976, Tilton 1973, Manhes & Allegre 1978, Tatsumoto et al 1973, 1976). These data were coupled, where possible, with Nd from the same meteorite; otherwise the data are from the same meteorite class [see Hart & Zindler (1986) for details and data sources]. The average U/Nd value for the five meteorite classes so derived is 0.01780, with a maximum range between classes of 13%. The use of the average of all five classes was felt to be preferable to using only C1 data, as only one isotope dilution analysis exists for C1's. There is no obvious trend of U/Nd with fractionation indicators such as Al/Si for the five classes; inclusion of CM2 and CV-CO meteorites, however, suggests a trend of decreasing U/Nd with increasing Al/Si. Use of the average U/Nd of 0.0178, coupled with Nd = 0.463 ppm in C1, gives U = 8.2 ppb, a value that compares well with that of 8.1 ppb for C1 from Anders & Ebihara (1982). Using U/Nd = 0.0178 and Nd in PUM = 1.17 ppm gives U in PUM = 0.0208 ppm.

highly volatile gaseous elements and Na, K, and P, and significant loss or noncondensation of Cr and Mn (column *B*, Table 2); (*b*) a partial loss or noncondensation of Si and Mg, elements that are moderately "volatile" in a solar nebula context (e.g. Grossman & Larimer 1974; see column *B*, Table 2); and (*c*) formation of an iron core that also incorporates Ni and Co, leaving enough Fe in the mantle to yield a MgO/(MgO + FeO) molar ratio of 0.90 (column *C*, Table 2). The core generated in this calculation falls slightly short of an earth core mass (30.47% vs 31.50%, column *D*, Table 2). If sufficient oxygen (or other volatile elements) is left in the core to bring it up to mass (column *E*, Table 2), the final core is composed of 90% Fe, 5.4% Ni, 4.7% O, and a trace of Co. The fraction of O is significantly lower than the light-element abundances usually predicted for the core (10–15%) on the basis of geophysical constraints (e.g. Ringwood 1979), but in line with the recent estimates of Jeanloz (1986). When the silicate component of C1 is recalculated to 100% after core removal (column *G*, Table 2), the abundances and ratios of refractory lithophile elements are in precise agreement with the PUM model developed earlier (Table 1), as they should be since Fe is the only element not preadjusted to be compatible with PUM in this calculation. Note that if Si and Mg were treated as strictly refractory elements in this calculation, it would require 17% light elements in the core to bring it up to mass, and many of the element abundances and ratios would be at variance with the PUM composition.

Thus, by starting with a basic C1 composition, removing all or most of the highly volatile elements (H_2O, C, S, Na, K, P, etc), and allowing for small but significant deficiencies in Si (19%) and Mg (8%), we can construct an Earth that not only has a proper size core with an appropriate composition, but also has absolute abundances of the refractory lithophile elements (2.47 times the average C1 value, as given by Anders & Ebihara 1982), which agree with our PUM estimate. We call this the LOSIMAG C1 Earth model.

The apparent success in estimating a primitive mantle composition based on upper-mantle lherzolites and associated depletion trends casts considerable doubt on models that call on gross chemical differences between the upper and lower mantle in order to accommodate inferred elastic properties for candidate mantle phases (e.g. Anderson & Bass 1985).

Is There a Pb Paradox?

The so-called "Pb paradox" refers to the observation that Pb in ocean islands and MORBs is, almost without exception, more radiogenic than a 4.55-b.y. bulk Earth "geochron" (Figure 9). If the age of the Earth is 4.55 b.y. (= chondrite age), the bulk Earth must lie somewhere on this geochron. Because Pb in oceanic basalts comprises most of what is known of the

Table 2 LOSIMAG C1 Earth model

	B[b]	C	D	E	F	G
SiO_2	18.58	18.58	31.96	18.58	31.48	45.96
Al_2O_3	1.640	1.64	2.82	1.64	2.78	4.06
FeO	24.49	3.05	5.25	3.05	5.17	7.55
MgO	15.27	15.27	26.26	15.27	25.87	37.77
CaO	1.298	1.298	2.23	1.298	2.20	3.21
Na_2O	0.134	0.134	0.230	0.134	0.227	0.331
K_2O	0.013	0.013	0.022	0.013	0.022	0.032
Cr_2O_3	0.189	0.189	0.325	0.189	0.320	0.467
MnO	0.053	0.053	0.091	0.053	0.090	0.131
TiO_2	0.073	0.073	0.126	0.073	0.124	0.181
NiO	1.39	0.112	0.193	0.112	0.190	0.277
CoO	0.064	0.0052	0.009	0.0052	0.009	0.013
P_2O_5	0.008	0.008	0.014	0.008	0.014	0.020
Silicate sum	63.202	40.425	69.530	40.425	68.50	100
HVE[a]	30.21	30.21	—	30.21	—	
MVE[a]	1.258	1.258	—	1.258	—	
SVE[a]	5.33	5.33	—	5.33	—	
Oxygen[a]	—	5.06	—	4.185	—	
	36.798	41.858	—	40.983	—	
Fe	—	16.665	28.664	16.116	28.239	89.648
Ni	—	1.004	1.727	1.004	1.700	5.397
Co	—	0.046	0.079	0.046	0.078	0.248
Oxygen[a]	—	—	—	0.875	1.483	4.708
Core sum		17.715	30.470	18.590	31.50	100
Sum all	100	100	100	100	100	

[a] HVE = highly volatile elements (H_2O, S, C, organics); MVE = moderately volatile elements (Na, K, P, Cr, Mn); SVE = slightly volatile elements (Mg, Si); oxygen is that obtained from FeO, NiO, CoO reduction.
[b] Column headings are as follows:
B—Major components of C1 chondrites from Anders & Ebihara (1982) all HVE separated from silicate components; partial separation of MVE, to leave MVE/Al ratios in silicate similar to those of Jagoutz et al (1979) mantle; partial separation of SVE, to leave SVE/Al ratios as in PUM, Table 1.
C—Separation of metallic core, leaving enough FeO to provide MgO/(MgO + FeO) = 0.90 (molar) and enough NiO and CoO to provide ratios to Al similar to those of Jagoutz et al (1979) in mantle.
D—Removal of HVE, MVE, and SVE and oxygen from C; remainder renormalized to 100%.
E—Column C, with partial oxygen budget left in core.
F—Removal of HVE, MVE, SVE and part of oxygen from E, remainder renormalized to 100%.
G—Silicate mantle and oxide core renormalized separately to 100%.

interactive portions of the mantle, we must look to some less obvious reservoir of unradiogenic Pb (which lies to the left of the geochron) to balance the oceanic mantle. The paradox arises from the fact that no such reservoir can be unambiguously identified, although several suggestions, including the lower continental crust (e.g. O'Nions et al 1979), DMM (Anderson 1982b), the core (Vidal & Dosso 1978, Oversby & Ringwood 1971, Allegre et al 1980, Vollmer 1977, Allegre 1982), and the lower mantle (Anderson 1983), have appeared in the recent literature. Before going on to evaluate these various suggestions, we summarize pertinent constraints on U and Pb budgets in the crust and mantle.

U AND Pb BUDGETS   A variety of Pb data for various average crustal rock types and composites has been compiled and is given in Table 3. There is a small but significant difference in Pb between upper and lower crustal rock types; this is largely related to the more basic character of lower crustal rocks and not to Pb depletion during granulite facies metamorphism. Relative to a PUM silicate Earth, the crust appears to contain an unusually large fraction of the silicate Earth's Pb inventory (37–70%; Table 3). This Pb enrichment is in contrast to values derived for crustal Sr and Nd inventories of 10% and 18%, respectively (Table 3), and appears inescapable.

Extreme variations of U in lower crustal (granulite facies) rocks hinder an accurate assessment of the U contents of, and the distribution within, the continental crust. Our estimates are given in Table 3 and are in reasonably good agreement with Taylor's (1982) estimates, though they have been compiled largely from independent data sources. Relative to a PUM silicate Earth, ~ 22–52% of Earth's U is in the continental crust.

Our estimates for U and Pb in the continental crust lead to rather low μ values ($\mu = {}^{238}U/{}^{204}Pb$), in terms of conventional wisdom, particularly for the upper crust (mean $\mu = 8.7$, range $= 6$–14; Table 3). In contrast, most data that exist for upper crustal materials strongly indicate the radiogenic nature of the upper crust. [${}^{206}Pb/{}^{204}Pb$ ratios for pelagic and terrigenous sediments range from 18.5 to 19.1 and from 18.8 to 19.8, respectively (Sun 1980, White et al 1985).] For crust with a mean age of 2.0 b.y. (Goldstein et al 1984, Allegre & Rousseau 1984), a range of μ from 10 to 14 is required to generate ${}^{206}Pb/{}^{204}Pb$ ranging from 18.5 to 19.8. While the estimated uncertainties in upper crustal U and Pb encompass μ values as high as 14, this is barely sufficient to generate the most radiogenic Pb observed in oceanic terrigenous sediments. In terms of the total crust estimate, the uncertainty in μ is large enough to tolerate a crustal reservoir lying either to the left or the right of the bulk earth geochron, though the best estimate will be well to the left of the geochron.

Insofar as U appears to be slightly more incompatible than Pb in most

Table 3 U-Pb budget for various terrestrial reservoirs

	U (ppm)	Pb (ppm)	μ	Nd (ppm)	Sr (ppm)	Sr/Nd	U/Nd	Pb/Nd
C1[a]	0.0200	6.12	0.146	1.13	19.5	17.3	0.0177	5.42
PUM[b]	0.0208	0.155	8.4	1.17	19.6	16.8	0.0178	0.132
MORB[c]	0.081	0.541	9.4	7.7	128	16.6	.0105	0.070
MORB mantle	0.0084	0.056	—	0.8	13.3	—	—	—
Upper crust[d]	2.4±0.5	18±4	8.7 (6–14)	32.5	300	9.2	.073	0.55
Lower crust[e]	0.7±0.5	11±3	3.8 (1–9)	32.5	300	9.2	.022	0.34
Total crust[f]	1.2±0.5	13.1±4	5.7 (3–12)	32.5±1.5	300±30	9.2	.037	0.40
Mass in crust	37%	54%	—	18%	10%	—	—	—
Mass in PUM								

[a] Element ratios are for C1 chondrites (Anders & Ebihara 1982); abundances from C1 chondrites multipled by factor of 2.47 (see Hart & Zindler 1986) as in the LOSIMAG C1 Earth model.
[b] From Table 1; Pb calculated for μ = 8.4 (see text).
[c] Average of 17 fresh MORB glasses and 4 leached whole rocks (Cohen & O'Nions 1982a,b, Cohen et al 1980, Vidal & Clauer 1981); average K for this suite is 1027 ppm. Twenty-two glasses from Jochum et al (1983) average 0.075 ppm U.
[d,e] Data sources: Allegre & Rousseau (1984), Dupuy et al (1979), Fahrig & Eade (1968), Goldstein et al (1984), Gray (1977), Gray & Oversby (1972), Heinrichs et al (1980), McLennan & Taylor (1980), Michard et al (1985), Montgomery (1977), Reilly & Shaw (1967), Shaw (1967), Shaw et al (1976), Taylor & McLennan (1981), Taylor et al (1983), Tilton & Barreiro (1980), White et al (1985).
[f] Calculated as 30% upper crust–70% lower crust.

igneous processes (Watson et al 1986), it might appear contradictory to speak of a crustal μ that is lower than that of the BSE, as the crust is the end result of a series of melting and crystallization processes. However, we must consider that arc processes are complex and appear to produce numerous other element fractionation patterns that contravene conventional wisdom (see, for example, Morris & Hart 1983), and thus they may also produce enrichments of Pb over U. This contention is supported by the relatively low μ values measured in many arc volcanics (e.g. Tatsumoto & Knight 1969, Oversby & Ewart 1972, Tatsumoto 1978).

Also shown in Table 3 is the average Pb content (and U, Sr, and Nd contents) of 17 fresh MORB glasses, compiled from the literature. While the various element ratios may approximate those of the MORB source mantle itself, the concentrations have obviously been raised by partial melting and fractional crystallization processes. Approximations to the contents of U and Pb in the MORB mantle have been derived by comparison to Nd estimates of 0.6–1.0 ppm (Hart & Zindler 1986). Using the U/Nd and Pb/Nd ratios in MORB (Table 3) would place the composition of MORB mantle (for 0.8 ppm Nd) at 13.3 ppm Sr, 0.0084 ppm U, and 0.056 ppm Pb. Thus, relative to PUM, the MORB mantle is depleted in Nd, Sr, U, and Pb by factors of 0.68, 0.68, 0.40, and 0.36, respectively (i.e. U and Pb behave more incompatibly than Sr and Nd).

DISCUSSION   The silicate portion of the Earth is clearly depleted in Pb relative to the LOSIMAG C1 chondrite model. This is an old observation and has been attributed to removal of Pb to the core or to volatilization or noncondensation of Pb during Earth accretion (e.g. Murthy 1975, Oversby & Ringwood 1971). In the event that any of these processes were synchronous with accretion, and provided that accretion took place rapidly at an age given by chondrites ($\sim$ 4.55 b.y.), then the 4.55-b.y. geochron accurately specifies the locus of possible BSE Pb isotope compositions. The Earth is also depleted in a variety of other volatile elements, many of which are not expected to be chalcophilic [that is, potentially core compatible (e.g., Sun 1982)], and thus there is no compelling reason, except for the Pb paradox itself, to assert that the depletion of Pb in the silicate Earth is due to "core pumping." However, it is clearly possible that the Pb depletion is a result of both nonaccretion and core pumping. In the event that some of the "missing" Pb was coprecipitated with the Fe-Ni core, and that core formation took a finite time to run to completion ($>$ 100 m.y.), then the bulk silicate earth is not required to lie on a 4.55-b.y. geochron. Various two-stage isochrons are shown in Figure 24 to illustrate this process.

While the $^{206}Pb/^{204}Pb$ ratio of PREMA is not well constrained, it is probably $\sim$ 18.2–18.5 (Figure 9); insofar as the continental crust is the total complement to PREMA, and insofar as the crust contains $\sim 50\%$ of the silicate Earth Pb budget (Table 3), then relative to PREMA, the crust must lie on the opposite side (and equidistant from) the relevant BSE isochron. If the average crustal μ is indeed as low as 5.7, as suggested earlier (Table 3), PREMA plus crust will lie on a 4.55-b.y. geochron, and little core pumping of Pb is required. If the crustal μ is the upper limit allowed (~ 12), then about 150 m.y. of core pumping is required. At this point, the most that can be said is that the missing Pb needed to explain the Pb paradox may be accommodated in either the crust or the core, or both.

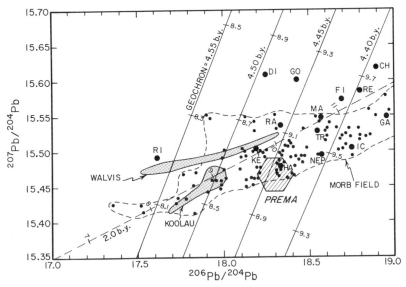

Figure 24 $^{207}Pb/^{204}Pb$-$^{206}PB/^{204}Pb$ plot, showing MORB data (small dots), selected averaged oceanic islands (large dots), and the field for Koolau (Oahu) and Walvis Ridge basalts, in comparison with various terrestrial isochrons. The true age (t_0) of the Earth (geochron) is taken to be 4.55 b.y.; the other isochrons represent possible core-mantle differentiation "ages." These isochrons are calculated for a first-stage $\mu = 0.21$ from t_0 to t, and second stage from t to present. Various closed-system μ values for the second stage are indicated as tick marks on the isochrons. Also shown for reference is a 2-b.y. secondary isochron (dashed), calculated with a first-stage $\mu = 8.1$ (t_0 to 2 b.y.)—second-stage μ values are shown as tick marks. This isochron may be used to estimate positions of various crust and mantle reservoirs having a ~ 2.0-b.y. mean age as a function of second-stage-reservoir μ value. Cross-hatched field indicates possible position for PREMA. Data references are given in Zindler et al (1982), Hart (1984b), and Hamelin & Allegre (1985b), Koolau data are from Stille et al (1983) and S. R. Hart (unpublished), and primordial Pb parameters are from Tatsumoto et al (1973). Oceanic islands: CH (Christmas), DI (Discovery), FI (Fiji), GA (Galapagos), GO (Gough), HA (Hawaii), IC (Iceland), KE (Kerguelen), MA (Marion), RA (Raratonga), RE (Reunion), RI (Rio Grande Rise), NEP (NE Pacific seamounts), TR (Tristan da Cunha).

We may also examine the other solutions to the Pb paradox introduced earlier. Can the depleted MORB mantle be the reservoir that "balances" the other oceanic Pb components? Certainly some MORBs have Pb isotope compositions lying to the left of the geochron (see Figure 24), although the great majority lie to the right. Relative to PUM, the MORB mantle only contains $\sim$ 11% of the silicate Earth Pb budget (if we assume that the MORB reservoir is 30% of the mantle mass), and DMM likely contains much less. Clearly the continental crust dominates the Pb budget of the Earth, and it is unlikely that an unradiogenic DMM component can be effective in balancing the mantle segment represented by the oceanic Pb array. (Here, we presume that much or all of the mantle, exclusive of the MORB reservoir, is characterized by radiogenic Pb similar to that observed in oceanic islands.) Obviously one can postulate that the lower mantle is unradiogenic (and also never sampled in erupted volcanics). For example, Anderson's (1983) model derives a lower-mantle μ value that is 25% lower than the μ of the upper mantle; however, his bulk silicate Earth also has a very high μ (10.5), which would demand more than 150 m.y. of core pumping of Pb.

We conclude that the Pb paradox is at least a "pink" herring, although we feel it is an important issue insofar as it may be one of the few ways available to constrain the time interval of core formation if, in fact, Pb did partition into the core. In this context, however, it seems that a modern Pb budget is not a promising approach, and that much better constraints are available by working with Pb isotope data from Archean rocks. Various works of this kind have been reported (Manhes et al 1979, Gancarz & Wasserburg 1977, Tera 1980); generally these give ages for the BSE (= core formation?) of 4.4–4.5 b.y., but with uncertainties that include the probable accretion age of 4.55 b.y.

Geochemical Mass Balance

The derivation of constraints on the relative volumes of geochemically distinct mantle components has received considerable attention in the recent literature (e.g. Jacobsen & Wasserburg 1979, DePaolo 1980, O'Nions et al 1979, Allegre et al 1980, Turcotte 1985). Most of these models are based only on Sr and Nd systematics and have assumed that the depleted MORB mantle is the sole complement of the enriched continental crust (the remainder of the mantle being BSE); thus, they have not dealt with much of the geochemical complexity known to characterize the mantle. In addition, many modelers have chosen to utilize input parameters without considering the effects of the substantial uncertainties known to be associated with them. Consequently, the conclusions based on these models (e.g. 32–45% of the mantle is depleted) are difficult to interpret at best.

Goldstein et al (1982) considered a more complex model, via inclusion of a HIMU component and U-Th-Pb systematics, and investigated the effects of input parameter uncertainties using a Monte Carlo approach. They concluded that the mass fraction of the depleted mantle is ill constrained by the present-day geochemical data base and could conceivably range from $< 20\%$ to $> 90\%$ of the mantle. Allegre et al (1983c) used an inversion method to investigate the effects of input uncertainties; even without the inclusion of a HIMU component, they concluded that the error bounds on the estimate of depleted mantle mass were large and could encompass values ranging from 30–90% of the entire mantle. Even with clearly oversimplified models, it has not been possible to delineate definitive constraints on the mass fraction of depleted mantle.

In this section, we derive constraints on the nature of mantle reservoirs by making simplifying assumptions that are consistent with the degree of chemical complexity known to characterize the mantle. It is appropriate here that we clarify our usage of the terms "reservoir" and "component." We have used component, in the preceding discussion, to refer to a geochemically distinct mantle segment that is homogeneous on scales sampled by the smallest effective melting volume. Obviously it may also be homogeneous on smaller or larger scales. Thus, a component may compose the entire upper mantle or be dispersed as meter- or even centimeter-scale veins or blobs. In contrast, we use the term reservoir to denote a portion of the mantle that is under discussion; it need not be chemically homogeneous nor physically contiguous. That is, we can speak of the MORB source reservoir, which is thought to be physically contiguous but chemically heterogeneous, as comprising HIMU and EM components in a DMM matrix, or we can refer to a relatively enriched reservoir that includes the sum of HIMU and EM components and is thus neither contiguous nor homogeneous.

As a starting point, we consider the simplest form of crust-mantle mass balance where the mantle and crust constitute complementary reservoirs. In addition, we consider only the Sm-Nd system, since (geochemically speaking) it constitutes the best-behaved parent-daughter pair. This approach does not deny that the mantle is very heterogeneous; it simply asks what the mean composition of this heterogeneous mantle is. The mass balance equation for this model may be rewritten (from Allegre et al 1983c) as

$$\mu_m = (0.0064\,Nd_c\mu_c - \mu_t Nd_t)/(0.0064\,Nd_c - Nd_t),$$

where the subscripted μ is $^{147}Sm/^{144}Nd$, subscripts c, m, and t refer to crust, depleted mantle, and bulk Earth, respectively, and Nd is the concentration in ppm. The factor 0.0064 is simply the mass ratio of continental crust to total mantle plus crust (Turcotte & Schubert 1982).

The selection of input parameters is given in the caption to Figure 25, with uncertainties that represent a subjective estimate of the maximum allowable range. The value for Nd_t is taken as that in PUM (see above). The calculated value for μ_m is $0.215 \pm 1.1\%$; the error in this value was propagated by allowing the errors in the individual parameters to combine in the worst possible way, and thus the error bounds are fairly conservative. Figure 25 also shows how μ_m varies as a function of Nd_c only. From simple mass balance, using the chosen parameters, we find that $\sim 18\%$ of the Earth's Nd budget is stored in the crust (Table 3), and thus the Nd

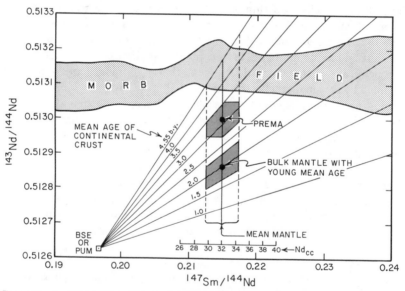

Figure 25   Sm/Nd isochron diagram showing various mean age isochrons and the value for $^{147}Sm/^{144}Nd$ of the mean depleted mantle (0.2148 ± 0.0024) calculated from a bulk Earth–continental crust depletion model. MORB field taken from literature values. Parameters and uncertainties chosen for model as follows: primitive mantle (bulk Earth) $Nd = 1.17 \pm 0.02$ ppm (PUM value; see text), $^{147}Sm/^{144}Nd = 0.1967 \pm 0.0003$ (PUM value taken from Jacobsen & Wasserburg 1984); continental crust $Nd = 32.5 \pm 1.5$ ppm, $^{147}Sm/^{144}Nd = 0.113 \pm 0.003$ (Goldstein et al 1984, Taylor et al 1983, Allegre & Rousseau 1984, Michard et al 1985, Allegre et al 1983c). The effect of other choices for crustal Nd concentration may be judged from the scale labeled Nd_{cc}. Shaded polygons delimit solutions for two choices of mean age $= 2.0 \pm 0.3$ b.y. and 3.1 ± 0.4 b.y. The lack of significant overlap between the mean depleted mantle field and that of MORB mantle simply means that the whole mantle is, on average, unlikely to be of MORB isotopic composition unless the mean age of depletion is significantly greater than 3.5 b.y. See text for discussion of error propagation.

concentration of mean depleted mantle is 0.96 ppm (18% depleted from 1.17 ppm).

This model value of μ_m can be used to derive the $^{143}Nd/^{144}Nd$ ratio of mean depleted mantle if we can estimate the mean age of the total crust-mantle system. If continental crust extraction has occurred at a constant rate over the age of the Earth, the mean age of the system will be ~ 2.28 b.y.; if crust extraction was more rapid during the Archean, when there was more heat available for differentiation (as argued for by Allegre & Rousseau 1984), then the mean age will be older. However, for either of these scenarios, significant recycling of average crustal material into the mantle, which is consistent with Sm-Nd systematics in modern and ancient sediments (Goldstein et al 1984, Allegre & Rousseau 1984), will tend to push the crust toward younger mean ages (perhaps as young as 1.7. b.y.). If all or a significant fraction of the continental crust was extracted early, as in continental big-bang models (e.g. Patterson & Tatsumoto 1964, Armstrong 1981), the mean age might be as old as 3.8 b.y. All things considered, there is an extremely large range of mean ages that have been proposed as characterizing the continental crust.

We have chosen to consider the consequences of two "end-member" mean-age ranges for the continental crust: 2.0 ± 0.3 b.y., corresponding to mean mantle $^{143}Nd/^{144}Nd$ of $0.51286 \pm ^8_6$; and 3.1 ± 0.4 b.y., with mean mantle $^{143}Nd/^{144}Nd$ of 0.51300 ± 5 (Figure 25). The latter estimate has the additional significance that it corresponds to the composition of PREMA. Based on the concentration of Nd and the $^{143}Nd/^{144}Nd$ ratio in the MORB reservoir (0.5–0.9 ppm, 0.51315; see Figure 14), we can now investigate mass and compositional relationships between the MORB reservoir and the remainder of the mantle, which we refer to as the "complementary reservoir." This complementary reservoir is enriched relative to the MORB reservoir, though it may be depleted or enriched relative to BSE or PUM. Figure 26 shows the relationship between the relative masses of the MORB and complementary reservoirs and the $^{143}Nd/^{144}Nd$ ratio of the complementary reservoir for the two mean mantle $^{143}Nd/^{144}Nd$ ratios estimated above. Contours for Nd concentration in the complementary reservoir are also shown in Figure 26 and document a positive correlation between this parameter and the mass of the MORB mantle.

If the complementary reservoir is primitive (BSE or PUM), as has been assumed in most geochemical models, then the relationships in Figure 26 can be used to constrain the mass fraction of this reservoir. This is done by ascertaining the mass of the complementary reservoir, which corresponds to the intersection of the PUM $^{143}Nd/^{144}Nd$ value with the PUM (1.17 ppm) Nd contour. The resulting mass fractions for a PUM reservoir are $\sim$ 30 and 48 wt%, respectively, for the models shown in Figures 26A and 26*B*. Taken at face value, this result suggests that if any primitive mantle volume survives today, it must constitute significantly less than the ~ 70% of the mantle that lies below the 670-km seismic discontinuity. This result lies within the permissible ranges calculated by Goldstein et al (1982) and Allegre et al (1983c); the higher precision and, therefore, enhanced significance of the current result are largely due to our improved understanding of Nd in PUM.

This result implies a material flux into and/or out of the lower mantle at some time during Earth history and, thus, is not compatible with the oft-proposed persistent, unbreachable chemical boundary layer at 670 km;

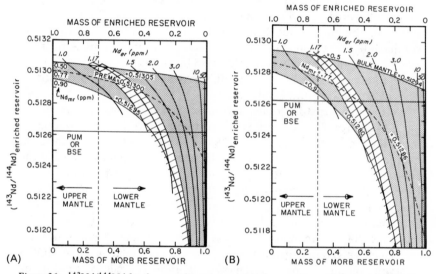

Figure 26 $^{143}Nd/^{144}Nd$ for the complementary enriched reservoir as a function of the mass of the MORB source reservoir (expressed as a mass fraction of the whole mantle). Panels *A* and *B* are for different mean mantle compositions corresponding to "old" and "young" mean ages for the continental crust, respectively (see Figure 25). In Panel *A*, the mean mantle has the composition of PREMA. "Upper" and "lower" mantle volumes are defined by the position of the 670-km seismic discontinuity. The shaded areas correspond to the uncertainties indicated in Figure 25 for the two mean mantle compositions in conjunction with Nd concentrations in the MORB reservoir (Nd_{mr}) ranging from 0.5–0.9 ppm. Nd concentrations in the complementary enriched reservoir (Nd_{er}) are shown as contours across the shaded areas. $^{143}Nd/^{144}Nd$ in PUM or BSE is shown as a horizontal line across each diagram. Approximate error bounds are shown for the $Nd_{er} = 1.17$ (= PUM) contour. The intersection of this error band with the PUM $^{143}Nd/^{144}Nd$ value defines the mass fraction of the complementary reservoir, if it comprises primitive, undifferentiated mantle. Resulting estimates for such a PUM reservoir are 26–38% and 42–53% of the mantle for models in Panels *A* and *B*, respectively. This suggests that if a primitive, undifferentiated mantle reservoir exists, it cannot constitute the entire lower mantle.

such a boundary may exist today, but it cannot have existed continuously since the earliest Archean. Furthermore, it seems likely that there is no convectively isolated reservoir for PUM today, and if PUM survives, it must exist as a dispersed component within a convecting medium. Hence, arguments citing the need for convective isolation of a PUM or BSE component (e.g. O'Nions & Oxburgh 1983) need to be reevaluated.

Earlier, we suggested that PREMA may be a physically distinct mantle component. If we now consider that the complementary reservoir is PREMA, then it must be volumetrically dominant, with the MORB reservoir constituting less than $\sim 10\%$ of the mass of the mantle (Figure 26A). Alternatively, if the complementary reservoir comprises PREMA + HIMU + EM I + EM II $\pm$ PUM (or BSE), then the relative masses of the MORB and complementary reservoirs are ill constrained.

These mass-balance considerations, together with component characterizations, suggest a mantle that is composed of DMM, HIMU, EM I, and EM II components and may also include PREMA and/or PUM. The MORB reservoir is volumetrically dominated by DMM (as a result of its highly depleted character) but contains various amounts of the enriched components, so that its mean composition is essentially that of average MORB. This heterogeneous MORB reservoir may be underlain by, or may grade downward into, mantle with an overall more enriched isotopic character. If the mean $^{143}Nd/^{144}Nd$ of this complementary reservoir is taken to be no lower than the most enriched OIB (~ 0.5123—Kerguelen), then the relationships in Figure 26 allow this reservoir to account for as little as 20% or (based on a mean isotopic character similar to the mean mantle composition) as much as $\sim 90\%$ of the mantle. The geochemical character of this relatively enriched reservoir may be satisfied by any number of scenarios involving variable fractions of HIMU, EM I, EM II, PREMA, PUM, and DMM. This relatively enriched reservoir could contain dispersed enriched components in the DMM matrix, differing from the MORB reservoir only in the relative abundance or representation of the enriched components, or it could represent a convectively isolated mantle domain.

Three factors tend to preclude derivation of definitive geochemical constraints on the placement of and physical relationships of the various geochemical components within the mantle: (a) Most observed isotopic compositions represent mixtures between two or more components; (b) studies of small seamounts, orogenic lherzolites, and ultramafic nodule suites document small-scale heterogeneities within the upper mantle; and (c) the number of components identified here (considered to be a minimum) makes it impossible to define uniquely mixing scenarios to explain isotope ratios observed in specific tectonic settings.

As geochemical philosophers, unaware of the various and sundry geophysical constraints on the structure of the mantle, we can objectively make and defend the following statements:

1. The MORB source reservoir, which is situated in the upper mantle, contains small-scale heterogeneities.
2. The MORB reservoir cannot constitute the whole mantle unless the mean age of the continental crust is on the order of 4.0 b.y.
3. The MORB source reservoir may either grade downward into mantle that is (on average) less depleted or abruptly contact such mantle across a midmantle chemical boundary layer. The possible existence of an enriched layer in the uppermost mantle, which overlies the MORB reservoir (Anderson 1982c), cannot be excluded.
4. Mantle of PUM composition cannot constitute more than half of the mantle. Some depleted mantle must be present below the 670-km discontinuity.
5. The mantle below the MORB reservoir may represent a mixture similar in style to the MORB reservoir, but with less DMM component, or it may be highly stratified with homogeneous or heterogeneous layers.
6. The Southern Hemisphere Dupal anomaly documents megascale lateral heterogeneity within the mantle.
7. Basalts with PREMA isotopic characteristics are found in a wide variety of oceanic, arc, and continental tectonic settings. These basalts must represent either the melting of a PREMA reservoir or a highly reproducible mixing between other components.
8. The constancy of $^3He/^4He$ ratios in MORBs likely results from a mixing between He in the various components that compose the MORB reservoir. High $^3He/^4He$ components may include PUM, EM I, and/or PREMA; low $^3He/^4He$ components may include DMM, EM II, and HIMU. The small range of $^3He/^4He$ ratios in MORB relative to the other isotope ratios of interest documents an enhanced mobility of He in the upper mantle, due either to high diffusion rates in the presence of melt or to migration of volatile-rich phases in the uppermost mantle.
9. The association of excess ^{129}Xe and high $^{40}Ar/^{36}Ar$ in some of the most depleted MORBS reflects the early (≥ 4.4 b.y.) and continuing degasssing of at least some portions of the MORB reservoir.
10. Mechanisms that account for the production of EM I, EM II, and HIMU components operate in conjunction with magmatism or metasomatism and associated tectonic activity within the mantle. Some or all of the proposed mechanisms may be operative in subduction environments, and thus some enriched mantle components (e.g. EM II) may incorporate or represent recycled crustal materials.

Together, these statements constitute a geochemist's view of the mantle, and without the benefit of geophysical contraints, they say disappointingly little regarding the physical structure of the mantle. We offer them as hopefully objective constraints that can be used to build mantle models in the context of the various geophysical constraints that are being developed. In the next section, we consider the interplay of these geochemical constraints with the geophysical constraints, as seen through the eyes of a geochemist.

The Nature of Mantle Convection

Perhaps the most fundamental question in solid-earth geophysics today pertains to the nature of the ~ 670-km seismic discontinuity (e.g. Dziewonski et al 1975, Dziewonski & Anderson 1981). Early workers (e.g. Anderson 1967, 1976, Ringwood 1975) attributed the discontinuity to a phase change in a homogeneous mantle. Recently, compelling arguments have been (re)made in support of a compositional boundary at 670 km. The existence of a chemical boundary at 670 km (with higher SiO_2 in the lower mantle) is expected to act as a barrier to convection and to limit mixing between upper- and lower-mantle reservoirs (e.g. Richter & McKenzie 1981, Olson & Yuen 1982), whereas convection across a phase boundary is likely permissible (e.g. Schubert & Turcotte 1971). Thus, resolution of this question is critical to the development of realistic models for the chemical and thermal evolution of the Earth.

Both whole-mantle and layered-mantle camps have argued persuasively for their respective mantle models. Unfortunately, individual authors have not always addressed apparently conflicting arguments made on the basis of similar observations by other authors. The current state of the art is, therefore, difficult to assess, particularly for the naive but well-intentioned geochemist. Hence, in the following paragraphs, we summarize the present state of geophysical arguments pertaining to the nature of the 670-km discontinuity.

UNDERSIDE REFLECTIONS   The high reflection coefficient inferred for the 670-km discontinuity on the basis of $P'670P'$ underside reflections (Richards 1972) suggests that the transition in physical properties at this boundary occurs over 5 km or less. Through comparison with calculated reflection coefficients, Lees et al (1983) have shown that observed reflectivities cannot be produced by phase transitions alone and require an associated discontinuous change in chemical composition. However, Muirhead (1985) has summarized arguments, largely from the works of J. R. Cleary, that permit an alternative interpretation for some of the $P'670P'$ data; this interpretation involves a core-mantle scattering phase. P. G. Richards (personal communication) acknowledges the viability of the

Muirhead-Cleary hypothesis for some of the "low quality" $P'670P'$ reflections, but he asserts that in other cases, the evidence for a midmantle reflector is incontrovertible.

The possibly discontinuous nature of the 670-km discontinuity is suggested by a comparison of two recent investigations of different subduction zones. Jordan (1977) and Creager & Jordan (1984) have compiled evidence for a high-velocity path extending to depths of at least 1000 km along an extension of the subducting slab in the Kuriles. These authors believe that this documents an aseismic portion of the slab itself that is extending into the lower mantle and contend that their initial results receive further support from their ongoing investigations (T. H. Jordan, personal communication). In contrast, Giardini & Woodhouse (1984) have shown that seismicity at Tonga documents the "en echelon" stacking of broken slab fragments, suggesting the impenetrable character of the 670-km discontinuity at this locality.

CONVECTIVE DESTRUCTION OF HETEROGENEITIES   The need to preserve chemical heterogeneities within the mantle has led a number of convection dynamicists to investigate the fate of such heterogeneities in a convecting mantle (e.g. Richter & Ribe 1979, Richter et al 1982, Olson et al 1984, Hoffman & McKenzie 1985). The results of these studies are in general accord (when heterogeneity and sampling scales are considered) and can be summarized as follows: (a) Large heterogeneities (similar in scale to the depth of the convecting layer) require very long time scales for dispersal and will likely persist for the age of the Earth; (b) small-scale heterogeneities will be mixed over "regional" volumes on time scales significantly shorter than the age of the Earth; and (c) total destruction of even the smallest scale heterogeneities depends on chemical diffusion. It is important to note that in all models, heterogeneities were considered to be passive (that is, indistinguishable from the ambient mantle matrix with regard to rheology and density). It is, therefore, difficult to generalize these results to a more realistic situation involving contrasting rheologies (such as subducted eclogite in a peridotite matrix). In such cases, the survival of heterogeneities may be fostered by the distinct rheologies, and in fact, density differences may produce a heterogeneity gradient with depth (e.g. Davies 1983). A further consequence of distinct chemical properties is a difference in solidus temperatures, which may facilitate sampling of even the smallest scale heterogeneities during melting (Zindler et al 1984, Sleep 1984). All in all, we see nothing in the modeling that precludes the long-term survival of heterogeneities within a convecting mantle domain, at least when very small-scale lengths are considered to be important. We therefore contend that convective isolation is not necessary to preserve ancient mantle heterogeneity.

PHYSICAL PROPERTIES OF MANTLE MINERALS   The elastic properties of mantle phases are being used to test the suitability of candidate bulk compositions to explain observed seismic velocities and inferred densities at various mantle levels (Lees et al 1983, Bass & Anderson 1984). Results from both of these studies agree that above 400 km, both olivine- and pyroxene-rich assemblages ("pyrolite" and "piclogite," respectively) can satisfy seismic constraints. While this may also be true for the lower mantle below 670 km, as recently pointed out by Wolf & Bukowinski (1985), D. L. Anderson (personal communication) and Knittle et al (1985) consider, respectively, that the relatively low *velocity* and low *density* of olivine in the lower mantle (perovskite plus magnesiowustite) may preclude an olivine-rich (e.g. pyrolite) lower mantle. Within the transition region, there are a number of problems: (*a*) The phases investigated by Lees et al (1983; olivine, orthopyroxene, and garnet) are too fast to satisfy PREM or PEM earth models (Dziewonski et al 1975, Dziewonski & Anderson 1981); and (*b*) the deeper phase transitions summarized by Jeanloz & Thompson (1983) and studied by Lees et al (1983) occur at substantially shallower depths than can be made consistent with a 670-km seismic discontinuity. Bass & Anderson (1984) and D. L. Anderson (personal communication) have suggested that these apparent enigmas may be explained if Ca-rich pyroxene persists in its low-pressure manifestation through much of the transition region and if increased alumina in the garnet and pyroxene component pushes the perovskite transition to higher pressures (via the ilmenite transition). The contentions of Bass & Anderson (1984), taken at face value, suggest that the transition region is rich in clinopyroxene, and therefore that piclogite is more appropriate than pyrolite as a bulk composition and that eclogite cannot penetrate the 670-km discontinuity because of its retarded transformation to perovskite (Bass & Anderson 1984, Anderson & Bass 1985). In contrast, Weidner (1985) has argued that a classic pyrolite composition is consistent with seismic data in and above the transition region.

Because cosmochemical constraints are seriously violated by a piclogite transition region *and* lower mantle, Anderson & Bass (1985) propose that the 670-km discontinuity represents a fundamental chemical boundary across which the piclogite transition region is in contact with an orthopyroxene-rich lower mantle. They suggest that this chemical discontinuity was formed as a result of extensive melting of the Earth during accretion, and that the enstatite-rich lower mantle is essentially devoid of incompatible trace elements. This leads, however, to absolute abundance levels of these elements in the upper mantle that are 2.5–3 times higher than the PUM value derived earlier. No upper-mantle materials are as yet known that both are this enriched in refractory lithophile elements and preserve chondritic abundance ratios. At present, the uncertainties associated with the basic physical properties data are significant and (as suggested by the transition region problems mentioned above) may be more important than is indicated by the experiments. Although a chemically layered mantle appears favored in current interpretations of the physical properties data, the data cannot preclude a more-or-less homogeneous composition (e.g. pyrolite or PUM) extending from the upper mantle, through the transition region, and into the lower mantle.

An interesting consequence of Anderson's hypothesized sub-670-km transformation of eclogite to perovskite, in conjunction with the density relationships shown by Lees et al (1983) and Bass & Anderson (1984), is that the transition region would act as a "density trap" for subducted eclogite. Having undergone the eclogite transformation, subducted oceanic crust will drop through the upper mantle into the transition region, where it becomes neutrally buoyant. Penetration into the lower mantle will be retarded by the alumina-induced late transformation to the perovskite structure (D. L. Anderson, personal communication). The production of eclogite through the generation and subduction of oceanic crust over the age of the Earth can account for as much as 12% of the mantle if the average rate of crust production has been about twice the present rate. The trapping of this eclogite in the transition region can satisfy the velocity constraints of the Bass & Anderson (1984) model without invoking fractional crystallization of an early magma ocean.

THERMAL STRUCTURE AND VISCOSITY OF THE MANTLE   Observations pertaining to postglacial rebound, nontidal acceleration, true polar wander, and changes in the degree-two component of the Earth's gravitational potential field all suggest that the viscosity of the mantle is constant (within about a factor of 5) across its entire thickness (Peltier & Andrews 1976, Peltier 1981, 1983). This condition is most easily met in the context of whole-mantle convection models. If viscosity is proportional to homologous temperature (T_m/T; e.g. Schubert & Spohn 1981, Spohn & Schubert 1982), constant viscosity across a midmantle thermal boundary layer implies that the melting temperatures (T_m) of the lower- and upper-mantle materials would have to fortuitously differ by an amount comparable to the temperature increase across the boundary layer ($\sim 500–700°C$; Jeanloz & Richter 1979). However, if we accept only a mantle with nearly constant viscosity, only *one* of the following conditions need hold: (*a*) There is no thermal boundary layer at 670 km; (b) the change in chemical properties associated with a chemical discontinuity at 670 km fortuitously results in a nearly constant homologous temperature across the boundary; or (c) the relationship between the homologous temperature and viscosity is not as simple as has been supposed. Although the contention that the mantle is isoviscous has been taken as support for whole-mantle convection (e.g. Peltier 1981, Spohn & Schubert 1982), this requirement is not absolute, since there are clearly other alternatives.

Assuming a constant homologous temperature throughout the mantle, Spohn & Schubert (1982) have further argued that the temperature increase across a midmantle thermal boundary layer predicts core-mantle temperatures significantly in excess of the 3200–3500 K estimated by Jeanloz & Richter (1979). R. Jeanloz (pesonal communication) has made preliminary measurements of the melting temperature of perovskite as a function of pressure that suggest little or no pressure dependence of T_m. In their calculation, Spohn & Schubert (1982) assumed a melting temperature gradient of $0.64°C$ km^{-1} for the lower mantle, and thus the new data of Jeanloz may be qualitatively consistent with an isoviscous lower mantle and reasonable core-mantle temperatures, even in the presence of a thermal boundary layer at 670 km. Essentially, this result predicts that the geotherm proposed by Jeanloz & Richter (1979) may be reconcilable with an isoviscous lower mantle.

MANTLE HEAT FLOW AND THE UREY RATIO   The ratio of heat flow to heat production in the Earth (the so-called Urey ratio) can potentially constrain the nature of mantle convection by offering a measure of the efficiency of heat removal from the mantle (Richter 1984). Using heat flow data from Sclater et al (1980) and estimates for PUM U (20.8 ppb; Table 1), with Th/U = 3.8 and K/U = 1.27×10^4, we obtain a Urey ratio of approximately 2.1. Furthermore, if we eliminate that fraction of the total Earth heat flow ($\sim 15\%$) and heat production ($\sim 37\%$) thought to originate in the continental crust, where thermal steady state is closely approached, a "reduced" Urey ratio for mantle heat flow to heat production is obtained of about 2.8. This suggests that mantle heat flow requires a significant component ($\sim 64\%$) of secular cooling. As discussed by Richter (1984), predictions from whole-mantle convection calculations suggest $< 20\%$ secular cooling (given no significant heat source within, or cooling of, the core) due to the efficiency of heat removal in these models. Layered models, on the other hand, are capable of accounting for much larger percentages of secular cooling ($\sim 50\%$). Typical convection calculations, however, ignore the insulating effects of continental lithosphere, a consideration that will tend to increase the secular component available in whole- and layered-mantle models. Even so, taken at face value, these arguments suggest either a layered mantle or a significant heat source in the core, given that no more than about 5–10% of the present-day heat flow can result from cooling of the core (Stacey 1969). Speculation regarding the nature and magnitude of possible core heat sources is not fruitful at this time other than to say that K has long been considered a possibility (Hall & Murthy 1971, Lewis 1971), and that more than about one silicate Earth budget of K in the core would violate cosmochemical constraints related to K depletion during accretion of the Earth (Sun 1982). One silicate Earth budget of K in the core would account for 15% of the present total Earth heat flow (at steady state), or up to 60% of the total heat flow if the core heat is fossil production from a few billion years ago. The above analysis assumes that the heat flow estimates of Sclater et al (1980) are representative of time-integrated values. If the rate of present-day volcanic activity at midocean ridges, and associated heat flow, is higher than the time-integrated average, then the appropriate mantle Urey ratio would be smaller than that calculated here. Further, F. M. Richter, W. R. Peltier & D. L. Turcotte (personal communication) all concur that when realistic uncertainties in the convection calculations are considered, the possibility of an extensive secular component in whole-mantle convection models cannot be excluded.

O'Nions & Oxburgh (1983) have argued that the relationship between the oceanic ^{4}He flux (Craig et al 1975) and heat flow supports the contention of a depleted upper mantle that is convectively isolated from a primitive lower mantle. Implicit in their reasoning are the assumptions that ^{4}He and heat are removed from the upper mantle with the same efficiency (i.e. that the entire oceanic crustal section, including both extrusive and intrusive components, is outgassed completely) and that radiogenic ^{4}He has a very short residence time in the upper mantle. This is in marked contrast to their contention that heat is transported across the 670-km boundary layer much more efficiently than is He. In principle, the He flux and heat flow data could be used to calculate a "fractionation factor" for heat and He at the ridge.

He arguments aside, the conclusions of O'Nions & Oxburgh (1983) are essentially based on the inferred depletion of U in the MORB source reservoir (a fact easily documented by examination of U in fresh MORBs) in conjunction with convection arguments similar to those summarized by Richter (1984). In this context, we note that the U budget of O'Nions & Oxburgh (1983) ($\sim$ 1.5 ppm U in the crust and 20 ppb U in the BSE) has about 47% of the Earth's U concentrated in the continental crust, and thus it precludes total convective isolation of a primitive lower mantle (since the mantle above 670 km constitutes only $\sim$ 30% of the total mantle).

DISCUSSION   The geophysical arguments summarized above do not clearly support either the whole-mantle or layered-mantle convection hypotheses; at some level, all observations could be reconciled with either hypothesis, although some special begging would be required. Based on the geochemical arguments presented in the preceding section, we contend that if the lower mantle is chemically distinct and convectively isolated from the upper mantle, it does not have a primitive or undifferentiated PUM (or BSE) composition. Thus, if there is an unbreachable chemical boundary layer at 670 km at present, it could not have existed for all of Earth history. At some time, this boundary must have been permeable in order to permit differentiation (depletion) of the lower mantle. We also know that the MORB source reservoir, which probably composes some portion of the upper mantle, cannot represent the mean composition of the mantle (see Figure 26), and therefore the present-day mantle must contain large-scale chemical gradients or discontinuities (though, as discussed earlier, these heterogeneities are largely documented by trace elements and isotopic ratios and need not imply significant major-element variations).

A compromise model that is compatible with many geochemical and geophysical observations might involve a chemical boundary layer at 670 km that is subject to periodic instability in time and/or space [similar to the model proposed by Herzberg (1983) and Herzberg & Forsythe (1983)]. Such a model offers a means to achieve some differentiation of the lower mantle, while at the same time providing for the current existence of a chemical boundary or gradient at 670 km. In fact, material exchange across the 670-km discontinuity may be necessary in order to effect the chemical differentiation between the upper and lower mantle that is necessary to stabilize a convective boundary layer. Heat and material flux across the boundary will serve to "buffer" the mean mantle temperature distribution near an adiabat and thus satisfy the viscosity predictions of Peltier (1981) and the thermal constraints discussed by Spohn & Schubert (1982). The high reflectivity of some of the $P'670P'$ reflections would then document stable portions of the boundary at a given time and place. We stress that with regard to geochemical budgets and mass balances, this compromise model has many of the same ramifications as whole-mantle convection.

5.   SUMMARY AND CONCLUSIONS

Chemical geodynamics, the study of the physical and chemical evolution of the Earth (Allegre 1982), is a field of inquiry that is clearly still in its infancy. However, great strides have been made over the past decade as solid-earth geochemists and geophysicists have begun more and more to work together in this quest. Although the present state of the art does not permit the absolute definition of broad and sweeping constraints on the physical and chemical structure of the Earth, we believe that the outlook for the near future is indeed rosy. We are currently on the steep part of a learning curve, and answers to many of the most intriguing problems in the geosciences appear to be almost within grasp.

The synthesis presented here explores three principal boundary conditions relating to the nature and development of chemical structure in the Earth's mantle: (a) inferred scale lengths for mantle chemical heterogeneities, (b) interrelationships of the various isotopic tracers, and (c) the bulk composition of the Earth. These boundary conditions are integrated with geophysical constraints in an attempt to evaluate models for the development of the physical and chemical structure of the mantle. The major points of the synthesis are the following:

1. Kilometer-size heterogeneities can survive diffusive equilibration for billions of years, even in the presence of a melt phase. The situation for the rare gases is not drastically different than that for other elements of interest.
2. The mantle is chemically heterogeneous on both very small (10 m) and very large (> 1000 km) scales. There is a clear positive correlation between the amplitudes of isotopic variations and the maximum scale lengths of observation.
3. Isotopic heterogeneities in the mantle require the existence of four "end-member" components (DMM, HIMU, EM I, and EM II) and are consistent with the existence of at least two additional components (BSE, PREMA).
4. Primitive undepleted mantle (BSE or PUM) can make up no more than about 55% of the total mantle, and convective isolation of such material is therefore precluded.
5. He isotope data cannot be simply interpreted in terms of the mantle components delineated by Sr, Nd, and Pb. Preeruptive degassing of He may explain high observed U/He ratios; consequent ingrowth of radiogenic ^{4}He may significantly perturb mantle He signals.
6. The composition of the primitive upper mantle (PUM) can be ascertained via comparison of mantle-derived lherzolites and chondrites using a procedure modified after Jagoutz et al (1979). Resulting abundances of lithophile refractory elements in PUM are constrained to better than $\pm 5\%$, and are equal to $2.51 \times C1$ chondrites (Anders & Ebihara 1982).
7. The Pb paradox is shown to be essentially nonparadoxical. Numerous

low-μ reservoirs are available to balance the ubiquitous high-μ oceanic basalt data.

8. The U content of the BSE is tightly constrained to be ~ 21 ppb. The fraction of this that now resides in the continental crust is less well constrained, but it lies between 22–52%. The reduced mantle Urey ratio (crustal heat flow and heat production subtracted) derived from these values is ~ 2.8; this high value suggests either a poor thermal efficiency for mantle convection or a significant component of heat deriving from the core.

ACKNOWLEDGMENTS

This work contains a myriad of old and new ideas that we have attempted to weave into a coherent self-consistent view of the Earth's mantle. We have endeavored to accurately cite appropriate references for old ideas, and we apologize in advance for where we have failed. The development of our new ideas has been immeasurably aided by conversations with many friends and colleagues. While we cannot hope to name them all, there are those whom we cannot fail to mention: Dave Graham, Don Anderson, Claude Allegre, Bill White, John Ludden, Jim Rubenstone, Charlie Langmuir, Ed Stolper, Jill Pasteris, Hubert Staudigel, Gerd Worner, Emil Jagoutz, Steve Goldstein, Laurie Reisberg, Ray Jeanloz, Frank Richter, Don Turcotte, and Dick Peltier. We would also like to thank Mark Kurz, Don Anderson, Ray Jeanloz, Claude Allegre, Fred Frey, Roger Hart, Dave Graham, and Gerd Worner for furnishing helpful preprints and unpublished data; Claude Allegre for his early role as a silent partner; and Emil Jagoutz for inspiring "PUM." AZ would like to thank the participants in his recent mantle geochemistry seminar for acting as a sounding board for earlier manifestations of much of this material, and to assure them that the learning process was always a two-way street. SRH acknowledges the denizens of the eleventh floor for abiding his absence during the writing of this manuscript, and MIT for paying him to stay home for a year. This manuscript was skillfully produced by Michelle Lezie, who suffered through untold versions while thoroughly crash-testing Proofwriter version 2.09, and Joan Totton, who provided timely help with the final staging. As usual, Patty Catanzaro did a superb job with the graphics, under no small amount of pressure from us. And finally, the work we present would not have been possible without the generous support of the National Science Foundation (grants EAR 82-11461, EAR 83-20066, and OCE 84-10615 to AZ, and EAR 83-08809 to SRH). This work represents Lamont-Doherty Geological Observatory Contribution No. 3941.

Literature Cited

Ahrens, L. H., Von Michaelis, H. 1969. The composition of stony meteorites. III. Some inter-element relationships. *Earth Planet. Sci. Lett.* 5: 395–400

Allegre, C. J. 1982. Chemical geodynamics. *Tectonophysics* 81: 109–32

Allegre, C. J., Rousseau, D. 1984. The growth of the continents through geological time studied by Nd isotope analysis of shales. *Earth Planet. Sci. Lett.* 67: 19–34

Allegre, C. J., Turcotte, D. L. 1985. Geodynamic mixing in the mesosphere boundary layer and the origin of oceanic islands. *Geophys. Res. Lett.* 12: 207–10

Allegre, C. J., Othman, D. B., Polve, M., Richard, P. 1979. The Nd-Sr isotopic correlation in mantle materials and geodynamic consequences. *Phys. Earth Planet. Inter.* 19: 293–306

Allegre, C. J., Brevart, O., Dupre, B., Minster, J.-F. 1980. Isotopic and chemical effects produced in a continuously differentiating convecting Earth mantle. *Philos. Trans. R. Soc. London Ser. A* 297: 447–77

Allegre, C. J., Staudacher, T., Sarda, P., Kurz, M. 1983a. Constraints on evolution of Earth's mantle from rare gas systematics. *Nature* 303: 762–66

Allegre, C. J., Dupre, B., Hamelin, B. 1983b. Geochemistry of oceanic ridge basalts explained by blob–upper mantle mixing, *Eos, Trans. Am. Geophys. Union* 64: 324

Allegre, C. J., Hart, S. R., Minster, J.-F. 1983c. Chemical structure and evolution of the mantle and continents determined by inversion of Nd and Sr isotopic data. II. Numerical experiments and discussion. *Earth Planet. Sci. Lett.* 66: 191–213

Allegre, C. J., Hamelin, B., Dupre, B. 1984. Statistical analysis of isotopic ratios in MORB: the mantle blob cluster model and the convective regime of the mantle. *Earth Planet. Sci. Lett.* 71: 71–84

Allegre, C. J., Staudacher, T., Sarda, P. 1986. Rare gas systematics: formation of the atmosphere, evolution and structure of the Earth's mantle. *Earth Planet. Sci. Lett.* In press

Anders, E., Ebihara, M. 1982. Solar system abundances of the elements. *Geochim. Cosmochim. Acta* 46: 2363–80

Anderson, D. L. 1967. Phase changes in the upper mantle. *Geophys. J. R. Astron. Soc.* 14: 135–64

Anderson, D. L. 1976. The 650-km mantle discontinuity. *Geophys. Res. Lett.* 5: 347–49

Anderson, D. L. 1979. Chemical stratification of the mantle. *J. Geophys. Res.* 84: 6297–98

Anderson, D. L. 1982a. Hotspots, polar wander, Mesozoic convection and the geoid. *Nature* 297: 391–93

Anderson, D. L. 1982b. Isotopic evolution of the mantle: the role of magma mixing. *Earth Planet. Sci. Lett.* 57: 1–12

Anderson, D. L. 1982c. Isotopic evolution of the mantle: a model. *Earth Planet. Sci. Lett.* 57: 13–24

Anderson, D. L. 1983. Chemical composition of the mantle. *J. Geophys. Res.* 88: B41–B52 (*Proc. Lunar Planet. Conf., 14th*)

Anderson, D. L. 1985. Hotspot magmas can form by fractionation and contamination of MORB. *Nature.* In press

Anderson, D. L., Bass, J. D. 1985. The transition region of the Earth's upper mantle. *Nature.* In press

Armstrong, R. L. 1981. Radiogenic isotopes: the case for crustal recycling on a near-steady-state no-continental growth Earth. *Philos. Trans. R. Soc. London Ser A* 301: 443–72

Bass, J. D., Anderson, D. L. 1984. Composition of the upper mantle: geophysical tests of two petrological models. *Geophys. Res. Lett.* 11: 237–40

Basu, A. R., Tatsumoto, M. 1979. Samarium-neodymium systematics in kimberlites and in the minerals of garnet lherzolite inclusions. *Science* 205: 398–401

Basu, A. R., Tatsumoto, M. 1980. Nd-isotopes in selected mantle-derived rocks and minerals and their implications for mantle evolution. *Contrib. Mineral. Petrol.* 75: 43–54

Batiza, R., Vanko, D. 1984. Petrology of young Pacific seamounts. *J. Geophys. Res.* 89: 11235–60

Birch, F., Schairer, J. F., Spicer, H. C. 1942. Handbook of physical constants. *Geol. Soc. Am. Spec. Pap.* 36. 325 pp.

Boynton, W. V. 1975. Fractionation in the solar nebula: condensation of yttrium and the rare earth elements. *Geochim. Cosmochim. Acta* 39: 569–84

Brueckner, H., Zindler, A., Petrini, R., Otonello, G., Bonatti, E. 1986. Nd and Sr isotope systematics of Zabargad Island, Red Sea. Submitted for publication

Carlson, R. W. 1984. Isotopic constraints on Columbia River flood basalt genesis and the nature of the subcontinental mantle. *Geochim. Cosmochim. Acta* 48: 2357–72

Carter, S. R., Evensen, N. M., Hamilton, P. J., O'Nions, R. K. 1978. Continental volcanics derived from enriched and depleted source regions: Nd and Sr isotope evidence. *Earth Planet. Sci. Lett.* 37: 401–8

Chase, C. G. 1981. Ocean island Pb: two-stage histories and mantle evolution. *Earth Planet. Sci. Lett.* 52: 277–84

Chen, J. H., Tilton, G. R. 1976. Isotopic lead

investigations on the Allende carbonaceous chondrite. *Geochim. Cosmochim. Acta* 40: 635–43

Cohen, R. S., O'Nions, R. K. 1982a. The lead, neodymium and strontium isotopic structure of ocean ridge basalts. *J. Petrol.* 23: 299–324

Cohen, R. S., O'Nions, R. K. 1982b. Identification of recycled continental material in the mantle from Sr, Nd and Pb isotope investigations. *Earth Planet. Sci. Lett.* 61: 73–84

Cohen, R. S., Evensen, N. M., Hamilton, P. J., O'Nions, R. K. 1980. U-Pb, Sm-Nd and Rb-Sr systematics of mid-ocean ridge basalt glasses. *Nature* 283: 149–53

Cohen, R. S., O'Nions, R. K., Dawson, J. B. 1984. Isotope geochemistry of xenoliths from East Africa: implications for development of mantle reservoirs and their interaction. *Earth Planet. Sci. Lett.* 68: 209–20.

Condomines, M., Gronvold, K., Hooker, P. J., Muehlenbachs, K., O'Nions, R. K., et al. 1983. Helium, oxygen, strontium and neodymium isotopic relationships in Icelandic volcanics. *Earth Planet. Sci. Lett.* 66: 125–36

Craig, H., Lupton, J. E. 1978. Helium isotope variations: evidence for mantle plumes at Yellowstone, Kilauea and the Ethiopian Rift Valley. *Eos, Trans. Am. Geophys. Union* 59: 1194 (Abstr.)

Craig, H., Rison, W. 1982. Helium 3: Indian Ocean hotspots and the East African Rift. *Eos, Trans. Am. Geophys. Union* 63: 1144 (Abstr.)

Craig, H., Rison, W. 1983. Helium isotopes and mantle heterogeneity. *Eos, Trans. Am. Geophys. Union* 64: 348

Craig, H., Clarke, W. B., Beg, M. S. 1975. Excess ³He in deep water on the East Pacific Rise. *Earth Planet Sci. Lett.* 26: 125–32

Creager, K. C., Jordan, T. H. 1984. Slab penetration into the lower mantle. *J. Geophys. Res.* 89: 3031–50

Dasch, E. J., Green, D. H. 1975. Strontium isotope geochemistry of lherzolite inclusions and host basaltic rocks, Victoria, Australia. *Am. J. Sci.* 275: 461–69

Davidson, J. P. 1983. Lesser Antilles isotopic evidence of the role of subducted sediment in island arc magma genesis. *Nature* 306: 253–56

Davies, G. F. 1983. Viscosity structure of layered convecting mantle. *Nature* 301: 592–94

Delaney, J. R., Muenow, D. W., Graham, D. G. 1978. Abundance and distribution of water, carbon and sulfur in the glassy rims of submarine pillow basalts. *Geochim. Cosmochim. Acta* 42: 581–94

DePaolo, D. J. 1980. Crustal growth and mantle evolution: inferences from models of element transport and Nd and Sr isotopes. *Geochim. Cosmochim. Acta* 44: 1185–96

DePaolo, D. J. 1981. Nd isotopic studies: some new perspectives on Earth structure and evolution. *Eos, Trans. Am. Geophys. Union* 62: 137–140

DePaolo, D. J., Wasserburg, G. J. 1976a. Inferences about magma sources and mantle structure from variations of ¹⁴³Nd/¹⁴⁴Nd. *Geophys. Res. Lett.* 3: 743–46

DePaolo, D. J., Wasserburg, G. J. 1976b. Nd isotopic variations and petrogenetic models. *Geophys. Res. Lett.* 3: 249–52

DePaolo, D. J., Wasserburg, G. J. 1977. The sources of island arcs as indicated by Nd and Sr isotopic studies. *Geophys. Res. Lett.* 4: 465–68

Dupre, B., Allegre, C. J. 1983. Pb-Sr isotope variation in Indian Ocean basalts and mixing phenomena. *Nature* 303: 142–46

Dupuy, C., Leyreloup, A., Vernierest, J. 1979. The lower continental crust of the massif central (Bournac, France)—with special references to REE, U and Th composition, evolution, heat-flow production. In *Origin and Distribution of the Elements, Physics and Chemistry of the Earth*, Vol. II, ed. L. H. Aherns, New York: Pergamon

Dziewonski, A., Anderson, D. L. 1981. Preliminary reference Earth models. *Phys. Earth Planet. Inter.* 25: 297–356

Dziewonski, A. M., Hales, A. L., Lapwood, E. R. 1975. Parametrically simple earth models consistent with geophysical data. *Phys. Earth Planet. Inter.* 10: 12–48

Fahrig, W. F. Eade, K. E. 1968. The chemical evolution of the Canadian shield, *Can. J. Earth Sci.* 5: 1247–52

Flemings, M. C. 1974. *Solidification Processing.* New York: McGraw-Hill.

Frey, F. A., Green, D. H. 1974. The mineralogy, geochemistry and origin of lherzolite inclusions in Victorian basanites. *Geochim. Cosmochim. Acta* 38: 1023–59

Frey, F. A., Suen, C.-Y. J. 1985. The Ronda high temperature peridotite: geochemistry and petrogenesis. *Geochim. Cosmochim. Acta* 49: 2469–91

Gancarz, A. J., Wasserburg, G. J. 1977. Initial Pb of the Amitsoq gneiss, West Greenland, and implications for the age of the Earth. *Geochim. Cosmochim. Acta* 41: 1283–1301.

Giardini, D., Woodhouse, J. H. 1984. Deep seismicity and modes of deformation in Tonga subduction zone. *Nature* 307: 505–9

Goldstein, S., Zindler, A., Jagoutz, E. 1982. Evolution of the mantle-crust system: quantitative geochemical modelling of

Pb, Sr, and Nd isotopic systematics. *Eos, Trans. Am. Geophys. Union* 63 : 460 (Abstr.)

Goldstein, S. L., O'Nions, R. K., Hamilton, P. J. 1984. A Sm-Nd isotopic study of atmospheric dusts and particulates from major river systems. *Earth Planet. Sci. Lett.* 70 : 221–36

Graham, D., Zindler, A., Reisberg, L., Kurz, M. D., Jenkins, W. J., Batiza, R. 1984. He, Sr and Nd isotopes in basaltic glasses from young Pacific seamounts. *Eos, Trans. Am. Geophys. Union* 65 : 1079 (Abstr.)

Graham, D., Zindler, A., Reisberg, L., Kurz, M., Jenkins, W., Batiza, R. 1986. He, Sr, and Nd isotope geochemistry of young Pacific seamounts. Submitted for publication.

Gray, C. M. 1977. The geochemistry of central Australian granulites in relation to the chemical and isotopic effects of granulite facies metamorphism, *Contrib. Mineral. Petrol.* 65 : 79–89

Gray, C. M., Oversby, V. M. 1972. The behaviour of lead isotopes during granulite facies metamorphism. *Geochim. Cosmochim. Acta* 36 : 939–52

Greenland, L. P., Rose, W. I., Stokes, J. B. 1985. An estimate of gas emissions and magmatic gas content from Kilauea volcano. *Geochim. Cosmochim. Acta* 49 : 125–29

Grossman, L., Larimer, J. W. 1974. Early chemical history of the solar system. *Rev. Geophys. Space Phys.* 12 : 71–101

Hall, H. T., Murthy, V. R. 1971. The early thermal history of the Earth. *Phys. Earth Planet. Inter.* 2 : 19–29

Hamelin, B., Allegre, C. J. 1985a. Lead isotopic composition of high-temperature peridotites from Lherz, Lanzo, Beni-Bousera and the genesis of isotopic heterogeneities in the earth's mantle. *Eos, Trans. Am. Geophys. Union* 46 : 1114

Hamelin, B., Allegre, C. J. 1985b. Large scale regional units within the depleted upper mantle : Pb-Sr-Nd study of the south-west Indian ridge. *Earth Planet. Sci. Lett.* In press

Hamilton, P. J., Evensen, N. M., O'Nions, R. K., Tarney, J. 1979. Sm-Nd systematics of Lewisian gneisses : implications for the origin of granulites. *Nature* 277 : 25–28

Hart, R., Hogan, L., Dymond, J. 1985. The closed system approximation for evolution of argon and helium in the mantle, crust, and atmosphere. *Isotope Geoscience* 52 : 45–73

Hart, R., Hogan, L., Dymond, J. 1986. Archaean degassing of the upper mantle and formation of the atmosphere. *Earth Planet. Sci. Lett.* In press

Hart, S. R. 1980. Diffusion compensation in natural silicates. *Geochim. Cosmochim. Acta* 45 : 279–91

Hart, S. R. 1984a. He diffusion in olivine. *Earth Planet. Sci. Lett.* 70 : 297–302

Hart, S. R. 1984b. A large-scale isotope anomaly in the Southern Hemisphere mantle. *Nature* 309 : 753–57

Hart, S. R., Zindler, A. 1986. In search of a bulk Earth composition. *Chem. Geol.* In press

Hawkesworth, C. J., Vollmer, R. 1979. Crustal contamination versus enriched mantle : $^{143}Nd/^{144}Nd$ and $^{87}Sr/^{86}Sr$ evidence from the Italian volcanics. *Contrib. Mineral. Petrol.* 69 : 151–69

Hawkesworth, C. J., Erlank, A. J., Marsh, J. S., Menzies, M. A., Van Calsteren, P. 1983. Evolution of the continental lithosphere : evidence from volcanics and xenoliths in southern Africa. In *Continental Basalts and Mantle Xenoliths*, ed. C. J. Hawkesworth, M. J. Norry, pp. 111–38. Cheshire, Engl : Shiva

Hawkesworth, C. J., Rogers, N. W., van Calsteren, P. W. C., Menzies, M. A. 1984. Mantle enrichment processes. *Nature* 311 : 331–35

Heinrichs, H., Schulz-Dobrick, B., Wedepohl, K. H. 1980. Terrestrial geochemistry of Cd, Bi, Tl, Pb, Zn and Rb. *Geochim. Cosmochim. Acta* 44 : 1519–33

Herzberg, C. T. 1983. Chemical stratification in the silicate Earth. *Earth Planet. Sci. Lett.* 67 : 249–60

Herzberg, C. T., Forsythe, R. D. 1983. Destabilization of a 650-km chemical boundary layer and its bearing on the evolution of the continental crust. *Phys. Earth Planet. Inter.* 32 : 352–60

Hoffman, N. R. A., McKenzie, D. P. 1985. The destruction of geochemical heterogeneities by differential fluid motions during mantle convection. *Geophys. J. R. Astron. Soc.* 82 : 163–206

Hofmann, A. W., Hart, S. R. 1978. An assessment of local and regional isotopic equilibrium in the mantle. *Earth Planet. Sci. Lett.* 38 : 44–62

Hofmann, A. W., Magaritz, M. 1977. Equilibrium and mixing in a partially molten mantle. In *Magma Genesis, Bull.* 96, ed. H. J. B. Dick, pp. 37–42. Portland : Oreg. Dep. Geol. Miner. Ind.

Hofmann, A. W., White, W. M. 1982. Mantle plumes from ancient oceanic crust. *Earth Planet. Sci. Lett.* 57 : 421–36

Jacobsen, S. B., Wasserburg, G. J. 1979. The mean age of mantle and crustal reservoirs. *J. Geophys. Res.* 84 : 7411–27

Jacobsen, S. B., Wasserburg, G. J. 1984. Sm-Nd isotopic evolution of chondrites and achondrites, II. *Earth Planet. Sci. Lett.* 67 : 137–50

Jagoutz, E., Palme, H., Badenhausen, H., Blum, K., Cendales, M., et al. 1979. The abundances of major, minor and trace elements in the Earth's mantle as derived from primitive ultramafic nodules. *Proc. Lunar Planet. Sci. Conf., 10th*, pp. 2031–50

Jagoutz, E., Carlson, R. W., Lugmair, G. W. 1980. Equilibrated Nd–unequilibrated Sr isotopes in mantle xenoliths. *Nature* 286: 708–10

Jeanloz, R. 1986. High pressure chemistry of the Earth's mantle and core. In *Mantle Convection*, ed. W. R. Peltier. New York: Gordon & Breach. In press

Jeanloz, R., Richter, F. M. 1979. Convection, composition, and the thermal state of the lower mantle. *J. Geophys. Res.* 84: 5497–5504

Jeanloz, R., Thompson, A. B. 1983. Phase transitions and mantle discontinuities. *Rev. Geophys. Space Phys.* 21: 51–74

Jeffrey, P. M., Anders, E. 1970. Primordial noble gases in separated meteoritic minerals—I. *Geochim. Cosmochim. Acta* 34: 1175–98

Jochum, K. P., Hofmann, A. W., Ito, E., Seufert, H. M., White, W. M. 1983. K, U, and Th in mid-ocean ridge basalt glasses. The terrestrial K/U and K/Rb ratios and heat production in the depleted mantle. *Nature* 306: 431–36

Jordan, T. H. 1977. Lithospheric slab penetration into the lower mantle beneath the Sea of Okhotsk, *J. Geophys.* 43: 473–96

Khitarov, N. I., Kadik, A. A. 1973. Water and carbon dioxide in magmatic melts and peculiarities of the melting process. *Contrib. Mineral. Petrol.* 41: 205–15

Knittle, E., Jeanloz, R., Smith, G. L. 1985. The thermal expansion of silicate perovskite and stratification of the Earth's mantle. *Nature* 319: 214–16

Kramers, J. D., Smith, C. B., Lock, N. P., Harmon, R. S., Boyd, F. R. 1981. Can kimberlites be generated from an ordinary mantle? *Nature* 291: 53–56

Kramers, J. D., Roddick, J. C. M., Dawson, J. B. 1983. Trace element and isotope studies on veined, metasomatic and "MARID" xenoliths from Bultfontein, South Africa. *Earth Planet. Sci. Lett.* 65: 90–106

Kurz, M. D. 1982. *Helium isotope geochemistry of oceanic volcanic rocks: implications for mantle heterogeneity and degassing.* PhD thesis. Mass. Inst. Technol./Woods Hole Oceanogr. Inst., Cambridge/Woods Hole, Mass. 281 pp.

Kurz, M. D., Jenkins, W. J. 1981. The distribution of helium in oceanic basalt glasses. *Earth Planet. Sci. Lett.* 53: 41–54

Kurz, M. D., Jenkins, W. J., Schilling, J. G., Hart, S. R. 1982. Helium isotopic systematics of ocean islands and mantle heterogeneity. *Nature* 297: 43–47

Kurz, M. D., Jenkins, W. J., Hart, S. R.,

Clague, D. 1983. Helium isotopic variations in volcanic rocks from Loihi Seamount and the Island of Hawaii. *Earth Planet. Sci. Lett.* 66: 388–406

Kurz, M. D., Meyer, P. S., Sigurdsson, H. 1985. Helium isotopic systematics within the neovolcanic zones of Iceland. *Earth Planet. Sci. Lett.* 74: 291–305

Langmuir, C. H., Vocke, R. D., Hanson, G. N. 1978. A general mixing equation with applications to Icelandic basalts. *Earth Planet. Sci. Lett.* 37: 380–92

Larimer, J. W. 1979. The condensation and fractionation of refractory lithophile elements. *Icarus* 40: 446–54

Lees, A. C., Bukowinski, M. S. T., Jeanloz, R. 1983. Reflection properties of phase transition and compositional change models of the 670-km discontinuity. *J. Geophys. Res.* 88: 8145–59

leRoex, A. P., Dick, H. J. B., Erlank, A. J., Reid, A. M., Frey, F. A., Hart, S. R. 1983. Geochemistry, mineralogy and petrogenesis of lavas erupted along the Southwest Indian Ridge between the Bouvet Triple Junction and 11 degrees east. *J. Petrol.* 24: 267–318

Lewis, J. S. 1971. Consequences of the presence of sulfur in the core of the Earth. *Earth Planet. Sci. Lett.* 11: 130–34

Loubet, M., Shimizu, N., Allegre, C. J. 1975. Rare earth elements in Alpine peridotites. *Contrib. Mineral. Petrol.* 53: 1–12

Lupton, J. E., Craig, H. 1975. Excess ^{3}He in oceanic basalts: evidence for terrestrial primordial helium. *Earth Planet. Sci. Lett.* 26: 133–39

Manhes, G., Allegre, C. J. 1978. Time differences as determined from the ratio of lead 207–206 in concordant meteorites. *Meteoritics* 13: 543–48

Manhes, G., Allegre, C. J., Dupre, B., Hamelin, B. 1979. Lead-lead systematics, the "age of the Earth" and the chemical evolution of our planet in a new representation space. *Earth Planet. Sci. Lett.* 44: 91–104

Manson, V. 1967. Geochemistry of basaltic rock: major elements. In *Basalts*, ed. H. H. Hess, A. Poldervaart, pp. 215–69. New York: Interscience

McKenzie, D., O'Nions, R. K. 1983. Mantle reservoirs and ocean island basalts. *Nature* 301: 229–31.

McLennan, S. M., Taylor, S. R. 1980. Th and U in sedimentary rocks: crustal evolution and sedimentary recycling. *Nature* 285: 621–24

Melson, W. G., Vallier, T. L., Wright, T. L., Byerly, G., Nelen, J. 1976. Chemical diversity of abyssal volcanic glass erupted along Pacific, Atlantic, and Indian ocean seafloor spreading centers. In *The Geophysics* of the Pacific Ocean Basin and its Margin, Geophys. Monogr., ed. G. H. Sutton, M. H. Manghnani, R. Moberly, pp. 351–68. Washington DC: Am. Geophys. Union

Menzies, M. 1983. Mantle ultramafic xenoliths in alkaline magmas: evidence for mantle heterogeneity modified by magmatic activity. In Continental Basalts and Mantle Xenoliths, ed. C. J. Hawkesworth, M. J. Norry, pp. 92–110. Cheshire, Engl: Shiva. 272 pp.

Menzies, M., Murthy, V. R. 1980a. Enriched mantle: Nd and Sr isotopes in diopsides from kimberlite nodules. Nature 283: 634–36

Menzies, M., Murthy, V. R. 1980b. Nd and Sr isotope geochemistry of hydrous mantle nodules and their host alkali basalts: implications for local heterogeneities in metasomatically veined mantle. Earth Planet. Sci. Lett. 46: 323–34

Menzies, M. A., Wass, S. Y. 1983. CO_2- and LREE-rich mantle below eastern Australia: a REE and isotopic study of alkaline magmas and apatite-rich mantle xenoliths from the Southern Highlands Province, Australia. Earth Planet. Sci. Lett. 65: 287–302

Michard, A., Gurriet, P., Soudant, M., Albarede, F. 1985. Nd isotopes in French Phanerozoic shales: external vs. internal aspects of crustal evolution. Geochim. Cosmochim. Acta 49: 601–10

Montgomery, C. P. W. 1977. Uranium-lead isotopic investigation of the Archean Imataca complex, Guayana shield, Venezuela. PhD thesis, Mass. Inst. Technol. 261 pp.

Morgan, W. J. 1971. Convection plumes in the lower mantle. Nature 230: 42–43

Morris, J. D. 1984. Enriched geochemical signatures in Aleutian and Indonesian arc lavas: an isotopic and trace element investigation. PhD thesis. Mass. Inst. Technol., Cambridge. 320 pp.

Morris, J. D., Hart, S. R. 1983. Isotopic and incompatible element constraints on the genesis of island arc volcanics from Cold Bay and Amak Island, Aleutians, and implications for mantle structure. Geochim. Cosmochim. Acta 47: 2015–30

Muirhead, K. 1985. Comments on "Reflection properties of phase transition and compositional change models of the 670-km discontinuity" by Alison C. Lees, M. S. T. Bukowinski, and Raymond Jeanloz. J. Geophys. Res. 90: 2057–59

Murthy, V. R. 1975. Composition of the core and the early chemical history of the Earth. In The Early History of the Earth, ed. B. F. Windley, pp. 21–31. London: Wiley. 619 pp.

Mysen, B. O., Arculus, R. J., Eggler, D. H. 1975. Solubility of carbon dioxide in melts

of andesite, tholeiite and olivine nephelenite composition to 3 kbar pressure. Contrib. Mineral. Petrol. 53: 227–39

O'Hara, M. J., Saunders, M. J., Mercey, E. L. P. 1975. Garnet peridotite, primary ultrabasic magma and eclogite: interpretation of upper mantle processes in kimberlite. In Physics and Chemistry of the Earth, 9: 571–604. New York: Pergamon

Olson, P., Yuen, D. A. 1982. Thermochemical plumes and mantle phase transitions. J. Geophys. Res. 87: 3993–4002

Olson, P., Yuen, D. A., Balsiger, D. 1984. Mixing of passive heterogeneities by mantle convection. J. Geophys. Res. 89: 425–36

O'Nions, R. K., Oxburgh, E. R. 1983. Heat and helium in the Earth. Nature 306: 429–31

O'Nions, R. K., Evensen, N. M., Hamilton, P. J. 1979. Geochemical modeling of mantle differentiation and crustal growth. J. Geophys. Res. 84: 6091–6101

O'Nions, R. K., Hamilton, P. J., Evensen, N. M. 1977. Variations in $^{143}Nd/^{144}Nd$ and $^{87}Sr/^{86}Sr$ ratios in oceanic basalts. Earth Planet. Sci. Lett. 34: 13–22

Oversby, V. M., Ewart, A. 1972. Lead isotopic compositions of Tonga Kermadec volcanics and their petrogenetic significance. Contrib. Mineral. Petrol. 37: 181–210

Oversby, V. M., Ringwood, A. E. 1971. Time of formation of the Earth's core. Nature 234: 463–65

Ozima, M., Podosek, F. A. 1983. Noble Gas Geochemistry. New York: Cambridge Univ. Press. 367 pp.

Ozima, M., Zashu, S. 1983. Noble gases in submarine pillow volcanic glasses. Earth Planet. Sci. Lett. 62: 24–40

Patchett, P. J., Tatsumoto, M. 1980. Hafnium isotope variations in oceanic basalts. Geophys. Res. Lett. 7: 1077–80

Patterson, C. C., Tatsumoto, M. 1964. The significance of lead isotopes in detrital feldspar with respect to chemical differentiation within the Earth's mantle. Geochim. Cosmochim. Acta 28: 1–28

Peltier, W. R. 1981. Ice age geodynamics. Ann. Rev. Earth Planet. Sci. 9: 199–225

Peltier, W. R. 1983. Constraint on deep mantle viscosity from Lageos acceleration data. Nature 304: 434–36

Peltier, W. R., Andrews, J. T. 1976. Glacial isostatic adjustment. I. The forward problem. Geophys. J. R. Astron. Soc. 46: 605–46

Polve, M., Allegre, C. J. 1980. Orogenic lherzolite complexes studied by ^{87}Rb-^{87}Sr: a clue to understand the mantle convection processes? Earth Planet. Sci. Lett. 51: 71–93

Poreda, R., Craig, H., Schilling, J.-G. 1980.

^{3}He/^{4}He variations along the Reykjanes Ridge. *Eos, Trans. Am. Geophys. Union* 61: 1158 (Abstr.)

Reilly, G. A., Shaw, D. M. 1967. An estimate of the composition of part of the Canadian shield in Northwestern Ontario. *Can. J. Earth Sci.* 4: 725–39

Reisberg, L., Zindler, A. 1985. The scale of mantle heterogeneity: isotopic evidence from Ronda. *Eos, Trans. Am. Geophys. Union* 66: 413 (Abstr.)

Reisberg, L., Zindler, A. 1986. Extreme isotopic variability in the upper mantle: evidence from Ronda. *Earth Planet. Sci. Lett.* In press

Richard, P., Allegre, C. J. 1980. Neodynium and strontium isotope study of ophiolite and orogenic lherzolite petrogenesis. *Earth Planet. Sci. Lett.* 47: 327–42

Richard, P., Shimizu, N., Allegre, C. J. 1976. ^{143}Nd/^{144}Nd, a natural tracer: an application to oceanic basalt. *Earth Planet Sci. Lett.* 31: 269–78

Richards, P. G. 1972. Seismic waves reflected from velocity gradient anomalies within the Earth's upper mantle. *J. Geophys.* 38: 517–27

Richardson, S. H., Erlank, A. J., Duncan, A. R., Reid, D. L. 1982. Correlated Nd, Sr and Pb isotope variation in Walvis Ridge basalts and implications for the evolution of their mantle source. *Earth Planet. Sci. Lett.* 59: 327–42

Richardson, S. H., Gurney, J. J., Erlank, A. J., Harris, J. W. 1984. Origin of diamonds in old enriched mantle. *Nature* 310: 198–202

Richardson, S. H., Erlank, A. J., Hart, S. R. 1985. Kimberlite-borne garnet peridotite xenoliths from old enriched subcontinental lithosphere. *Earth Planet. Sci. Lett.* 75: 116–28

Richter, F. M. 1984. Regionalized models for the thermal evolution of the Earth. *Earth Planet. Sci. Lett.* 68: 471–84

Richter, F. M., McKenzie, D. P. 1981. On some consequences and possible causes of layered mantle convection. *J. Geophys. Res.* 86: 6133–42

Richter, F. M., Ribe, N. M. 1979. On the importance of advection in determining the local isotopic composition of the mantle. *Earth Planet. Sci. Lett.* 43: 212–22

Richter, F. M., Daly, S. F., Nataf, H. C. 1982. A parameterized model for the evolution of isotopic heterogeneities in a convecting system. *Earth Planet. Sci. Lett.* 60: 178–94

Ringwood, A. E. 1975. *Composition and Petrology of the Earth's Mantle.* New York: McGraw-Hill. 618 pp.

Ringwood, A. E. 1979. *Origin of the Earth and Moon.* New York: Springer-Verlag. 295 pp.

Rison, W. 1981. Loihi seamount: mantle volatiles in the basalts. *Eos, Trans. Am. Geophys. Union* 62: 1083 (Abstr.)

Rison, W., Craig, H. 1982. Helium-3: coming of age in Samoa. *Eos, Trans. Am. Geophys. Union* 63: 114 (Abstr.)

Roden, M. K., Hart, S. R., Frey, F. A., Melson, W. G. 1984. Sr, Nd and Pb isotopic and REE geochemistry of St. Paul's Rocks: the metamorphic and metasomatic development of an alkali basalt mantle source. *Contrib. Mineral. Petrol.* 85: 376–90

Sarda, P., Staudacher, T., Allegre, C. J. 1985. ^{40}Ar/^{36}Ar in MORB glasses: constraints on atmosphere and mantle evolution. *Earth Planet. Sci. Lett.* 72: 357–75

Schilling, J.-G. 1973. Icelandic mantle plume: geochemical evidence along the Reykjanes Ridge. *Nature* 242: 565–71

Schubert, G., Spohn, T. 1981. Two-layer mantle convection and the depletion of radioactive elements in the lower mantle. *Geophys. Res. Lett.* 8: 951–54

Schubert, G. D., Turcotte, D. L. 1971. Phase changes and mantle convection. *J. Geophys. Res.* 76: 1424–32

Sclater, J. G., Jaupart, C., Galson, D. 1980. The heat flow through oceanic and continental crust and the heat loss of the earth. *Rev. Geophys. Space Phys.* 18: 269–311

Shaw, D. M. 1967. U, Th, and K in the Canadian Precambrian shield and possible mantle compositions. *Geochim. Cosmochim. Acta* 31: 1111–13

Shaw, D. M., Dostal, J., Keays, R. R. 1976. Additional estimates of continental surface Precambrian shield composition in Canada. *Geochim. Cosmochim. Acta* 40: 73–83

Sleep, N. H. 1984. Tapping of magmas from ubiquitous mantle heterogeneities: an alternative to mantle plumes? *J. Geophys. Res.* 89: 10,029–41

Sneeringer, M., Hart, S. R., Shimizu, N. 1984. Strontium and samarium diffusion in diopside. *Geochim. Cosmochim. Acta* 48: 1589–1608

Spohn, T., Schubert, G. 1982. Modes of mantle convection and the removal of heat from the Earth's interior. *J. Geophys. Res.* 87: 4682–96

Stacey, F. D. 1969. *Physics of the Earth.* New York: Wiley. 414 pp.

Staudacher, T., Allegre, C. J. 1982. Terrestrial xenology. *Earth Planet. Sci. Lett.* 60: 389–406

Staudigel, H., Zindler, A., Hart, S. R., Leslie, T., Chen, C. Y., Clague, D. 1984. The isotope systematics of a juvenile intraplate volcano: Pb, Nd and Sr isotope ratios of basalts from Loihi Seamount, Hawaii. *Earth Planet. Sci. Lett.* 69: 13–29

Stille, P., Unruh, D. M., Tatsumoto, M. 1983. Pb, Sr, Nd and Hf isotopic evidence of multiple sources for Oahu, Hawaii basalts. *Nature* 304: 25–29

Stosch, H. G., Lugmair, G. W. 1984. Evolution of lower continental crust as recorded by granulite facies xenoliths from the Eifel/Germany. *Eos, Trans. Am. Geophys. Union* 65: 230 (Abstr.)

Stosch, H. G., Carlson, R. W., Lugmair, G. W. 1980. Episodic mantle differentiation: Nd and Sr isotopic evidence. *Earth Planet. Sci. Lett.* 47: 263–71

Stueber, A. M., Ikramuddin, M. 1974. Rubidium, strontium and the isotopic composition of strontium in ultramafic nodule minerals and host basalts. *Geochim. Cosmochim. Acta* 38: 207–16

Suen, C. J. 1978. *Geochemistry of peridotites and associated mafic rocks, Ronda Ultramafic Complex, Spain.* PhD thesis. Mass. Inst. Technol., Cambridge. 305 pp.

Sun, S.-S. 1980. Lead isotopic study of young volcanic rocks from mid-ocean ridges, ocean islands and island arcs. *Philos. Trans. R. Soc. London Ser. A* 297: 409–45

Sun, S.-S. 1982. Chemical composition and origin of the Earth's primitive mantle. *Geochim. Cosmochim. Acta* 46: 179–92

Sun, S.-S., Hanson, G. N. 1975. Evolution of the mantle: geochemical evidence from alkali basalts. *Geology* 3: 297–302

Tatsumoto, M. 1978. Isotopic composition of lead in oceanic basalt and its implication to mantle evolution. *Earth Planet. Sci. Lett.* 38: 63–87

Tatsumoto, M., Knight, R. J. 1969. Isotopic composition of lead in volcanic rocks from central Honshu—with regard to basalt genesis. *Geochem. J.* 3: 53–86

Tatsumoto, M., Knight, R. J., Allegre, C. 1973. Time difference in the formation of meteorites as determined from the ratio of lead-207 to lead-206. *Science* 180: 1278–83

Tatsumoto, M., Unruh, D. M., Desborough, G. A. 1976. U-Th-Pb and Rb-Sr systematics of Allende and U-Th-Pb systematics of Orgueil. *Geochim. Cosmochim. Acta* 40: 617–34

Taylor, S. R. 1982. Lunar and terrestrial crusts: a contrast in origin and evolution. *Phys. Earth Planet. Inter.* 29: 233–41

Taylor, S. R., McLennan, S. M. 1981. The composition and evolution of the continental crust: rare earth element evidence from sedimentary rocks. *Philos. Trans. R. Soc. London Ser. A* 301: 381–99

Taylor, S. R., McLennan, S. M., McCulloch, M. T. 1983. Geochemistry of loess, continental crustal composition and crustal model ages. *Geochim. Cosmochim. Acta* 47: 1897–1905

Tilton, G. R. 1973. Isotopic lead ages of chondritic meteorites. *Earth Planet. Sci. Lett.* 19: 321–29

Tilton, G. R., Barreiro, B. A. 1980. Origin of lead in Andean calcalkaline lavas, Southern Peru. *Science* 210: 2145–47

Turcotte, D. L. 1985. Chemical geodynamic models. In preparation

Turcotte, D. L., Schubert, G. 1982. *Geodynamics.* New York: Wiley. 435 pp.

Vidal, P., Clauer, N. 1981. Pb and Sr isotopic systematics of some basalts and sulfides from the East Pacific Rise at 21°N (project RITA). *Earth Planet. Sci. Lett.* 55: 237–46

Vidal, P., Dosso, L. 1978. Core formation: catastrophic or continuous? Sr and Pb isotope geochemistry constraints. *Geophys. Res Lett.* 5: 169–72

Vollmer, R. 1977. Terrestrial lead evolution and formation time of the Earth's core. *Nature* 270: 144–47

Wasserburg, G. J., DePaolo, D. J. 1979. Models of Earth structure inferred from neodymium and strontium isotopic abundances. *Proc. Natl. Acad. Sci. USA* 76: 3594–98

Watson, E. B., Othman, D. B., Luck, J. M., Hofmann, A. W. 1986. Partitioning of U, Pb, Cs, Yb, Hf, Re and Os between chromian diopsidic pyroxene and haplobasaltic liquid. *Geochim. Cosmochim. Acta* In press

Weidner, D. J. 1985. A mineral physics test of a pyrolite mantle. *Geophys. Res. Lett.* 12: 417–20

White, W. M. 1985. Sources of oceanic basalts: radiogenic isotopic evidence. *Geology* 13: 115–18

White, W. M., Hofmann, A. W. 1982. Sr and Nd isotope geochemistry of oceanic basalts and mantle evolution. *Nature* 296: 821–25

White, W. M., Patchett, J. 1984. Hf-Nd-Sr isotopes and element abundances in island arcs: implications for magma origins and crust mantle evolution. *Earth Planet. Sci. Lett.* 67: 167–85

White, W. M., Dupre, B., Vidal, P. 1985. Isotope and trace element geochemistry of sediments from the Barbados Ridge–Demerara Plain region, Atlantic Ocean. *Geochim. Cosmochim. Acta* 49: 1875–86

Wolf, G. H., Bukowinski, M. S. T. 1985. Ab initio structural and thermoelastic properties of orthorhombic $MgSiO_3$ perovskite. *Geophys. Res. Lett.* 12: 809–12

Worner, G., Zindler, A., Staudigel, H. 1986. Sr, Nd, and Pb isotope geochemistry of Tertiary and Quaternary volcanics from W. Germany. Submitted for publication

Yoder, H. S. Jr. 1976. *Generation of Basaltic Magma.* Washington, DC: Natl. Acad. Sci. 265 pp.

Zindler, A., Hart, S. R. 1986. Helium: problematic primordial signals. *Earth Planet. Sci. Lett.* In press

Zindler, A., Jagoutz, E. 1986. Mantle cryptology. *Geochim. Cosmochim. Acta* In press

Zindler, A., Hart, S. R., Frey, F. A., Jakobsson, S. P. 1979. Nd and Sr isotope ratios and rare earth elements abundances in Reykjanes Peninsula basalts: evidence for mantle heterogeneity beneath Iceland. *Earth Planet. Sci. Lett.* 45: 249–62

Zindler, A., Jagoutz, E., Goldstein, S. 1982. Nd, Sr and Pb isotopic systematics in a three-component mantle: a new perspective. *Nature* 58: 519–23

Zindler, A., Staudigel, H., Batiza, R. 1984. Isotope and trace element geochemistry of young Pacific seamounts: implications for the scale of upper mantle heterogeneity. *Earth Planet. Sci. Lett.* 70: 175–95

Zindler, A., Staudigel, H., Hart, S. R., Endres, R., Goldstein, S. 1983. Nd and Sr isotopic study of a mafic layer from Ronda ultramafic complex. *Nature* 304: 226–30

SUBJECT INDEX

A

Aardvarks, 77
Accretionary prisms
  and subduction zones, 314
Advection
  and conduction in the Earth,
    399–402
  as a heat-transfer mechanism,
    378, 384–85, 388
  and mantle geotherms, 389–96
Aftershock areas
  expansion patterns of, 310–13
Air bubbles
  in ice cores, 204
Air-sea exchange
  of CO_2, 213–14, 218
Alaska
  earthquakes in, 293
  land rising in, 258
  subduction zones in, 297
  uplift of the shorelines in, 300
Alaskan earthquake
  in 1964, 295, 296
    aftershock area of, 311
    data from, 11
    waveforms of, 308–9
Albumin
  in fossils, 80
Aleutian Islands
  earthquakes in the, 293,
    306
  subduction zones in the, 297
Amino acid(s)
  changes of
    in proteins, 72–73
  and DNA, 72
  in fossils, 79–80
Ammonoids
  and the Triassic, 100
Amphibole
  and calc-alkaline, 439–42
Andesites
  defined, 419
  petrogenesis of, 417–51
  and subduction zones, 299
Andesitic volcanism
  and subduction zones, 319
Angayucham terrane
  chert in the, 476
Antarctic Bottom Water
  CO_2 in, 218
  tritium in, 214
Antarctic ice
  CO_2 in air bubbles in, 204
Antarctic Intermediate Water,
  218
Archean
  heat production decreased
    since, 410

Archean geotherms, 378, 402–5
Arctic Ocean drift ice, 221–22
Armoring
  and grain entrainment sorting,
    133–34
Arrhenius, S., 202
Artiodactyls, 77
Asama Volcano, 271
Asia
  biostratigraphic work in, 95–
    98
Asperity
  and subduction-zone earth-
    quakes, 305–10, 313–15,
    319
Atlantic Ocean
  CO_2 in the, 217, 219
  see also North Atlantic Ocean
Atmosphere
  CO_2 in the, 201–10, 220
  along the equator in the Pacif-
    ic, 47
  future carbon levels in the,
    207–8
  and ocean
    in Bjerknes' hypothesis,
      47–48, 53
    a two-way coupling be-
      tween, 44, 46
  and ocean interaction model of
    El Niño events, 62–65
  Tyndall on the gases in the,
    201–2
Atmospheric changes
  and El Niño, 53–54
Atmospheric response
  during ENSO, 60–61
Australasian strewn field, 324,
  326–27, 338
Australasian tektites, 331–32,
  344
Australia
  biostratigraphic work in, 95–
    98
  Ordovician conodont faunas
    of, 97–98

B

Back-arc basin
  and subduction zones, 299,
    319
Back-arc opening
  and seismicity, 304
Baltoscandia
  and Ordovician conodont zon-
    al biostratigraphy, 93
Baltoscandic conodont succession, 93–95
  in China, 97

Basaltic magma
  and andesite petrogenesis,
    417–51 passim
Basalts
  and mantle geochemical
    heterogeneities, 498–502
  oceanic
    isotope systematics in, 532–
      35
    and subduction zones, 299
  see also Midocean ridge
    basalt; Ocean island
    basalt
Beach
  dissipative, 242, 246–47, 261
  reflective, 242, 246, 261
Beach erosion, 237, 239, 258,
  261
  and rip currents, 249–52, 261
  storm waves and wave run-up
    and, 239–43
Beach morphology
  and edge waves, 246–49
Bedding
  cyclic
    in pelagic and hemipelagic
      sediments, 363–69
  rhythmic
    in platform limestones,
      369–70
Bedding style
  and diagenetic processes, 463–
    65
  of radiolarian chert, 462
Benioff-zone dip angle(s), 301
  in subduction zones, 299, 319
Biosphere
  and the carbon cycle, 208
  CO_2 from the, 209–10
Biostratigraphic correlation
  and conodonts, 85–108
  explained, 86–88
Biostratigraphic scheme
  the Mohawkian-Cincinnatian,
    92
Biostratigraphic zonation
  for rocks, 455
Biostratigraphy
  conodonts in Paleozoic and
    Triassic, 86
  of the high-latitude faunal
    realm, 93–95
  of the low-latitude faunal
    realm, 89–93
  radiolarian, 480–84, 486
Biozonal schemes
  traditional, 105–8
Biozones, 93
  in biostratigraphic correlation,
    87

573

conodont-based
in the Devonian, 99
of the lowermost
Carboniferous, 99
of the Triassic, 100
Bjerknes hypothesis, 47–48, 53
for the growth of ENSO
anomalies, 65
modification of, 62
Boundary layer
cold
at the top of the mantle,
393
see also Thermal boundary
layer(s)
Bovids
immune phylogeny of, 77–79
Brittle infrastructure
and faulting, 164, 171
Bulk silicate Earth (BSE), 504–
63 passim

C

C1 chondrite
Earth as a, 541–42
Calc-alkaline andesites, 418
and fractionation accompanied
by crustal contamination,
442–47, 451
and fractionation involving
amphibole, 439–42
and fractionation involving anhydrous silicates, 437–39
Calc-alkaline rock series, 419–
21, 431–33, 447
Calc-alkaline trend, 436–37,
449, 451
Calendar frequency band, 351
Cambrian
and conodont-based
biostratigraphy, 88
Carbon
in the oceans, 211
distribution of, 212–13
Carbon cycle, 207–8
Carbon dioxide
in air bubbles in Antarctic ice,
204
atmospheric, 201–10
consequences of rising,
220–24
the largest natural perturbation of, 224
a steady rise in, 206–7, 230
from the biosphere, 209–10
and chemical geodynamics,
517–22
the increase of
and fossil fuel, 208–10
and possible effects on climate, 201–30

modeling future increases,
207–10, 230
in the oceans, 204, 213–20
preindustrial level, 203–6,
225, 229
removal and disposal of, 228–
29
^{13}C record
in tree rings, 205, 210
Carbonate data
direct oceanic, 216–20
Carbonate system
in the oceans, 211–13
Carboniferous
and conodont zonal
biostratigraphy, 99–100
cyclothems, 370–71
Carnivores, 77
Carranza, Dr. Luis, 43
Cataclastic flow
and faulting, 156–57
Cetaceans, 77
Chalcopyrite, 186–87
Chamberlin, T. C., 202
Chemical geodynamics, 493–
563
Chert
bedded, 456
sedimentary structures in,
462–63
and silica diagenesis, 460–
61
chert-bearing assemblages,
479–80
in the circum-Pacific rim,
478–79
color of, 465
dating, 480–86
defined, 456
lithologic associations and
tectonic settings of, 470–
73
lithologic attributes
composition, 456–61
physical characteristics,
461–66
primary and secondary, 456
radiolarian
Paleozoic and Mesozoic,
455–88
sponge spicule, 456
temporal distribution, 469
unfossiliferous, 458
Chert association
continental margin, 472
island-arc, 471
mélange, 472–73
ophiolitic, 470–71, 474–75
seamount, 471
Chile
plate motion in, 316
subduction zones in, 297

time functions for earthquakes
from, 309
uplift of the shorelines in, 300
Chilean earthquake
in 1960, 295, 296
data from, 11
China
biostratigraphic work in, 95–
98
Chulitna terrane
and ophiolitic chert, 475–76
Cincinnati Region
conodont species in the, 90–
91
Cincinnatian
and Mohawkian
biostratigraphic scheme,
92–93
Circulation anomalies
midlatitude
and tropical Pacific SST, 46
Circum-Pacific rim
chert-bearing terranes in, 478–
79
Climate
and CO_2 increase in the
atmosphere and oceans,
201–30
changes in, 227
Climatic rhythms
recorded in strata, 351–73
Coal
a ban on, 227
burning, 209
Coal-fired plants, 228
Coast Range ophiolite
and chert, 476–77
Coastal erosion
and sea level, 256–62
Coastal processes
and shoreline erosion, 237–62
Codons
and DNA, 72, 74
Collagen(s)
fossil, 79–80
glycine in, 73
Colo Volcano, 288–89
Columbia
subduction zones in, 297
Columbia earthquake
data from, 11
in 1906, 295
Colorado
the Cretaceous of, 356
see also Denver
Conduction
and advection in the Earth,
399–402
as a heat transfer mechanism,
378, 384–85, 388, 393
and lithosphere geotherms,
396–98

Conodonts
  and biostratigraphic correla-
    tion, 85–108
  as biostratigraphic indices, 86
  Cambrian, 88
Continental drift, 3–4, 12
Continental margin chert association, 472
Convection
  in the core, 12–13
  within the Earth, 378, 385–
    94, 405
  isoviscous, 407
  layered, 409
  modeling of, 12
  whole-mantle, 409–10
Convective flow model
  of the mantle, 304–5
Copper
  in Mississippi-Valley type de-
    posits, 184–87
Core
  convection in the, 12–13
  heat flow out of the, 390–91
  outer
    temperature in the, 379–
      81
  the radius of the, 8
Coriolis effect
  and winds along the South
    American coast, 50
Coriolis force
  and the equator, 55
Coulomb-Navier failure criterion, 25
Cretaceous
  the atmospheric CO_2 level in
    the, 220
  of Colorado, 356
Cretaceous/Tertiary boundary
  Ir anomaly at, 340
Crust
  geotherm in the
    constraints on, 379–84
  temperature distribution in the,
    377–411
  uplift and erosion of, 399–402
Crustal contamination
  and calc-alkaline andesites,
    442–47, 451
Crustal structure
  seismic refraction studies of,
    2–3, 5, 7, 11–12, 19
Crystallization
  fractional
    and basaltic magma, 417–
      19
    controls, 422–37
    and tholeiitic and calc-
      alkaline trends, 420–22
Cyclothems
  Carboniferous, 370–71

Cytochrome c
  a family tree from, 73
  phylogeny of, 74
  sequence data for, 71

D

Dating
  isotopic
    of radiolarian chert, 485–86
  magnetostratigraphic
    of chert, 484–85
  radiolarian chert, 480–86
De Saussure, H., 201
Deep magnetic soundings, 9, 12
Deforestation, 203
  and the CO_2 problem, 210
  large-scale, 205, 229
  see also Forests
Deformation
  coseismic (γ-phase), 159–60
  in crustal fault zones, 149–71
  and fault rocks, 164–66
  postseismic (δ-phase), 160–61
  preseismic (β-phase), 159
Deformation processes
  in fault zones, 154–61
Denver, Colorado
  earthquakes in, 27
  induced seismicity at, 22
  seismicity triggered at, 39
Devonian
  biozonal boundaries in the,
    108
  and conodont zonal
    biostratigraphy, 99
Diagenetic overprints, 364–65
Diagenetic processes
  and bedding style, 463–65
Diamond(s)
  3.5 Gyr old, 404
  in xenoliths, 383
Dinosaurs
  reconstruction of, 81
Dilution cycles
  and stratification, 364
Discontinuities
  the 200-km, 10
    refractions from below, 19
  the 400-km, 382–83
  the 670-km, 552, 554–58
  the M, 3, 11–12
Dislocation creep
  and faulting, 157–58
Dissolution cycles
  and stratification, 364
DNA
  and the eukaryotic genome, 72
  mitochondrial
    as a molecular clock, 74–75
    as a molecular clock, 73–75

  as a newly deciphered lan-
    guage, 72
  and quagga systematics, 81
  sequencing, 71
Downdropping
  of the crust, 399–402
Dugong, 80

E

Earth
  as a Cl chondrite, 541–42
  temperature in the lower man-
    tle of, 381–82
  temperature in the outer core,
    379–81
  thermal evolution of the, 378–
    79
  and structure of the mantle,
    405–11
Earthquake(s)
  in crustal fault zones, 149–71
  cycle
    regularity in the, 31–32
  expansion patterns of after-
    shock areas, 310–13
  forecasting, 267, 269
  foreshock-mainshock-
    aftershock sequences,
    153–54, 159–61
  interseismic healing, 161
  at Long Valley caldera, 284–
    85
  "man-made," 28
  beneath Mount St. Helens,
    272
  at the Pozzuoli caldera, 282–
    83
  prediction of, 32
  quantification of, 293–97
  regional variation of, 293
  rupture controls, 171
  stress cycle
    and deformation, 151–53,
      159–61
  subduction-zone
    rupture process, 293–319
  triggered, 21–40
    and increased seismic risk,
      35–40
    mitigation of risk from, 38–
      40
    and stress cycles on faults,
      31–34
  types of, 26–31
Easterlies
  and El Niño, 52–53, 67
Eccentricity cycles
  the Earth's, 353–55, 363,
    366–67, 372
Eccentricity-driven bundles

precession-caused couplets in, 365–68
Edentates, 77
El Niño, 43–68
 and climate, 43–44
 the composite, 50–54, 56
 event, 52–53, 67
 explained, 260
 the interannual temperature variability associated with, 49
 interval between events, 54
 mature phase of, 53
 ocean-atmosphere interaction model, 62–65
 oceanic changes during, 54–57
 oceanography of, 48
 onset of, 52
 prelude to, 52
 sea-level changes produced by, 260–62
 and SST, 48, 57–61
 and Southern Oscillation, 44–47, 68
El Niño Current, 52
El Niño-Southern Oscillation (ENSO), 44, 47
 atmospheric circulation during, 58
 atmospheric response during, 60–61
 cycle, 65–68
 cyclic nature of, 224
Electrical conductivity
 of the Earth, 9
Elephants
 relatives of the, 80
End-member-depleted MORB mantle (DMM), 504–62 passim
Engineering
 and seismic hazards, 35–38
Enriched mantle components(EMs), 507–35 passim, 553–54, 562
ENSO
 see El Niño-Southern Oscillation
Entrainment sorting
 and grain behaviors, 131–33
 of grains, 117, 123–34
Eocene lacustrine record
 and lake varves, 359–60
Euconodonts
 in the upper Upper Cambrian, 88
Eukaryotic genome, 72–73
Eutherians, 75
Evaporite varves, 362–63
Evolution
 molecular and morphological, 82
Evolutionary relationships, 71

Excess trench sediments (ETS), 314
 see also Sediment(s)

F

Failure criterion
 Coulomb-Navier, 25
 Mohr-Coulomb, 23, 24
Far-field time function(s), 307–9
 the frequency spectrum of, 317–19
Fault asperities, 305–6, 319
Fault heterogeneity
 and high-frequency seismic waves, 317
Fault model
 and far-field body-wave time function, 307–8
Fault plane
 and the asperity model of earthquakes, 305–6, 309, 319
Fault rocks
 and deformation, 164–66
Fault zone(s)
 conceptual models, 150, 169
 continental model, 161–69
 crustal
  earthquakes and rock deformation in, 149–71
 deformation processes in, 154–61
 evolution of, 150–51, 170
 geometry, 169–70
 rheological models for, 170–71
Faulting
 and great earthquakes, 295–96
Faults
 stress cycles on
  and triggered earthquakes, 31–34
Faunal realm
 high-latitude
  biostratigraphy of, 93–95
 low-latitude
  biostratigraphy of, 89–93
Faunas
 conodont
  high-latitude and low-latitude, 88
Fibrinopeptides, 73
Flooding
 historic occurrences of, 258
Flow zones
 in stream channels, 117–19
Fluid flow
 during formation of Mississippi Valley–type deposits, 190
Fluid injection

 and triggered seismicity, 26–27, 39
Fluid migration
 gravity-driven, 192–95
 out of sedimentary basins, 192–95
Forests
 cutting down of, 220
 large-scale clearing of, 202
 see also Deforestation
Fossil fuel(s)
 burning of, 202, 220
 consumption of, 205, 223
 continued use of, 229
 and the cumulative increase of CO_2, 208–10
 reduction of use of, 225–27
Fossil-fuel CO_2
 injected into the ocean, 228
Fossil record
 and taxonomic information, 75
Fossils
 amino acids in, 79–80
 and biostratigraphic correlation, 87
 index, 86, 88, 108
 Shaw's graphic correlation of, 87
Fourier, J., 201
Fourier's law, 393
Free oscillation, 17
 inversion, 15
Frictional regime
 in faulting, 164–66

G

Galena, 186–87
 and Mississippi Valley–type deposits, 177, 182–83, 196
 solubilities of, 189
 sulfur isotopic composition of, 187–88
Garnet
 in the mantle, 381–82
Gas
 natural
  the recoverable supply of, 209
General circulation model (GCM)
 atmospheric, 220–21
Genetic code
 and the eukaryotic genome, 72
Genetic information
 from fossils, 79–80
Genome
 eukaryotic, 72–73
Geochemistry
 of bedded radiolarian chert, 459–60
 isotope, 5–6

of Mississippi Valley–type deposits, 181–88
of tektites and impact glasses, 323–44
Geodynamics, 402–11
chemical, 493–563
Geology
of Mississippi Valley–type deposits, 181–88
Geomagnetic reversals, 340
Geotherm(s)
Archean, 378, 402–5
in the crust and mantle
constraints on, 379–84
estimates of, 378
lithosphere
and conduction, 396–98
mantle
and advection, 389–96
through the oceanic lithosphere, 400
Geothermometer
the two-pyroxene, 383
Geothermometer/geobarometer
calibrations, 383–84
Glaciers
advances and retreats of, 258
melting, 262
see also Ice sheet(s)
Globins
sequence data for, 71
Glycine
in collagen, 73
Golconda terrane
chert associations in the, 477–78
Gold, 113
in open framework gravels, 134
in the Ventersdorp Contact placer, 131–32
Grain boundary diffusion
and faulting, 158
Grains
in the binomial rock classification system, 458
entrainment of, 117, 123–34
mineral
in placer sites, 116
settling of, 117, 120–23
the shearing of, 117, 138–39
in stream channels, 117–20
transport sorting of, 117, 134–38
Graphic correlation
of fossils, 87
of Upper Ordovician strata, 90
Gravity-driven fluid migration, 192–95
Great Lakes
erosion rates in the, 260, 262
Greenhouse effect
explained, 202

mitigating, 225–29
Greenhouse warming
delaying, 224–29
Gulf Coast
sinking, 258

H

Hadley circulation, 47
Hale cycle
the 22-yr, 351, 363, 372
Hawaiian Volcano Observatory, 269
Heat loss
out of the Earth, 390–91
Heat transfer
by conduction, 378, 384–85, 388, 393
Heat transfer mechanisms
in the Earth, 384–402
Hekla Volcano, 272–73
Hiding
and grain entrainment sorting, 134
High U/Pb mantle composition (HIMU), 504–35 passim, 549, 553–54, 562
Holocene shoreline
around the Pacific, 300
Horst-and-graben structures (HGS), 314–15
Hsinfengkian (China) dam
earthquake at, 21
Hydrofracturing
to measure stress in rock, 27
Hyraxes, 77

I

Ice cores
sampling air bubbles in, 204
Ice rhythms
Pleistocene, 355–56
Ice sheet(s)
Pleistocene
the melting of, 386
West Antarctic
disappearance of the, 221–22
the melting of the, 223
the world, 223, 225, 230
Impact glasses
geochemistry of, 323–44
Indonesian Low, 48
Industrial Revolution
and the atmospheric CO_2 content, 203, 229
and CO_2 in surface water, 204
Infragravity component
and beach erosion, 241
Infragravity motions
and beach erosion, 242, 247, 261

and edge waves, 243–46
Infrasonic array
in Australia, 17–18
International Statigraphic Guide, 91, 93
Intertropical Convergence Zone, 48, 50, 54
displacement of the, 222
and El Niño, 52
Ion probe, 14
Iron
in the Earth's core, 379–80
Island-arc chert association, 471
Isotope systematics
in oceanic basalts, 532–35
Isotopic characterization
of mantle components, 502–35
Isotopic dating
of radiolarian chert, 485–86
Ivory Coast strewn field, 324
Ivory Coast tektites, 329–31, 338, 341, 343
Izalco Volcano, 274

J

Jalisco, Mexico earthquake
in 1932, 316
Juan de Fuca subduction zone, 314–16
Jurassic times
and the reversal of the Earth's magnetic field, 7

K

Kamchatka
earthquakes in, 293
subduction zones in, 297
Kashmir
biozones in, 106
Kelvin waves
and El Niño, 54–57
Kilauea Volcano, 269, 288
eruption pattern of, 274–76
Komatiites
and mantle temperatures in the Archean, 403–4
the occurrence of, 405
Koyna Reservoir, 30
the largest induced earthquake at, 35
a triggered earthquake at, 21
Kurile Islands
earthquakes in the, 306, 308–9
plate motion in the, 316

L

Lagomorphs, 76–77
Lake varves, 359–60

Lava flow(s)
  in Hawaii, 269, 277–78
  human diversion of, 288
Lead
  Pb paradox
    and the mantle, 542–48
    and zinc deposits
      Mississippi Valley–type,
        177–96
    see also High U/Pb mantle
      composition (HIMU)
Libyan Desert Glass (LDG),
  339, 342
Limestones
  chert nodules in, 456
  rhythmic bedding in, 369–70
Liquid line(s) of descent
  explained, 418
  high-pressure, 436–37
  low-pressure
    and magmas of contrasting
      bulk composition, 425–
      33
Lithosphere
  continental
    heat transfer through, 378,
      404
  oceanic
    geotherms through, 400
    heat transfer through, 378,
      404
    and time-dependent conduc-
      tion, 398–99
Lithosphere geotherms
  and conduction, 396–98
Lonar glass, 342
Long Valley, California
  caldera at, 284–86
Longshore currents
  and sand transport, 253–55

M

Magma(s)
  basaltic
    and andesite petrogenesis,
      417–51 passim
  and chemical geodynamics,
    498–530 passim
  of contrasting bulk composi-
    tion
    and liquid lines of descent,
      425–33
  in Mauna Loa, 277–78
  in Mount St. Helens, 278–79
  at Rabaul, Pozzuoli, and Long
    Valley calderas, 285
  and volcanic eruptions, 269–
    70
Magma intrusion
  at the Long Valley caldera,
    285
  at the Pozzuoli caldera, 283

  at the Rabaul caldera, 280
Magmatism
  and tectonics, 417
Magnetic field
  Earth's, 380–81
Magnetostratigraphic dating
  of radiolarian chert, 484–85
Mammals
  family trees of, 75–77
Mammoths
  and radioimmunoassay, 80–
    81
Manatees, 80
Mantle
  and body wave travel times,
    15–17
  chemical and physical struc-
    ture of, 535–61
  chemical heterogeneity of,
    494–502
  the cold boundary layer at the
    top of the, 393
  components
    isotopic characterization of,
      502–35
    composition of the lower and
      upper, 408, 411
  convection in the, 388, 555–
    61
  convective flow model of the,
    304–5
  discontinuities in the, 19
  electrical conductivity of the,
    9
  geotherm in the
    constraints on, 379–84
  geotherms
    and advection, 389–96
  He signals, 510–23
  heat flow and the Urey ratio,
    535, 559–61, 563
  layered, 408–11
  lower
    temperature in the, 381–82
  mappings of the, 18
  minerals, 557–58
  and seismic travel times, 8
  structure
    and the thermal evolution of
      the Earth, 405–11
  surrounding the core, 380–81
  temperature distribution in the,
    377–411
  thermal boundary layer at the
    base of the, 13, 391
  thermal structure and viscosity
    of the, 558–59
  upper
    dominant minerals of the,
      381–82
Marianas
  plate motion in the, 316
  subduction zones in the, 297

  uplift of the shorelines in the,
    300
Marin Headlands, California
  and ophiolitic chert, 474–75
Marine varves, 360, 362
Marsupials, 75–77
Martinsburg Formation
  conodont species in, 90–91
Mastodons
  and radioimmunoassay, 80–81
Mauna Loa
  gas analyzers near, 206
Mauna Loa Volcano, 276–78
  eruption pattern of, 274–75
  repose period between erup-
    tions of, 272
Maxwell relaxation time, 385–86
Mélange chert association, 472–
  73
Mesozoic
  the warmth of the, 223
Mesozoic radiolarian chert, 455–
  88
Metal-bisulfide complexing
  for Mississippi Valley–type
    ore-forming fluids, 191–
    92
Metal-chloride complexing
  alternatives to, 191–92
  and Mississippi Valley–type
    deposits, 189–91, 195
Metal ratios
  in Mississippi Valley–type de-
    posits, 184–87
Meteorological patterns
  northern Pacific, 261
Mexican subduction zone, 312
Mexico
  subduction zones in, 297
Microfossils
  in cherty rocks, 455
Microtektites, 339–41
Midocean ridge basalt (MORB),
  499–561 passim
  source of, 525–27
  see also Basalts
Milankovitch frequency band,
  351, 372
Milankovitch rhythmicity, 371–
  72
  in the pre-Pleistocene record,
    356–57
Mill(s)
  dressing, 123, 141
  geomorphological, 116–17,
    139
Mineral(s)
  creep-deformation of, 385
  entrainment, 127
  in placer sites, 116
  in a pulp, 122, 134
  in radiolarian-bearing chert
    sequences, 486

transport velocities of, 137–38
in a turbulent flow, 121
Mines
planning the excavation process of, 39
Mining
and triggered seismicity, 26–28, 33
Mississippi Valley–type deposits
characteristics of, 177–79
fluid flow during the formation of, 190
genesis of, 177–96
geology and geochemistry of, 181–88
lead and sulfur isotopic geochemistry of, 187–88, 195–96
and metal-chloride complexing, 189–91, 195
alternatives to, 191–92
metal ratios in, 184–87
wall-rock alteration in, 183–84
Miyagi-Oki, Japan, earthquake in 1978
aftershock area of, 311
Mohawkian-Cincinnatian biostratigraphic scheme, 92–93
Mohr-Coulomb failure criterion, 23–24
Moldavite strewn field, 324
Moldavites, 324–27, 329–30, 338, 344
Molecular clocks
proteins and DNA as, 73–75
Moment release rate (MRR)
for subduction zones, 301–2
Monitoring techniques
of volcanoes, 275–76
Monotremes, 75–76
Monsoon
the Asian, 50
Monsoon fluctuations, 44
Mont Pelée
the eruption of, 268–69
Mount Fuji, 276
Mount Nuovo, 280
the eruption of, 282
Mount Rainier, 276
Mount St. Helens, 278–80
Mount St. Helens Volcano, 271–73, 276
Mudstones
siliceous, 456
Muong Nong–type tektites, 326, 336–38

N

Natural disasters
forecasting, 267

Niigata earthquake
the 1964, 296
waveforms of, 308–9
New Hebrides
aftershock expansion ratios in the, 312–13
subduction zones in the, 297
time functions for earthquakes from the, 309
uplift of the shorelines in the, 300
North America
Baltoscandic conodont zones in, 95
the Ordovician in, 92
western
radiolarian chert in, 473–78
North American Midcontinent
biostratigraphy for the Middle and Upper Ordovician of the, 90
conodont faunas of, 97
graphic correlation and biozones in the, 87
Ordovician rocks in the, 89
North American plate, 316
North American strewn field, 324
North American tektites, 324, 327, 332, 338
North Atlantic Deep Water, 214
North Atlantic Ocean
CO_2 in the, 211, 217, 219
tritium in the, 215
North Equatorial Countercurrent
and El Niño, 60
Northern Hemisphere
atmospheric CO_2 content for the, 203
Nuclear bomb tests
of 1958–1962, 213–14
Nusselt number, 393, 406–7
and Rayleigh number, 406

O

Obliquity cycle
the Earth's, 353–54, 357, 363
Obliquity-driven oscillations, 367–68
Ocean(s)
and atmosphere
in Bjerknes' hypothesis, 47–48, 53
a two-way coupling between, 44, 46
carbon in the
distribution of, 212–13
CO_2 in the, 204, 213–20
carbonate system in the, 211–13
fossil-fuel CO_2 injected into the, 228

model for the carbon cycle in the, 207–8
as a sink for excess CO_2, 202, 211, 230
vertical mixing rates in the, 214–16
Ocean-atmosphere interaction model
of El Niño events, 62–65
and the ENSO cycle, 65–68
Ocean island basalt (OIB), 499, 511, 514, 522–23, 526, 531, 533, 535
see also Basalts
Oceanic changes
and El Niño, 46, 54–57
Oceanic lithosphere
geotherms through the, 400
heat transfer through the, 378, 404
and time-dependent conduction, 398–99
Oil
shale, 209, 227
Olivine
in the mantle, 381–82
Ophiolitic chert association, 470–71, 474–75
Orbital variations
the Earth's, 353–55
Ordovician
and conodont-based biostratigraphy, 88–98
Middle
biozonal boundaries in, 108
Ore
in Mississippi Valley–type deposits, 177–79
Ore formation
the basinal brine hypothesis of, 179–81, 194–95
Ore-forming constituents
transport and precipitation of, 189–92
transportation of, 179–81
models, 180–81
Ore-forming fluids
and aquifer lithologies, 186–87
lead and sulfur isotopic characteristics of, 187–88
Oregon coast
erosion along the, 260, 262
rising, 258
sea level on the, 261
Organometallic complexing
for Mississippi Valley–type ore-forming fluids, 191–92
Overpassing
and grain entrainment sorting, 132–33

P

Pacific Ocean
  CO_2 concentrations in the, 213
  and El Niño, 44–67, 260–61
  the Holocene shoreline around
    the, 300
  SST along the equator in the,
    47
  subduction zones in the, 312
  tropical
    climatology of, 48–50
    warming of, 46
Pakistan
  biozones in, 105–6
  conodont-based biostratigraphy
    in, 100–5
Paleobiogeography
  of radiolarians, 468–69
Paleomagnetism, 3–5
  of Stormberg lavas, 5, 7
Paleozoic radiolarian chert, 455–
    88
Pander Society
  biostratigraphy volume of the,
    85, 87–88, 94, 100
Pangolin, 77
Paraconodonts
  and biostratigraphy, 88
Parametric Earth model (PEM),
    15
Peclet number, 388–89
Peridotite
  the fusion curve of, 384
  mantle, 382, 417, 420
  melting, 383–84
Permian
  and conodont zonal
    biostratigraphy, 99
Peru
  time functions for earthquakes
    from, 309
Peru Current, 50
Petrogenesis
  of andesites, 417–51
Petrology
  of Mississippi Valley–type de-
    posits, 181–83
Pezet, Federico Alfonso, 43, 68
Phase relations
  and elevated pressure, 433–37
  and one-atm melting ex-
    periments on basalts,
    423–33
  to predict crystallization paths,
    422–23
Philippines
  aftershock expansion ratios in
    the, 312–13
Photosynthesis
  enhanced, 225
  on land and in the oceans, 209

Phylogenetics
  molecular, 71–82
Phylogeny
  immunological
    of bovids, 77–79
    of elephants and sea cows,
      81
Pilanesberg dikes, 4
Placer(s)
  conditions for development of,
    116–17
  and differential transport, 138
  enrichments of, 139–42
  explained, 113
  and free settling of grains,
    122–23
  and hindered settling of
    grains, 123
  mining, 140
    the earliest, 113
  origin of, 113–43
  water-laid, 114–16
Plate coupling
  and high-frequency seismic
    waves, 317
  and subduction-zone earth-
    quakes, 297, 299, 302,
    313–15, 319
Plate motion
  aseismic and seismic, 316–
    17
  and seismicity, 300–305
Plate tectonic motions
  and convection in the mantle,
    388
Plate tectonics, 12–13, 378
  in the Archean, 404–5
Plate velocities
  at the Earth's surface, 406
Plates
  oceanic, 398–99
Pleistocene ice rhythms, 355–56
Pleistocene ice sheets, 386
Polar Front
  tritium near the, 214
Pore pressure
  compaction-induced changes
    in, 40
  and reservoir loading, 29–30
  and triggered earthquakes, 26–
    27, 33
Pouillet, C., 201
Pozzuoli, Italy
  caldera at, 280–84
  magma at, 285
Precession
  the Earth's, 354–57, 363,
    366, 372
Precession-caused couplets, 365–
    68
Precession index, 355, 357,
    366–67, 369

Preliminary reference Earth mod-
    el (PREM), 16, 19
Prevalent mantle composition
    (PREMA), 504–62 passim
Primitive upper mantle (PUM),
    536–41
Proboscidea, 80–81
Productivity cycles
  and stratification, 364
Prokaryotes
  and taxonomic information, 75
Protein(s)
  amino acid changes in, 72–73
  and DNA, 72
  evolution, 73
  in fossils, 80
  as a molecular clock, 73–75
Protoconodonts
  and biostratigraphy, 88
Pyroxene
  in the mantle, 381–82
Pyroxene geothermometer, 383

Q

Quagga systematics, 81
Quarrying operations
  and triggered seismicity, 26–
    28
Quartz
  in chert, 456, 458
Quasi-plastic regime
  in faulting, 166–67

R

Rabaul caldera, 280
  magma at, 285
Radiation
  as a heat transfer mechanism,
    384–85
Radioactive heat production,
    402–3, 411
Radioimmunoassay (RIA)
  of bovids, 77–79
  of fossils, 80
Radiolarian biostratigraphy, 480–
    84, 486
Radiolarian chert
  geochemistry of, 459–60
  Paleozoic and Mesozoic, 455–
    88
  in western North America,
    473–78
Rainfall
  and El Niño, 54
  and the Southern Oscillation,
    44
Rangely, Colorado
  triggered seismicity at, 27, 39
Rat Islands earthquake, 308–9

Rayleigh number(s)
  and convection within the
    Earth, 386–87, 392
  of the Earth's mantle, 388
  in the mantle of terrestrial
    planets, 387
  and Nusselt number, 406–7
Redox cycles
  and stratification, 364, 366–67
*Report on the National Shoreline
  Study, A*, 237
Reservoir impounding
  and triggered earthquakes, 28–
    31
Reservoir impoundment, 37
Reservoir-induced seismicity,
  22–23, 33, 35–36
Reservoir load
  and tectonic environment, 25
Reservoir loading
  effects of, 29
  and triggered earthquakes, 33
Reservoirs
  seismicity related to, 34, 37–
    40
Reynolds number
  boundary
    in stream channels, 119,
      125–26
  grain, 120–21
Rheological properties
  and temperature, 385
Rheology
  temperature-dependent, 406–8
Richter's earthquake magnitude
  scale, 295
Rip currents
  and beach erosion, 249–52,
    261
River drainage systems
  and the effects of global
    warming, 224
Rivera plate, 316
RNA bases
  and taxonomic information, 75
Rock deformation
  in crustal fault zones, 149–71
Rock failure
  and induced stress changes,
    23–26
  in underground mines, 27–28
Rock series
  and parent magmas, 420–22,
    433, 436–37
Rock type
  and the binomial classification
    system, 457–58
Rocks
  creep-deformation of, 385
  sedimentary
    fine-grained siliceous, 456–
      59, 461

general classes of, 455
    ultramafic, 495–98
Rocky Mountain Arsenal
  injection of fluid wastes at, 22
Rossby waves
  and El Niño, 55
Rupture controls
  and earthquakes, 171
Rupture process
  of subduction zone earth-
    quakes, 293–319
Ryukyu
  aftershock expansion ratios in,
    312

S

Sand transport
  and longshore currents, 253–
    55
  offshore, 243, 261, 263
Sanriku earthquake, 316–17
Scythian conodont zones, 105
Scythian rocks
  conodont distribution in, 100
Sea cows
  and radioimmunoassay, 80–81
Sea level
  and CO_2-induced warming,
    223
  and coastal erosion, 256–62
  data
    the work of Wyrtki in, 48,
      54
  and El Niño, 52–57, 67, 260–
    62
  a 4–6 m rise in, 223
  the long-term rise in 237, 239
  in the Pacific, 49
Sea-surface temperature (SST)
  and El Niño, 44, 46, 48, 52–
    53, 57–61, 66–67
  equatorial Pacific, 47, 49–50
  and the ocean-atmosphere in-
    teraction model of El
    Niño, 62–65
Seamount chert association, 471
Sediment(s)
  and allocyclic stratigraphic
    oscillations, 352–53
  baked, 4
  compaction of, 192–93
  deep-sea
    microtektites in, 339–41
  movement of
    in stream channels, 119
    in natural streams, 117–19
  oscillations
    and beach erosion, 243
  pelagic and hemipelagic
    bedding in, 363–69
  trench

and subduction zones, 313–
    14
Sedimentary structures
  in bedded chert, 462–63
Sedimentation
  of the crust, 399–402
Seismic body waves, 305–6, 308
Seismic moment, 296–97
  and the far-field time function,
    307
Seismic potential
  evaluation of, 315–16
Seismic refraction
  and gravity measurements pro-
    gram, 6
Seismic refraction experiments
  in the United States, 9–10
Seismic refraction studies
  of crustal structure, 2–3, 5, 7,
    11–12, 19
Seismic risk
  and triggered earthquakes, 35–
    40
Seismic swarms
  in Mount St. Helens, 278
  and volcanic eruptions, 276
Seismic waves
  radiation of high-frequency,
    317–19
Seismicity
  beneath Mauna Loa, 277
  at Mount St. Helens, 278–86
  and plate motion, 300–5
  at the Rabaul caldera, 280
  reservoir-induced, 22–23
  and subduction zones, 297–
    300
Seismograph
  the Wood-Anderson, 295
Settling velocities
  of grains, 120–23, 132–33
Shear resistance profiles
  and faulting, 167–69
Shear stress
  tectonic
    and crustal fault mechanics,
      170
Shearing
  of grains, 117, 138–39
  mechanisms allowing steady,
    155–59
  steady aseismic, 171
Shearing flow
  and crustal fault zones, 156
Shields parameter
  and grain entrainment, 129–31
Shoreline erosion
  and coastal processes, 237–62
  computer models of, 255–56
Shoreline(s)
  Holocene, 300
  uplift rate of, 300

Siberian Platform
  Ordovician deposits of the, 95
Silica diagenesis
  and bedded chert, 460–61
  and diagenetic features, 463
Silicate perovskite
  in the lower mantle, 382
Silicates
  anhydrous
    and calc-alkaline andesites,
      437–39
  of the lower mantle, 381
  see also Bulk silicate Earth
    (BSE)
Silurian
  and conodont-based
    biostratigraphy, 98–99
Sirenia, 80–81
Solar magnetic polarity, 351
Solomon Islands
  earthquakes in the, 305–6,
    312
  surface waves of earthquakes
    in the, 305
  time functions for earthquakes
    from the, 309
Sorting mechanisms
  hydraulic, 116–39
South America
  earthquakes in, 293
  Ordovician conodont faunas
    of, 97
South American coast, 49
  and El Niño, 54
  mean upwelling at the, 58
  sea level along the, 58
  winds along the, 50
South American subduction
  zones
    aftershock expansion ratios in,
      312
South Equatorial Current, 50, 53
South Pacific Convergence
  Zone, 48
  and El Niño, 52
South Pole
  gas analyzers near the, 206
Southern Hemisphere
  atmospheric CO_2 content for
    the, 203
Southern Oscillation, 67
  and El Niño, 44–47, 68
  see also El Niño–Southern
    Oscillation
Species
  extinct
    molecular systematics of,
      79–82
Spectroscopic data
  CO_2 analysis of, 205
Sphalerite, 186–87
  in Mississippi Valley–type de-
    posits, 177, 181–83, 196

solubilities of, 189
sulfur isotopic composition of,
  187–88
Stokes' law
  and grain-settling velocities,
    120–21
Strabo, 113
Stratigraphic oscillations
  allocyclic, 352–53
Stratigraphic sequences
  autocyclic, 351–52
Stream channels
  flow zones in, 117–19
Streams
  placer deposits formed by,
    114
Stress
  elastic
    and mining and quarrying
      operations, 27
    in mining operations, 28
    and reservoir loading, 29,
      30
    and triggered earthquakes,
      27
Stress change(s)
  induced
    and rock failure, 23–26
    and seismic risk assessment,
      37–38, 40
  produced by reservoirs, 28
Stress cycles
  on faults
    and triggered earthquakes,
      31–34
Subduction-zone earthquakes
  rupture process of, 293–319
Subduction zones
  mantle temperature distribu-
    tions along, 395
Subtropical Front, 214
Sunspot hemicycle
  the 11-yr, 351, 363, 372
Superplastic flow
  and faulting, 158–59
Surface-wave magnitude
  in seismology, 295–96
Surface-wave studies, 6, 11
Surface waves
  of Solomon Island earth-
    quakes, 305

T

Taxonomy
  mammalian, 76–77
Tectonic features
  and subduction zones, 297–
    300, 319
Tectonics
  and magmatism, 417
Tektites
  age, 338–39

geochemistry of, 323–44
  and isotopes, 338–39
  major element chemistry of,
    326–31
  and microtektites, 340
  mineralogical and petrograph-
    ical characteristics, 325–
    26
  Muong Nong–type, 326, 336–
    38
  strewn fields, 324–25
  trace elements of, 331–36
  water content of, 339
Temperature
  atmospheric
    changes attributable to in-
      creasing CO_2, 222
  and convection within the
    Earth, 386–88, 390–94
  in the crustal mantle, 377–
    411
  global
    increase in the, 220–21,
      225–27
  and its effects on rheological
    properties, 385
  in the lower mantle, 381–82
  in the outer core, 379–81
  surface
    and heat flux in the Earth,
      384
Thermal boundary layer(s)
  at the base of the mantle, 13,
    391
  in the Earth, 393, 405
  modeling heat transfer in the
    Earth's, 378
Thermocline
  and El Niño, 52, 55–56, 58
  in the Pacific, 49–50
Tholeiitic andesites, 418
  see also Basaltic magma
Tholeiitic rock series, 419–21,
  427–28, 430–31, 433, 447
Tides
  perigean spring, 258
Tonga
  aftershock expansion ratios in,
    312–13
  subduction zones in, 297
  uplift of the shorelines in, 300
Tonga earthquake, 319
Tournaisian
  and conodont zonal
    biostratigraphy, 99–100
Trade winds, 67
  and El Niño, 53
  in the Pacific, 48–50, 56
Transport sorting
  of grains, 117, 134–38
Trapping
  and grain entrainment, 131–32
Tree growth

and solar magnetic polarity,
   351
Tree ring ^{13}C record, 205, 210
Tremadocian
  Baltoscandian
     conodont succession in, 94
Trench sediments
  and subduction zones, 313–14
Triassic
  and conodont-based
     biostratigraphy, 100
  Lower
     biozonal boundaries in the,
        108
Triassic lacustrine record
  and lake varves, 360
Trophoblast, 76
Tyndall, J., 201–2

U

Uplift
  of crust, 399–402
Uplift rate
  of shorelines, 300
Upper Mississippi Valley
  CO_2 contents of fluid in-
     clusions from the, 190
  Mississippi Valley–type de-
     posits in, 195–96
  sphalerite in the, 181–83
  sulfur isotopic studies of the,
     187
Urey ratio, 410
  and mantle heat flow, 535,
     559–61, 563

V

Varve counts, 356–63
Varves
  evaporite, 362–63
  lake, 359–60
  marine, 360, 362
Ventersdorp Contact placer,
   131–32
Vertebrate nuclei
  and taxonomic information, 75

Viruses
  and taxonomic information,
     75
Volcanic eruptions
  forecasting, 267–89
  based on eruptive record,
     270–75
  monitoring approach, 275–
     86
  an ideal forecast of, 269
Volcanic explosions
  triggering, 229

W

Walker, Sir Gilbert, 44
Walker Circulation, 47–48, 61
Wall-rock alteration
  in Mississippi Valley–type de-
     posits, 183–84
Warming
  global, 223
  social-economic-political
     consequences of, 224
Washington-Oregon coast
  subduction zone off the, 314–
     16
Water resources
  redistribution of, 224
Wave run-up
  and beach erosion, 239–43,
     258, 261
Wave set-up
  and beach erosion, 239–42,
     249, 261
Waves
  at an angle to the shoreline,
     253–55
  edge
     and beach morphology,
        246–49
     and infragravity motions,
        243–46
  equatorial
     and El Niño, 55–56
  and nearshore currents, 249–
     52
  Kelvin
     and El Niño, 54–57

Rossby
  and El Niño, 55
storm
  and beach erosion, 239–43,
     258, 261
Weddell Sea
  CO_2 in the, 211, 219
Weddell Sea Bottom Water, 218
West Antarctic ice sheet, 221–23
West Antarctica, 223
Wheat belt
  migration of the, 224
Wind anomalies
  and the ocean-atmosphere in-
     teraction model of El
     Niño, 64–66
Wind changes
  and El Niño, 54–55
Winnowing
  and grain entrainment sorting,
     132–33
Witwatersrand
  seismicity at, 28
Witwatersrand Carbon Leader
  Reef, 123
Witwatersrand mines, 4
Witwatersrand paleoplacers, 113
Witwatersrand System, 5–6

X

Xenoliths
  from the mantle, 402
  melting, 383–84
  ultramafic, 495–96

Y

Yalin number
  and grain entrainment, 126
Yangtze Gorges Region
  and the Ordovician, 97

Z

Zhamanshin crater, 324–25,
   330–32, 340–41, 343
Zinc
  in Mississippi Valley–type de-
     posits, 184–87

CUMULATIVE INDEXES

CONTRIBUTING AUTHORS VOLUMES 1-14

A

Abelson, P. H., 6:325-51
Alyea, F. N., 6:43-74
Anderson, D. E., 4:95-121
Anderson, D. L., 5:179-202
Andrews, J. T., 6:205-28
Anhaeusser, C. R., 3:31-53
Apel, J. R., 8:303-42
Arculus, R. J., 13:75-95
Armstrong, R. L., 10:129-54
Arnold, J. R., 5:449-89
Axford, W. I., 2:419-74

B

Bada, J. L., 13:241-68
Bambach, R. K., 7:473-502
Banerjee, S. K., 1:269-96
Banks, P. M., 4:381-440
Barnes, I., 1:157-81
Barry, R. G., 6:205-28
Barth, C. A., 2:333-67
Barton, P. B. Jr., 1:183-211
Bassett, W. A., 7:357-84
Bathurst, R. G. C., 2:257-74
Behrensmeyer, A. K., 10:39-60
Benninger, L. K., 5:227-55
Benson, R. H., 9:59-80
Bergström, S. M., 14:85-112
Bhattacharyya, D. B., 10:441-57
Birch, F., 7:1-9
Bishop, F. C., 9:175-98
Black, R. F., 4:75-94
Blandford, R., 5:111-22
Bodnar, R. J., 8:263-301
Bonatti, E., 3:401-31
Bottinga, Y., 5:65-110
Brewer, J. A., 8:205-30
Brown, L., 12:39-59
Browne, P. R. L., 6:229-50
Brownlee, D. E., 13:147-73
Bryan, K., 10:15-38
Buland, R., 9:385-413
Bullard, E., 3:1-30
Burdick, L. J., 7:417-42
Burke, D. B., 2:213-38
Burke, K., 5:371-96
Burnett, D. S., 11:329-58
Burnham, C. W., 1:313-38

Burnham, L., 13:297-314
Burns, J. A., 8:527-58
Burns, R. G., 4:229-63; 9:345-83
Burst, J. F., 4:293-318
Busse, F. H., 11:241-68

C

Cane, M. A., 14:43-70
Carpenter, F. M., 13:297-314
Carter, S. R., 7:11-38
Castleman, A. W. Jr., 9:227-49
Champness, P. E., 5:203-26
Chapman, C. R., 5:515-40
Chase, C. G., 3:271-91; 13:97-117
Chave, K. E., 12:293-305
Chen, C.-T. A., 14:201-35
Chou, L., 8:17-33
Clark, D. R., 5:159-78
Clark, G. R. II, 2:77-99
Claypool, G. E., 11:299-327
Cluff, L. S., 4:123-45
Coroniti, F. V., 1:107-29
Crompton, A. W., 1:131-55
Crough, S. T., 11:165-93
Crutzen, P. J., 7:443-72
Cruz-Cumplido, M. I., 2:239-56
Cunnold, D. M., 6:43-74

D

Daly, S. F., 9:415-48
Damon, P. E., 6:457-94
Decker, R. W., 14:267-92
Dieterich, J. H., 2:275-301
Domenico, P. A., 5:287-317
Donahue, T. M., 4:265-92
Donaldson, I. G., 10:377-95
Drake, E. T., 14:201-35
Duce, R. A., 4:187-228
Durham, J. W., 6:21-42

E

Eaton, G. P., 10:409-40
Eugster, H. P., 8:35-63
Evans, B. W., 5:397-447
Evensen, N. M., 7:11-38

F

Farlow, N. H., 9:19-58
Filson, J., 3:157-81
Fischer, A. G., 14:351-76
Fripiat, J. J., 2:239-56

G

Garland, G. D., 9:147-74
Gay, N. C., 6:405-36
Gibson, I. L., 9:285-309
Gierlowski, T. C., 13:385-425
Gieskes, J. M., 3:433-53
Gilluly, J., 5:1-12
Gingerich, P. D., 8:407-24
Graf, D. L., 4:95-121
Grant, T. A., 4:123-45
Grey, A., 9:285-309
Gross, M. G., 6:127-43
Grossman, L., 8:559-608
Grove, T. L., 14:417-54
Gueguen, Y., 8:119-44
Gulkis, S., 7:385-415

H

Haggerty, S. E., 11:133-63
Hales, A. L., 14:1-20
Hallam, A., 12:205-43
Hamilton, P. J., 7:11-38
Hanson, G. N., 8:371-406
Hargraves, R. B., 1:269-96
Harms, J. C., 7:227-48
Harris, A. W., 10:61-108
Hart, S. R., 10:483-526; 14:493-571
Hay, W. W., 6:353-75
Heirtzler, J. R., 7:343-55
Helmberger, D. V., 7:417-42
Hem, J. D., 1:157-81
Herron, E. M., 3:271-91
Hinze, W. J., 13:345-83
Hoffman, E. J., 4:187-228
Holman, R. A., 14:237-65
Holton, J. R., 8:169-90
Housen, K. R., 10:355-76
Howell, D. G., 12:107-31
Hsü, K. J., 10:109-28
Hubbard, W. B., 1:85-106
Hulver, M. L., 13:385-425

Hunt, G. E., 11:415–59
Hunten, D. M., 4:265–92
Huppert, H. E., 12:11–37

J

James, D. E., 9:311–44
Javoy, M., 5:65–110
Jeanloz, R., 14:377–415
Jeffreys, H., 1:1–13
Jenkins, F. A. Jr., 1:131–55
Johns, W. D., 7:183–98
Johnson, M. E., 7:473–502
Johnson, N. M., 12:445–88
Johnson, T. C., 12:179–204
Johnson, T. V., 6:93–125
Jones, D. L., 12:107–31; 14:455–92
Jones, K. L., 5:515–40

K

Kanamori, H., 1:213–39; 14:293–322
Karig, D. E., 2:51–75
Keesee, R. G., 9:227–49
Kellogg, W. W., 7:63–92
Kinzler, R. J., 14:417–54
Kirkpatrick, R. J., 13:29–47
Kistler, R. W., 2:403–18
Koeberl, C., 14:323–50
Komar, P. D., 14:237–65
Koshlyakov, M. N., 6:495–523
Kröner, A., 13:49–74
Ku, T.-L., 4:347–79
Kvenvolden, K. A., 3:183–212; 11:299–327

L

Langevin, Y., 5:449–89
Leckie, J. O., 9:449–86
Lerman, A., 6:281–303
Lerman, J. C., 6:457–94
Levine, J., 5:357–69
Levy, E. H., 4:159–85
Lewis, B. T. R., 6:377–404
Lick, W., 4:49–74; 10:327–53
Lilly, D. K., 7:117–61
Lindsay, E. H., 12:445–88
Lindzen, R. S., 7:199–225
Lion, L. W., 9:449–86
Lister, C. R. B., 8:95–117
Long, A., 6:457–94
Lottes, A. L., 13:385–425
Lowe, D. R., 8:145–67
Lowenstein, J. M., 14:71–83
Lupton, J. E., 11:371–414

M

Macdonald, K. C., 10:155–90
Malin, M. C., 12:411–43

Margolis, S. V., 4:229–63
Mavko, G. M., 9:81–111
McBirney, A. R., 6:437–56; 12:337–57
McConnell, J. C., 4:319–46
McConnell, J. D. C., 3:129–55
McGarr, A., 6:405–36
McKerrow, W. S., 7:473–502
McNally, K. C., 11:359–69
Melosh, H. J., 8:65–93
Mendis, D. A., 2:419–74
Mercer, J. H., 11:99–132
Millero, F. J., 2:101–50
Miyashiro, A., 3:251–69
Mogi, K., 1:63–94
Mohr, P., 6:145–72
Molnar, P., 12:489–518
Monin, A. S., 6:495–523
Morris, S., 14:377–415
Munk, W. H., 8:1–16
Murase, T., 12:337–57
Murchey, B., 14:455–92
Murthy, V. R., 13:269–96
Mysen, B. O., 11:75–97

N

Nairn, A. E. M., 6:75–91
Navrotsky, A., 7:93–115
Naylor, R. S., 3:387–400
Ness, N. F., 7:249–88
Newburn, R. L. Jr., 10:297–326
Nicolas, A., 8:119–44
Nier, A. O., 9:1–17
Nixon, P. H., 9:285–309
Normark, W. R., 3:271–91
Norton, D. L., 12:155–77
Nozaki, Y., 5:227–55
Nunn, J. A., 8:17–33

O

Okal, E. A., 11:195–214
Oliver, J. E., 8:205–30
O'Nions, R. K., 7:11–38
Opdyke, N. D., 12:445–88
Ostrom, J. H., 3:55–77

P

Pálmason, G., 2:25–50
Palmer, A. R., 5:13–33
Park, C. F. Jr., 6:305–24
Parker, R. L., 5:35–64
Pasteris, J. D., 12:133–53
Peltier, W. R., 9:199–225
Philander, S. G. H., 8:191–204
Phillips, R. J., 12:411–43
Pilbeam, C. C., 3:343–60
Pittman, E. D., 7:39–62
Poag, C. W., 6:251–80
Pollack, H. N., 10:459–81

Pollack, J. B., 8:425–87
Prinn, R. G., 6:43–74

R

Raitt, W. J., 4:381–440
Reeburgh, W. S., 11:269–98
Ressetar, R., 6:75–91
Richet, P., 5:65–110
Richter, F. M., 6:9–19
Ridley, W. I., 4:15–48
Riedel, W. R., 1:241–68
Roden, M. F., 13:269–96
Rodgers, J., 13:1–4
Roedder, E., 8:263–301
Rogers, N. W., 9:285–309
Ross, R. J. Jr., 12:307–35
Rothrock, D. A., 3:317–42
Rowley, D. B., 13:385–425
Rubey, W. W., 2:1–24
Rudnicki, J. W., 8:489–525
Rumble, D. III, 10:221–33
Russell, D. A., 7:163–82

S

Sachs, H. M., 5:159–78
Saemundsson, K., 2:25–50
Sahagian, D. L., 13:385–425
Saleeby, J. B., 11:45–73
Savage, J. C., 11:11–43
Savin, S. M., 5:319–55
Schermer, E. R., 12:107–31
Schnitker, D., 8:343–70
Schopf, J. W., 3:213–49
Schopf, T. J. M., 12:245–92
Schramm, S., 13:29–47
Schubert, G., 7:289–342
Schumm, S. A., 13:5–27
Schunk, R. W., 4:381–440
Scotese, C. R., 7:473–502
Sekanina, Z., 9:113–45
Sharp, R. P., 8:231–61
Sheldon, R. P., 9:251–84
Shimizu, N., 10:483–526
Shoemaker, E. M., 11:461–94
Sibson, R. H., 14:149–75
Simpson, D. W., 14:21–42
Simpson, G. G., 4:1–13
Skinner, B. J., 1:183–211
Sleep, N. H., 8:17–33
Slingerland, R., 14:113–47
Sloss, L. L., 12:1–10
Smith, E. J., 7:385–415
Smith, J. V., 9:175–98
Smith, K. A., 13:29–47
Smith, N. D., 14:113–47
Southam, J. R., 6:353–75
Sparks, R. S. J., 12:11–37
Squyres, S. W., 12:83–106
Stacey, F. D., 4:147–57
Stanley, S. M., 3:361–85
Stevenson, D. J., 10:257–95

Stewart, R. H., 12:61–82
Suzuki, Z., 10:235–56
Sverjensky, D. A., 14:177–99
Sweet, W. C., 14:85–112

T

Thomas, G. E., 6:173–204
Thomas, W. A., 13:175–99
Thompson, A. B., 2:179–212
Thompson, G. A., 2:213–38
Thomsen, L., 5:491–513
Thorne, R. M., 1:107–29
Timur, A., 13:315–44
Toksöz, M. N., 2:151–77;
    13:315–44
Toon, O. B., 9:19–58
Turcotte, D. L., 10:397–408
Turekian, K. K., 5:227–55
Turner, G., 13:29–47

U

Ubaghs, G., 3:79–98

V

Vaišnys, J. R., 3:343–60
Van der Voo, R., 10:191–220
Van Houten, F. B., 1:39–61;
    10:441–57
Van Schmus, W. R., 13:345–83
Veblen, D. R., 13:119–46
Verhoogen, J., 11:1–9
Veverka, J., 8:527–58
von Huene, R., 12:359–81
Vonnegut, B., 1:297–311

W

Walcott, R. I., 1:15–37
Walker, R. M., 3:99–128
Ward, W. R., 10:61–108
Warren, P. H., 13:201–40
Watts, A. B., 9:415–48
Webb, T. III, 5:159–78
Weertman, J., 3:293–315;
    11:215–40
Weertman, J. R., 3:293–315

Wetherill, G. W., 2:303–31
Whipple, F. L., 6:1–8
Wilkening, L. L., 10:355–76
Wilson, J. T., 10:1–14
Windley, B. F., 9:175–98
Wofsy, S. C., 4:441–69
Wood, D. S., 2:369–401
Woolum, D. S., 11:329–58
Wright, H. E. Jr., 5:123–58

Y

Yang, W.-H., 13:29–47
Yeomans, D. K., 10:297–326
York, D., 12:383–409
Yung, Y. L., 8:425–87

Z

Zartman, R. E., 5:257–86
Zen, E., 2:179–212
Ziegler, A. M., 7:473–502;
    13:385–425
Zindler, A., 14:493–571

CHAPTER TITLES VOLUMES 1–14

PREFATORY CHAPTERS
Developments in Geophysics   H. Jeffreys   1:1–13
Fifty Years of the Earth Sciences—A
   Renaissance   W. W. Rubey   2:1–24
The Emergence of Plate Tectonics: A Personal
   View   E. Bullard   3:1–30
The Compleat Palaeontologist?   G. G. Simpson   4:1–13
American Geology Since 1910—A Personal
   Appraisal   J. Gilluly   5:1–12
The Earth as Part of the Universe   F. L. Whipple   6:1–8
Reminiscences and Digressions   F. Birch   7:1–9
Affairs of the Sea   W. H. Munk   8:1–16
Some Reminiscences of Isotopes,
   Geochronology, and Mass Spectrometry   A. O. Nier   9:1–17
Early Days in University Geophysics   J. T. Wilson   10:1–14
Personal Notes and Sundry Comments   J. Verhoogen   11:1–9
The Greening of Stratigraphy 1933–1983   L. L. Sloss   12:1–10
Witnessing Revolutions in the Earth Sciences   J. Rodgers   13:1–4
Geophysics on Three Continents   A. L. Hales   14:1–20

GEOCHEMISTRY, MINERALOGY, AND PETROLOGY
Chemistry of Subsurface Waters   I. Barnes, J. D. Hem   1:157–82
Genesis of Mineral Deposits   B. J. Skinner, P. B. Barton Jr.   1:183–212
Order-Disorder Relationships in Some
   Rock-Forming Silicate Minerals   C. W. Burnham   1:313–38
Low Grade Regional Metamorphism: Mineral
   Equilibrium Relations   E. Zen, A. B. Thompson   2:179–212
Clays as Catalysts for Natural Processes   J. J. Fripiat, M. I. Cruz-Cumplido   2:239–56
Phanerozoic Batholiths in Western North
   America: A Summary of Some Recent
   Work on Variations in Time, Space,
   Chemistry, and Isotopic Compositions   R. W. Kistler   2:403–18
Microstructures of Minerals as Petrogenetic
   Indicators   J. D. C. McConnell   3:129–55
Advances in the Geochemistry of Amino
   Acids   K. A. Kvenvolden   3:183–212
Volcanic Rock Series and Tectonic Setting   A. Miyashiro   3:251–69
Metallogenesis at Oceanic Spreading Centers   E. Bonatti   3:401–31
Chemistry of Interstitial Waters of Marine
   Sediments   J. M. Gieskes   3:433–53
Multicomponent Electrolyte Diffusion   D. E. Anderson, D. L. Graf   4:95–121
Argillaceous Sediment Dewatering   J. F. Burst   4:293–318
The Uranium-Series Methods of Age
   Determination   T.-L. Ku   4:347–79
Interactions of CH and CO in the Earth's
   Atmosphere   S. C. Wofsy   4:441–69
A Review of Hydrogen, Carbon, Nitrogen,
   Oxygen, Sulphur, and Chlorine Stable
   Isotope Fractionation Among Gaseous
   Molecules   P. Richet, Y. Bottinga, M. Javoy   5:65–110
Transmission Electron Microscopy in Earth
   Science   P. E. Champness   5:203–26

588   CHAPTER TITLES

Geochemistry of Atmospheric Radon and
Radon Products
K. K. Turekian, Y. Nozaki,
L. K. Benninger   5:227–55

Transport Phenomena in Chemical Rate
Processes in Sediments   P. A. Domenico   5:287–317
Metamorphism of Alpine Peridotite and
Serpentinite   B. W. Evans   5:397–447
Hydrothermal Alteration in Active Geothermal
Fields   P. R. L. Browne   6:229–50
Temporal Fluctuations of Atmospheric ^{14}C:
Causal Factors and Implications   P. E. Damon, J. C. Lerman, A.
Long   6:457–94

Geochemical and Cosmochemical Applications
of Nd Isotope Analysis   R. K. O'Nions, S. R. Carter, N. M.
Evensen, P. J. Hamilton   7:11–38
Calorimetry: Its Application to Petrology   A. Navrotsky   7:93–115
Clay Mineral Catalysis and Petroleum
Generation   W. D. Johns   7:183–98
Geochemistry of Evaporitic Lacustrine
Deposits   H. P. Eugster   8:35–63
Geologic Pressure Determinations from Fluid
Inclusion Studies   E. Roedder, R. J. Bodnar   8:263–301
Rare Earth Elements in Petrogenetic Studies of
Igneous Systems   G. N. Hanson   8:371–406
Depleted and Fertile Mantle Xenoliths from
Southern African Kimberlites   P. H. Nixon, N. W. Rogers, I. L.
Gibson, A. Grey   9:285–309

The Combined Use of Isotopes as Indicators of
Crustal Contamination   D. E. James   9:311–44
Intervalence Transitions in Mixed-Valence
Minerals of Iron and Titanium   R. G. Burns   9:345–83
The Role of Perfectly Mobile Components in
Metamorphism   D. Rumble III   10:221–33
Phanerozoic Oolitic Ironstones—Geologic
Record and Facies Model   F. B. Van Houten, D. B.
Bhattacharyya   10:441–57

Applications of the Ion Microprobe to
Geochemistry and Cosmochemistry   N. Shimizu, S. R. Hart   10:483–526
The Structure of Silicate Melts   B. O. Mysen   11:75–97
Radioactive Nuclear Waste Stabilization:
Aspects of Solid-State Molecular
Engineering and Applied Geochemistry   S. E. Haggerty   11:133–63
In Situ Trace Element Microanalysis   D. S. Burnett, D. S. Woolum   11:329–58
Terrestrial Inert Gases: Isotope Tracer Studies
and Clues to Primordial Components in the
Mantle   J. E. Lupton   11:371–414
Double-Diffusive Convection Due to
Crystallization in Magmas   H. E. Huppert, R. S. J. Sparks   12:11–37
Applications of Accelerator Mass Spectrometry   L. Brown   12:39–59
Kimberlites: Complex Mantle Melts   J. D. Pasteris   12:133–53
Theory of Hydrothermal Systems   D. L. Norton   12:155–77
Rheological Properties of Magmas   A. R. McBirney, T. Murase   12:337–57
Cooling Histories from ^{40}Ar/^{39}Ar Age Spectra:
Implications for Precambrian Plate Tectonics   D. York   12:383–409
Solid-State Nuclear Magnetic Resonance
Spectroscopy of Minerals   R. J. Kirkpatrick, K. A. Smith, S.
Schramm, G. Turner, W.-H. Yang   13:29–47
Evolution of the Archean Continental Crust   A. Kröner   13:49–74
Oxidation Status of the Mantle: Past and
Present   R. J. Arculus   13:75–95
Direct TEM Imaging of Complex Structures
and Defects in Silicates   D. R. Veblen   13:119–46
Mantle Metasomatism   M. F. Roden, V. R. Murthy   13:269–96

Petrogenesis of Andesites	T. R. Grove, R. J. Kinzler	14:417–54
Chemical Geodynamics	A. Zindler, S. Hart	14:493–571

GEOPHYSICS AND PLANETARY SCIENCE

Structure of the Earth from Glacio-Isostatic Rebound	R. I. Walcott	1:15–38
Interior of Jupiter and Saturn	W. B. Hubbard	1:85–106
Magnetospheric Electrons	F. V. Coroniti, R. M. Thorne	1:107–30
Theory and Nature of Magnetism in Rocks	R. B. Hargraves, S. K. Banerjee	1:269–96
Geophysical Data and the Interior of the Moon	M. N. Toksöz	2:151–77
Solar System Sources of Meteorites and Large Meteoroids	G. W. Wetherill	2:303–31
The Atmosphere of Mars	C. A. Barth	2:333–67
Satellites and Magnetospheres of the Outer Planets	W. I. Axford, D. A. Mendis	2:419–74
Interaction of Energetic Nuclear Particles in Space with the Lunar Surface	R. M. Walker	3:99–128
Array Seismology	J. Filson	3:157–81
High Temperature Creep of Rock and Mantle Viscosity	J. Weertman, J. R. Weertman	3:293–315
The Mechanical Behavior of Pack Ice	D. A. Rothrock	3:317–42
Mechanical Properties of Granular Media	J. R. Vaišnys, C. C. Pilbeam	3:343–60
Petrology of Lunar Rocks and Implications to Lunar Evolution	W. I. Ridley	4:15–48
Paleomagnetism of Meteorites	F. D. Stacey	4:147–57
Generation of Planetary Magnetic Fields	E. H. Levy	4:159–85
Hydrogen Loss from the Terrestrial Planets	D. M. Hunten, T. M. Donahue	4:265–92
The Ionospheres of Mars and Venus	J. C. McConnell	4:319–46
The Topside Ionosphere: A Region of Dynamic Transition	P. M. Banks, R. W. Schunk, W. J. Raitt	4:381–440
Understanding Inverse Theory	R. L. Parker	5:35–64
Discrimination Between Earthquakes and Underground Explosions	R. Blandford	5:111–22
Composition of the Mantle and Core	D. L. Anderson	5:179–202
Laser-Distance Measuring Techniques	J. Levine	5:357–69
The Evolution of the Lunar Regolith	Y. Langevin, J. R. Arnold	5:449–89
Theoretical Foundations of Equations of State for the Terrestrial Planets	L. Thomsen	5:491–513
Cratering and Obliteration History of Mars	C. R. Chapman, K. L. Jones	5:515–40
Mantle Convection Models	F. M. Richter	6:9–19
The Galilean Satellites of Jupiter: Four Worlds	T. V. Johnson	6:93–125
The Interstellar Wind and its Influence on the Interplanetary Environment	G. E. Thomas	6:173–204
Evolution of Ocean Crust Seismic Velocities	B. T. R. Lewis	6:377–404
The Magnetic Fields of Mercury, Mars, and Moon	N. F. Ness	7:249–88
Subsolidus Convection in the Mantles of Terrestrial Planets	G. Schubert	7:289–342
The Diamond Cell and the Nature of the Earth's Mantle	W. A. Bassett	7:357–84
The Magnetic Field of Jupiter: A Comparison of Radio Astronomy and Spacecraft Observations	E. J. Smith, S. Gulkis	7:385–415
Synthetic Seismograms	D. V. Helmberger, L. J. Burdick	7:417–42
Cratering Mechanics—Observational, Experimental, and Theoretical	H. J. Melosh	8:65–93
Heat Flow and Hydrothermal Circulation	C. R. B. Lister	8:95–117
Seismic Reflection Studies of Deep Crustal Structure	J. A. Brewer, J. E. Oliver	8:205–30
Geomorphological Processes on Terrestrial Planetary Surfaces	R. P. Sharp	8:231–61

Origin and Evolution of Planetary
  Atmospheres   J. B. Pollack, Y. L. Yung   8:425–87
The Moons of Mars   J. Veverka, J. A. Burns   8:527–58
Refractory Inclusions in the Allende Meteorite L. Grossman   8:559–608
Rotation and Precession of Cometary Nuclei   Z. Sekanina   9:113–45
The Significance of Terrestrial Electrical
  Conductivity Variations   G. D. Garland   9:147–74
Ice Age Geodynamics   W. R. Peltier   9:199–225
Free Oscillations of the Earth   R. Buland   9:385–413
Long Wavelength Gravity and Topography
  Anomalies   A. B. Watts, S. F. Daly   9:415–48
Dynamical Constraints of the Formation and
  Evolution of Planetary Bodies   A. W. Harris, W. R. Ward   10:61–108
Pre-Mesozoic Paleomagnetism and Plate
  Tectonics   R. Van der Voo   10:191–220
Interiors of the Giant Planets   D. J. Stevenson   10:257–95
Halley's Comet   R. L. Newburn Jr., D. K. Yeomans 10:297–326
Regoliths on Small Bodies in the Solar System K. R. Housen, L. L. Wilkening   10:355–76
Heat and Mass Circulation in Geothermal
  Systems   I. G. Donaldson   10:377–95
Magma Migration   D. L. Turcotte   10:397–408
The Heat Flow from the Continents   H. N. Pollack   10:459–81
Strain Accumulation in Western United States J. C. Savage   11:11–43
Oceanic Intraplate Seismicity   E. A. Okal   11:195–214
Creep Deformation of Ice   J. Weertman   11:215–40
Recent Developments in the Dynamo Theory
  of Planetary Magnetism   F. H. Busse   11:241–68
Seismic Gaps in Space and Time   K. C. McNally   11:359–69
The Atmospheres of the Outer Planets   G. E. Hunt   11:415–59
Asteroid and Comet Bombardment of the
  Earth   E. M. Shoemaker   11:461–94
The History of Water on Mars   S. W. Squyres   12:83–106
Tectonic Processes Along the Front of Modern
  Convergent Margins—Research of the Past
  Decade   R. von Huene   12:359–81
Tectonics of Venus   R. J. Phillips, M. C. Malin   12:411–43
The Geological Significance of the Geoid   C. G. Chase   13:97–117
Cosmic Dust: Collection and Research   D. E. Brownlee   13:147–73
The Magma Ocean Concept and Lunar
  Evolution   P. H. Warren   13:201–40
Downhole Geophysical Logging   A. Timur, M. N. Toksöz   13:315–44
The Midcontinent Rift System   W. R. Van Schmus, W. J. Hinze   13:345–83
Triggered Earthquakes   D. W. Simpson   14:21–42
Forecasting Volcanic Eruptions   R. W. Decker   14:267–92
Rupture Process of Subduction-Zone
  Earthquakes   H. Kanamori   14:293–322
Geochemistry of Tektites and Impact Glasses   C. Koeberl   14:323–50
Temperature Distribution in the Crust and
  Mantle   R. Jeanloz, S. Morris   14:377–415

OCEANOGRAPHY, METEOROLOGY, AND PALEOCLIMATOLOGY
Electrical Balance in the Lower Atmosphere   B. Vonnegut   1:297–312
The Physical Chemistry of Seawater   F. J. Millero   2:101–50
Numerical Modeling of Lake Currents   W. Lick   4:49–74
Features Indicative of Permafrost   R. F. Black   4:75–94
Chemical Fractionation at the Air/Sea Interface R. A. Duce, E. J. Hoffman   4:187–228
Pacific Deep-Sea Manganese Nodules: Their
  Distribution, Composition, and Origin   S. V. Margolis, R. G. Burns   4:229–63
Quaternary Vegetation History—Some
  Comparisons Between Europe and America   H. E. Wright Jr.   5:123–58
The History of the Earth's Surface
  Temperature During the Past 100 Million
  Years   S. M. Savin   5:319–55

Photochemistry and Dynamics of the Ozone
  Layer   R. G. Prinn, F. N. Alyea,
      D. M. Cunnold   6:43–74
Effects of Waste Disposal Operations in
  Estuaries and the Coastal Ocean   M. G. Gross   6:127–43
Glacial Inception and Disintegration During
  the Last Glaciation   J. T. Andrews, R. G. Barry   6:205–28
Synoptic Eddies in the Ocean   M. N. Koshlyakov, A. S. Monin   6:495–523
Influences of Mankind on Climate   W. W. Kellogg   7:63–92
The Dynamical Structure and Evolution of
  Thunderstorms and Squall Lines   D. K. Lilly   7:117–61
Atmospheric Tides   R. S. Lindzen   7:199–225
The Role of NO and NO_2 in the Chemistry of
  the Troposphere and Stratosphere   P. Crutzen   7:443–72
The Dynamics of Sudden Stratospheric
  Warnings   J. R. Holton   8:169–90
The Equatorial Undercurrent Revisited   S. G. H. Philander   8:191–204
Satellite Sensing of Ocean Surface Dynamics   J. R. Apel   8:303–42
Particles Above the Tropopause   N. H. Farlow, O. B. Toon   9:19–58
Nucleation and Growth of Stratospheric
  Aerosols   A. W. Castleman Jr., R. G. Keesee   9:227–49
The Biogeochemistry of the Air-Sea Interface   L. W. Lion, J. O. Leckie   9:449–86
Poleward Heat Transport by the Oceans:
  Observations and Models   K. Bryan   10:15–38
Thirteen Years of Deep-Sea Drilling   K. J. Hsü   10:109–28
The Transport of Contaminants in the Great
  Lakes   W. Lick   10:327–53
Cenozoic Glaciation in the Southern
  Hemisphere   J. H. Mercer   11:99–132
Oceanography from Space   R. H. Stewart   12:61–82
Pre-Quaternary Sea-Level Changes   A. Hallam   12:205–43
El Niño   M. A. Cane   14:43–70
Carbon Dioxide Increase in the Atmosphere
  and Oceans and Possible Effects on Climate   C.-T. A. Chen, E. T. Drake   14:201–35

PALEONTOLOGY, STRATIGRAPHY, AND SEDIMENTOLOGY
  Origin of Red Beds: A Review—1961–1972   F. B. Van Houten   1:39–62
Mammals from Reptiles: A Review of
  Mammalian Origins   A. W. Crompton, F. A. Jenkins Jr.   1:131–56
Cenozoic Planktonic Micropaleontology and
  Biostratigraphy   W. R. Riedel   1:241–68
Growth Lines in Invertebrate Skeletons   G. R. Clark II   2:77–99
Marine Diagenesis of Shallow Water Calcium
  Carbonate Sediments   R. G. C. Bathurst   2:257–74
The Origin of Birds   J. H. Ostrom   3:55–77
Early Paleozoic Echinoderms   G. Ubaghs   3:79–98
Precambrian Paleobiology: Problems and
  Perspectives   J. W. Schopf   3:213–49
Adaptive Themes in the Evolution of the
  Bivalvia (Mollusca)   S. M. Stanley   3:361–85
Biostratigraphy of the Cambrian System—A
  Progress Report   A. R. Palmer   5:13–33
Paleoecological Transfer Functions   H. M. Sachs, T. Webb III,
      D. R. Clark   5:159–78
The Probable Metazoan Biota of the
  Precambrian as Indicated by the Subsequent
  Record   J. W. Durham   6:21–42
Stratigraphy of the Atlantic Continental Shelf
  and Slope of the United States   C. W. Poag   6:251–80
Chemical Exchange Across Sediment-Water
  Interface   A. Lerman   6:281–303
Organic Matter in the Earth's Crust   P. H. Abelson   6:325–51
Quantifying Biostratigraphic Correlation   W. W. Hay, J. R. Southam   6:353–75

Volcanic Evolution of the Cascade Range   A. R. McBirney   6:437–56
Recent Advances in Sandstone Diagenesis   E. D. Pittman   7:39–62
The Enigma of the Extinction of the Dinosaurs   D. A. Russell   7:163–82
Primary Sedimentary Structures   J. C. Harms   7:227–48
Paleozoic Paleogeography   A. M. Ziegler, C. R. Scotese,
W. S. McKerrow, M. E. Johnson,
R. K. Bambach   7:473–502
Archean Sedimentation   D. R. Lowe   8:145–67
Quaternary Deep-Sea Benthic Foraminifers and
Bottom Water Masses   D. Schnitker   8:343–70
Evolutionary Patterns of Early Cenozoic
Mammals   P. D. Gingerich   8:407–24
Form, Function, and Architecture of Ostracode
Shells   R. H. Benson   9:59–80
Metamorphosed Layered Igneous Complexes
in Archean Granulite-Gneiss Belts   B. F. Windley, F. C. Bishop,
J. V. Smith   9:175–98
Ancient Marine Phosphorites   R. P. Sheldon   9:251–84
The Geological Context of Human Evolution   A. K. Behrensmeyer   10:39–60
Rates of Biogeochemical Processes in Anoxic
Sediments   W. S. Reeburgh   11:269–98
Methane and Other Hydrocarbon Gases in
Marine Sediment   G. E. Claypool, K. A. Kvenvolden   11:299–327
Sedimentation in Large Lakes   T. C. Johnson   12:179–204
Rates of Evolution and the Notion of "Living
Fossils"   T. J. M. Schopf   12:245–92
Physics and Chemistry of Biomineralization   K. E. Chave   12:293–305
The Ordovician System, Progress and
Problems   R. J. Ross Jr.   12:307–35
Blancan-Hemphillian Land Mammal Ages
and Late Cenozoic Mammal Dispersal
Events   E. H. Lindsay, N. D. Opdyke, N.
M. Johnson   12:445–88
Patterns of Alluvial Rivers   S. A. Schumm   13:5–27
Amino Acid Racemization Dating of Fossil
Bones   J. L. Bada   13:241–68
The Geological Record of Insects   F. M. Carpenter, L. Burnham   13:297–314
Paleogeographic Interpretation: With an
Example From the Mid-Cretaceous   A. M. Ziegler, D. B. Rowley, A. L.
Lottes, D. L. Sahagian, M. L.
Hulver, T. C. Gierlowski   13:385–425
Molecular Phylogenetics   J. M. Lowenstein   14:71–83
Conodonts and Biostratigraphic Correlation   W. C. Sweet, S. M. Bergström   14:85–112
Occurrence and Formation of Water-Laid
Placers   R. Slingerland, N. D. Smith   14:113–47
Genesis of Mississippi Valley–Type Lead-Zinc
Deposits   D. A. Sverjensky   14:177–99
Coastal Processes and the Development of
Shoreline Erosion   P. D. Komar, R. A. Holman   14:237–65
Climatic Rhythms Recorded in Strata   A. G. Fischer   14:351–76
Geologic Significance of Paleozoic and
Mesozoic Radiolarian Chert   D. L. Jones, B. Murchey   14:455–92

TECTONOPHYSICS AND REGIONAL GEOLOGY
Rock Fracture   K. Mogi   1:63–84
Mode of Strain Release Associated with Major
Earthquakes in Japan   H. Kanamori   1:213–40
Iceland in Relation to the Mid-Atlantic Ridge   G. Pálmason, K. Saemundsson   2:25–50
Evolution of Arc Systems in the Western
Pacific   D. E. Karig   2:51–75
Regional Geophysics of the Basin and Range
Province   G. A. Thompson, D. B. Burke   2:213–38
Earthquake Mechanisms and Modeling   J. H. Dieterich   2:275–301

Current Views of the Development of Slaty
  Cleavage   D. S. Wood   2:369–401
Precambrian Tectonic Environments   C. R. Anhaeusser   3:31–53
Plate Tectonics: Commotion in the Ocean and
  Continental Consequences   C. G. Chase, E. M. Herron,
                                                    W. R. Normark   3:271–91
Age Provinces in the Northern Appalachians   R. S. Naylor   3:387–400
Radar Imagery in Defining Regional Tectonic
  Structure   T. A. Grant, L. S. Cluff   4:123–45
Geochronology of Some Alkalic Rock
  Provinces in Eastern and Central United
  States   R. E. Zartman   5:257–86
Aulacogens and Continental Breakup   K. Burke   5:371–96
Paleomagnetism of the Peri-Atlantic
  Precambrian   R. Ressetar   6:75–91
Afar   P. Mohr   6:145–72
Critical Mineral Resources   C. F. Park Jr.   6:305–24
State of Stress in the Earth's Crust   A. McGarr, N. C. Gay   6:405–36
The North Atlantic Ridge: Observational
  Evidence for Its Generation and Aging   J. R. Heirtzler   7:343–55
Platform Basins   N. H. Sleep, J. A. Nunn, L. Chou   8:17–33
Deformation of Mantle Rocks   Y. Gueguen, A. Nicolas   8:119–44
Seismic Reflection Studies of Deep Crustal
  Structure   J. A. Brewer, J. E. Oliver   8:205–30
Fracture Mechanics Applied to the Earth's
  Crust   J. W. Rudnicki   8:489–525
Mechanics of Motion on Major Faults   G. M. Mavko   9:81–111
Cordilleran Metamorphic Core
  Complexes—From Arizona to Southern
  Canada   R. L. Armstrong   10:129–54
Mid-Ocean Ridges: Fine Scale Tectonic,
  Volcanic and Hydrothermal Processes
  Within the Plate Boundary Zone   K. C. Macdonald   10:155–90
Earthquake Prediction   Z. Suzuki   10:235–56
The Basin and Range Province: Origin and
  Tectonic Significance   G. P. Eaton   10:409–40
Accretionary Tectonics of the North American
  Cordillera   J. B. Saleeby   11:45–73
Hotspot Swells   S. T. Crough   11:165–93
The Origin of Allochthonous Terranes:
  Perspectives on the Growth and Shaping of
  Continents   E. R. Schermer, D. G. Howell,
                                                    D. L. Jones   12:107–31
Structure and Tectonics of the Himalaya:
  Constraints and Implications of Geophysical
  Data   P. Molnar   12:489–518
The Appalachian-Ouachita Connection:
  Paleozoic Orogenic Belt at the Southern
  Margin of North America   W. A. Thomas   13:175–99
Earthquakes and Rock Deformation in Crustal
  Fault Zones   R. H. Sibson   14:149–75

Annual Reviews Inc. | ORDER FORM |

A NONPROFIT SCIENTIFIC PUBLISHER
4139 El Camino Way, Palo Alto, CA 94306-9981, USA • (415) 493-4400

Annual Reviews Inc. publications are available directly from our office by mail or telephone (paid by credit card or purchase order), through booksellers and subscription agents, worldwide, and through participating professional societies. Prices subject to change without notice.

- **Individuals:** Prepayment required on new accounts by check or money order (in U.S. dollars, check drawn on U.S. bank) or charge to credit card — American Express, VISA, MasterCard.
- **Institutional buyers:** Please include purchase order number.
- **Students:** $10.00 discount from retail price, per volume. Prepayment required. Proof of student status must be provided (photocopy of student I.D. or signature of department secretary is acceptable). Students must send orders direct to Annual Reviews. Orders received through bookstores and institutions requesting student rates will be returned.
- **Professional Society Members:** Members of professional societies that have a contractual arrangement with Annual Reviews may order books through their society at a reduced rate. Check with your society for information.

Regular orders: Please list the volumes you wish to order by volume number.
Standing orders: New volume in the series will be sent to you automatically each year upon publication. Cancellation may be made at any time. Please indicate volume number to begin standing order.
Prepublication orders: Volumes not yet published will be shipped in month and year indicated.
California orders: Add applicable sales tax.
Postage paid (4th class bookrate/surface mail) **by Annual Reviews Inc.** Airmail postage extra.

ANNUAL REVIEWS SERIES		Prices Postpaid per volume USA/elsewhere	Regular Order Please send:	Standing Order Begin with:
			Vol. number	Vol. number
Annual Review of ANTHROPOLOGY (Prices of Volumes in brackets effective until 12/31/85)				
[Vols. 1-10	(1972-1981)	**$20.00/$21.00**]		
[Vol. 11	(1982)	**$22.00/$25.00**]		
[Vols. 12-14	(1983-1985)	**$27.00/$30.00**]		
Vols. 1-14	(1972-1985)	**$27.00/$30.00**		
Vol. 15	(avail. Oct. 1986)	**$31.00/$34.00**	Vol(s). _____	Vol. _____
Annual Review of ASTRONOMY AND ASTROPHYSICS (Prices of Volumes in brackets effective until 12/31/85)				
[Vols. 1-2, 4-19	(1963-1964; 1966-1981)	**$20.00/$21.00**]		
[Vol. 20	(1982)	**$22.00/$25.00**]		
[Vols. 21-23	(1983-1985)	**$44.00/$47.00**]		
Vols. 1-2, 4-20	(1963-1964; 1966-1982)	**$27.00/$30.00**		
Vols. 21-23	(1983-1985)	**$44.00/$47.00**		
Vol. 24	(avail. Sept. 1986)	**$44.00/$47.00**	Vol(s). _____	Vol. _____
Annual Review of BIOCHEMISTRY (Prices of Volumes in brackets effective until 12/31/85)				
[Vols. 30-34, 36-50	(1961-1965; 1967-1981)	**$21.00/$22.00**]		
[Vol. 51	(1982)	**$23.00/$26.00**]		
[Vols. 52-54	(1983-1985)	**$29.00/$32.00**]		
Vols. 30-34, 36-54	(1961-1965; 1967-1985)	**$29.00/$32.00**		
Vol. 55	(avail. July 1986)	**$33.00/$36.00**	Vol(s). _____	Vol. _____
Annual Review of BIOPHYSICS AND BIOPHYSICAL CHEMISTRY (Prices of Vols. in brackets effective until 12/31/85)				
(*Formerly* Annual Review of Biophysics and Bioengineering)				
[Vols. 1-10	(1972-1981)	**$20.00/$21.00**]		
[Vol. 11	(1982)	**$22.00/$25.00**]		
[Vols. 12-14	(1983-1985)	**$47.00/$50.00**]		
Vols. 1-11	(1972-1982)	**$27.00/$30.00**		
Vols. 12-14	(1983-1985)	**$47.00/$50.00**		
Vol. 15	(avail. June 1986)	**$47.00/$50.00**	Vol(s). _____	Vol. _____
Annual Review of CELL BIOLOGY				
Vol. 1	(1985)	**$27.00/$30.00**		
Vol. 2	(avail. Nov. 1986)	**$31.00/$34.00**	Vol(s). _____	Vol. _____
Annual Review of COMPUTER SCIENCE				
Vol. 1	(avail. late 1986)	**Price not yet established**	Vol. _____	Vol. _____
Annual Review of EARTH AND PLANETARY SCIENCES (Prices of Volumes in brackets effective until 12/31/85)				
[Vols. 1-9	(1973-1981)	**$20.00/$21.00**]		
[Vol. 10	(1982)	**$22.00/$25.00**]		
[Vols. 11-13	(1983-1985)	**$44.00/$47.00**]		
Vols. 1-10	(1973-1982)	**$27.00/$30.00**		
Vols. 11-13	(1983-1985)	**$44.00/$47.00**		
Vol. 14	(avail. May 1986)	**$44.00/$47.00**	Vol(s). _____	Vol. _____

ANNUAL REVIEWS SERIES		Prices Postpaid per volume USA/elsewhere	Regular Order Please send:	Standing Order Begin with:

Annual Review of **ECOLOGY AND SYSTEMATICS** (Prices of Volumes in brackets effective until 12/31/85)

[Vols. 1-12	(1970-1981)................	**$20.00/$21.00]**		
[Vol. 13	(1982).....................	**$22.00/$25.00]**		
[Vols. 14-16	(1983-1985)................	**$27.00/$30.00]**		
Vols. 1-16	(1970-1985)................	**$27.00/$30.00**		
Vol. 17	(avail. Nov. 1986).............	**$31.00/$34.00**	Vol(s). _____	Vol. _____

Annual Review of **ENERGY** (Prices of Volumes in brackets effective until 12/31/85)

[Vols. 1-6	(1976-1981)................	**$20.00/$21.00]**		
[Vol. 7	(1982).....................	**$22.00/$25.00]**		
[Vols. 8-10	(1983-1985)................	**$56.00/$59.00]**		
Vols. 1-7	(1976-1982)................	**$27.00/$30.00**		
Vols. 8-10	(1983-1985)................	**$56.00/$59.00**		
Vol. 11	(avail. Oct. 1986).............	**$56.00/$59.00**	Vol(s). _____	Vol. _____

Annual Review of **ENTOMOLOGY** (Prices of Volumes in brackets effective until 12/31/85)

[Vols. 9-16, 18-26	(1964-1971; 1973-1981)........	**$20.00/$21.00]**		
[Vol. 27	(1982).....................	**$22.00/$25.00]**		
[Vols. 28-30	(1983-1985)................	**$27.00/$30.00]**		
Vols. 9-16, 18-30	(1964-1971; 1973-1985)........	**$27.00/$30.00**		
Vol. 31	(avail. Jan. 1986).............	**$31.00/$34.00**	Vol(s). _____	Vol. _____

Annual Review of **FLUID MECHANICS** (Prices of Volumes in brackets effective until 12/31/85)

[Vols. 1-5, 7-13	(1969-1973; 1975-1981)........	**$20.00/$21.00]**		
[Vol. 14	(1982).....................	**$22.00/$25.00]**		
[Vols. 15-17	(1983-1985)................	**$28.00/$31.00]**		
Vols. 1-5, 7-17	(1969-1973; 1975-1985)........	**$28.00/$31.00**		
Vol. 18	(avail. Jan. 1986).............	**$32.00/$35.00**	Vol(s). _____	Vol. _____

Annual Review of **GENETICS** (Prices of Volumes in brackets effective until 12/31/85)

[Vols. 1-15	(1967-1981)................	**$20.00/$21.00]**		
[Vol. 16	(1982).....................	**$22.00/$25.00]**		
[Vols. 17-19	(1983-1985)................	**$27.00/$30.00]**		
Vols. 1-19	(1967-1985)................	**$27.00/$30.00**		
Vol. 20	(avail. Dec. 1986).............	**$31.00/$34.00**	Vol(s). _____	Vol. _____

Annual Review of **IMMUNOLOGY**

Vols. 1-3	(1983-1985)................	**$27.00/$30.00**		
Vol. 4	(avail. April 1986).............	**$31.00/$34.00**	Vol(s). _____	Vol. _____

Annual Review of **MATERIALS SCIENCE** (Prices of Volumes in brackets effective until 12/31/85)

[Vols. 1-11	(1971-1981)................	**$20.00/$21.00]**		
[Vol. 12	(1982).....................	**$22.00/$25.00]**		
[Vols. 13-15	(1983-1985)................	**$64.00/$67.00]**		
Vols. 1-12	(1971-1982)................	**$27.00/$30.00**		
Vols. 13-15	(1983-1985)................	**$64.00/$67.00**		
Vol. 16	(avail. August 1986)............	**$64.00/$67.00**	Vol(s). _____	Vol. _____

Annual Review of **MEDICINE** (Prices of Volumes in brackets effective until 12/31/85)

[Vols. 1-3, 5-15, 17-32	(1950-52; 1954-64; 1966-81)......	**$20.00/$21.00]**		
[Vol. 33	(1982).....................	**$22.00/$25.00]**		
[Vols. 34-36	(1983-1985)................	**$27.00/$30.00]**		
Vols. 1-3, 5-15, 17-36	(1950-52; 1954-64; 1966-85)......	**$27.00/$30.00**		
Vol. 37	(avail. April 1986).............	**$31.00/$34.00**	Vol(s). _____	Vol. _____

Annual Review of **MICROBIOLOGY** (Prices of Volumes in brackets effective until 12/31/85)

[Vols. 18-35	(1964-1981)................	**$20.00/$21.00]**		
[Vol. 36	(1982).....................	**$22.00/$25.00]**		
[Vols. 37-39	(1983-1985)................	**$27.00/$30.00]**		
Vols. 18-39	(1964-1985)................	**$27.00/$30.00**		
Vol. 40	(avail. Oct. 1986).............	**$31.00/$34.00**	Vol(s). _____	Vol. _____

Annual Review of **NEUROSCIENCE** (Prices of Volumes in brackets effective until 12/31/85)

[Vols. 1-4	(1978-1981)................	**$20.00/$21.00]**		
[Vol. 5	(1982).....................	**$22.00/$25.00]**		
[Vols. 6-8	(1983-1985)................	**$27.00/$30.00]**		
Vols. 1-8	(1978-1985)................	**$27.00/$30.00**		
Vol. 9	(avail. March 1986)............	**$31.00/$34.00**	Vol(s). _____	Vol. _____